# Biological Invasions

Economic and Environmental Costs of
Alien Plant, Animal, and Microbe Species

## Second Edition

# Biological Invasions

## Economic and Environmental Costs of Alien Plant, Animal, and Microbe Species

## Second Edition

Edited by David Pimentel

CRC Press
Taylor & Francis Group
Boca Raton   London   New York

CRC Press is an imprint of the
Taylor & Francis Group, an **informa** business

Cover photograph of Johnsongrass used with permission from the Kansas Department of Agriculture.

CRC Press
Taylor & Francis Group
6000 Broken Sound Parkway NW, Suite 300
Boca Raton, FL 33487-2742

---

**Library of Congress Cataloging-in-Publication Data**

---

Biological invasions : economic and environmental costs of alien plant, animal, and microbe species / [edited by] David Pimentel. -- 2nd ed.
    p. cm.
    Summary: "A revised, expanded, and updated second version to the successful <EM>Biological Invasions: Economic and Environmental Costs of Alien Plant, Animal, and Microbe Species</EM>, this reference discusses how non-native species invade new ecosystems and the subsequent economic and environmental effects of these species. With nine new chapters, this text provides detailed information on the major components of the invasive-species problem from six continents, including impacts on human health and livestock. The book examines ways in which non-native species destroy vital crops and forests; damage ecosystem dynamics, which leads to plant and animal biodiversity losses; and cause soil erosion and water loss"-- Provided by publisher.
    Summary: "Some 10 million species of plants, animals, and microbes are thought to inhabit the earth, but so far only about 1.5 million of these have been identified. A mere 15 of the approximately 250,000 known plant species provide the world's human population with about 90 percent of its food.1 These crops are wheat, rice, corn, rye, barley, soybeans, and common millet. Although these crops are now grown in nearly every nation, only one or two of these crop species originated in any specific country. Among animals, eight species currently provide the bulk of the meat, milk, and eggs consumed by humans. These leading livestock species are cattle, buffalo, sheep, goats, horses, camels, chickens, and ducks. Farms in the United States feed approximately 100 million cattle, 7 million sheep, and 9 billion chickens each year"-- Provided by publisher.
    Includes bibliographical references and index.
    ISBN 978-1-4398-2990-5 (hardback)
    1. Biological invasions--Economic aspects. 2. Biological invasions--Environmental aspects. I. Pimentel, David, 1925- II. Title.

QH353.B57 2011
577'.18--dc22
                                                        2010051444

---

**Visit the Taylor & Francis Web site at**
**http://www.taylorandfrancis.com**

**and the CRC Press Web site at**
**http://www.crcpress.com**

# Contents

# Acknowledgments

I wish to express my sincere gratitude to the Cornell Association of Professors Emeriti for the partial support of our research through the Albert Podell Grant Program. In addition, I wish to thank Mike Burgess for his valuable assistance in the preparation of our book. He did an exceptional job with editing the chapters.

# Editor

**David Pimentel** is a professor of ecology and agricultural science at Cornell University, Ithaca, New York (e-mail: dp18@cornell.edu). He received his BS in 1948 from the University of Massachusetts Amherst and his PhD from Cornell University in 1951. From 1951 to 1954, he was chief of the Tropical Research Laboratory, U.S. Public Health Service, San Juan, Puerto Rico. From 1954 to 1955, he was a postdoctoral research fellow at the University of Chicago; from 1960 to 1961, OEEC Fellow at Oxford University (England); and 1961 NSF Scholar at Massachusetts Institute of Technology. He was appointed assistant professor of insect ecology at Cornell University in 1955 and associate professor in 1961. In 1963, he was appointed professor and head of the Department of Entomology and Limnology. He served as department head until 1969 when he returned to full-time research and teaching as professor of ecology and agricultural sciences.

Nationally, he served on the President's White House Science and Technology Program, Washington, D.C., 1969, and on a Presidential Commission on the Environment. He also served on numerous National Academy of Sciences Committees and Boards, including chairing the Board of Ecology. He has served on committees in the U.S. Department of Health, Education and Welfare; U.S. Department of Energy; U.S. Department of Agriculture; U.S. Congressional Office of Technology Assessment; and the U.S. State Department. He served as president of the Rachel Carson Council and as an elected member of the National Audubon Society and the American Institute of Biological Sciences.

Dr. Pimentel's honors and achievements include being a fellow at Oxford University (England); being appointed honorary professor, Institute of Applied Ecology, China; receiving an Honorary Degree of Science, University of Massachusetts; receiving the Distinguished Service Award from the Rural Sociology Society; and serving on the Board of Directors, International Institute of Ecological Economics, Royal Swedish Academy of Science. He is now serving as editor-in-chief of the journal *Environment, Development and Sustainability*.

Dr. Pimentel has authored nearly 700 scientific publications, written three books, and edited 30 others. His research spans the fields of energy, biological control, biotechnology, land and water conservation, and environmental policy.

# Contributors

**David Pimentel**
College of Agriculture and Life Sciences
Cornell University
Ithaca, New York

## Australia

**Deon Canyon**
Disaster Health and Crisis
    Management Unit
Anton Breinl Centre for Public Health and
    Tropical Medicine
James Cook University
Townsville, Australia

**Richard H. Groves**
CSIRO Plant Industry and CRC Australian
    Weed Management
Canberra, Australia

**Ian Naumann**
Australian Government Department of
    Agriculture Fisheries and Forestry
Canberra, Australia

**Rick Speare**
Disaster Health and Crisis
    Management Unit
Anton Breinl Centre for Public Health
    and Tropical Medicine
James Cook University
Townsville, Australia

**Ken Winkel**
Department of Pharmacology
University of Melbourne
Parkville, Victoria, Australia

## Brazil

**Helena Godoy Bergallo**
Department of Ecology
Institute of Biology
Universidade do Estado do Rio
    de Janeiro
Rio de Janeiro, Brazil

**Rosana Mazzoni**
Department of Ecology
Institute of Biology
Universidade do Estado do Rio
    de Janeiro
Rio de Janeiro, Brazil

**Carlos Frederico D. Rocha**
Department of Ecology
Institute of Biology
Universidade do Estado do Rio
    de Janeiro
Rio de Janeiro, Brazil

## British Isles

**Adriana E. S. Ford-Thompson**
Environment Department
University of York
York, England

**Stephen Harris**
School of Biological Sciences
University of Bristol
Bristol, England

**Carolyn J. Snell**
Department of Social Policy and
    Social Work
University of York
York, England

**Piran C. L. White**
Environment Department
University of York
York, England

**Mark Williamson**
Department of Biology
University of York
York, England

*Europe*

**Marie-Laure Desprez-Loustau**
INRA, UMR BioGeCo INRA-Université
Cestas, France

**Salit Kark**
The Biodiversity Research Group
Department of Evolution,
    Systematics and Ecology
Silberman Institute of Life Sciences
Hebrew University of Jerusalem
Jerusalem, Israel

**Alain Roques**
INRA UR 633
Zoologie Forestière
Orléans, France

**Anne-Sophie Roy**
European and Mediterranean Plant
    Protection Organization (EPPO)
Paris, France

**Ivan Sache**
INRA, UMR INRA-AgroParisTech
    Bioger-CPP
Campus AgroParisTech
Thiverval-Grignon, France

**Susan M. Shirley**
Department of Forest Ecosystems
    and Society
Oregon State University
Corvallis, Oregon

**Frédéric Suffert**
INRA, UMR INRA-AgroParisTech
    Bioger-CPP
Campus AgroParisTech
Thiverval-Grignon, France

*India*

**T. N. Ananthakrishnan**
Zoological Survey of India (ret.)

**Daizy R. Batish**
Department of Botany
Panjab University
Chandigarh, India

**R. K. Kohli**
Department of Botany
Panjab University
Chandigarh, India

**H. P. Singh**
Department of Environment and
    Vocational Studies
Panjab University
Chandigarh, India

*New Zealand*

**M. N. Clout**
Centre for Biodiversity and Biosecurity
School of Biological Sciences
Tamaki Campus
University of Auckland
Auckland, New Zealand

**Susan M. Timmins**
Department of Conservation
Wellington, New Zealand

**Peter A. Williams**
Landcare Research
Nelson, New Zealand

## South Africa

**Christian T. Chimimba**
Mammal Research Institute (MRI) &
    DST-NRF Centre of Excellence for
    Invasion Biology (CIB)
Department of Zoology and Entomology
University of Pretoria
Pretoria, South Africa

**Sarah J. Davies**
DST-NRF Centre of Excellence for
    Invasion Biology (CIB)
Faculty of Science
Stellenbosch University
Matieland, South Africa

**W. J. de Lange**
Natural Resources and the Environment
CSIR
Stellenbosch, South Africa

**D. C. Le Maitre**
Natural Resources and the Environment
CSIR
Stellenbosch, South Africa

**Faansie Peacock**
DST-NRF Centre of Excellence for
    Invasion Biology
Department of Zoology and Entomology
University of Pretoria
Pretoria, South Africa

**D. M. Richardson**
DST-NRF Centre of Excellence for
    Invasion Biology (CIB)
Department of Botany and Zoology
Stellenbosch University
Matieland, South Africa

**Dian Spear**
DST-NRF Centre of Excellence for
    Invasion Biology (CIB)
Department of Botany and Zoology
Stellenbosch University
Matieland, South Africa

**Berndt J. van Rensburg**
Department of Zoology and Entomology
University of Pretoria
Pretoria, South Africa

**B. W. van Wilgen**
Natural Resources and the Environment
CSIR
Stellenbosch, South Africa

**Nicola J. van Wilgen**
DST-NRF Centre of Excellence for Invasion
    Biology (CIB)
Department of Botany and Zoology
Stellenbosch University
Matieland, South Africa

**Olaf L. F. Weyl**
South African Institute for Aquatic
    Biodiversity
Grahamstown, South Africa

**R. M. Wise**
Economics and Policy Research Branch
Policy and Strategy Group
Department of Primary Industries
Melbourne, Australia

## United States

**Michael L. Avery**
USDA Animal and Plant Health
    Inspection Service
National Wildlife Research Center
    (NWRC)
Gainesville, Florida

**Tyler A. Campbell**
USDA Animal and Plant Health
    Inspection Service
National Wildlife Research Center
    (NWRC) Texas Field Station
Texas A&M University
Kingsville, Texas

**Peter J. Egan**
Armed Forces Pest Management Board
Forest Glen Section-WRMC
Washington, D.C.

**Richard M. Engeman**
USDA/APHIS/Wildlife Services
National Wildlife Research Center
Fort Collins, Colorado

**Michael W. Fall**
USDA/APHIS/Wildlife Services
National Wildlife Research Center
Fort Collins, Colorado

**William C. Pitt**
USDA/APHIS/WS/NWRC Hilo HI
    Field Station
Hilo, Hawaii

**Stephanie A. Shwiff**
National Wildlife Research Center
    Wildlife Services
Animal and Plant Health Inspection
    Service
USDA
Fort Collins, Colorado

**Gary W. Witmer**
USDA/APHIS/Wildlife Services
National Wildlife Research Center
Fort Collins, Colorado

# chapter one

# Introduction

## Nonnative species in the world

**David Pimentel**

### Contents

Some 10 million species of plants, animals, and microbes are believed to inhabit the Earth, but so far only about 1.5 million of these have been identified. A mere 15 of the approximately 250,000 known plant species provide the world's human population with about 90% of its food.[1] These crops are wheat, rice, corn, rye, barley, soybeans, and common millet. Although these crops are now grown in nearly every nation, only one or two of these crop species originated in any specific country.

Among animals, eight species currently provide the bulk of the meat, milk, and eggs consumed by humans. These leading livestock species are cattle, buffalo, sheep, goats, horses, camels, chickens, and ducks. Farms in the United States feed approximately 100 million cattle, 7 million sheep, and 9 billion chickens each year.[2]

Although much is known about the world's major food sources, relatively little is known about the vast number of plant, animal, and microbe species that have migrated throughout the world and invaded new ecosystems. Every nation now has thousands of nonnative, introduced species inhabiting their ecosystems. Many crop and livestock species were intentionally introduced into these ecosystems because native plants and livestock could not provide sufficient food for a country's needs; other species were either intentionally or accidentally introduced into a nation's ecosystems, along with human invasions.

The invasion of nonnative species into new ecosystems is accelerating as the world's human population multiplies and goods are transported ever more rapidly on an increasingly global scale. Several of these nonnative plant, animal, and microbe species were originally introduced for use in agriculture but have since become major pests. In the

United States, for example, these include Johnson grass, which was introduced for live-stock grazing, and cats, which were introduced for mouse control.

The impact of invasive species is second only to that of human population growth and associated activities as a cause of the loss of biodiversity throughout the world. In the United States, invasions of nonnative plants, animals, or microbes are believed to be responsible for 42% of the decline of native species now listed as endangered or threat-ened.[3] The loss of biodiversity caused by invasive species is the result of competition from invasive species and the resulting displacement of native species, as well as by predation and hybridization.

Several decades ago the British ecologist Charles Elton[4] investigated invading species worldwide and the widespread environmental damages they caused. He became aware of the need to assemble information about such invasive organisms, including their ecologi-cal effects and the difficulty in controlling those that become pests.

The contributors to this book have built on Elton's early studies and share in these pages their investigations into the environmental and economic impacts of invading spe-cies. They compare the number of native and nonnative species for several regions of the world. Where possible, information is provided on how nonnative species invaded an eco-system, as well as the environmental and economic consequences.

Contributing scientists from Australia, Brazil, the British Isles, Europe, India, New Zealand, South Africa, and the United States share their expertise in this book. Several factors were involved in selecting the nations discussed here, as will be explained next.

## 1.1   Australia

Australia's relative geographic isolation has not protected the continent from the influx of invasive species. Groves, in his investigation of invasive plants in Australia, reports that the number of introduced plant species is believed to be roughly equal to the number of native species—about 25,000 each.

Groves estimates the number of alien plant species that have been established in Australian natural habitats at 2681. A few of the major weed pests include wild oats, skeleton weed, Mexican feather grass, Spanish thistle, serrated tussock grass, and Paterson's curse. The most costly damage inflicted by invasive weeds is to crop systems, which suffer an estimated damage of AU$1.271 billion each year. Damage to pasture land accounts for another AU$494 million per year, while the horticultural industry bears a cost of AU$213 million each year.

Bomford and Hart[5] expand the knowledge of invasive vertebrate species in Australia and indicate that more than 80 species of nonindigenous vertebrates have become estab-lished in Australia. Of these species, more than 30 have become serious pests, among them the European rabbit, feral pigs, feral cats, the dingo dog, feral goats, the European starling, and the cane toad. The direct economic losses caused to agriculture by these introduced vertebrate pests are an estimated US$420 million per year. Control costs borne by the gov-ernment and landholders represent an additional US$60 million per year, while another US$20 million or so is spent on related research.

Although no estimate is reported here as to the overall number of invertebrates that have been introduced into Australia, several of the major nonnative invertebrate pests are discussed by Canyon, Naumann, Speare, and Winkel in Section I. These species include the mosqui-toes *Aedes aegypti* and *Culex gelidus*, both of which transmit serious diseases; honeybees and wasps, which cause human deaths; red fire ants, which cause human, livestock, and wildlife problems; the cattle tick; screw-worm fly complex; the red-legged earth mite, which damages crops; and the European wood wasp, which attacks forests. The invasive species investigated by the contributors indicate that invertebrates in Australia are responsible for more than $5.3 billion in annual damage and control costs. One table listing several of the major pests estimated damages totaling AU$4.7 billion per year from this group of pests alone.

## 1.2  Brazil

Rocha, Bergallo, and Mazzoni report in Section II that Brazil has a high biodiversity of vertebrates, a total of 6623 species; the proportion of invasive species (2.06% or 137 species) cannot be considered negligible especially considering their total impacts. Fish invaders, which number 109 (4.2%), and mammals (2.45%) are groups of vertebrates that presently have a higher proportion of invasive species among the known living species in Brazil. Of the 850 amphibian species known to occur in Brazil, only 3 (0.35%) species are invasive species in some areas. Among reptiles, of the 709 species recorded in Brazil (including lizards, amphisbaenians, snakes, turtles and crocodiles), 5 species (0.71%) are invasive. Presently, in Brazil, 1825 bird species have been recorded, and of these, 4 (0.22%) are invasive species. Pooling these vertebrate groups, the data show that at least 137 (2.06%) of the living vertebrate species in Brazil are invasive exotics. The most serious invasive mammal species include rats, cats, and pigs.

## 1.3  British Isles

In his study of invasive plants in the British Isles, Williamson found that the number of native plant species is about 1500, whereas the number of known alien plant species is 1642. The number of alien plants that have become well established in natural ecosystems is estimated to range from 210 to 558 species. Most of the damage and control costs, which range from £200 to £300 million per year, are associated with the impact of alien species on crops.

P. C. L. White, Ford-Thompson, Snell, and Harris report in Section III that Britain's native vertebrate species, other nonvertebrate species, and the environment have been affected by the introductions of alien vertebrate species. The number of invasive species introduced into Britain are as follows: mammals, 22; bird species, 21; fish species, 13; and reptilian and amphibian species, 11.

These introduced alien species are estimated to cost Britain about £2 billion per year. Alien rabbits are costing about £529,000 per year through attempts to control the rabbits to protect Britain's forests. About £334,000 is invested in controlling the introduced U.S. gray squirrel that damages trees. It is interesting to note that people, including children, in cities like the gray squirrel. The gray squirrel is also causing a decline in the native red squirrel in Britain.

## 1.4  Europe

The assessment of the impacts of alien vertebrate in Europe carried out by Shirley and Kark reports that at least 88 species of alien mammals have been introduced into Europe. Among the 140 bird species introduced into Europe, economic impacts were reported for 56 species, whereas biodiversity and human health impacts were reported for 27 and 10 species, respectively. Of the bird species, Canada geese are reported to cause the greatest damage to agricultural crops, mostly grain crops. A total of 55 reptile and amphibian species have been successfully introduced into Europe.

A total of 1590 alien arthropod species were reported by Roques as of June 2010. Insects make up 94% of these species, and most (90%) were introduced with ornamental plants. The introduced invertebrates show a strong affinity for environments disturbed by human activities. The insects in greenhouses do remain in these structures, but they are not escape proof. Overall, it is estimated that only 14% of the alien invertebrates have a negative ecological effect on the environment.

The invading plant pathogens of Europe are reported on by Sache, Roy, Suffert, and Desprez-Loustau. Fungi are the most serious group of plant pathogens attacking plants in Europe, and this trend holds for the world. The most serious plant pathogens (fungi) in Europe are those attacking grapes and potatoes. The list of alien fungi contains 688 species and the plant pathogens make up 77% of this list. The plant pathogens attacking crops in the British Isles are estimated to cause US$2 billion in damages per year. Two pathogens, downy mildew and powdery mildew, both introduced from America, are major threats to grapes. The most notable case of a pathogen causing problems in Europe was the fungal attack of potato late blight in Ireland in the mid-nineteenth century that caused massive starvation and mass emigration from Ireland.

## 1.5   India

According to Batish, Kohli, and Singh, invasive plants pose a serious threat to native eco- systems in India by altering plant composition, reducing biodiversity, changing soil struc- ture, and affecting public health, costing at least US$91 billion per year. An estimated 18% of the Indian flora are alien species and are causing severe environmental problems. In the Kashmir Himalayas, there are an estimated 571 alien plant species. These plants are having major impacts on the native flora. Some of the plants that have spread widely and are serious pests are billy goat weed, lantana, and water hyacinth. However, the Indian government is proposing to introduce another serious pest plant, jatropha, for oil produc- tion. Australia already rejected this plant. For oil production, a worker has to work all day (8 hours) to collect US$2.86 worth of nuts for a small quantity of oil.

An enormous number of invertebrate species have been introduced into India, as reported by Ananthakrishnan. These introductions have caused major economic dam- age. For example, the eriophyid mite is causing serious damage to coconut as well as the *Brontispa* beetle. The cotton mealy bug is devastating cotton production in several regions of India. The papaya mealy bug is causing economic damage to a wide array of crops. Another introduced invertebrate, the golden apple snail is a serious pest of rice and is the intermediate host of a nematode parasite on humans.

## 1.6   New Zealand

New Zealand, a historically isolated ancient landmass, has suffered severe damage from invasive species. According to Williams and Timmins, the number of native plant species in New Zealand is about 2000 species, while an estimated 1800 species of alien plant species have invaded the island nation. New Zealand's primary industries of agri- culture, horticulture, and forestry are based on a total of 140 species, most of which are introduced. The cost to New Zealand of defending its borders against new weeds and managing or controlling those that already exist amounts to about NZ$276 million per year. Those species that are not successfully controlled and that directly affect the nation's productive output cost a further NZ$302 million per year.

Clout reports that an ecological catastrophe followed the arrival of humans and alien mammals on New Zealand. Maori settlers brought dogs and rats with them. At least 58 species of endemic bird species were lost in this initial settlement phase, including several flightless bird species. European settlers have successfully introduced more than 90 species of alien vertebrates, including 32 mammals, 36 birds, 19 fish, and 4 species of frogs and reptiles. Several of the vertebrates, including cattle and sheep, have been highly valuable to New Zealand.

## 1.7   South Africa

South Africa suffers from a large number of nonindigenous species. In a detailed analysis of plants introduced into South Africa, LeMaitre, de Lange, Richardson, Wise, and van Wilgen report that more than 9000 plant species have invaded the vast South African ecosystem and about 1000 of these are self-sustaining. Of this number, about 161 species now rank as serious pest weeds. These invasive weeds are causing loss of natural biodiversity, water shortages, loss of crop and forest production, and increased soil erosion. The authors estimate that the annual environmental loss is equal to 2.5% of South Africa's gross domestic product and is just over US$1 billion per year for all weeds. Biological control is proving to be one effective way to control the invading weeds.

The report on invasive vertebrates of South Africa was prepared by van Rensberg, Weyl, Davies, van Wilgen, Peacock, Spear, and Chimimba. South Africa is an alien freshwater fish hot spot. At least 21 species have invaded South Africa. Nearly 300 species of reptiles have been imported in pet stores, but only 3 species have been reported as established. One of the most common is the flowerpot snake imported from the East Indies more than 200 years ago, but it is not causing any significant damage. Only a few species of amphibians have become established, but they remain relatively rare. To date, 77 alien bird species have been recorded in South Africa. Only 12 of the 77 species have potential to be problems, and these include the house sparrow, common starling, common myna, rock dove, rose-ringed parakeet, Indian house crow, mallard duck, and red-billed quelea. The two starling species are pests of fruits and cereal crops. The most serious bird problem is the red-billed quelea, with an estimated population of 1.5 billion, with 190 million in South Africa. The average size flock is 400,000 queleas, and this size flock can consume 1.6 tons of grain in one day with a value of US$128,000, which places an enormous burden on South Africa. Only 50 or so mammals have been introduced, including cattle, sheep, and goats. Those that are pests include the European rabbit, several species of common rats, mice, and feral cats. One of the serious mammal pests is the common pig.

## 1.8   United States

Pimentel reports that more than 50,000 species of plants, animals, and microbes have been introduced either accidentally or intentionally into the United States in the past 100 years. Among these are 128 crop species that were intentionally introduced into the United States but have since become annoying weeds or serious pests of agriculture and horticulture. One such pest is Johnson grass, which was introduced as a forage grass but now is a major pest weed throughout the southern United States. The melaleuca tree, intentionally introduced as an ornamental tree, is now spreading rapidly throughout Florida and other southern states, where it displaces native trees and other vegetation and is removing vital moisture from the Everglades and other ecosystems.

The spread of invasive weeds causes an estimated $34 billion in damage and control cost in the United States each year. When invasive plants displace native vegetation, the native animals and microbes associated with the native vegetation species are greatly reduced in number. Most of the damage from invading plants in the United States occurs to natural ecosystems, primarily in the South and the West.

Vertebrate species introduced to the United States cause an estimated $46.8 billion in damage and control costs each year, with rats and cats being responsible for the majority of the problems and losses. Meanwhile, invading invertebrate species, such as pest insects, destroy some $14.7 billion worth of the U.S. crops and forests each year.

Invading plant pathogens attack crops and forest, causing an estimated $13.1 billion worth of damage and control costs annually in the United States. An additional $91.6 billion is spent in the United States to deal with introduced microbes, such as HIV/AIDS and influenza viruses.

An assessment of the impacts of invading vertebrates in the United States was conducted by Fall, Avery, Campbell, Egan, Engeman, Pimentel, Pitt, Shwiff, and Witmer, and they reported on nearly 20 species of alien vertebrates that have been introduced into the United States. These animals are estimated to cause more than $47 billion in damages and control costs per year. Rodents (rats) cause the most damage, including the Norway rat, roof rat, Asian house rat, and Polynesian rat. Of the group, the Norway rat is the most destructive and dangerous due to diseases it carries. Mice can be and are frequently a problem. Mice are short lived but can reproduce at an enormous rate. For example, 20 mice can increase to 2000 in just 8 months! Swine are a growing problem in the southern United States. Pigeons are a problem in cities and towns, and starlings are a problem to agriculture. The Burmese python is a growing problem in Florida and is spreading. The noisy coqui frog is a problem in Hawaii and Florida. The European and Asian carp are growing problems in many regions of the United States.

## 1.9   World overview

In a preliminary investigation, Pimentel et al.[6] summarize the economic and environmental damage caused by alien plant, animal, and microbe species in the United States, the British Isles, Australia, Europe, South Africa, India, and Brazil. They report that more than 120,000 nonnative species of plants, animals, and microbes not only have invaded these nations but also have become well established in the new ecosystems. The invasion of these nonnative organisms causes more than $300 billion per year in damage and control costs in those key regions.

Kim reports on the number of humans infected by invading organisms in Australia, Brazil, the British Isles, India, New Zealand, South Africa, and the United States. Surprisingly, little is known about the origins and the spread of several pathogenic diseases that affect human health.

One of the most recent invading infectious organisms, and now one of the best known, is the HIV virus, which causes AIDS. In the seven nations studied, nearly 9 million people are currently infected with HIV/AIDS, with about 7.6 million infected initially in South Africa and India. The World Health Organization (WHO) estimates that $7 billion per year is needed to fight against HIV/AIDS.

Worldwide, about 2 billion people are currently infected with tuberculosis (TB), and 2.4 billion are infected with malaria. These two diseases are causing enormous economic hardships and a great many deaths each year. The WHO reports that several billion dollars are needed to control these two major diseases. In India alone, TB costs $3 billion each year in terms of deaths, lost work, and medical treatment. AIDS, influenza, and syphilis claim the lives of more than 40,000 people each year in the United States and treatment costs for these diseases plus syphilis total more than $90 billion per year and do not include the other exotic diseases.

The information provided in this book reconfirms the diverse and unpredictable roles that nonnative species assume as they invade new ecosystems. They often attack vital crops and forests, and they may cause major damage to ecosystems that result in loss of biodiversity, soil erosion, and water loss. In addition, major human and livestock diseases have invaded many countries, resulting in significant health and economic problems.

Alien species invasions will be an ongoing problem in the future as the human population multiplies and becomes increasingly mobile. The increasing movement of goods associated with globalization will also tend to accelerate the spread of alien species as never before.

## Acknowledgment

We wish to express our sincere gratitude to the Cornell Association of Professors Emeriti for the partial support of our research through the Albert Podell Grant Program.

## References

1. Pimentel, D., and M. Pimentel. 1996. *Food, Energy, and Society*. Rev. ed. Niwot, CO: University Press of Colorado.
2. USDA. 2000. *Agricultural Statistics*. Washington, DC: U.S. Department of Agriculture.
3. Nature Conservancy. 1996. *America's Least Wanted: Alien Species Invasions of U.S. Ecosystems*. Arlington, TX: The Nature Conservancy.
4. Elton, C. S. 1958. *The Ecology of Invasions by Animals and Plants*. London: Methuen.
5. Bomford, M., and Q. Hart. 2002. Non-indigenous vertebrates in Australia. In *Biological Invasions: Economic and Environmental Costs of Alien Plant, Animal, and Microbe Species*, ed. D. Pimentel, 25. Boca Raton, FL: CRC Press.
6. Pimentel, D., S. McNair, J. Janecka, J. Wightman, C. Simmonds, C. O'Connell, E. Wong, L. Russel, J. Zern, T. Aquino, and T. Tsomondo. 2002. Economic and environmental threats of alien plant, animal, and microbe invasions. In *Biological Invasions: Economic and Environmental Costs of Alien Plant, Animal, and Microbe Species*, ed. D. Pimentel, 307. Boca Raton, FL: CRC Press.

*section one*

---

*Australia*

*chapter two*

# The impacts of alien plants in Australia

*Richard H. Groves*

## Contents

## 2.1  Introduction

A large number of alien plant species have been introduced to Australia, both accidentally and deliberately. One publication on Australian plants of horticultural significance[1] lists some 30,000 plant names as being available from 450 nurseries in all states and territories of Australia. This listing includes not just plant species that have been deliberately introduced but also some native plants that are used in horticulture, together with their synonyms and cultivar names. A subsequent publication[2] cites more than 27,000 taxa known to be present in the alien flora of Australia—a total number slightly more than the estimated number of higher plant species native to Australia. Not all of the ca. 27,000 alien species have become naturalized, however. The most recently published listing of the naturalized alien flora of Australia[3] gives a total of 2681 plant species that are known to be naturalized and to have voucher specimens lodged in Australian herbaria. In other words, about 10%–15% of the total Australian alien flora is naturalized. A minority of these alien and naturalized plant species affect, or are perceived to affect, human activities in some way and may be regarded as weeds. This chapter considers some of these 2681 alien naturalized plant species and their impacts on Australian ecosystems, but the coverage will not be limited to the smaller proportion of naturalized aliens that are regarded as weeds.

Some cosmopolitan species may be regarded as either alien or native.[4–6] For Australia as a whole, it was estimated[3] that this uncertain status applies to only 34 plant species, that is, 1.1%, which is a small proportion of the total alien flora. These relatively few plant species typically occupy either wetlands or beach strandlines, where bird- or water-dispersed species predominate. Although these cosmopolitan species may be numerically significant among most insular floras, including those of Australian islands, they comprise only a trivial proportion of the flora of the large land mass of continental Australia.

The proportion of the total Australian flora that is alien varies from region to region and from ecosystem to ecosystem. For instance, offshore islands have a high proportion of alien species (60% on Norfolk Island[7] and 48% on Lord Howe Island[7,8]), whereas the floras of some arid (Uluru National Park[9]) and alpine (Kosciuszko National Park[10]) areas are only about 5%–7% alien.[11] Although the percentage of alien species may not have changed greatly over time since European settlement, the number of naturalized alien species has increased inexorably since the first state floras were compiled. Specht[12] showed a fairly constant rate of increase of about five species per year per state for Queensland, New South Wales, Victoria, and South Australia for the period from 1870 to 1980. More recently, Groves et al.[13] provided evidence that, nationally, this rate may have increased over the period from 1981 to 1995. Certainly, they could find no evidence that the proportion of naturalized alien plants in the Australian flora had decreased, despite slightly more than 100 years of quarantine legislation that regulates the entry of alien plant species to Australia.

In this chapter, I will discuss the impacts of this increasing number of alien plant species on the Australian community from the perspective of the economics of crop and pasture enterprises, native plant diversity, and human and animal health. Although in some cases, the effects of alien plants on some agricultural systems have been quantified and the cost–benefit ratios of managing them calculated, few such quantitative estimates are available regarding the impacts on natural ecosystems in terms of native plant and animal diversity or human and animal health aspects. Some recommendations are made for further research on the impacts of alien species on Australian ecosystems and the native plant diversity present in those ecosystems.

## 2.2   *Impacts on agricultural ecosystems*

Alien plants influence crop and pasture ecosystems in many ways. The crop systems themselves consist largely of alien economic plants, as few native Australian plants have been domesticated. The plant species that form the basis of pasture ecosystems in southern Australia are also alien, having been introduced mainly from Mediterranean Europe. On the other hand, most of the plants that form the basis of northern (summer-wet) and central (semi-arid, rangeland) grazing systems in Australia are native to those regions. The negative impacts of alien species on agricultural ecosystems in southern Australia will be stressed, because that is where the available data are concentrated. However, it should also be recognized that some alien species, such as *Trifolium subterraneum*, whilst useful in pasture ecosystems may impact negatively on crop systems.

### 2.2.1   *Economic aspects*

The negative impacts of alien plants on crop and pasture systems throughout Australia in general have been estimated. The incidence of alien plants leads to the need to cultivate land for crops or to resow pastures, or to spray with herbicides, or both. The presence of these aliens is associated directly with reductions in crop or pasture yield and with product contamination. Some alien plant species may poison animals or lead to poor animal performance.

Each aspect incurs a financial cost. For Australian crop systems as a whole, Combellack[14] was the first to estimate the financial costs of each aspect. For the financial year 1981–1982, cultivation to control alien plants cost AU$592 million and purchase of herbicides cost $137 million, plus $34 million (all amounts in this chapter are in Australian dollars (AU$)) to apply them. In the same year, losses in crop yields were estimated to be $422 million and product contamination to cost $86 million. These estimates gave a total annual cost of alien plants in crop systems of $1.271 billion for that year. Financial estimates of the negative impacts of alien plants in pasture systems for the same year were $494 million, in horticulture $213 million, and in "noncrop" areas $119 million. For all agricultural systems, Combellack's estimates (based on 1981–1982 statistics) totaled $2.1 billion, which translated into $3.3 billion in 1995-dollar terms (see Jones et al.[15] for questions on the validity of such an extrapolation). The results of a subsequent study showed that product losses and expenditure on control of aliens at current infestation levels in crop systems amounted to $1.133 billion for the financial year 1988–1999.[15] Sinden et al.[16] updated these previous estimates of the costs of weeds to Australia and arrived at a figure of between $3.554 billion and $4.532 billion per year. These latter authors were unable to estimate the impacts of alien plants in urban areas and the cost of weeds to human health (see Section 2.4). Whatever the accounting system used, alien plants annually cost Australian agricultural producers a substantial amount of money, and the costs are also borne by consumers.

The direct financial cost of a few individual weeds in the crop and pasture systems has been estimated. The alien species complex called wild oats (*Avena* spp.) in grain crops was estimated to cost $42 million for the financial year 1987–1988.[17] Nevertheless, this estimate was conservative, because it did not include the cost of grain contamination, increased opportunity provided by the *Avena* to host pathogens, or increased resistance to control methods. The financial impact of skeleton weed (*Chondrilla juncea*) on wheat crops was estimated at $20 million[18,19] for the financial year 1972–1973, of which $18.5 million was attributable to lost productivity and $1.5 million to spraying costs. These two examples suffice to show that the costs associated with the presence of some alien species among crop systems are considerable. With a pattern of increasing resistance to herbicides shown by several of these alien species, especially the annual grasses group, the costs of aliens in crop systems will increase.

Some other negative impacts of alien plants on agricultural systems may add to the annual costs. For instance, attempts to prevent the incursion of two alien plant species (*Nassella tenuissima*, or Mexican feather grass, and *Onopordum nervosum*, a Spanish thistle) that have serious potential to modify pasture systems were estimated to generate benefits of $83 million to producers in 2000–2001.[20] This estimate was based on a reduction in the probability of these weeds becoming naturalized, and thereby a reduction of potential costs they would impose should they ever become established. In 2001, both species were available only in the nursery industry (as plants or seeds, respectively, for landscaping) and were not yet known to have escaped cultivation, let alone become naturalized. In 2010, the situation with *N. tenuissima* was less certain; some material may still be found in suburban gardens, and search attempts continue, although there is still no record of its naturalization.

Alien plants also directly impact pasture ecosystems. Serrated tussock (*Nassella trichotoma*) is a perennial grass of South American origin that reduces the livestock-carrying capacity of southern Australian pastures. Its presence incurs an annual cost of $40 million in New South Wales[21] and, in 1997, about $5.1 million in Victoria.[22] If the weed is not contained in Victoria, the cost estimate could increase to $15 million or more per year by 2010.[22] The same species is now spreading in Tasmania as well, representing an additional but hitherto unquantified cost to southern Australian pasture production.

Financial estimates of the costs of other individual alien species in southern Australia pastures are also available for two cases in which biological control of the species was proposed but faced opposition from some sectors of the Australian community. In each case, there were demonstrable conflicts of interest arising from the fact that some alien species have both negative and positive effects. For instance, costs of the alien species were estimated as part of the overall decision to allow the release of biological control agents. The first case concerns Paterson's curse (*Echium plantagineum*), which produces alkaloids that affect liver function in grazing animals (especially sheep), but which also produces honey with a pale color preferred by exporters to the Japanese market. Further, though Paterson's curse is a serious pasture weed in most parts of southern Australia, it may be considered useful fodder for animals in some semiarid rangelands, especially in northern South Australia, where its common name is, appropriately, "salvation Jane." An independent inquiry into the negative and positive aspects of biological control of this weed recommended the release of insects to control the growth and flowering of Paterson's curse on the basis of an economic analysis of the costs ($30 million annually) and benefits ($2 million annually) to Australia.[23]

My second case concerns blackberry (*Rubus fruticosus* agg.). Data gathered in the 1980s considered the additional costs to Tasmanian berry growers and honey producers for biologically controlling blackberry with a rust proposed for release, balanced against the benefits of increasing pasture production by controlling blackberry. The data collected in the early 1980s indicate a total annual cost of $41.5 million to Australia.[24] Subsequently, both the negative and positive impacts of blackberry infestations have been itemized by James and Lockwood.[25] These authors stressed the need to collect much more information on current distribution and impact valuation before an up-to-date economic analysis can be made for the 8.8 million hectares that blackberry occupies in southern Australia.

These examples collectively show that alien plants can directly and significantly impact southern Australian crop and pasture systems. In economic terms, the negative impacts of alien species far outweigh any positive ones. Continuing research that leads to improved levels of control of such aliens that negatively impact agriculture is usually highly cost-effective.[18,20] Less attention has been paid to the effects of such alien species on the sustainability of southern Australian agriculture (and specifically its profitability), although with steadily increasing groundwater salinity and the increasing resistance of crop and pasture weeds to herbicides, these effects are urgently in need of increased research attention.

## 2.3  Impacts on natural ecosystems

### 2.3.1  Biodiversity aspects

The impacts of alien plants on natural ecosystems are complex and vary with human attitudes and knowledge. Impact assessment in these systems can be highly subjective. For instance, a few people express zero tolerance for alien plants in natural ecosystems. To them, any alien species in a nature reserve lessens the quality of the natural environment. Other individuals may tolerate some alien plants, such as those with brightly colored flowers in the ground layer, but will be intolerant of spiny shrubs or rampant vines that may prevent access to waterways or viewpoints. At the other extreme are those individuals who do not even recognize some species as being alien to Australia, such as willows (*Salix* spp.) or poplars (*Populus* spp.), in part because they appear so frequently in early paintings of the Australian landscape; some Australians in fact believe such aliens to be native species. The impacts that alien plants in natural ecosystems may have on people vary far more than do alien plants in agricultural systems, where alien species are usually identified more accurately and their costs estimated more realistically.

A consideration of alien plants in natural ecosystems is also made more complex because the same plant species may affect both agricultural and natural ecosystems. For instance, blackberry is a major weed of pastures, but it is an equally major weed in natural ecosystems, especially along waterways in southern Australia. Furthermore, blackberry is strongly weedy in the establishment phase of forest plantations. The weediness of St. John's wort (*Hypericum perforatum*) was first recognized as a weed of dairy pastures. Land use changed, as a result of this weed status, from pasture to forest plantations of *Pinus radiata* in some regions. Currently, the same species occurs mainly in natural ecosystems and roadsides along which it spreads, although it continues to be weedy in pine plantations.

A further example is provided by horehound (*Marrubium vulgare*). This native of the Mediterranean region was introduced to Australia as a source of herbal compounds. It spread to become a weed of sheep-grazed pastures in relatively high-rainfall regions, where it is still a problem plant because of its unpalatability. More recently, it has increased in dominance in some semiarid areas, where its fruits are spread not only by sheep but also by native animals such as kangaroos. In northwestern Victoria's Wyperfeld National Park, it is common to see horehound as a major weed in areas where kangaroos congregate and rest overnight.

From these few examples, it is clear that the distinction between alien plants in agricultural and natural ecosystems is far from rigid, and many widespread alien plants impact both systems, albeit in different ways. What's more, their relative impacts on each system may change with time as the same species come to have less effect on agricultural systems and more on natural ones.

As with agricultural systems, alien plants impacting natural ecosystems do so either negatively or positively, and some may have no apparent effect. Adair and Groves[26] proposed four hypothetical models for assessing the relationships possible between alien infestations and the biodiversity of natural ecosystems (Figure 2.1). Such models were able to relate levels of biodiversity (e.g., native species richness) to some measure of alien plant infestation. Such models require further development and testing, however, before they become generally acceptable, especially to managers of land affected by alien plants.

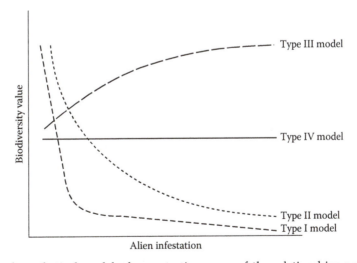

*Figure 2.1* Four hypothetical models demonstrating some of the relationships possible between biodiversity value (e.g., number of native plant species) and alien infestation (e.g., weed density). (Redrawn from Adair, R. J., and R. H. Groves. 1998. Impact of environmental weeds on biodiversity: A review and development of a methodology. National Weeds Program, Environment Australia, Occasional Paper, Canberra.)

Consider the following two examples of different impacts of alien plants on some Australian natural ecosystems. The first is *Mimosa pigra,* a native of Central America, which has been introduced to the tropical wetlands of the Darwin region of northern Australia. Negative aspects of this leguminous shrub on the ecosystem include the formation of monospecific thickets of shrubs that replace the native sedgeland, on which the endangered magpie goose (*Anseranas semipalmata*) has historically depended for nesting sites and food. Overall, bird abundance was reduced as a result of mimosa infestation, as was lizard abundance.[27] On mimosa-infested sites, there was also less herbaceous vegetation and fewer native tree seedlings than in noninvaded natural vegetation. All these indices of biodiversity were affected negatively by the presence of high densities of *M. pigra.* On the other hand, frog numbers seemed to be unaffected by mimosa density—an example of a neutral impact. In the same ecosystem, the presence of *M. pigra* was associated with increased numbers of the rare marsupial mouse *Sminthopsis virginiae,* presumably because of the increased high-quality food supply provided by *M. pigra* seeds and the increased shelter from predators provided by the dense thickets of the alien shrub. The latter is clearly a positive impact if one measures only *Sminthopsis* numbers as an index of biodiversity. Depending on which measure of biodiversity is chosen, impacts may be negative, neutral, or even positive, and three of the four models proposed[26] may apply to this one alien when measured by different indices of biodiversity.

Much the same mix of impacts seems to apply to the second example, which concerns the invasion of an arid river system in central Australia by the tree species tamarisk (*Tamarix aphylla*), a Eurasian species.[28] The banks of the Finke River in central Australia were originally dominated naturally by river red gum (*Eucalyptus camaldulensis*), but after serious flooding of the river system in 1974, tamarisk seeds were washed downstream from homesteads where the trees had been planted for amenity and shade. Within 15 years of the flood, the eucalypt riverine woodland had changed to one dominated by tamarisk, and various indices of biodiversity had changed markedly. Regeneration of the previously dominant native tree was reduced, the floristic composition of the ground vegetation was changed, and the numbers of reptiles and most birds were reduced—all negative impacts. But, while most bird species declined, some aerial insectivorous species increased—a positive impact—and there seemed to be no effect on the number of granivorous bird species, a neutral effect. The increase in some aerial insectivorous birds may, in turn, lead to positive or negative effects on other aspects of the ecosystem.[27]

For both examples, the impacts are overwhelmingly negative in human terms. Mimosa infestations and the potential spread of this alien to nearby World Heritage-listed Kakadu National Park threaten the traditional food-gathering patterns of the resident Aborigines. In addition, the tourist industry is potentially threatened by the associated loss in regional ecosystem diversity. The value of production from pastoral areas in the region adjoining Kakadu is compromised, and the rounding up of cattle on these properties is made more difficult. The negative effects of mimosa on human values justify the large amount of research funds already spent on mimosa control, whereas the impacts of mimosa on biodiversity are mixed; so, too, with tamarisk in relation to its ability to increase the salinity of the invaded region, although less has been spent on its control in the Finke River system. A replay of the scenario that happened in the Finke River may be happening currently in the lower Gascoyne River, near Carnarvon, in Western Australia. This situation may eventually bring increased attention to the control of tamarisk in other regions of semiarid and arid Australia. My second example is matched by an analogous situation that has developed in the southwestern United States and northwestern Mexico, where the closely related *T. ramosissima* is having similarly strong negative effects on the salinity of river systems and native fish populations in these semiarid contiguous regions.[29]

The next two examples of alien species' impacts on natural ecosystems concern individual species, rather than the ecosystems of which the species form a part. The alien climbing species bridal creeper (*Asparagus asparagoides*), native to South Africa, has been shown to directly affect the populations of two rare or threatened native plant species.

The first native species is the sandhill greenhood orchid, *Pterostylis arenicola*, which is known to be indigenous to only several sites in South Australia. The orchid is terrestrial; it emerges from root tubers in late autumn each year and forms a small, flat rosette of leaves over winter, then flowers in spring before senescing in late spring.[30] The phenology of the native orchid is matched almost exactly by that of the alien bridal creeper, which sprouts annually from a mat of perennial tubers in autumn, overtops native vegetation, flowers in spring, and forms berries in late spring in summer-dry areas of southern Australia. Thus, the cover of the alien is at its peak when the native orchid rosettes are present, which means that the latter are shaded and rendered less competitive. Sorensen and Jusaitis[30] showed that with bridal creeper absent, the number of orchid rosettes present was about 40 per m$^2$, whereas with the alien present, the number of rosettes was less than 10 per m$^2$. In this instance, there seem to be no positive or neutral impacts on the ecosystem. This example represents one of the few in Australia in which the effect of an alien on the population of an endangered native plant species has been quantified.

The low shrub *Pimelea spicata* is a minor but once-common component of the shrub layer in Cumberland Plain Woodland to the southwest of Sydney. At one site where its numbers are greatest (several thousands), it co-occurs with bridal creeper. Again, the phenology of the native matches that of the alien. *Pimelea spicata* has a thick perennial taproot, from which new shoots emerge after drought, fire, or other natural disturbances. Shoots elongate and then flower any time between spring and the following autumn, depending on summer rainfall patterns (the Sydney Basin has year-round rain compared to regions farther south or inland). Bridal creeper's effect on this uncommon native plant is both to smother its shoots through winter and spring and to compete with it for water and nutrients when its shoot canopy has senesced.[31] Once again, there seem to be no positive effects on the part of the alien from this species–species interaction in the woodland ecosystem in which both occur.

These two examples of the impacts of bridal creeper illustrate perhaps the most appropriate way to explore the direct interactions between alien and native species, both in controlled conditions in a greenhouse and experimentally in the field. Interactions between alien species and natural ecosystems are more complex and often indirect, and different types of impacts (negative, positive, and neutral) and combinations of those impact types are possible. A further complexity is that alien plants may provide food and refuge for aliens from other taxonomic groupings, whether they be vermin (such as foxes or pigs), pests (e.g., insects), or pathogens (crop diseases and the like). The few generalizations that can be drawn from this limited number of examples need further testing as more examples are documented. In terms of bridal creeper examples, the relationships between alien density and biodiversity value are probably best represented by Type I or II models.[26] The previous examples of the impacts of alien plants on ecosystems are more complex and involve mixes of Type II, III, and IV response models (Figure 2.1).

A start has been made recently to assess the number of native species and plant communities at risk from continued expansion of bitou bush (*Chrysanthemoides monili-fera* subsp. *rotundata*),[32] the results of which show a surprisingly high number of species (34 plant species and 4 animal species) and five ecological communities at risk from bitou bush and boneseed (*C.m.* subsp. *monilifera*) invasion in New South Wales. A similar methodology[33] has been employed more recently to prioritize alien species on their threat and ability to impact biodiversity in New South Wales. The modeling process identified three "extreme" alien species (bitou bush, lantana [*Lantana camara*], and Madeira vine [*Anredera cordifolia*]) and a further 19 "very high-priority" alien species with respect to their ability to negatively impact biodiversity.[33] The impacts of lantana (*Lantana camara*) on native plant species and plant community types in New South Wales and Queensland have also been assessed recently.[34] Despite this recent increased effort on several major alien plant species, it remains surprising that the impacts on species richness (as one measure of biodiversity) of the majority of Australia's major alien plants are still unknown,[26] even though they are recognized as weeds of national significance, and major programs for their control are underway.

## 2.3.2   Economic aspects

Data on the economic impacts of alien species on natural ecosystems are few and indirect. Panetta and James[35] presented a strategy for collecting and analyzing such data to overcome this deficiency. Several attempts have been made to obtain data by assessing the cost-effectiveness of control programs for alien species. For instance, Sinden et al.[16] estimated that at least $19.6 million was spent each year controlling alien plants in national parks and other areas listed in National Heritage Trust agreements; furthermore, this level of spending was rising rapidly each year. The financial costs of managing broom (*Cytisus scoparius*) in Barrington Tops National Park in central coastal New South Wales have been studied.[36,37] Use of a detailed bioeconomic model showed that intervention in the management of broom in natural ecosystems is clearly economically justified and that a combination of control measures, rather than any single measure, is almost always justified on economic grounds.[36] Furthermore, a combination of controls that targets both alien plant density and the seed bank is important in the longer term, because being a leguminous shrub, broom has a long-lived seed bank.[37]

A cost–benefit analysis for the control of bitou bush, an alien from South Africa now dominating coastal vegetation in eastern Australia, involves both the plant's effect on biodiversity and its threat to public access to beaches. A control program for bitou bush has been implemented that involves strategic use of herbicides, hand-pulling of plants by volunteers, release of a number of highly specific insects from South Africa for long-term biological control, and some revegetation with competitive native plant species. A preliminary economic analysis of the cost-effectiveness of this program arrived at a benefit-to-cost ratio of about 20.[20] With the knowledge recently obtained concerning the impact of bitou bush on biodiversity values,[32] a survey using choice modeling to better measure the economic impact of aliens such as bitou bush now needs to be instituted to assess the economic value the Australian public places on biodiversity. Only then can the impacts of aliens on biodiversity be analyzed economically. Given that there are few well-documented examples of the influences of aliens on the biodiversity of Australian ecosystems, and even fewer on the financial costs of those impacts, it is clear that more examples are needed. This is particularly true for major alien species in northern Australia, such as rubber vine (*Cryptostegia grandiflora*) and prickly acacia (*Acacia nilotica*). Adair and Groves[26] have suggested that, given active control programs for these aliens, it may be more important to determine threshold levels for declines in biodiversity and to identify management barriers to invasion (or reinvasion), rather than simply measuring impacts in some generalized manner.

## 2.4    Impacts on human health

As with the impacts on agriculture and native biodiversity, aliens may have both positive and negative effects on human health. A relatively small number of the alien plant species introduced to Australia were imported deliberately for their putative herbal properties. I have already mentioned the case of horehound (*Marrubium vulgare*), but others include St. John's wort (*Hypericum perforatum*), variegated thistle (*Silybum marianum*), and, possibly, intentional introductions of dandelion (*Taraxacum officinale*) and pennyroyal (*Mentha pulegium*). Some species that were introduced accidentally are now found to have some benefit to human health. For instance, Paterson's curse (*Echium plantagineum*) is now cultivated, albeit in England, because its seeds are high in omega-3 acids (S. Lloyd, pers. comm.).

Many plants contain compounds that cause physiological reactions in people that negatively affect their well-being and quality of life, and alien plants in Australia are no exception. Although the examples chosen here apply only to aliens present in Australia, I stress that they apply equally to the same plants present in other countries, whether they be considered alien or native to those environments.

Parthenium weed (*Parthenium hysterophorus*) is native to the southern United States and Mexico, as well as to Central and South America. The North American variant of this species has come to have a major impact on humans in central Queensland.[38] This variant contains parthenin, a sesquiterpene lactone that can cause allergic dermatitis in humans who have repeated contact with parts of the plant, especially the flowers or the trichomes on its leaves.[38] Cases of dermatitis have been reported in the United States, where the parthenin-containing variant is native, although most such cases, including some deaths,[38–40] have been reported in India, where the plant is an alien. The problem is less acute at this time in central Queensland, but contact dermatitis has been recorded in that region as well.[41] The dermatitis seems confined to adult males in most cases reported, presumably because of parthenin's interaction with male sex hormones. The human health problem caused by parthenium weed could become greater, given its potential to spread further in Australia.[42]

Skin irritation (urticaria) and allergic rhinitis have been reported to occur in Australia after repeated contact with Paterson's curse (*Echium plantagineum*).[43] The causative ingredient is unknown; it may be one of the pyrrolizidine alkaloids that are known to affect animal health[44] and are contained in the plant hairs or other particulate matter. The closely related *Echium vulgare* has been recorded as causing dermatitis, but not urticaria.[44] Other alien plant species known to cause forms of skin irritation include some of the brassicas (*Brassica alba* and *B. napus*), the nettles (*Urtica* spp.), *Erigeron* spp., stinkweed (*Inula graveolens*), and the garden plant *Rhus toxicodendron*,[45] which is not yet known to be naturalized in Australia.

A large number of Australians are affected chronically by hay fever (allergic rhinitis) and chronically or acutely by asthma (allergic bronchitis) as a result of inhaling allergenic pollen produced mainly in spring by a wide range of alien plants. Many of the introduced grasses (especially *Lolium* spp.) are a major source of such pollen, as are radiata pine (*Pinus radiata*), the ragweeds (*Ambrosia* spp.) in southern Queensland, pellitory (*Parietaria judaica*) in urban Sydney, and more widely the privets (*Ligustrum* spp.), the olives (*Olea europaea* and *O. cuspidata*), the poplars (*Populus* spp.), and peppercorn (*Schinus molle*).[45] Medical statistics on the prevalence of this condition are confounded, however, because some native plants, such as *Atriplex spongiosa* and *Allocasuarina* spp.,[45] also produce allergenic pollen.

A few alien plants contain poisonous compounds, which if ingested may lead to serious illness and death. Examples include thorn apples (*Datura* spp.), arum lily (*Zantedeschia aethiopica*), and hemlock (*Conium maculatum*). While contact with leaves of oleander (*Nerium oleander*) may cause eczema, ingestion of its leaves or flowers can cause death because the toxic glucosides it contains have a digitalis-like action in humans. Gardner and Bennetts report that people have even "been fatally poisoned by eating meat when oleander twigs were used as skewers or spits during its cooking (p. 152)."[45]

It is questionable whether the presumed positive effects of alien plants on human health by way of increasing recognition of the value of herbals for well-being in Australia will ever outweigh the decrease in that same well-being caused by chronic allergenicity. For the present purposes, however, both are significant aspects of the overall impact of alien (and native) plants on the health and well-being of the Australian public and the national economy.

## 2.5   Impacts on animal health

Many alien plants contain chemical compounds that affect animal health to varying extents, and hence affect agricultural productivity. Animal health and even animal survival after ingestion of such chemicals depend on many factors, including past grazing history, stage of plant growth, whether the diet is mixed or monospecific, and the type of animal (whether monogastric or not), in addition to the level and nature of the actual toxic constituents in the alien plants. Different cultivars of the same plant species may contain different levels of the active compounds, as in subterranean clover (*Trifolium subterraneum*).[46] The following examples illustrate that the impacts on animal health can be acute or chronic and negative or positive, depending on the particular situation.

Much of southern Australian animal production depends on pastures that contain the alien species subterranean clover (*Trifolium subterraneum*), phalaris (*Phalaris aquatica*), and ryegrasses (*Lolium* spp.), all of which are Mediterranean in origin. Although such species form the very basis of animal productivity in southern Australia and have a strongly positive impact, under certain circumstances the same species can strongly and negatively influence animal health. Subterranean clover plants contain estrogens that can cause abortions and infertility in sheep that graze pastures dominated by this species.[46] If sheep are forced to eat only phalaris or ryegrass in their diet, especially if they are suddenly moved to pastures in which these species are actively growing, they can suffer and even die from a nervous condition known as "staggers" that is caused by the alkaloids present in phalaris or ryegrass.[46] In these cases, a mixed diet appears to overcome such drastic negative effects.

Other alien plants, such as variegated thistle (*Silybum marianum*), may contain large amounts of nitrate ions in spring, which if ingested in sufficient amounts can cause blood poisoning.[46] The alien St. John's wort (*Hypericum perforatum*), widespread in southern Australia, contains hypericin, which if ingested in sufficient quantities can cause photosensitization in animals, especially sheep. Because the flowers contain the highest concentrations of hypericin, grazing in pastures dominated by St. John's wort in late spring can lead to a suite of debilitating symptoms that reduce animal condition.[47] Affected animals recover when they no longer ingest St. John's wort. The equally widespread alien plant Paterson's curse (*Echium plantagineum*) contains eight pyrrolizidine alkaloids that can interfere with liver function in grazing animals, especially sheep and horses. These alkaloids can cause cumulative liver damage and even death if eaten in large quantities over a long enough period in the spring.[47] The animal health problem is exacerbated if the same animals have access to spring-germinating plants such as heliotrope (*Heliotropium europaeum*), which also contain such alkaloids. In all these examples, symptoms usually can be avoided and animal health maintained by careful pasture and animal management.

   Some aliens can cause poisoning in animals, especially those that produce cyanogenic glycosides and glucosinolates, such as members of the Brassicaceae family. In some cases, these toxic compounds can be broken down in the rumen. Aliens in the Brassicaceae, Oxalidaceae, and Polygonaceae families produce oxalates that may be acutely toxic. Many types of poisoning attributable to many alien plants containing these compounds and some of the factors known to moderate chronic or acute symptoms are discussed in several texts.[46,47] Anecdotal accounts of the effects of potentially poisonous plants straddle the boundary between fact and fiction. I repeat the words on the cover of one of Connor's[48] texts: "Fact and myth conflict in the realm of poisonous plants; a false reputation for toxicity may, over the years, build up around a harmless plant, but true reports of poisoning, though made public, are sometimes overlooked."

   The impacts of alien plants on animal health, thus, may be strongly negative (when the veterinary symptoms are acute) or weakly so (chronic states). On the other hand, the very basis of animal production, at least in southern Australia, depends on the positive effects of alien plants introduced from Mediterranean Europe in terms of the availability of high-quality forage, especially in winter, when native grasses are inadequate to sustain introduced livestock. Many of the alien plants that have negative effects on animal health and production evolved in the same Mediterranean region and their properties have been selected for, either deliberately or inadvertently. It should come as no surprise that negative impacts in their native region are replicated when the same species are introduced to another region, such as southern Australia.

## 2.6   Conclusions

The impacts of alien plants in Australia vary according to the ecosystem and the index considered. The monetary costs of aliens to the Australian economy are high, especially in terms of losses to agricultural productivity and human well-being. With the prospect of more alien species naturalizing[49] and in view of increased resistance to herbicides in some species that are already naturalized, the future appears worrisome. Negative impacts on the diversity of Australian plants, animals, and ecosystems are many, but are largely unquantified scientifically, let alone in economic terms. An increased effort in this regard has commenced[32–34] of the type that will be essential to the provision of better information to decision makers. Positive effects of alien species are reflected in increases in export markets and domestic growth in economic terms, as well as increases in quality of life for humans and their enjoyment of native plants and animals, natural landscapes, and ecosystems. It is possible that the "services" provided by such natural landscapes may be valued more effectively in the near future, especially in relation to the provision of water and the amelioration of salinity. The balance between these two sets of impacts—negative and positive—will influence the future habitability of Australia for its people.

   As mentioned earlier in Section 2.1 of this chapter, the number of naturalized aliens in the Australian flora represents about 10%–15% of the total number of introduced species. I conclude, on the basis of the limited evidence available currently, that only a small proportion of the total of 2681 species have a known impact on the Australian economy, either directly or indirectly by way of effects on agricultural production, native biodiversity, or human or animal welfare. Further, from the limited number of examples cited in this chapter, the impacts of still fewer alien plant species have been documented, and these refer only to those having major, chiefly negative, impacts. If more examples were available, any bias toward negative impact could be tested more validly and a more balanced appraisal arrived at for the overall impact of alien plants in Australia, such as is available for two weed candidates for biological control, Paterson's curse (*Echium plantagineum*) and blackberry (*Rubus fruticosus* agg.; see Section 2.2.1). Furthermore, the examples cited have all referred to single species of aliens, whereas, at least in southern Australia, aliens usually occur as a group of taxonomically diverse species and the impacts of a group of species need to be considered to better reflect the field situation; this shift in thinking has yet to be tackled at a landscape scale, either in an economic sense or in terms of biodiversity loss (but see Downey et al.[50] for a first attempt in this direction for conservation of biodiversity).

Research results, if acted upon, have the potential to reduce the negative effects of alien species and to increase any positive aspects of indigenous species in natural ecosystems. Research sometimes occurs only when the impacts of aliens have been recognized and even quantified in some way. Future research and ecosystem management should be aimed equally at those species only recently introduced or naturalized, before their negative or positive effects are expressed fully. This latter approach has gained some momentum recently in Australia, with the publication of a so-called alert list of alien species[51] and attempts to eradicate some alien species recently detected in Australia and of known major impact elsewhere (e.g., *Kochia scoparia* in Western Australia and *Chromolaena odorata* in coastal Queensland). Increased collation of knowledge of alien species worldwide would help to identify many species not yet known to be present in Australia on which quarantine and research should focus. After all, about the only generalization presently tenable is that if the alien has a negative impact elsewhere, it will most probably have a similarly negative effect if introduced to Australia. International efforts such as this book will help to refine such hypotheses, and thereby reduce the negative impacts of aliens on Australian ecosystems and the people who inhabit and manage them and derive a living or enjoyment from them.

## Acknowledgments

A draft of this chapter for the first edition of this book benefitted greatly from the comments of Jeremy Burdon, Dick Mack, Dick Medd, Trudi Mullett, Dane Panetta, Paul Weiss, and Tony Willis, to all of whom I am grateful. I further acknowledge the help of Paul Downey and Jack Sinden especially in updating information for this second-edition chapter on the environmental and economic impacts of several major weeds of natural ecosystems of southeastern Australia. The results of their recent and continuing research have advanced the subject considerably.

## References

1. Hibbert, M. 2000. *The Aussie Plant Finder 2000/2001*. Glebe: Florilegium.
2. Randall, R. P. 2002. *A Global Compendium of Weeds*. Melbourne: R. G. & F. J. Richardson.
3. Groves, R. H. et al. 2003. *Weed Categories for Natural and Agricultural Ecosystem Management*. Canberra: Department of Agriculture, Fisheries and Forestry.
4. Groves, R. H. 1986. Plant invasions of Australia: An overview. In *Ecology of Biological Invasions: An Australian Perspective*, ed. R. H. Groves and J. J. Burdon, 137. Canberra: Australian Academy of Science.
5. Kloot, P. M. 1984. The introduced elements of the flora of southern Australia. *J Biogeog* 11:63.
6. Michael, P. W. 1994. Alien plants. In *Australian Vegetation*. 2nd ed., ed. R. H. Groves, 44. Cambridge, UK: Cambridge University Press.
7. Green, P. S. 1994. Norfolk Island & Lord Howe Island. In *Flora of Australia*. Vol. 49, *Oceanic Islands 1*, ed. A. E. Orchard, 1. Canberra: Australian Biological Resources Study, AGPS Press.

8. Pickard, J. 1984. Exotic plants on Lord Howe Island: Distribution in space and time, 1853–1981. *J Biogeog* 11:181.

9. Griffin, G. E., and D. J. Nelson. 1988. *Vegetation Survey of Selected Land Units in the Uluru (Ayers–Mt. Olga) National Park.* A report to the Australian National Parks and Wildlife Service, Canberra.

10. Costin, A. B. et al. 1979. *Kosciusko Alpine Flora.* Melbourne: CSIRO/Collins.

11. Humphries, S. J., R. H. Groves, and D. S. Mitchell. 1991. Plant invasions of Australian ecosystems. A status review and management directions. *Kowari* 2:1.

12. Specht, R. L. 1981. Major vegetation formations in Australia. In *Ecological Biogeography of Australia*, ed. A. Keast, 163–297. The Hague: Dr. W. Junk.

13. Groves, R. H. et al. 1998. Recent incursions of weeds to Australia, 1971–1995. Technical Series No. 3, CRC for Weed Management Systems, Adelaide.

14. Combellack, J. H. 1987. Weed control pursuits in Australia. *Chem and Ind* (April, 1987): 273–280.

15. Jones, R. et al. 2000. The distribution, density and economic impact of weeds in the Australian annual winter cropping system. Technical Series No. 4, CRC for Weed Management Systems, Adelaide.

16. Sinden, J. et al. 2004. The economic impact of weeds in Australia. Technical Series No. 8, CRC for Australian Weed Management, Adelaide.

17. Medd, R. W., and S. Pandey. 1990. Estimating the cost of wild oats (*Avena* spp.) in the Australian wheat industry. *Plant Prot Q* 5:142.

18. Marsden, J. S. et al. 1980. *Returns on Australian Agricultural Research.* Melbourne: CSIRO.

19. Cullen, J. M. 1985. Bringing the cost benefit analysis of biological control of *Chondrilla juncea* up to date. In *Proceedings of the VI International Symposium on Biological Control of Weeds*, August 1984, Vancouver, ed. E. S. Delfosse, 145. Ottawa: Agriculture Canada.

20. CIE (Centre for International Economics). 2001. The CRC for weed management systems: An impact assessment. Technical Series No. 6, CRC for Weed Management Systems, Adelaide.

21. Jones, R. E., and D. T. Vere. 1998. The economics of serrated tussock in New South Wales. *Plant Prot Q* 13:70.

22. Nicholson, C., A. Patterson, and L. Miller. 1997. The cost of serrated tussock control in central western Victoria. A report to the Department of Natural Resources and Environment, Victoria, Melbourne.

23. IAC (Industries Assistance Commission). 1985. Biological control of *Echium* species (including Paterson's curse/salvation Jane). Industries Assistance Commission Report No. 371, Australian Government Publishing Service, Canberra.

24. Field, R. P., and E. Bruzzese. 1984. *Biological Control of Blackberry, Report 1984/2.* Frankston: Keith Turnbull Research Institute, Department of Conservation, Forests and Lands.

25. James, R., and M. Lockwood. 1998. Economics of blackberries: Current data and rapid valuation techniques. *Plant Prot Q* 13:175.

26. Adair, R. J., and R. H. Groves. 1998. Impact of environmental weeds on biodiversity: A review and development of a methodology. National Weeds Program, Environment Australia, Occasional Paper, Canberra.

27. Braithwaite, R. W., W. M. Lonsdale, and J. A. Estbergs. 1989. Alien vegetation and native biota in tropical Australia: Impact of *Mimosa pigra. Biol Conserv* 48:189.

28. Griffin, G. E. et al. 1989. Status and implications of the invasion of Tamarisk (*T. aphylla*) on the Finke River, Northern Territory. *J Environ Manage* 29:297.

29. Loope, L. L. et al. 1988. Biological invasions of arid land reserves. *Biol Conserv* 44:95.

30. Sorensen, B., and M. Jusaitis. 1995. The impact of bridal creeper on an endangered orchid. In *Weeds of Conservation Concern*, ed. D. Cooke and J. Choate, 27. South Australia, Adelaide: Department of Environment and Natural Resources and Animal and Plant Control Commission.

31. Willis, A. J., J. A. Matarczyk, and R. H. Groves. Competitive interactions between an endangered shrub, *Pimelea spicata*, and a threatening weed, *Asparagus asparagoides. Biol Conserv* in review.

32. Coutts-Smith, A. J., and P. O. Downey. 2006. The impact of weeds on threatened biodiversity in New South Wales. Technical Series No. 11, CRC for Australian Weed Management, Adelaide.

33. Downey, P. O., T. J. Scanlon, and J. R. Hosking. 2010. Prioritizing weed species on their threat and ability to impact on biodiversity: A case study from New South Wales. *Plant Prot Q* 25:111–126.
34. Turner, P. J., and P. O. Downey. 2010. Ensuring invasive alien plant management delivers biodiversity conservation: Insights from a new approach using *Lantana camara*. *Plant Prot Q* 25:102–110.
35. Panetta, F. D., and R. F. James. 1997. Weed control thresholds: A useful concept in natural ecosystems? *Plant Prot Q* 14:68.
36. Odom, D. et al. 2005. Economic issues in the management of plants invading natural environments: Scotch broom in Barrington Tops National Park. *Biol Invasions* 7:445.
37. Odom, D. et al. 2003. Policies for the management of weeds in natural ecosystems: The case of scotch broom (*Cytisus scoparius* L.) in an Australian national park. *Ecol Econ* 44:119.
38. Navie, S. C. et al. *Parthenium hysterophorus* L. In *The Biology of Australian Weeds*, vol. 2, ed. F. D. Panetta, R. H. Groves, and R. C. H. Shepherd, 157. Melbourne: R. G. & F. J. Richardson.
39. Lonkar, A., J. C. Mitchell, and C. D. Calnan. 1974. Contact dermatitis from *Parthenium hysterophorus*. *Trans St Johns Hosp Dermatol Soc* 60:45.
40. Subba Rao, P. V. et al. 1977. Clinical and immunological studies on persons exposed to *Parthenium hysterophorus* L. *Experientia* 33:1387.
41. Towers, G. H. N. 1981. Allergic eczematous-contact dermatitis from parthenium weed (*Parthenium hysterophorus*). In *Proceedings of the 6th Australian Weeds Conference, Gold Coast, Queensland*, ed. B. J. Wilson and J. T. Swarbrick, 143. Broadbeach: Queensland Weed Society.
42. Williams, J. D., and R. H. Groves. 1980. The influence of temperature and photoperiod on growth and development of *Parthenium hysterophorus* L. *Weed Res* 20:47.
43. Burdon, J. J., and J. G. W. Burdon. 1983. Allergy associated with Paterson's curse. *Med J Aust* 2:87.
44. Parsons, W. T., and E. G. Cuthbertson. 2001. *Noxious Weeds of Australia*. 2nd ed. Melbourne: CSIRO Publishing.
45. Gardner, C. A., and H. W. Bennetts. 1956. *The Toxic Plants of Western Australia*. Perth: West Australian Newspapers Ltd.
46. Everist, S. L. 1974. *Poisonous Plants of Australia*. Sydney: Angus and Robertson.
47. Bourke, C. A. 1997. Effects of *Hypericum perforatum* (St. John's wort) in animal health and production. *Plant Prot Q* 12:91.
48. Connor, H. E. 1977. *The Poisonous Plants of New Zealand*. 2nd ed. Wellington: Government Printer.
49. Caley, P., R. H. Groves, and R. Barker. 2008. Estimating the invasion success of introduced plants. *Divers Distrib* 14:196.
50. Downey, P. O. et al. 2010. Managing alien plants for biodiversity outcomes—the need for triage. *Invasive Plant Sci Manag* 3:1–11.
51. DEH (Department of Environment and Heritage). 2000. National environmental alert list. Canberra: Department of Environment and Heritage, www.weeds.gov.au/weeds/lists/alert.html (accessed 28 January 2010).

*chapter three*

# Environmental and economic costs of invertebrate invasions in Australia

*Deon Canyon, Ian Naumann, Rick Speare, and Ken Winkel*

## Contents

## 3.1 Introduction

Many exotic invertebrate pests have arrived on Australian shores due to human and animal migration, the transportation of goods, and weather patterns. The very first exotic species most likely were human ectoparasites, such as body, head, and pubic lice.

While many of these introductions have been innocuous, some have had significant environmental and economic effects. A clear picture of costs associated with these pests is not available in the literature, and there are few publications that have directly addressed exotic invasions in Australia. Of these, few have presented related costs involved, and medical pests are not usually included.[1] The second known major wave of human colonization by Westerners in the late 1780s coincided with the introduction of a number of exotic invertebrate organisms. These pests colonized Australia via ship from various ports around the world. Outbreaks of disease caused by the importation of mosquitoes and other insects, such as lice, proliferated throughout tropical regions especially in prospecting townships where squalid conditions prevailed. Food stores and timbers contaminated with exotic insects added to the local insect population capable of significantly affecting future agricultural and forestry operations. But this is nothing compared to the vast sea of global traffic witnessed in this day and age in which exotic insects can regularly be found and via which many organisms have become well established in foreign countries.

The traditional barriers of Australia, the sea and the Great Dividing Range, create distinct climatic zones and have long served to limit pest populations. However, as long as humans create habitats for themselves or for their crops, exotic insect pests find a way to exploit them. For decades the pattern has been set in which "civilization" and advances in basic hygiene play the most important role in ridding countries of imported vector-borne disease. The question remains as to whether this will continue to provide protection in the face of increasing global tourism and traffic. There are already indications that increased population movement is changing the global distribution of insect vectors and their related diseases. This is occurring through the movement of people, products, and animals as a result of world trade agreements and the decreased time taken to travel between countries, which favors invertebrates with shorter life spans.

This chapter examines exotic invertebrates in four main areas: medical, veterinary, agricultural, and marine. Each section focuses on several important species and outlines the current situation. Economic costs relating to the introduction of these species are related where possible from cost analyses, but in some cases, the best guesstimate is presented. Environmental costs are difficult to determine for most pests due to unknown impacts. How, for instance, would one estimate the damage done to the environment by the practice of spraying insecticide over a mangrove mosquito breeding site, apart from simply monitoring local animal populations or the extent of fish breeding?

While local effects may be easier to obtain, broader effects are confounded by too many factors to enable a reasonable level of certainty in any conclusion. Thus, environmental costs are stated where possible; however, these costs are not always in terms of dollars. It is very difficult to compare medical costs with other costs because they are complex. Medical costs register quite far below agricultural costs, but the intangible factors such as those relating to suffering and psychological effects may translate to a lifetime of lost production and/or social damage.

The estimated costs in some areas are divided by the number of years they pertain to so that an annual amount is generated. Control costs are used for agricultural items since that figure is more comparable to the figures generated in the medical section. From the information collated in this chapter, a conservative estimate on the cost of exotic invertebrates in Australia is in the range of AU$1 billion per annum, whereas an estimate including production loss and other intangibles would be around $5–$8 billion annually. No figure is presented for potential losses due to several recently introduced invertebrates, although their impact is expected to be considerable in the years to come.

## 3.2   Invasions of medical importance

### 3.2.1   Aedes aegypti

Dengue fever, dengue hemorrhagic fever (DHF), and dengue shock syndrome are various forms in which the dengue virus manifests itself in humans. The peridomestic mosquitoes, *Aedes aegypti* and *Ae. albopictus*, are responsible for biological transmission of four serotypes. Cross-protective immunity lasts for about 2 months,[2] and immunity to a particular serotype is lifelong.[3] Dengue viruses are particularly effective because they are able to replicate to a high level in mosquitoes and produce a high viremia in humans, which facilitates other mosquitoes becoming infected. Globally, *Aedes aegypti* is responsible for most urban infections.

Dengue is advancing on a geographic basis, and the World Health Organization (WHO) has placed dengue on the agenda of its infectious arm, the Committee for Tropical Disease Research (TDR). The WHO estimates that every year 100 million cases of dengue fever and 500,000 cases of DHF occur with an average case fatality rate of 5%. Thus, 25,000–30,000 fatalities are caused by DHF each year. In Puerto Rico, the disability adjusted life years lost per million people increased by 25% from 1984 to 1994, placing the economic impact of the disease in the same order of magnitude as malaria, tuberculosis, hepatitis, STDs (excluding AIDS), the childhood cluster (polio, measles, pertussis, etc.), or the tropic cluster (Chagas, schistosomiasis, and filariasis).[4]

#### 3.2.1.1   History of epidemics

Since 1879, dengue has manifested itself in epidemic form in Australia. A general infection rate of 75% was proposed for all areas experiencing dengue up until the 1953–1955 epidemic, with infection rates since this date ranging from 2% to 38% depending on geographic area.[5] The mortality rate varies substantially. Typically, 1–7 DHF cases would result from 100 dengue fever cases, and prior to the development of modern and adequate hospital management, 50% of DHF cases would die.[2]

The earliest known dengue epidemics occurred from 1885 to 1901 and spread throughout most of Thursday Island, Townsville, Cairns, Cooktown, Port Douglas, Charters Towers, Normanton, Mackay, Ingham, and Bowen, with cases inland at Hughenden, Barcaldine, and so on. This widespread epidemic penetrated into New South Wales in 1898 and at least three deaths were reported in Brisbane.[6] Based on an infection rate of 75% and a population of around 500,000 in 1900,[7] 375,000 people were likely infected with dengue. Cases continued to a lesser degree until 1904–1906, when the virus travelled north to infest Thursday Island, and south to Townsville where nine deaths occurred and down to Brisbane where a large outbreak caused 94 deaths. One death was also reported in Sydney. Thus, an estimated 190 DHF cases are likely, with a maximum of 19,000 cases. However, if only 15% actually reported ill to a health clinic,[5] a probable 126,730 people were infected in and around the Brisbane region. Interestingly, the population of Brisbane at this time was around 126,000.[6] Thus, the rate of 1 death to 1000 possible cases seems plausible. From 1885 to 1923, 52 deaths were recorded in the Townsville region,[8] which arose from around 52,000 probable infections. From 1916 to 1919 and from 1924 to 1926, New South Wales and Queensland were broadly struck with two epidemics, which produced a similar infection rate. The number of infected people was estimated at around 600,000[6,8,9] in each of the two epidemics. From 1938 to 1939, dengue was reintroduced and eventually caused another large epidemic in 1941–1943. This epidemic swept from Queensland down to Brisbane with up to 85% infection rates in some towns.[8,9] In Townsville alone, 5,000 cases were reported with 25,000 probable infections. Judging from past records and taking other areas into account, this figure could be doubled. This epidemic also swept north to Darwin and initiated a highly successful campaign to eliminate *Aedes aegypti* from the Northern Territory. Dengue struck again in 1953–1955, infecting 10%–85% of the population with an estimated 15,000 cases.[10] In 1981–1983 dengue returned to Queensland and was confirmed in 458 people. Using the recent notification rate of 15%,[5] a possible 3100 people were infected in this epidemic. From 1991 to 2008, 4,747 confirmed cases have been reported translating to a probable 32,000 infections.

Cumulatively, this leads to an estimated figure of 1,855,000 dengue infections in Australia since the introduction of *Aedes aegypti* and dengue, which is certain to be conservative due to a lack of information on numerous places that experienced epidemics. Based on this estimation, 1,819,340 people were infected prior to the 1980s with an infection rate of 75%, and 35,000 infections have resulted since the 1980s with an infection rate of 15%.

### 3.2.1.2   Recent developments

Dengue was reintroduced in North Queensland on November 11, 2008 by an Australian tourist returning from Indonesia. In accordance with the Dengue Fever Action Plan for North Queensland 2005–2010, the Dengue Action Response Team (DART) responded.[11] Despite using lethal ovitraps, the development of which had received more than $750,000 in national grant funding, the team failed to control the outbreak. This was attributed to a shortening of the expected transmission cycle from 14 days to 9–10 days. On January 22, 2009, a historic decision was made. The senior director of the Tropical Population Health Unit in Cairns activated the Emergency Response Plan that was specifically developed for disaster response. Unfortunately, this plan had not been updated for several years, which slowed its activation and subsequent response. It took a further week before the chief medical officer responsible for the response actually met with responders in Cairns. Following this, an incident management team was created and they assumed responsibility. However, due to their lack of specific technical expertise, a lot of time was wasted and a lot of goodwill was lost due to power plays. The team was innovative and appeared to be successful, but cases only declined in May 2009, which may have been the natural end of the epidemic or control efforts. The arrival of H1N1 was managed using this new approach to infectious diseases, and an emergency preparedness position was established. The economic costs associated with this disaster response approach to infectious disease outbreaks are not yet available but are likely to be considerable.

### 3.2.1.3   Cost estimation

Meltzer et al.[4] determined a figure of AU$80 per capita for the 1977 Puerto Rico epidemic, which included medical costs, control efforts, lost work, and lost tourism.[12] If this figure is used to calculate the cost of all Australian dengue epidemics, the result is an all-inclusive estimated total cost of $148 million or $1.3 million per annum.

The costs appear to be higher in Australia, which may be related to the population structure in North Queensland. The average time lost through illness in the 1992–1993 Charters Towers epidemic was calculated to be 10.5 days.[13] Using the total number of infected people, the result is 19,477,500 man-days being lost in total or 177,068 days per annum. With each day valued at AU$96, according to an average income of $35,000, the annual cost since the introduction of dengue is almost $17 million in current terms. However, epidemics are much smaller these days, infection rates have changed, and the average wage has risen to $45,000, so it is only appropriate that the current situation should be separated from the past. Prior to 1990, the cost of work lost in today's dollars is close to $2 billion. In the last 18 years alone, 32,000 infections with 336,000 lost days have equated to $41.3 million or $2.3 million per annum in lost time alone.

Judging from correspondence from several local city councils, labor costs for the control of exotic mosquitoes and related diseases range from $2000 to $6000 per year with brief major jumps during epidemics caused by reallocation of current staff to vector control operations. The cost of insecticides for exotic mosquito control is minimal during non-epidemic years but ranges from a few thousand to nearly a million dollars per council per epidemic. The Charters Towers City Council determined the cost of vector control (insecticides and staff) for its 1992–1993 epidemic to be $750,000 for a population of 8,500, resulting in a cost of $88 per capita. In a similarly sized epidemic, the Townsville City Council estimated direct costs in 2000 to be at least $500,000 for a population of 110,000, resulting in a cost of $5 per capita. Thus, it is problematic to use the per capita method in the modern environment where epidemics cause similar numbers of infections with similar costs regardless of the population size. The population in North Queensland is comparatively widely spread and small, and dengue-related expenses can be considerable even though large populations are not involved. The epidemic costs in Townsville and Cairns, including annual maintenance costs, averaged at least $200,000 per annum during the 1990s.

When the Tropical Public Health Unit in Cairns dealt with a number of small epidemics lasting over 3 years from 1997 to 1999, they formed a vector control unit called DART. In 2002, they guesstimated an annual cost of $200,000 equating to $2 per capita since the formation of this team. Thus, over the last decade, approximately 32,000 infections have occurred within an area containing a human population of not more than 300,000 at a control cost of around $400,000 per annum. This same figure was quoted by Ritchie in the study by McMichael et al.[14] as the minimum amount spent per annum on dengue management in Far North Queensland, but the source of the amount was not acknowledged, and derivation of the amount was not explained. This figure also did not include health or economic costs.

If all epidemics are taken into account, the cost of the introduction of *Aedes aegypti* to Australia, including lost work and control costs, has been considerable at around $17 million per annum. Since 1990, however, costs were more reasonable averaging out at around $2.7 million per annum. These figures do not include the intangible costs to individuals and society that can greatly detract from the quality of life and general well-being. Intangibles are perhaps similar in nature to environmental costs where quality is almost impossible to measure except in great leaps and bounds. Judging from the replies from city councils in North Queensland and pesticide companies, the costs to control *non-exotic* mosquitoes far exceed the costs to control exotic mosquitoes.

Since 2000, Australia has experienced a surge in viremic importations and dengue epidemics due largely to the construction of a new international airport in Cairns. The Dengue Fever Management Plan for North Queensland 2005–2010 renewed the government's commitment to lowering dengue incidence by reducing vector breeding through education programs, encouraging greater awareness of the disease among general practitioners, and improving physical and serological surveillance and mosquito control. However, this has not been entirely successful. While Australia's medical entomologists argue that dengue is not endemic in Australia, the figures make this ever more difficult to justify.

On a final point, the state and federal governments have advocated the collection and storage of rain water in domestic tanks. While this is not of immediate concern, a lack of maintenance and deterioration of the tank screens over time will allow *Ae. aegypti* to breed in larger numbers. This would enhance the distribution and population of mosquitoes, which would impact any resurgence of dengue. Policing this potential problem will have further economic ramifications.

### 3.2.2   Aedes albopictus

*Aedes albopictus* has established itself across the world and in some ways is more of a threat than *Ae. aegypti* since it can tolerate a broader range of temperatures, transmit a wider variety of pathogens, and is more competitive. Fortunately, it is less anthropogenic and so it is less of a threat where alternative blood sources are available.

Globally, *Ae. albopictus* is responsible for the rural infection cycle of dengue. During the 1900s, dengue was not endemic in Australia, and until the introductions of *Ae. albopictus* in 2004, *Ae. aegypti* was the only vector. Since its introduction into the United States in 1980, this species has achieved a global distribution. Preventing the establishment of this species in Australia was a major focus of vector control in the north. This strategy failed in 2004 when *Ae. albopictus* was detected in the Torres Strait off northern Australia and was allowed to spread and establish itself on many islands.[15] There is little doubt that the species will establish itself on the mainland regardless of control strategies because from 1997 to 2005, the importation of *Ae. albopictus* was detected and prevented 28 times by the Australian Quarantine Inspection Service and other authorities at Australian international seaports (including Darwin, Cairns, Townsville, Brisbane, Sydney, and Melbourne).[16]

The impact of this species is difficult to gauge, and only time will tell. Since *Ae. albopictus* has effectively displaced *Ae. aegypti* and other species in many countries, all indications are that it will displace similar species in Australia. In fact, several medical entomologists have often spoken of this possibility and its potential effects. It is hypothesized that, since it is not known as an effective urban vector of dengue, its introduction may result in a decline in dengue-related costs due to a lower number of cases. But since it is an aggressive day-biter, increases in control for nuisance biting may be expected. Alternatively, *Ae. albopictus* is known to have caused dengue epidemics in the absence of *Ae. aegypti* in Japan and China,[17–20] Seychelles,[21] and Hawaii.[22] The capacity of this mosquito to transmit other pathogens, such as yellow fever virus, Japanese encephalitis virus (JEV), and Ross River virus (RRV)[19] is an additional concern that may have economic implications in the future.

*Ae. aegypti* has caused dengue epidemics as far south as New South Wales, but *Ae. albopictus* has the potential to inhabit the entire country. Thus, areas that have never been exposed to these tropical diseases may begin to experience them. No doubt this will be facetiously attributed to climate change when it happens.

### 3.2.3   Aedes (Aedimorphus) vexans vexans

*Aedes vexans* has become established on mainland Australia and is now distributed throughout Australasia and the Pacific.[16,23,24] It was detected during collections in the Kimberly region of Western Australia that took place in the wet seasons from 1996 to 2003. The mode of introduction is hypothesized to be by wind currents from the Indonesian archipelago or via aircraft from endemic islands arriving into Kununurra.[23] Its subsequent presence in annual collections indicates that it is here to stay.

### 3.2.4   Culex gelidus

In 1995, an outbreak of Japanese encephalitis (JE) occurred on the Torres Strait Islands in northern Australia. JE is a serious disease with an average hospital stay of 14 days and a mortality rate of 10%–50%. Forty percent of survivors experience mental or physical crippling and require 1–5 years of rehabilitation, whereas 10% require chronic care.[12]

During a 3-week period, three residents of the outer island Badu (population 700) manifested typical symptoms of acute illness with headaches, fever, convulsions, depressed level of consciousness, and coma, with two deaths.[25] A seroprevalence survey confirmed JE infection in 35 Badu people (16%), 20 other outer island people (1.5%–11%), and 63 pigs (70%).[25] There is a vaccine available that is 95% protective. In this epidemic, the majority of inhabitants of northern Torres Strait (3500 people) were vaccinated by Queensland Health in the same year. Sentinel pigs were established in 1996, and almost all had seroconverted by March that year, in addition to most horses tested. In early 1998, an adult male working on a boat at the mouth of the Mitchell River on the west coast of the Cape York Peninsula and a 12-year-old unvaccinated child from Badu were diagnosed as having JE. This is the first time that JE has been recorded on mainland Australia, and there is some concern because several seroconversions of wild pigs on the mainland have occurred.[26]

The Queensland Health Tropical Public Health Unit in Cairns, the most active responsible state health authority, declined to provide an estimate on the cost of JE to Australia. Their curious reticence in providing this information was attributed to the difficulty in obtaining data from the various agencies involved and confidentiality issues. Suffice to say that several million dollars are probably involved, including efforts such as serological surveillance, vaccinating most of the population, and building a new piggery to act as a permanent sentinel station.

Through viral isolations, the mosquito responsible for these outbreaks was believed to be *Culex annulirostris*, the vector of Murray Valley encephalitis and a common native swamp breeder.[25,27] However, a recent revelation has cast doubt on this. A misidentified alien mosquito species, *Culex gelidus*, widely distributed in the Torres Strait, mainland Queensland, and the Northern Territory has now been found, and JE has been isolated in it.[28] This exotic mosquito is capable of transmitting JE in addition to the Batai, Getah, and Tembusu viruses and increases the likelihood that JE will become more prevalent on the mainland. Northern Territory medical entomologist Peter Whelan said that this mosquito could become a threat as it breeds around piggeries, dairies, sewage treatment works, and abattoirs. Whelan believed that it was no longer possible to consider the eradication of this mosquito species.[29] The potential now exists for much larger, more widespread epidemics with severe impacts and increased costs.

In a recent study on the vector competence of *Cx. gelidus*, it was found to be refractory to infection with Barmah Forest virus and only 25% were capable of transmitting RRV.[30] Since the RRV result is similar to *Ochlerotatus vigilax*, *Cx. gelidus* may be considered a significant vector of this alphavirus. *Cx. gelidus* was more susceptible to flavivirus infection with transmission rates of 96%, 95%, and 41% for JEV, Kunjun virus, and Murray Valley encephalitis virus, respectively.[30] In this study, these rates were higher than those for *Cx. annulirostris*, the primary vector of these diseases in Australia. *Cx. gelidus* is thus a vector of significant public health concern, but it is too early to estimate costs at this time.

While there are control costs for *Ae. vexans* and *Cu. gelidus*, there are no obvious changes in the rates of vector-borne disease in the National Notifiable Disease Surveillance System and so there are no associated health costs despite fears to the contrary.[31]

## 3.2.5 Honeybees and wasps

Relatively little information is available on the economic costs of exotic venomous invertebrates in Australia. A review of the literature reveals that the greatest calculable economic impact is attributable to bees and wasps. Indeed, no information is available, for example, on the impact of exotic arachnids. Note that the attribution of specific health costs to bee stings, as distinct from wasp stings, is complicated by the failure of the current health classification system to resolve the two diagnoses. This is further discussed below.

Since its arrival in 1822, the European honeybee (*Apis mellifera*) has become widespread throughout all the states and territories of Australia. By 1998 more than 670,000 hives were officially registered.[32] Apart from the considerable income generated by honeybees, their stings are a leading cause of death due to venomous bites and stings in Australia.[33,34] For example, 63 bee-sting-related fatalities were registered by the Australian Bureau of Statistics and the National Coroners Information System from 1979–2010. This grouping includes at least 45 definite bee stings and at least 9 definite wasp stings, giving a 5:1 ratio of bee to wasp fatalities. As a group, bee and wasps were second among venomous bite and sting deaths in this period only to snake-bite fatalities. Using those cases directly attributable to honeybee stings gave a mortality rate of 0.12 per million population per year.[35]

Similarly, bee stings are a leading cause of human morbidity as demonstrated by emergency department (ED) presentations and hospitalization data. Under the current diagnostic system, bee and wasp stings are coded as a single category (International Classification of Disease, Version 10, External Cause Code X23). National hospitalization data for the year from July 2002 to June 2005, revealed that bee and wasp stings were responsible for 3547 new inpatient cases.[36] This included 2754 bee sting cases and 793 wasp sting cases, a nearly identical ratio to that seen with fatalities. This hospitalization rate was second only to spider bites over the same period. Similarly, an analysis of ED presentations in Victoria revealed that bee and wasp stings accounted for 41% of all presentations due to venomous bites and stings.[37]

Assuming a similar rate of ED presentations in the United States and costs per presentation and hospitalization episode, Australian bee and wasp stings result in at least $10 million in direct hospital expenditure annually. While as yet no data is available on the extent of less severe morbidity in humans and the impact of bee stings on domestic animals and livestock, it appears that the effects of honeybees on native plants and animals are minor.[32] Clearly, then, the economic impact of honeybees in Australia is overwhelmingly positive.

In contrast to the net positive value of the honeybee, exotic wasps, notably the European wasp, *Vespula germanica*, inflict damage without any benefits. A native of Europe, western Asia, and northern Africa, the European wasp was first introduced into Australia in 1954 but only became established in 1959 in Hobart, Tasmania.[38,39] This vespid arrived on the mainland in 1977 and, lacking any natural predators, has rapidly expanded its range ever since. By 1991, tens of thousands of nests were estimated as being destroyed in metropolitan Melbourne annually,[38] with wasp densities of up to 40 per $km^2$. These wasps are now found in Tasmania, Victoria, New South Wales, the Australian Capital Territory, and South Australia.[39]

Indeed the surge in numbers of *Vespula germanica* in southeastern Australia during the summer of 1997–1998 prompted the Victorian government to call for a national control strategy.[40] However, a recent analysis of wasp-sting mortality in Australia, driven by concern about the lethal potential of *V. germanica*, failed to detect any human fatalities attributable to this wasp during the last 20 years.[41,42] Research into the morbidity attributable to these wasps has been limited by the disease classification system that combines bee and wasp stings in one category. Such data was presented in the previous discussion on the honeybee. An attempt to calculate the economic and health impact of this wasp conservatively estimated the cost to Victoria alone at greater than $2 million annually.[43] This included the effects on horticultural industries, health care, national parks, and tourism as well as the direct cost of nest destruction.

### 3.2.6 Red imported fire ants

Two exotic ant species of potential medical and ecological significance have been found in Australia. The tropical fire ant, *Solenopsis geminata* (Fabricius), is estimated to have been introduced sometime before 1987 into the Northern Territory, and its current distribution is limited to northern coastal areas.[44] Although it does not appear to have caused significant ecological damage in Australia, this species has become a serious problem elsewhere in southeast Asia and the Pacific, especially Okinawa and Guam.[45] In 2001, the South American fire ant, commonly known as the red imported fire ant (RIFA), *S. invicta* Buren, was detected in southern Queensland.[46] The initial incursion undoubtedly had taken place several years earlier and most likely was via sea cargo.

The United States' experience with two imported fire ant species (*S. invicta* and *S. richteri* Forel) gave little cause for optimism in Australia. These species have developed resistance to natural and chemical control methods and have continued to cause significant ecological and agricultural damage and a variety of health problems for people living in southeastern states. The health risks range from sting site pustules, secondary infections, and large late-phase responses, to life-threatening anaphylaxis.[47,48] Occasionally, skin grafts or the amputation of an affected limb are called for.[49] Stings may occur indoors and outdoors. In areas endemic for RIFA, the most commonly reported cause of Hymenoptera venom allergy is now RIFA allergy.[50] It has also been estimated that RIFA sting more than 50% of persons living in endemic areas each year.[51]

As pests of agriculture, fire ants (1) damage or remove seeds; (2) damage roots, tubers, stems, and fruit; (3) protect injurious plant-sucking Hemiptera; (4) interfere with biological control; (5) are hazardous to hand labor; (6) damage irrigation systems; (7) build mounds that interfere with mechanical harvesters; and (8) harass stock, especially young animals. There are also a host of additional effects, such as damage to electrical equipment and structural damage due to undermining. Estimates of the monetary impact of fire ants on agriculture vary enormously. A recent (and perhaps conservative) analysis estimates the combined value of production losses and control costs in North America to be US$246 million.[52] Less substantiated estimates place annual costs in excess of $2 billion.

An estimate of US$2.4 million per annum has been made of the direct health costs attributable to RIFA in the state of South Carolina, where all 46 counties are now infested. These costs relate to the estimated 660,000 sting cases and 33,000 medical consultations.[53] Regional RIFA control programs were discontinued due to cost and environmental chemical concerns, and these ants now infest more than 310 million acres in the United States and Puerto Rico. Moreover, evolutionary changes have facilitated their expansion northward and westward, heightening public health concerns.[54]

The potential impact of *S. invicta* in Australia over 30 years was assessed assuming simple radial spread from the Brisbane incursion area and that the ant dispersed to occupy 60% of the area to which it was ecologically suited (as indicated by CLIMEX modeling).[55] Principal costs were estimated based on market and survey data from the United States but applied to Australia. These costs included damage to households (based on repair costs and domestic control measures), losses in property values, damage to livestock and agriculture (using the cattle industry as a proxy), costs of medical treatment to humans, loss of work days due to stings, damage to schools and costs of control on school premises, damages to electrical equipment, and damages to golf courses. These costs alone amount to a sobering $8.9 billion. Environmental costs were not included. Furthermore, the radial spread model was quite conservative and, for example, predicted that the ant would not reach the Sydney metropolitan area for about 15 years. Since *S. invicta* is transported easily by road or rail, the ant more than likely would reach southern population centers in less than 15 years. This would mean that the greatest urban costs would begin to accrue even earlier and that over 30 years the total cost would be greater than $8.9 billion.

An eradication program commenced almost immediately following the first detections and grew quickly to include baiting, nest destruction, surveillance, domestic quarantine measures, and intensive community engagement. Although the ant was successfully confined to the Brisbane area and eliminated from many urban and periurban areas, low-density populations over rural areas near to Brisbane and the continuing discovery of outlier nests bedevil the eradication program. Fortunately, only a single case of fire ant sting-related anaphylaxis has been reported in the Australian medical literature to date.[56] Operational costs in the order of $174 million have been incurred, and the eradication efforts are ongoing.

## 3.3   Invasions of veterinary importance

In the last half of the twentieth century, Australia has had the most stringent importation requirements for vertebrate animals of any country globally. The requirements have been particularly stringent for livestock, domestic pets, and wildlife and less so for fish and amphibians. One aspect of these requirements has been that any imported animal should be free of ectoparasites. This policy has been highly successful. No ectoparasites of significance to livestock or domestic pets have become established in Australia in the last 50 years.

### 3.3.1   Cattle tick

Early introductions of arthropod pests into Australia have been enormously costly. Cattle tick, *Boophilus microplus*, has been the most expensive. The cattle tick was introduced into Australia in 1872 by the importation of 12 Brahman cattle from Batavia.[57] They first appeared in Queensland in 1891, Western Australia in 1895, and New South Wales in 1906. Cattle ticks were introduced to Victoria in 1914 on horses from Queensland en route to the war in Egypt. However, *B. microplus* did not become established in Victoria, and this state along with Tasmania and South Australia has been free of cattle tick.[57] The distribution of the tick is determined by low temperature and humidity, and is hence confined in Australia to northern Western Australia, the northern half of the Northern Territory, coastal Queensland, and northern New South Wales. Two blood protozoan parasites, *Babesia bovis, B. argentina*, and a blood-borne bacterium, *Anaplasma marginale*, use the tick as vector and cause economic impact. The economic costs of the cattle tick are due to the direct effects of the ticks on cattle (loss of condition, anemia and death, susceptibility to drought, damage to hides, slow growth rate), effects of dipping on cattle (loss of body weight, loss of milk production, deaths during drought, loss of young calves, toxicity), and control costs (increased stock handling, additional capital expenditure, costs of acaricides), market effects (restrictions on movements), costs of tick-borne diseases (deaths, slow growth, vaccine costs, treatment costs, handling costs). Davis[58] presented cost estimates in past and current terms. Current terms are as of 1997 and are given in parentheses: 1959: £9,579,000 ($87 million); 1973: $15,500,000 ($87 million); 1995: $132 million ($134 million). The earlier estimates did not take into account government costs associated with control strategies and costs of dipping yards. On average, acaricides accounted for 11% of costs, additional labor for 35%, and production losses and deaths for 32%. An estimate of the cost of the cattle tick to dairy farming in Queensland, excluding tick fever costs, estimated that 49% was due to costs of control and 51% due to decreased production.[59] A quarantine barrier on the New South Wales/Queensland border was created to halt the southward spread of the cattle tick and is maintained at an annual cost of around $3.3 million. The savings and benefits from this tick line were estimated at $41.5 million per annum by Davis.[58] Sutherst[60] has predicted that climate change will affect pests and diseases with significant impacts on control costs productivity. Models have predicted an increase in costs ranging from $18 million to $192 million, associated with a change in global distribution. It was suggested that insect pests, with their high reproductive rates, short generation times, efficient dispersal rates, and ability to rapidly adapt, will be affected early on in the event of climate change events resulting in large costs.

### 3.3.2    Screw-worm fly

Australia is the only continent with tropical regions that does not have the screw-worm fly. The larval stages of screw-worm fly cause cutaneous myiasis. Both the Old and New World screw-worm flies are globally notifiable diseases to the World Organization for Animal Health (OIE). The major species of concern to Australia is the Old World screw-worm fly, *Chrysomya bezziana*, found in Papua New Guinea (PNG), Southeast Asia, India, parts of the Middle East, and Africa. Other species (New World screw-worm flies) such as *Dermatobia hominis* from South America, *Cochliomyia hominovorax* from Central America, and *Cordylobia arthropophaga* from Africa are of less concern owing to Australia's quarantine restrictions. *C. bezziana* is an obligate parasite of all warm-blooded animals. Female flies are attracted to open wounds in the skin and lay eggs on the wound edges. The eggs hatch in 12–24 hours, and the larvae move into the wound and feed for 5–7 days, after which they drop off onto the ground to pupate. During the larval feeding phase, the wound enlarges in diameter and depth. The economic cost of screw-worm fly is due to occasional deaths, drops in production, damage to hides and underlying muscle, costs of insecticides, and costs of additional labor for treatment and management protocols. In Australia, where much of the cattle industry in the tropics involves minimal inspection of cattle, the introduction of screw-worm fly would require a marked change in management practices with frequent inspections for management of unstruck wounds and fly treatment of wounds already infected by the larvae.

If *C. bezziana* was introduced and became endemic, it would occupy a large area of northern Australia. A high probability of establishment exists year-round in tropical regions, except in areas around the Gulf of Carpentaria and in the Northern Territory where dry weather mid-year would limit the survival.[61] Low temperatures make establishment in temperate areas unlikely. The potential area of permanent colonization in Australia extends south to the mid-coast of New South Wales. Comparison of areas suitable for permanent establishment with the potential for summer distribution indicates that large additional areas, carrying most of the continent's livestock, could be colonized in the summer months.[62] Since screw-worm fly can infect all warm-blooded animals, its economic impact depends on the juxtaposition of the fly, the climate, and suitable hosts. In northern Australia, the cattle industry would suffer the major impact, with some impact on sheep in more inland areas. Other species including goats, horses, domestic pets, and native and feral mammals would also suffer cutaneous myiasis from the fly. In 1979, it was estimated that the economic loss to the sheep and cattle industry if screw-worm fly was allowed to spread unchecked would be AU$101 million annually.[63]

*C. bezziana* could be introduced to Australia by the illegal importation of infected animals, importation on infected people, adult flies flying into Australia, and flies being carried into Australia in boats and aircraft. Since the southern coast of the Western Province of PNG is only 3 km from Saibai, the northern most Australian island in Torres Strait, the possibility of *C. bezziana* being introduced from PNG is considered a significant threat.

*C. bezziana* flies labeled with a radioactive tracer have been shown in PNG to deposit eggs a median distance of 10.8 km from their point of release with a maximum distance of 100 km.[64] It seems feasible therefore that adult flies from the PNG mainland could arrive unaided on the top islands of Torres Strait. In addition, the traditional visitors treaty between Australia and PNG allows for free movement of people between coastal regions of the Western Province and the top islands of Torres Strait for the purposes of trade and social interaction. Recent restrictions on movement of animals make introduction of infected animals from PNG less likely, but policing is difficult.

Screw-worm flies have arrived in Australia in boats and aircraft[65] and as cutaneous myiasis in people. In 1988, *C. bezziana* flies were found on a vessel in Darwin Harbor.[65] There are no known introductions on animals. The cases on humans have involved the South American fly, *Dermatobia hominis*, the tumbu fly, *Cordylobia arthropophaga*, from Africa, and the New World screw-worm fly imported on a traveler from Argentina and Brazil.[66–68] These cases on humans are low risk owing to the small numbers of larvae involved in these species where there is only one larva per lesion. However, wounds infected with *C. bezziana* can contain thousands of larvae,[62] and the risk of one person or animal bringing in sufficient numbers to establish the fly in Australia is much higher.

The economic costs of an eradication program even when detected early may be quite high. When the New World screw-worm fly, *Cochliomyia hominovorax*, a species very similar in biology to *C. bezziana*, was introduced to Libya in 1988, the eradication campaign cost approximately US$75 million.[69] The annual regional benefit of eradicating this invasion was estimated to be US$480 million at a benefit-cost ratio of 50:1.[69] The same species in the United States in 1960 cost US$100 million annually, and elimination from the southern United States and Mexico took more than 20 years and cost nearly US$700 million.[70] The cost–benefit ratio for this eradication program was 1:10.[70] The economic cost of an invasion by *C. bezziana* depends on the point of entry; for Brisbane, it has been estimated at AU$282 million per year in 1992 for producers and AU$775 million annually for both producers and consumers.[71]

In Australia, the approach is one of risk reduction, early detection, and preparedness.[63] Risk reduction involves quarantine requirements for animals imported formally into Australia, prohibiting the informal movement of animals from PNG to the Australian Torres Strait Islands and restricting movement of animals between islands in Torres Strait, insecticidal sprays prior to arrival for aircraft and ships entering Australia, a cattle-free zone in Cape York Peninsula, and attempts to reduce feral animals on the Torres Strait Islands. Early detection involves education to alert Torres Strait Islanders and communities on Cape York to screw-worm fly, submission of diagnostic specimens from struck animals, trapping of flies in traps baited with swormlure, and a sentinel wounded animal scheme and trapping was instituted, but is now reserved for use to map the distribution of any introductions. Since female screw-worm flies mate only once, the main control method is the release of sterile males once screw-worm fly is detected. A factory to produce sterile male screw-worm flies was established in Port Moresby (PNG), but is mothballed at the moment. The sterile insect release method was used effectively as a major tool to eliminate flies in the Libyan outbreak and to eradicate screw-worm flies from the southern United States. The cost of prevention over a 20-year period was estimated in 1979 to be AU$20,230,000.[63] Modeling on a sterile insect release program showed the sterile insect release program to be biologically and economically feasible.[70] However, when the lowered competitiveness of captive raised sterile males was used in an alternative modeling study, the time for eradication extended beyond a decade, potentially making use of sterile males uneconomic.[72]

Monitoring in Torres Strait has shown that the risk of introduction via Torres Strait is low.[70] Importation of infected dogs through legal channels could potentially result in the accidental introduction of *C. bezziana* into Australia if clinical assessment is not detailed.[73] However, the major risk is the illegal introduction of an infested animal. Ongoing monitoring is recommended. Hence, screw-worm fly has a significant economic cost to Australia even though it has never invaded the continent.

## 3.4   *Invasions of importance to agriculture and forestry*

Surprisingly, the statistics compiled annually to describe the economic value of agricultural and forest production in Australia do not reveal the general economic impact of arthropod pests and certainly give little indication of the particular impact of exotic pest species. Some individual appraisals have identified massive impacts. For example, the introduced red-legged earth mite, *Halotydeus destructor* Tucker (Acarina: Penthaleidae) is believed to cause more than $200 million worth of damage each year to Australian pastures, and stored grain pests collectively possibly account for $100 million in losses annually (unpublished).[74] Even sporadic outbreaks of introduced pests can be extremely damaging. In South Australian plantations, the European wood wasp, *Sirex noctilio* Fabricius (Hymenoptera: Siricidae), killed more than 5 million *Pinus radiata* trees between 1987 and 1989 with a royalty value of $10–$12 million.[75] The wood wasp damage occurred despite the presence of effective biological control agents in Australia.

What are we to make of these and other scattered, oft-repeated estimates of impacts when the methods of reckoning generally are not explained? Furthermore, in the absence of systematic, direct measures of losses at the farm gate, mill, or market, or accurate costs of control measures, how are we to obtain a quantitative impression of the monetary impact of the full range of exotic pests?

This section describes two different approaches to the question. The first uses an estimation technique based on total production values and assumptions regarding percentage crop losses. The second approach embodies more formal economic analysis and is based on more precise data on losses and costs of control measures. The first approach gives no better than a first approximation of impacts, but it does allow the sketching of the broad picture. The second approach, though more rigorous, calls for data, which are available for few industries, commodities, or pests.

### 3.4.1   *Estimation from production values*

Table 3.1[76] summarizes an application of the estimation technique and lists 48 insect and mite species introduced accidentally into Australia between 1971 and 1995. Each species has been assigned a pest status (major, sporadic, minor) based on the performance of the species in other countries and on Australian experience postintroduction. Admittedly, the approach is subjective.

Exactly 50% of the species listed in Table 3.1 are classified as major pests. A major pest causes economic loss over a large part of the distribution of the crop and requires control measures most of the time.[77] Clarke[78] assumed that in the absence of control measures, a major pest would cause a loss of 10% or more in value of the commodity. For many major pests, losses are potentially massive. The introduced codling moth, *Cydia pomonella* Linnaeus (Lepidoptera: Tortricidae), whose larvae tunnel in fruit, can infest and render unmarketable up to 100% of apples in an unprotected Australian orchard.[76]

**Table 3.1** Estimated Economic Costs of Production Losses and Control Costs ($,000)
Associated with the Introduction of Insects 1971–1995

| Scientific name | Pest status | Industry[a] | Production loss | Control cost |
|---|---|---|---|---|
| Mango psyllid | Unknown | Mango | N/A | N/A |
| Spotted clover aphid | Major | Pastoral | 500,000 | 50,000 |
| *Abacarus hystrix* | Major | Pastoral | 500,000 | 50,000 |
| *Acyrthosiphon kondoi* | Sporadic | Pastoral | 100,000 | 10,000 |
| *Acyrthosiphon pisum* | Sporadic | Pastoral/peas | 100,000 | 10,000 |
| *Aleurocanthus spiniferus* | Major | Citrus | 6,000 | 600 |
| *Aleurodicus dispersus* | Sporadic | Horticultural | 10,000 | 1,000 |
| *Ametastegia glabrata* | Sporadic | Raspberry/grape | 7,600 | 760 |
| *Apis cerana* | Major | Beekeeping | 3,500 | 350 |
| *Aulacaspis tegalensis* | Major | Sugarcane | 85,000 | 8,500 |
| *Bactrocera frauenfeldi* | Minor | Horticultural | N/A | N/A |
| *Bactrocera papayae* | Major | Horticultural | 200,000 | 75,000 |
| *Bemisia tabaci* biotype B | Major | Horticultural | 500,000 | 150,000 |
| *Bombus terrestris* | Innocuous | N/A | N/A | N/A |
| *Brevennia rehi* | Major | Rice/sorghum | 25,000 | 2,500 |
| *Brontispa longissima* | Major | Coconut/palms | 1,000 | 100 |
| *Chilo terrenellus* | Major | Sugarcane | 85,000 | 8,500 |
| *Coptotermes formosanus* | Major | Timber | 600,000 | 60,000 |
| *Deanolis albizonalis* | Major | Mango | 6,800 | 680 |
| *Dysaphis aucupariae* | Innocuous | N/A | N/A | N/A |
| *Eriophyes hibisci* | Sporadic | Hibiscus | 50 | 5 |
| *Eumetopina flavipes* | Minor | Sugarcane | N/A | N/A |
| *Frankliniella occidentalis* | Major | Horticultural | 250,000 | 100,000 |
| *Hercinothrips femoralis* | Major | Horticultural | 75,000 | 10,000 |
| *Heteropsylla cubana* | Major | *Leucaena* | 5,000 | 500 |
| *Hylotrupes bajulus* | Major | Softwood timber | 200,000 | 20,000 |
| *Hypothenemus californicus* | Minor | Wheat | N/A | N/A |
| *Hypurus bertrandi* | Innocuous | N/A | N/A | N/A |
| *Idioscopus clypealis* | Sporadic | Mango | 1,400 | 140 |
| *Idioscopus niveosparsus* | Sporadic | Mango | 1,400 | 140 |
| *Ithome lassula* | Major | *Leucaena* | 5,000 | 500 |
| *Lachesilla quercus* | Nuisance | Bulk grain | N/A | 100 |
| *Melittiphis alvearius* | Innocuous | N/A | N/A | N/A |
| *Metopolophium dirhodum* | Major | Rose/barley | 6,500 | 650 |
| *Phenacoccus parvus* | Major | Vegetables | 7,000 | 700 |
| *Phyllonorycter messaniella* | Minor | Environmental | N/A | N/A |
| *Pyrrhalta luteola* | Major | Environmental | 100,000 | 1,000 |
| *Oxycanthae* | Sporadic | Fruit/grain | 80,000 | 8,000 |
| *Ribautiana ulmi* | Sporadic | Environmental | N/A | 500 |
| *Scapteriscus didactylus* | Sporadic | Pastoral | 100,000 | 10,000 |
| *Scaptomyza flava* | Minor | Horticultural | N/A | N/A |

***Table 3.1*** Estimated Economic Costs of Production Losses and Control Costs ($,000) Associated with the Introduction of Insects 1971–1995 (*Continued*)

| Scientific name | Pest status | Industry[a] | Production loss | Control cost |
|---|---|---|---|---|
| *Scolytus multistriatus* | Sporadic | Environmental | N/A | 500 |
| *Temnorrhynchus retusus* | Innocuous | N/A | N/A | N/A |
| *Therioaphis trifolii* f. *maculata* | Major | Pastoral | 500,000 | 50,000 |
| *Thrips palmi* | Major | Horticultural | 200,000 | 75,000 |
| *Trogoderma variabile* | Major | Bulk grain | 400,000 | 40,000 |
| *Varroa jacobsoni* | Major | Apiary/honey | 4,000 | 400 |
| *Vespula germanica* | Nuisance | Environmental | N/A | 1,000 |
| Total | | | 4,665,250 | 747,125 |

*Source:* Clarke, G. M, *Exotic Insects in Australia: Introductions, Risks and Implications for Quarantine*, Bureau of Resource Sciences, Canberra, 1996. With permission.

[a] Values for the pastoral industry include loss in seed production and production losses associated with grazing and may be very conservative.

Approximately 23% of the species listed in Table 3.1 are classified as sporadic pests. These are pests that are usually unimportant, perhaps controlled by natural enemies or weather conditions, but occasionally cause economic damage. A sporadic pest can cause damage equivalent to that of a major pest but on average only once every 5 years. For such pests, annualized yield losses would amount to 2%. The dock sawfly, *Ametastegia glabrata* Fallen (Hymenoptera: Tenthredinidae), is an example of a newly introduced, sporadic pest. Its larvae feed on a range of herbaceous weeds, and the species has become widespread in southeastern Australia. It can become a significant pest in orchards where populations build up on dock, growing as a weed beneath apple trees. On maturation, larvae of the sawfly abandon the dock and tunnel into fruit in search of pupation sites.[79]

Ten percent of the species introduced between 1971 and 1995 are classified as minor pests. A minor pest feeds or oviposits on a valuable host plant but does not inflict economically significant damage. For example, larvae of the introduced oak leaf miner, *Phyllonorycter messaniella* Zeller (Lepidoptera: Gracillariidae) tunnel in the leaves of ornamental oak trees (*Quercus* spp.) and can be very abundant. However, the mines have no discernible effect on the performance of oaks as shade trees, and control measures are not called for.

Table 3.1 gives the host plant or commodity affected by the introduced species. Annual production values for hosts or commodities were obtained from publications of the Australian Horticultural Council, the Australian Bureau of Statistics, and the Australian Bureau of Agricultural and Resource Economics (ABARE). Losses due to each introduced pest were estimated by calculating 10% or 2% of these production values, depending on the status of the pest. Where expert opinion was available these estimates were revised upward or downward to reflect more precise knowledge of the impact of particular pests.

In reality, the individual losses and the aggregate annual loss of nearly $4.7 billion are, at best, estimates of potential losses. Control measures generally are applied, with varying degrees of success and at varying costs. In some grain crops, control measures against a key pest can cost 1% of total production value with 4% of yield still lost to the pest.[80] Effective control of horticultural pests can cost as much as 30% of crop value.[81] If we were to use the low value (i.e., effective control for 1% of the value of the crop) and assume no residual damage, total control costs would be as listed in the final column of Table 3.1, summing to nearly $750 million per annum. Of course, the economics of pest management are complex. Decisions on the commitment of resources to pest control are made using various economic threshold, optimization, or decision theory models, or may be made without regard to economics at all.[82]

Table 3.1 does not include the many serious pests that were introduced into Australia before 1971. For example, many of the stored product pests came with the First Fleet in 1788, and many others were introduced progressively (and probably repeatedly) during the nineteenth century. Also since 1995, additional exotic species have been introduced accidentally to Australia. Thus, Table 3.1 depicts the rate of increase of the economic losses due to exotic species over the last quarter of the twentieth century rather than the total annual loss due to all exotic species. Table 3.1 does not take into account the impact of plant diseases, nor the impact of weeds, which has been estimated at $3.9 billion annually.[83] Nevertheless, it is indicative of the economic impact of exotic pests on Australian agriculture and forestry.

### 3.4.2   Papaya fruit fly

Papaya fruit fly (PFF), *Bactrocera papayae* Drew and Hancock (Diptera: Tephritidae) is one of the very few exotic arthropods for which economic impact and the costs of control in Australia have been documented in any detail. The insect, which attacks a wide variety of tropical and temperate fruit and vegetables, was detected on mainland Australia for the first time in October 1995. PFF is a well-known and widely feared polyphagous, horticultural pest. When the species was detected in North Queensland, a number of Australia's trading partners promptly imposed trade bans on susceptible fruit and vegetables originating in North Queensland or Australia generally.

Following the initial discovery of the species in the Cairns area, a quarantine zone was established to prevent the spread of the pest to other parts of Australia. A program involving extensive surveillance and toxic baiting within the quarantine zone began. A cost–benefit analysis of proposals to eradicate the fly from the quarantine zone was performed,[81] and between 1996 and 1999, an eradication program, based on toxic baiting, was undertaken. The cost–benefit analysis highlighted the extreme complexity of assessing economic impact.

First, the choice of analytical technique is critical. General equilibrium analysis yields information not only on the industry or commodity directly affected by the pest but also information on the consequences for other industries and macroeconomic effects. General equilibrium analysis requires large amounts of data, for example, to describe interindustry effects. Partial equilibrium analysis requires less input data. It takes into account changes in the price and availability of commodities and, thus, the effect on consumers of the commodity. It gives a more realistic picture than do partial budgeting techniques, which focus largely on the effects on the particular industry or production system affected by the pest. Partial budgeting requires the least input data.

It is important to recognize that an industry may have several strategies in the face of a new and damaging exotic pest. For example, with the advent of PFF, three options were available to Australian fruit and vegetable growers:

1. They could simply accept the pest and redirect their exports to countries already infested by PFF. It was estimated that PFF-free markets offered a premium of $9 million per annum, which would then be lost to the Australian producers.
2. They could continue to export to premium, PFF-free markets, but only after fruit had been disinfested, at a cost of $7.59 million per annum.

*Table 3.2* Estimated Annual Costs of Papaya Fruit Fly in Australia[a]

| Source of costs | Within quarantine zone $ million | For remainder of Australia[b] $ million | Australia-wide $ million |
|---|---|---|---|
| Economic losses on exports | 0.08 | 7.51 | 7.59 |
| Cost of insecticide treatments | 0.37 | 52.88 | 53.25 |
| Cost of disinfestation for domestic market | 12.67 | 0 | 12.67 |
| Total | 13.12 | 60.39 | 73.51 |

*Source:* ABARE, *Papaya fruit fly: Cost-benefit analysis of proposed eradication program*, ABARE, Canberra, 1995. With permission.

[a] Undiscounted real values.
[b] Assumes 100% probability of infestation spreading from the quarantine zone to all other suitable regions in Australia.

3. They could redirect their exports to the domestic market, resulting in a $28.97-million loss in producer economic surplus. If the PFF dispersed throughout Australia, there would be no disinfestation costs. If the fly did not spread, then disinfestation costs would amount to $12.67 million per annum.

Clearly, the second alternative is the most attractive option for producers.

Many of the fruits and vegetables susceptible to PFF are also susceptible to the native Queensland fruit fly and are protected with regular pesticide sprays. However, the presence of PFF would require additional spray treatments. For example, bananas would require an additional 6 sprays each year at a cost of $46 per hectare per spray and tomatoes would require an additional 10 sprays, each at a cost of $27 per hectare. Australia-wide, additional spray treatments for all susceptible fruits and vegetables would amount to an additional impost of $53.25 million per annum. Table 3.2 depicts the most likely cost scenario.

In summary, the ABARE analysis indicated that annual cost to Australia as a result of the PFF incursion was likely to be approximately $74 million, which is significantly less than the figures given for PFF (*B. papayae*) in Table 3.1. By 1999, PFF had been eradicated at an actual cost of approximately $35 million,[84] which is another estimate of the cost of the incursion of this exotic species into Australia.

### 3.4.3  Citrus canker

Citrus canker is another invader for which precise, modern-day costs are available. The disease is caused by the bacterium *Xanthomonas axonopodis* (Hasse) and is recognized on citrus trees (oranges, grapefruit, limes, etc.) by lesions on the stems, leaves, and fruits; by premature leaf drop; and by a general loss of plant vitality. It is particularly pernicious because the unsightly fruit, though edible, are virtually unsalable. The disease originated in Asia, with *Citrus* and its relatives, and has spread to the New World, the Middle East, and the Pacific region. Dispersal can occur via infected planting material, on the wind, or on contaminated equipment. The disease was detected in the Northern Territory of Australia in the 1900s and eradicated by removal of host trees.

A fresh incursion of citrus canker was detected in 2004 in the Emerald region of central Queensland.[85] Three commercial orchards were found to be infected, but the source of the incursion was never established. The response program was based on removal of all commercial and noncommercial citrus trees and native hosts (such as the native lime, *Citrus glauca* Lindl. Burkill), restrictions on replanting and the introduction of potential hosts, and an intensive surveillance program. Some 490,000 citrus trees and 170,000 native citrus trees were destroyed over a quarantine area of some 3,000 km$^2$. Eradication was declared in February 2009. The reliably documented cost of the incursion comprised approximately $17 million for operational, eradication activities, a compensation program of $4.6 million for growers who were obliged to destroy healthy but susceptible trees, and an initial income support program of $1.5 million for affected growers. Since the incursion was detected in 2004, these costs were met entirely by federal and state governments, following long-established national practice. Since 2005, however, the costs of responses to outbreaks of exotic plant pests and diseases have been shared by the government and the private sector, under the terms of what is known as the Emergency Plant Pest Response Deed.[85] The deed is an agreement between the government and industry bodies, which recognizes the "beneficiary pays" principle. The deed also makes the contributions to response programs from the public purse subject to industry adopting reasonable pre-emptive measures to minimize the risk of incursions by invasive species. A similar agreement covers incursions affecting livestock, but neither this agreement nor the Emergency Plant Pest Response Deed encompass weeds, invasive species affecting freshwater and marine environments, or vertebrate pests.

### 3.4.4   Banana skipper

A cost–benefit analysis of a biological control program against the banana skipper, *Erionota thrax* Linnaeus (Lepidoptera: Hesperiidae) similarly provides insight into both the potential costs to Australia of an invasion by this pest species and the cost of control.[86] Biological control of this species has been demonstrated by a $700,000 program based in PNG.

In the absence of biological control agents banana skipper could be expected to cause production losses of up to $65.9 million per annum in Australia. However, in the presence of biological control agents, these losses could be expected to shrink to approximately $3 million annually. Thus, the invasion of Australia by banana skipper would cost the country a one-off sum of $700,000 (an estimate of the research and development costs of the biological control program) and a recurring annual amount of $3 million. The analysis by Waterhouse et al.[86] relied largely on partial budgeting.

### 3.4.5   European house borer

The European house borer, *Hylotrupes bajulus* Linnaeus, is a destructive pest of a range of coniferous timbers used in construction. It was detected in Perth in 2004. In 2007, a national control program commenced, and the borer remains confined to a handful of outer Perth suburbs. To assist decision making within the control program, economic outcomes of three alternative response scenarios were estimated.[87] If nothing were done, Australian home and business owners would be left to deal with the pest at an estimated cost of $120 million over 30 years. If the Western Australia government banned the use of untreated softwoods for structural purposes, compliance costs would amount to $697 million. If the government bans applied only to areas where the borer was known to occur (and the insect did not infest new areas), compliance costs would reduce to $37 million, but $1 million in damage could be expected.

### 3.4.6  Beneficial exotic arthropods

Apart from biological control agents, there are few examples of arthropod introductions to Australia that have created major economic benefits for agriculture or forestry. The honeybee, *Apis mellifera* Linnaeus (Hymenoptera: Apidae), and leafcutter bee, *Megachile rotundata* Fabricius (Hymenoptera: Megachilidae), stand out in this respect. The former is the mainstay of the Australian beekeeping industry, which has a gross production value between $60 and $65 million per annum, largely from honey, wax, and queen bees, which are a valuable export commodity.[32,88] Industry costs, which are principally labor and transportation, run at about 80% of revenue. However, the major and most pervasive economic impact of the honeybee is as a crop pollinator, with crops such as apples, cotton, citrus, onions, and mangoes being particularly dependent. Benefits to Australia of up to $1.2 billion per annum have been claimed.[32] Indeed, this estimate might be conservative. The nitrogen enrichment of New Zealand soils by pasture legumes, all pollinated by honeybees and leafcutter bees, has been valued at $1.87 billion. Clearly, the economic benefits of introduced bees are substantial, but even these come with some cost as related in the section on medical impacts.

The Asian bee, *Apis cerana* Fabricius, on the other hand, constitutes a major threat to the Australian bee industry due to its ectoparasitic mites belonging to the genus *Varroa*. Since 1980, these mites have been globally distributed via the illicit exportation of infested queen bees. The mites have been known for centuries as pests of Asian bees in India, China, Korea, and Southeast Asia and now commonly infest *Apis mellifera*. From these locations it was introduced into northern Europe and South America and is now established in the United States, Indonesian Papua, and PNG.[89] The introduction of *Varroa* to the Australian European bee population would lead to significant reductions in crop yields and pollination success.

It is anticipated that if *Varroa* were to establish in Australia it would devastate unmanaged colonies of *A. mellifera* and that free pollination of agricultural crops by feral honeybees effectively would cease. Managed, commercial hives can be treated for *Varroa*, but at a cost of perhaps $40–$50 per hive, per year. Treatment also increases the risk of chemical residues in honeybee products. In a scenario in which many more producers were obliged to pay for pollination services and the unit cost of these services were increased by the need to control *Varroa*, Australian plant industries overall could face additional costs of between $21 and $50 million over a 30-year period.[90] In southern Africa, resistance to *Varroa* emerged in African *A. mellifera* over a 10-year period. If this were to occur in Australia, the estimated cost to plant industries would be discounted substantially.

In fact, an incursion of *A. cerana* was detected in the Cairns area of tropical Queensland in 2007. *Varroa* seems not to have accompanied the incursion, but an eradication response was warranted and initiated. Three years on, it was estimated that a modest but effective surveillance, destruction, and community awareness program, directed at eradication of *A. cerana*, would cost at least $800,000 per year.

## 3.5  Invasions of marine importance

Australia has a large coastline with many ports of call, which leaves it particularly exposed to invasions of invertebrate marine species primarily via international shipping but also through mechanisms such as discharge of ballast water, attachment to vessel hulls, importation for the aquarium trade and fish farming, deliberate introduction, movement of fisheries products, and transportation in fishing equipment or anchors. Surprisingly, surveillance and action against marine invaders in Australia are a recent development.

In 1995, the National Introduced Marine Species Port Survey Program was initiated to provide baseline information on the status of introduced species in trading and other coastal ports.[91] To date, 170 exotic species have been found, although it is said that the scale of marine invasions is really not known in Australian waters.[92] Three of the identified exotic invertebrate species have been marked as likely to make significant contributions to economic and environmental costs in the years to come. They are the black-striped mussel (*Mytilopsis* sp.), the northern Pacific seastar (*Asterias amurensis*), the sabellid fan worm (*Sabella spallanzanii*), and the toxic dinoflagellate (*Gymnodinium catenatum*). The costs and many other factors relating to the other exotic species are unknown.

### 3.5.1   Black-striped mussels

The black-striped mussel invasion of northern Australia was detected in 1999 in three marinas in Darwin, and 400 infested vessels have since been tracked. This was the first known incursion of a serious marine pest into Australian tropical waters, and a great deal of concern was expressed due to the potential for considerable economical and ecological damage. In a month-long eradication operation involving 250 people, over 100 tons of chlorine and 10 tons of copper sulfate were dumped into infested waters at a cost of $2 million. Concern stems from the fact that this mussel is a close relative of the zebra mussel, *Dreissena polymorpha*, which invaded the U.S. Great Lakes system in the 1980s with an economic impact of over $600 million per annum. In India, the black-striped mussel has impacted in a similar manner to the zebra mussel by fouling all intertidal and sublittoral structures and vessels in large numbers. In Australia, predictions include infestation of marine oyster farms, marine pumping facilities (ballast and cooling systems), and recreational and inshore vessels and all port facilities, and costs are expected to be similar to the United States and India experience. The potential environmental impact of this organism is predicted to be substantial, with the possibility of vast monocultures in low estuarine habitats.

### 3.5.2   Northern Pacific seastar

In 1986, the first *Asterias amurensis* (Lutken) seastar specimen was discovered in southern Australian waters near Hobart.[92] Its natural habitats include cooler coasts ranging from the Bering Straits down to Canada and Japan. A decade later it had become well established in the lower Derwent River and parts of several other estuaries and bays with two specimens also found in Port Philip Bay.[93] In 1974, the Commonwealth Scientific and Research Organization (CSIRO) reported that the Derwent estuary was the most polluted river in the world and advised against eating anything in it. In 1999, the CSIRO released a media statement to the effect that there was a link between the pollution and the seastar population, which had now reached 30 million. Port Philip Bay recorded 50 specimens in early 1998 and 12 million in 1999.[94] Due to the link with pollution, it is possible that the seastars in the Derwent estuary have made a negligible contribution to environmental costs. It is still too soon to estimate control and economic costs, but these are on the rise because these incidents involving invasive marine invertebrates have prompted the federal government to endorse a $5 million dollar management program designed to address the threat.[95] This seastar is a well-adapted predator with a predilection for shellfish, but is capable of consuming any animal tissue encountered and will dig for buried prey. Research into impacts and costs have been initiated, however, no conclusive data has yet emerged. Considerable losses, in the range of millions of dollars per year, have been sustained by the shellfish mariculture industries in Japan, which is of concern to wild shellfish fisheries in Australia. The northern Pacific seastar is now recognized globally as a significant pest capable of causing great damage to the marine environment, aquaculture, and commercial and recreational fisheries.[96]

### 3.5.3   European fan worm

The European or sabellid fan worm, *Sabella spallanzanii*, was first identified in Western Australian waters in 1965. It has since made its way into Eastern waters, and in 1992 it was the dominant organism in the polluted Port Philip Bay.[97] Although there is no direct evidence in Western Australia to suggest that this species is negatively impacting any fisheries or native species, it is having a significant impact in Port Philip Bay where scallop farmers are under threat. Seagrass beds have been overgrown, and competition for food has been detrimental to native oysters and other shellfish. *Sabella spallanzanii* is an efficient filter feeder with a greater capacity to feed on phytoplankton than seagrass.[98] This results in a significant and detrimental reduction in the amount of food in the system thus impacting on the entire ecology. No effort has been made to quantify the impact costs of this organism.

   Management measures for ballast water transfer, the suspected transfer mechanism in many cases, are progressing on both the national and international levels due to the ability of marine organisms to transcend all boundaries and the necessity of limiting their distribution. While the International Marine Organization established an obligatory international framework for ballast water management in 2000, mandatory reporting has been in place in Australia since 1998. In addition, the National System for the Prevention and Management of Marine Pest Incursions was agreed to by all Australian governments in late 2003. It aims to control existing exotic pests and prevent further incursions.[97]

### 3.5.4   New Zealand screw shell

The distribution and impact of the New Zealand screw shell, *Maoricolpus roseus*, are currently being researched by the University of Tasmania and CSIRO. In the 1920s, this invasive marine pest was inadvertently introduced to southeastern Tasmania. To date, it has colonized more territory than any other exotic benthic pest in Australia, and its tolerance of temperature and depth makes further spread likely.[99]

   Bax[100] stated that this pest had become established in vast beds in the northern Bass Strait and off the coasts of eastern Tasmania, Victoria, and New South Wales. He estimated that it was distributed over a sea floor area the size of Tasmania. Barrett and colleagues from the Tasmanian Aquaculture and Fisheries Institute (TAFI) are mapping marine habitats off the coast of Freycinet, Bruny Island, and the Tasman Peninsula using an autonomous underwater vehicle. This process discovered that the New Zealand screw shell occurs in great numbers on sandy sediments around Tasmania's east coast. "We knew that this screw shell species had formed extensive cover in parts of the D'Entrecasteaux Channel, but were not aware of similar densities in other areas of the east coast," Dr. Barrett said.[101]

   Since few predators can break the hard shell of *M. roseus*, there is concern for its control and potential impact on other mollusk species, including scallops and native screw shells.[100] The pest breeds so prolifically that its live and dead shells smother the sea floor to a depth of 80 m along the continental shelf. While some researchers believe the dead shells create new habitats that allow other species, such as sponges, to settle and grow,[101] Bax believes that the substantial beds of dead shells are detrimental to other animals on the sea floor.[101]

At this point, it is not known how far the screw shell will be able to distribute or what impact it will have on other species that are required to maintain local ecosystems. It is important to manage the risk of this species being spread by shipping since genetic markers for *M. roseus* have been identified in the plankton and water available for ships' ballasts.[99] Australia has the richest fauna of screw shells or Turritellids in the world, so the impact on biodiversity may be considerable. Until the environmental impact is known, it is not possible to predict the economic impact.

## References

1. New, T. R. 1994. *Exotic Insects in Australia*. Adelaide: Gleneagles Publishing.
2. Hayes, E. B., and D. J. Gubler. 1992. Dengue and dengue hemorrhagic fever. *Pediatr Infect Dis* 11:311.
3. Halstead, S. B. 1980. Dengue hemorrhagic fever: A public health problem and a field for research. *Bull WHO* 58:1.
4. Meltzer, M. I., J. G. Rigau-Pérez, G. G. Clark, P. Reiter, and D. J. Gubler. 1998. Using disability-adjusted life years to assess the economic impact of dengue in Puerto Rico: 1984–1994. *Am J Trop Med Hyg* 59:265.
5. Kay, B. H., P. Barker-Hudson, N. D. Stallman, M. A. Wiemers, E. N. Marks, P. J. Holt, M. Muscio, and B. M. Gorman. 1984. Dengue fever. Reappearance in northern Queensland after 26 years. *Med J Aust* 140:264.
6. Cleland, J. B., and B. Bradley. 1918. Dengue fever in Australia. *J Hyg* 16:317.
7. Cameron, D. 1997. Well nigh beneath contempt! Urbanisation and the development of manufacturing in Queensland, 1860–1930. In *Images of the Urban Conference*, Sunshine Coast University College, University of Queensland.
8. Lumley, G. F., and F. H. Taylor. 1943. *Dengue*. School of Public Health and Tropical Medicine, Service Publication No. 3. Sydney: University of Sydney & Commonwealth Department of Health, Australasian Medical Publishing Company.
9. Walker, A. S., E. Meyers, A. R. Woodhill, and R. N. McCulloch. 1942. Dengue fever. *Med J Aust* 12:223.
10. Doherty, R. L. 1957. Clinical and epidemiological observations on dengue fever in Queensland, 1954–1955. *Med J Aust* 1:753.
11. QH. 2004. *Dengue Fever Management Plan for North Queensland 2005–2010*. Brisbane: Queensland Health, Queensland Government.
12. Shepard, D. S., and S. B. Halstead. 1993. Dengue (with notes on yellow fever and Japanese encephalitis). In *Disease Control Priorities in Developing Countries*, ed. D. T. Jamison, 303. New York: Oxford University Press Inc.
13. McBride, W. J. H., H. Mullner, R. Muller, J. Labrooy, and I. Wronski. 1998. Determinants of dengue 2 infection among residents of Charters Towers, Queensland, Australia. *Am J Epidemiol* 148:1111.
14. McMichael, A., R. Woodruff, P. Whetten, K. Hennessy, N. Nichols, S. Hales, A. Woodward, and T. Kjellstrom. 2003. *Human Health and Climate Change in Oceania: A Risk Assessment 2002*. Canberra: Commonwealth of Australia.
15. Ritchie, S. A., P. Moore, M. Carruthers, C. R. Williams, B. L. Montgomery, P. Foley, S. Ahboo et al. 2006. Discovery of a widespread infestation of *Aedes albopictus* in the Torres Strait, Australia. *J Am Mosq Control Assoc* 22:358.
16. Russell, R. C., C. R. Williams, R. W. Sutherst, and S. A. Ritchie. 2005. *Aedes* (Stegomyia) *albopictus*—a dengue threat for southern Australia? *Commun Dis Intell* 29:296.
17. Gratz, N. 2004. Critical review of the vector status of *Aedes albopictus*. *Med Vet Entomol* 18:215.
18. Shroyer, D. A. 1986. *Aedes albopictus* and arboviruses: A concise review of the literature. *J Am Mosq Control Assoc* 2:424.
19. Mitchell, C. J. 1995. Geographic spread of *Aedes albopictus* and potential for involvement in arbovirus cycles in the Mediterranean Basin. *J Vect Ecol* 20:44.

20. Almeida, A. P. G., S. S. S. G. Baptista, C. A. G. C. C. Sousa, M. T. L. M. Novo, H. C. Ramos, N. A. Panella, M. Godsey et al. 2005. Bioecology and vectorial capacity of *Aedes albopictus* (Diptera: Culicidae) in Macao, China, in relation to dengue virus transmission. *J Med Entomol* 42:419.

21. Metselaar, D., C. R. Grainger, K. G. Oei, D. G. Reynolds, M. Pudney, C. J. Leake, P. M. Tukei, R. M. D. Offay, and D. I. Simpson. 1980. An outbreak of type 2 dengue fever in the Seychelles probably transmitted by *Aedes albopictus*. *Bull WHO* 58:937.

22. Effler, P. V., L. Pang, P. Kitsutani, V. Vorndam, M. Nakata, T. Avers, J. Elm et al. 2005. Dengue fever, Hawaii, 2001–2002. *Emerg Infect Dis* 11:742.

23. Johansen, C. A., M. D. A. Lindsay, S. A. Harrington, P. I. Whelan, R. C. Russell, A. K. Broom. 2005. First record of *Aedes* (Aedimorphus) *vexans vexans* (Meigen) in Australia. *J Am Mosq Control Assoc* 21:222.

24. Russell, R. C. 2009. "Dramas" down-under: Changes and challenges in Australia. In *Vector Biology, Ecology and Control*, ed. P. W. Atkinson, 81. London: Springer.

25. Hanna, J. N., S. A. Ritchie, D. A. Phillips, J. Shield, M. C. Bailey, J. S. Mackenzie, M. Poidinger, B. J. McCall, and P. J. Mills. 1996. An outbreak of Japanese encephalitis in the Torres Strait, Australia, 1995. *Med J Aust* 165:256.

26. Preslar, D. 1998. Japanese encephalitis—Australia [2]. ProMED-mail post. http://osi.oracle.com:8070/promed/promed.folder.home?p_cornerid=61 (accessed March 24, 2010).

27. Hanna, J. N., S. A. Ritchie, D. A. Phillips, J. M. Lee, S. L. Hills, A. F. van den Hurk, A. T. Pyke, C. A. Johansen, and J. S. Mackenzie. 1999. Japanese encephalitis in north Queensland, Australia, 1998. *Med J Aust* 170:533.

28. Cosgriff, M. 2000a. Japanese encephalitis virus, mosquitoes—Australia. ProMED-mail. http://osi.oracle.com:8070/promed/promed.folder.home?p_cornerid=61 (accessed March 24, 2010).

29. Cosgriff, M. 2000b. Culex gelidus—Australia. ProMED-mail. http://osi.oracle.com:8070/promed/promed.folder.home?p_cornerid=61 (accessed March 24, 2010).

30. Johnson, P. H., S. Hall-Mendelin, P. I. Whelan, S. P. Frances, C. C. Jansen, D. O. Mackenzie, J. A. Northill, and A. F. van den Hurk. 2009. Vector competence of Australian *Culex gelidus* Theobald (Diptera: Culicidae) for endemic and exotic arboviruses. *Aust J Entomol* 48:234.

31. NNDSS. Number of notifications for all diseases by year, Australia, 1991 to 2009 and year-to-date notifications for 2010. Canberra: National Notifiable Diseases Surveillance System, Department of Health and Aging. http://www9.health.gov.au/cda/source/Rpt_2.cfm?RequestTimeout=500 (accessed March 24, 2010).

32. Gibbs, D. M. H., and I. F. Muirhead. 1998. The economic value and environmental impact of the Australian beekeeping industry. A report prepared for the Australian Beekeeping Industry, Australian Honeybee Industry Council, Maroubra.

33. Harvey, P., S. Sperber, F. Kette, R. J. Heddle, and P. J. Roberts-Thomson. 1984. Bee sting mortality in Australia. *Med J Aust* 140:209.

34. Levick, N. R., J. O. Schmidt, J. Harrison, G. S. Smith, and K. Winkel, 2000. Review of bee and wasp sting injuries in Australia and the USA. In *The Hymenoptera: Evolution, Biodiversity and Biological Control*, ed. A. D. Austin and M. Dowton, 437–447. Melbourne: CSIRO Publishing.

35. McGain, F., and K. D. Winkel. 2000. Bee and wasp sting related fatalities in Australia. In *XIIIth World Congress of the International Society on Toxinology*, Abstract Book, Paris, L122.

36. Bradley, C. 2008. Chapter 4. Wasps and bees. In *Venomous bites and stings in Australia to 2005*, 44. Australian Injury Research and Statistics Series No. 40. Australian Institute of Health and Welfare.

37. Winkel, K., G. Hawdon, and K. Ashby. 1998. Venomous bites and stings. *Aust J Emerg Care* 5:13.

38. Crosland, M. W. J. 1991. The spread of the social wasp, *Vespula germanica*, in Australia. *NZ J Zool* 18:375.

39. Spradbery, J. P., and G. F. Maywald. 1992. The distribution of the European or German wasp in Australia, past, present and future. *Aust J Zool* 40:495.

40. Levick, N. R., K. D. Winkel, and G. S. Smith. 1997. European wasps: An emerging hazard in Australia. *Med J Aust* 167:650.

41. McGain, F., J. Harrison, and K. D. Winkel. 2000. Wasp sting mortality in Australia. *Med J Aust* 173:198.

42. McGain, F., Harrison, J., and Winkel, K. D. 2001. Wasp sting mortality in Australia. *Med J. Aust* 174:255.

43. Honan, P. 1997. In *Proceedings of the European Wasp Strategy Meeting*, September 11, Victorian Department of Natural Resources and the Environment.

44. Anderson, A. N. 2000. *The Ants of Northern Australia*. East Melbourne: CSIRO.

45. Hoffman, D. R. 1997. Reactions to less common species of fire ants. *J Allergy Clin Immunol* 100:679.

46. Queensland Department of Primary Industries. 2001. Animal and plant health: Red imported fireants. http://www.dpi.qld.gov.au/health/3125.html (accessed March 24, 2010).

47. Stafford, C. T. 1996. Hypersensitivity to fire ant venom. *Ann Allergy Asthma Immunol* 77:87.

48. Prahlow, J. A., and J. J. Barnard. 1998. Fatal anaphylaxis due to fire ant stings. *Am J Forensic Med Pathol* 19:137.

49. Taber, S. W. 2000. *Fire Ants*. College Station, TX: Texas A&M University.

50. Freeman, T. M. 1997. Hymenoptera hypersensitivity in an imported fire ant endemic area. *Ann Allergy Asthma Immunol* 78:369.

51. DeShazo, R. D., D. F. Williams, and E. S. Moak. 1999. Fire ant attacks on residents in health care facilities: A report of two cases. *Ann Int Med* 131:424.

52. Semevsky, F. N., L. C. Thompson, and S. M. Semenov. 1998. An economic evaluation of the impact of fire ants on agricultural plant production in the southeastern USA. In *Proceedings 10th All-Russian Myrmecological Symposium*, Moscow, 144.

53. Caldwell, S. T., S. H. Schuman, and W. M. Simpson. 1999. Fire ants: A continuing community health threat in South Carolina. *JSC Med Assoc* 95:231.

54. Kemp, S. F., R. D. deShazo, J. E. Moffitt, D. F. Williams, and W. A. Buhner. 2000. Expanding habitat of the imported fire ant (*Solenopsis invicta*): A public health concern. *J Allergy Clin Immunol* 105:683.

55. Kompas, T., and N. Che. 2001. An economic assessment of the potential costs of red imported fire ants in Australia. Australian Bureau of Agricultural and Resource Economics Report for Department of Primary Industries, Queensland, Canberra.

56. Solley, G. O., Vanderwoude, C., and Knight, G. K. 2002. Anaphylaxis due to Red Imported Fire Ant sting. *Med J Aust* 176:521.

57. Bureau of Agricultural Economics. 1959. *The Economic Importance of Cattle Tick in Australia*. Canberra: BAE.

58. Davis, R. 1998. A cost benefit analysis of the removal of the tick-line in Queensland. In *42nd Annual Conference of the Australian Agricultural and Resource Economics Society*, University of New England, Armidale. http://www.uq.edu.au/~ecrdavis/Thesis/PDF_Files/cbatline.pdf (accessed March 24, 2010).

59. Jonsson, N. N., R. Davis, and M. De Witt. 2008. An estimate of the economic effects of cattle tick (*Boophilus microplus*) infestation on Queensland dairy farms. *Aust Vet J* 79:826.

60. Sutherst, R. W. 1996. Impacts of climate change on pests, diseases and weeds in Australia. Report of an International Workshop, Brisbane 1995, CSIRO Division of Entomology, Canberra.

61. Atzeni, M. G., D. G. Mayer, and M. A. Stuart. 1997. Evaluating the risk of the establishment of screw-worm fly in Australia. *Aust Vet J* 75:743.

62. Spradbery, J. P. 1991. *A Manual for the Diagnosis of Screwworm Fly*. Canberra: Australian Government Publishing Service.

63. Australian Bureau of Animal Health. 1979. *Screw-Worm Fly: Possible Prevention and Eradication Policies for Australia*. Canberra: Australian Government Printing Service.

64. Spradbery, J. P., R. J. Mahon, R. Morton, and R. S. Tozer. 1995. Dispersal of the Old World screw-worm fly *Chrysomya bezziana*. *Med Vet Entomol* 9:161.

65. Rajapaksa, N., and J. P. Spradbery. 1989. Occurrence of the Old World screwworm fly *Chrysomya bezziana* on livestock vessels and commercial aircraft. *Aust Vet J* 66:94.

66. Ng, S. O., and M. Yates. 1997. Cutaneous myiasis in a traveller returning from Africa. *Aust J Derm* 38:38.

67. Rubel, D. M., B. K. Walder, A. Jopp-McKay, and R. Rosen. 1993. Dermal myiasis in an Australian traveler. *Aust J Derm* 34:45.

68. Searson, J., L. Sanders, G. Davis, N. Tweddle, and P. Thornber. 1992. Screw-worm myiasis in an overseas traveler—case report. *Commun Dis Intell* 16:239–40.

69. FAO. 1992. *Eradicating the Screwworm*. Rome: Food and Agriculture Organization.

70. ARMCANZ. 1996. AUSVETPLAN—Australian veterinary emergency plan. Enterprise Manual, Dairy Processing, Department of Primary Industries and Energy, Agriculture and Resource Management Council of Australia and New Zealand, Canberra.

71. Anaman, K. A., M. G. Atzeni, D. G. Mayer, M. A. Stuart, D. G. Butler, R. J. Glanville, J. C. Walthall, and I. C. Douglas. 1993. Economic assessment of the expected producer losses and control strategies of a screwworm fly invasion of Australia. Department of Primary Industries, Brisbane.

72. Mayer, D. G., M. G. Atzeni, M. A. Stuart, K. A. Anaman, and D. G. Butler. 1998. Mating competitiveness of irradiated flies for screwworm fly eradication campaigns. *Prev Vet Med* 36:1.

73. Chemonges-Nielsen, S. 2003. *Chrysomya bezziana* in pet dogs in Hong Kong: A potential threat to Australia. *Aust Vet J* 81:202.

74. CSIRO. 1993. *Division of Entomology Report of Research*. Canberra: CSIRO.

75. Elliott, H. J., C. P. Ohmart, and F. R. Wylie. 1998. *Insect Pests of Australian Forests*. Melbourne: Inkata.

76. Geier, P. W. 1981. The codling moth, *Cydia pomonella* (L.): Profile of a key pest. In *The Ecology of Pests*, ed. R. L. Kitching and R. E. Jones, 109. Melbourne: CSIRO.

77. Hill, D. S. 1987. *Agricultural Insect Pests of the Tropics and Their Control*. Cambridge, UK: Cambridge University Press.

78. Clarke, G. M. 1996. *Exotic Insects in Australia: Introductions, Risks and Implications for Quarantine*. Canberra: Bureau of Resource Sciences.

79. Malipatil, M. B., I. D. Naumann, and D. G. Williams. 2007. First record of dock sawfly *Ametastegia glabrata* (Fallén) in Australia (Hymenoptera: Tenthredinidae). *Aust J Entomol* 34:95.

80. Hughes, R. D. 1997. Predicted consequences of the establishment of the Russian wheat aphid (*Diuraphis noxia*) in Australia. Bureau of Resources Sciences, Canberra, 1992. *Immunology* 100:679.

81. ABARE. 1995. *Papaya Fruit Fly: Cost-Benefit Analysis of the Proposed Eradication Program*. Canberra: ABARE.

82. Mumford, J. D., and G. A. Norton. 1984. Economics of decision making in pest management. *Ann Rev Entomol* 29:157.

83. Sinden, J. A. 2004. The economic impact of weeds in Australia: Report to the CRC for Australian Weed Management. CRC for Australian Weed Management, Technical Series No. 8, Adelaide.

84. De Barro, P. 1999. A penny spent is a pound saved: Pre-emptive approaches to managing quarantine threats to primary industries. In *Plant Health in the New Global Trading Environment: Managing Exotic Insects, Weeds and Pathogens*, ed. C. F. McRae and S. M. Dempsey, 151. Canberra: National Office of Animal and Plant Health.

85. Plant Health Australia. 2009. *National Plant Health Status Report (07/08)*. Canberra: Plant Health Australia.

86. Waterhouse, D., B. Dillon, and V. Vincent. 2000. Economic benefits to Papua New Guinea and Australia from the biological control of banana skipper (Erionota thrax). ACIAR Impact Assessment Series 12, Canberra.

87. Cook, D. C., S. Liu, and W. L. Procter. Deliberative methods for assessing utilities. ACERA report. http://www.acera.unimelb.edu.au/materials/endorsed/0803-final-report.pdf (accessed March 24, 2010).

88. Gill, R. 1997. Beekeeping and secure access to public land—how it benefits the industry and society. A report for the Rural Industries Research and Development Corporation and the Honeybee Industries Research and Development Council of Australia, RIRDC Research Paper Series 97/16, Canberra.

89. AQIS. 1999. Varroa mite. AQIS Public Relations, Commonwealth of Australia. http://www.aqis.gov.au:80/docs/schools/rl/8043.htm (accessed March 24, 2010).

90. CSIRO. 2008. Inquiry into the future development of the Australian honeybee industry. Submission no. 33, p. 10, House of Representatives Standing Committee.

91. CSIRO. 1999. Black-striped mussel. CSIRO Marine Research, Hobart. http://www.csiro.au/page.asp?type=faq&id=BlackStripedMussel (accessed March 24, 2010).

92. CSIRO. 1998. The northern Pacific seastar. CSIRO Marine Research, Hobart.

93. Australian Nature Conservation Agency. 1996. The introduced northern Pacific seastar, *Asterias amurensis* (Lutken), in Tasmania. ANCA, Canberra.

94. Bureau of Rural Sciences. 1999. Aquatic pests and diseases. http://www.brs.gov.au:80/fish/status99/aquatic.html (accessed March 24, 2010).

95. AQIS. 2000. National arrangements for invasive marine species. AQIS Public Relations Bulletin, Commonwealth of Australia. http://www.aqis.gov.au:80/docs/bulletin/ab800_6.htm (accessed March 24, 2010).

96. Fisheries Western Australia. 2000. Introduced marine aquatic invaders: Northern Pacific seastar. http://www.wa.gov.au/westfish/hab/broc/marineinvader/marine01.html (accessed March 24, 2010).

97. Clapin, G., and Evans, D. R. 1995. The status of the introduced marine fan worm *Sabella spallanzanii* in WA. CSIRO Technical Report 2, Division Fish, CSIRO.

98. Lemmens, J. W. T. J., G. Clapin, P. Lavery, and J. Cary. 1996. Filtering capacity of seagrass meadows and other habitats of Cockburn Sound, Western Australia. *Mar Ecol Prog Ser* 143:187.

99. Gunasekera, R. M., J. G. Patil, F. R. McEnnulty, and N. J. Bax. 2005. Specific amplification of mt-COI gene of the invasive gastropod *Maoricolpus roseus* in planktonic samples reveals a free-living larval life-history stage. *Mar Freshw Res* 56:901.

100. Bax, N. 2000. Screw shell's marine marathon. CSIRO media release. Ref. 2000/287. http://www.csiro.com/files/mediaRelease/mr2000/prScrewShell.htm (accessed May 13, 2010).

101. Tasmanian Aquaculture and Fisheries Institute, University of Tasmania. 2009. Remote seabed mapping reveals marine pests spread. http://www.utas.edu.au/events/Media%20Releases/2009/Mapping%20sea%20floor.pdf (accessed March 24, 2010).

*section two*

---

*Brazil*

# chapter four

# Invasive vertebrates in Brazil

*Carlos Frederico D. Rocha, Helena Godoy Bergallo,*
*and Rosana Mazzoni*

## Contents

## 4.1 Introduction

Brazil, one of the world's megadiverse countries, is presently facing different challenges to conserve its rich biodiversity. Such a large country encloses different biomes, such as the Amazon forest, the Cerrado (savannah-like vegetation located in central Brazil), the Caatinga (semiarid vegetation of the northeastern region), the Pantanal (wetlands in midwestern region), the Pampas (open fields in southern Brazil), and the Atlantic rainforest in the eastern portion of the country. All of these biomes presently experience a set of disturbances that strongly modify the natural landscape, and as a result, significant portions of them and their biological diversity have been lost.[1–6] The two main causes of biodiversity loss in Brazilian ecosystems are similar to those in most regions of the world: habitat fragmentation and presence of invasive exotic species. Increasing human commercial activities and the improvement of modes of transport experienced by civilizations, especially since the age of the great discoveries (after the fourteenth century), brought a massive interchange of fauna and flora leading to an increasing homogenization of the world's biota.[7,8] The process of homogenization of the world's biota[9] has been recognized as the second most important cause of biological diversity in many areas of the world.[7,10–13]

During the last approximately 500 years since European colonization, most natural ecosystems in Brazil have continuously received an influx of introduced (i.e., exotic) plant, animal, and microbial species from other ecosystems.[4,14,15,6] These species are understood as exotic species, coming from different biomes, ecosystems, or regions. Here, while it is important to realize that a species coming from a different biome can be understood to be an exotic invader species,[16] for some very geographically extensive biomes (like the

Brazilian Atlantic rainforest that extends along approximately 5000 km of the eastern Brazilian coastal region), plant and faunal movement across considerably distant regions of the same biome can also effectively constitute introductions, since distance along the biome gradient produces continual differences in communities' compositions.

An exotic species effectively establishing populations in new natural environments can be considered an invasive species (or invasive alien species [IAS]). The Convention on Biological Diversity (CBD, 1992) signed by 175 countries stated that an invasive species "is one introduced species that spreads without human assistance causing threat to natural or semi-natural environments outside its original range." The results of invasion may not only cause a reduction of the original local biological diversity but may also tend to promote significant environmental, economical, and social impacts.[17] Consideration for that in Brazil, a consequence of the negative impact of invasive exotic species on natural environments and their biological diversity has been the founding of CBD (held in Rio de Janeiro, Brazil, in 1992). Since then, the concern in Brazil regarding exotic species and their potential for becoming invasive species has been increasingly integrated into thinking and actions to protect biological diversity.

The first well-documented reports of the invasive species in Brazil dealt with the spread of the African Malaria mosquito *Anopheles gambiae* in northeastern Brazil during the last 30 to 40 years.[18–20] In terms of vertebrates, the first published records of invasion of natural environments by exotic species come from the study by Myers.[21] For fishes, the first known case of introduction occurred during the end of the nineteenth century, with the arrival of the first lot of *Cyprinus carpio* and its effective use in fish culture.[22]

Recent important efforts in Brazil to establish a data set of exotic species invading natural environments have resulted in data sets from governmental organizations, such as Ministério do Meio Ambiente (http://www.mma.gov.br/sitio/index.php?ido=conteudo.monta&idEstrutura=174), Instituto Chico Mendes de Conservação da Biodiversidade (http://www.icmbio.gov.br/), and nongovernmental organizations (NGOs), such as Instituto Hórus (http://www.institutohorus.org.br/) and Exoticfish (http://exoticfish.bio.br/lista.htm). Clearly, we still need to continually progress toward a more comprehensive understanding of which species are invasive vertebrates, where they are located, the economic issues regarding their presence in natural environments, and how to cope with them.

In this chapter, we seek to provide additional information regarding the actual *status* of the invasive vertebrate species in Brazil (exclusively freshwater Osteichthyes, Amphibia, Reptilia, Aves, and Mammalia), analyze the general distribution of these main vertebrate groups throughout the country, and provide an annotated checklist based on all known cases.

To obtain a view of the present *status* of invasive vertebrates in Brazil, we recorded data available in the literature and on the Internet regarding the occurrence and actual situations (e.g., only exotic or if also invasive) of the exotic and invasive vertebrate species, supplemented with the field records of the authors. These records composed a data set from which we extracted the information hereafter presented. The total number of living species of each vertebrate group occurring in Brazil was obtained from comprehensive and consistent catalogs or checklists that are widely accepted. In terms of fishes for the purpose of this study, we considered only exclusively freshwater species and based data for living species on the catalog of freshwater fishes from Brazil (Catálogo das Espécies de Peixes de Água Doce do Brasil).[23] For amphibians and reptiles, the total living species were obtained from the List of Amphibians and Reptiles of Brazil (2009) of the Brazilian Society of Herpetology (at http://www.sbherpetologia.org.br/). For living bird species in Brazil, we used the database of the Brazilian Committee of Ornithological Records (Comitê Brasileiro de Registros

Ornitológicos at http://www.cbro.org.br/CBRO/num.htm), and for mammals, the total living species were obtained from *Mamíferos do Brasil* by Reis et al.[24]

We also considered the exotic invasive species in terms of its Neotropical or extra-Neotropical (ENT) origin. For a Neotropical origin, we considered the species introduced into Brazil from another area (or biome) in the Neotropics, and for an ENT origin, those introduced species that come from other zoogeographic regions. We also aimed to estimate the number of invasive species of each vertebrate group (and also the pooled number of vertebrates) reported to be registered in each Brazilian state, except for freshwater fishes, to which we considered the number of species for the main Brazilian hydrographic regions (as established by the Conselho Nacional de Recursos Hídricos [CNRH][25]). According to the CNRH (resolution no. 32/2003), freshwater fishes in Brazil are distributed in 12 hydrographic regions: Amazônica, Tocantins/Araguaia, Atlântico Nordeste Ocidental, Parnaíba, Atlântico Nordeste Oriental, São Francisco, Atlântico Leste, Atlântico Sudeste, Paraná, Paraguai, Uruguai, and Atlântico Sul. Nonetheless, these hydrographic regions can be divided into many different ecoregions[26] following geologic design or barriers. Thus, in this study, species invasions toward other ecoregions in the same hydrographic region are considered introductions. This decision follows the recognition of freshwater ecoregions as large areas encompassing one or more freshwater systems with a distinct assemblage of freshwater species.[27]

## 4.2 Invasive vertebrates in Brazil

As a result of the increasing concern regarding invasive species, a set of introductory studies and actions has been made during the last few years to identify the invasive exotic species in Brazil. These studies and actions have resulted in an approximate knowledge of which introduced species remain exotic species and which can be considered invasive.

Based on the data available, we found that the proportion of invasive species in each vertebrate group in Brazil in relation to the total number of living species for each vertebrate group in the country can vary from 0.2% up to more than 4%, depending on the group, with the most invasions recorded being freshwater fishes (Table 4.1).

Osteichthyes (4.25%) and mammals (2.45%) are the groups of vertebrates that presently have the highest proportion of invasive species among the known living species in Brazil (Table 4.1). Of the 850 amphibian species known to occur in Brazil (including Anura, Caudata, and Gimnophiona), 3 species (0.35%) are invasive species (Table 4.1). Among reptiles, of the 709 species recorded in Brazil (including lizards, amphisbaenians, snakes, turtles and crocodiles), 5 species (0.71%) are invasive. Presently, in Brazil, there are 1825 recorded bird species (including 1019 passerine and 826 nonpasserine species); of these, 4 (0.22 %) constitute invasive species. Considering these vertebrate groups pooled, the data show that at least 138 (2.08%) of the living vertebrate (nonmarine) species in Brazil are invasive exotic (Table 4.1).

### 4.2.1 Fishes

Fishes compose the most diversified vertebrate group in the world totaling approximately 50% of all known species. Among all fish species, 9% (2587 species) are native to Brazilian freshwater systems. Such species diversity is widespread in many different ecosystems, reflecting a high phenotypic plasticity.[28]

A total of 109 species of freshwater Osteichthyes have been registered as invasive in Brazil (Tables 4.1 and 4.2). The motivation for these introductions depended on the fish

*Table 4.1* Number of Living Species, Number of Recorded Invasive Species, and Their Respective Proportion in Each Vertebrate Group in Brazil

| Vertebrate group | Living species in Brazil[a] | Invasive vertebrate species in Brazil (IVSB) | Proportion of invasive species (%) |
|---|---|---|---|
| Osteichthyes | 2587 | 110 | 4.25 |
| Amphibians | 850 | 3 | 0.35 |
| Reptiles | 709 | 5 | 0.71 |
| Aves | 1825 | 4 | 0.22 |
| Mammalia | 652 | 16 | 2.45 |
| Total | 6623 | 138 | 2.08 |

[a]*Sources:* Data from Freshwater fishes (Buckup, P. A., N. A. Menezes, and M. S. Ghazzi. 2007. *Catálogo das espécies de peixes de água doce do Brasil.* Série Livros 23. Rio de Janeiro: Museu Nacional.); amphibians and reptiles: Brazilian Society of Herpetology, List of Amphibians and Reptiles of Brazil (http://www.sbherpetologia.org.br); aves: Comitê Brasileiro de Registros Ornitológicos (http://www.cbro.org.br/CBRO/num.htm); mammals (Reis, N. R., A. L. Peracchi, W. A. Pedro, and I. P. Lima. 2006. *Mamíferos do Brasil.* Londrina: Ed. UEL.)

*Note:* The number of invasive vertebrate species in Brazil (IVSB) indicates solely the occurrence of the species as invasive in Brazil in at least one area, independent of the number of places it occurs as invasive.

group as well as on economical and ecological issues. The main causes for exotic fish introduction are intentional or nonintentional introduction for the improvement of fish culture ponds and aquaria; intentional introduction (stocking) for sportive fishing and biological control (e.g., mosquito control), and nonintentional elimination of natural barriers due to hydropower impoundments. Nonetheless, many cases of introduction of unknown causes still occur.

The ability to invade and establish viable populations in freshwater fishes depends on life-history attributes of the species involved. Ahead, we discuss, by family, the most numerous fish groups introduced in the Brazilian freshwater systems.

Characidae is one of the most heterogeneous fish groups arranged in many *incertae sedis* genera,[29] conferring a high ability to invade and establish viable populations in many different environments. This is the most abundant family, with 17 invasive species in Brazil, 70% of them which are invasive in the upper Paraná basin, both in the lentic habitat of Itaipú reservoir and in the lotic rivers and adjacent streams.[27,30] The ability to invade both lotic and lentic habitats is a consequence of the large variability and the exceptional swimming capability of this fish group.[31] In Brazil, all invasive species among the characids are of Neotropical origin.

Cichlids from the family Cichlidae occur on almost all continents in the world[32] and are registered as the second most important group with invasive species in the Brazilian freshwaters, with 15 species. Cichlids are mainly lentic species, and their introduction into Brazilian freshwater environments followed economic (fish culture and aquariology) and fishery interests.

The Cyprinidae is one of the two largest families of vertebrates and is widely distributed around the world because of its desirable attributes for fisheries and aquaculture. Cyprinids are native to Eurasia, Africa, and North America without native representatives in the Neotropical region.[33] Thus, all of the Cyprinidae species registered in Brazil were introduced for aquaculture purposes. This was the third most numerous fish group of invasive species in Brazil, with 14 species.

A total of 26 (23.8%) invasive fish species in Brazil were of ENT origin and all were introduced as a consequence of their use as fish stock for sportive fishing or fish culture

*Table 4.2* Summary of Recorded Invasive Vertebrates in Brazil

| Species/family | NT/ENT | Common name in Brazil | Originated from | Where it is invasive in Brazil | References |
|---|---|---|---|---|---|
| **Fishes** | | | | | |
| **Family Clupeidae** | | | | | |
| *Platanichthys platana* (Regan, 1917) | NT | Sardinha | South America: just north of Rio de Janeiro, Brazil to Uruguay and Argentina | Upper Paraná River in the Itaipú reservoir and confluence areas of small tributaries in the reservoir. Hydrographic regions L–M. | 30 |
| **Family Cyprinidae** | | | | | |
| *Aristichthys nobilis* (Richardson, 1845) | NT | Carpa cabeçona | Asia: China | Upper Paraná River in marginal areas of the Itaipú reservoir and confluence areas of small tributaries in the reservoir. Hydrographic regions L. | 30 |
| *Carassius auratus* (Linnaeus, 1758) | ENT | Japonês | Asia | Paraíba do Sul River. Hydrographic region H. | http://exoticfish.bio.br/lista.htm |
| *Ctenopharyngodon idella* (Valenciennes, 1844) | NT | Carpa da china | Asia: China to eastern Siberia | Upper Paraná River in marginal areas of the Itaipú reservoir and confluence areas of small tributaries in the reservoir and and Rio Grande do Sul at Lagoa Mirim and Lagoa dos Patos. Hydrographic regions I–L. | 30 |

*(Continued)*

*Table 4.2* Summary of Recorded Invasive Vertebrates in Brazil (*Continued*)

| Species/family | NT/ENT | Common name in Brazil | Originated from | Where it is invasive in Brazil | References |
|---|---|---|---|---|---|
| *Cyprinus carpio* (Linnaeus, 1757) | ENT | Carpa comum | Europe to Asia: Europe, Russia, China, India and Southeast Asia | Minas Gerais in the municipality of Muriaé and zona da Mata; Rio de Janeiro at Paraíba do Sul basin; reservoirs at Paraíba do Sul basin; Paraná, in the municipality of Londrina at Tibagi River/tributaries; upper Paraná River in rivers and streams and Rio Grande do Sul at Lagoa Mirim and Lagoa dos Patos. Hydrographic regions I-L-H. | 30 |
| *Danio frankei* (Meinken, 1963) | ENT | Danio-leopardo | Asia | Paraíba do Sul River. Hydrographic region H. | http://exoticfish.bio.br/lista.htm |
| *Danio malabaricus* (Jerdon, 1849) | ENT | Danio | Asia | Paraíba do Sul River. Hydrographic region H. | http://exoticfish.bio.br/lista.htm |
| *Danio rerio* (Hamilton, 1822) | ENT | Paulistinha | Asia | Paraíba do Sul River. Hydrographic region H. | http://exoticfish.bio.br/lista.htm |
| *Hypophthalmichthys molitrix* (Valenciennes, 1844) | ENT | Carpa prateada | Asia | Doce River. Hydrographic region H. | http://exoticfish.bio.br/lista.htm |
| *Hypophthalmichthys nobilis* (Richardson, 1845) | ENT | Carpa cabeçuda | Asia | Doce and Grande rivers. Hydrographic regions H-F. | http://exoticfish.bio.br/lista.htm |
| *Puntius conchonius* (Hamilton, 1822) | ENT | Barbo-conchônio | Asia | Paraíba do Sul River. Hydrographic region H | http://exoticfish.bio.br/lista.htm |
| *Puntius nigrofasciatus* (Günther, 1868) | ENT | Barbo-rubi | Asia | Paraíba do Sul River. Hydrographic region H. | http://exoticfish.bio.br/lista.htm |

| Species | Status | Common name | Origin | Basin | Location | Reference |
|---|---|---|---|---|---|---|
| *Puntius semifasciolatus* (Günther, 1868) | ENT | Barbo-ouro | Asia | | Paraíba do Sul River. Hydrographic region H. | http://exoticfish.bio.br/lista.htm |
| *Puntius tetrazona* (Bleeker, 1855) | ENT | Barbo-sumatrano | Asia | | Paraíba do Sul River. Hydrographic region H. | http://exoticfish.bio.br/lista.htm |
| *Tanichthys albonubes* (Lin, 1932) | ENT | Tanictis | Asia | | Paraíba do Sul River. Hydrographic region H. | http://exoticfish.bio.br/lista.htm |
| **Family Acestrorhynchidae** | | | | | | |
| *Acestrorhynchus pantaneiro* (Menezes, 1992) | NT | | Paraná-Uruguay basin | | Upper Paraná River in marginal areas of the Itaipú reservoir and confluence areas of small tributaries in the reservoir; Laguna dos Patos drainage. Hydrographic regions M-L. | 27 |
| **Family Anostomidae** | | | | | | |
| *Leporinus macrocephalus* (Garavello and Britski, 1988) | NT | Piavuçú | Paraná-Uruguay basin | | Upper Paraná River in marginal areas of the Itaipú reservoir and confluence areas of small tributaries in the reservoir; Doce and Paraíba do Sul basins. Hydrographic regions M-L-H. | 27, 30, http://exoticfish.bio.br/lista.htm |
| *Leporinus octofasciatus* (Steindachner, 1915) | NT | Aracú-pintado | Paraná-Uruguai basin | | São Francisco River Hydrographic region F. | 35 |
| *Schizodon borellii* (Boulenger, 1900) | NT | | Paraná-Uruguay basin | | Upper Paraná River in marginal areas of the Itaipú reservoir and confluence areas of small tributaries in the reservoir. Hydrographic regions M-L. | 27 |

*(Continued)*

**Table 4.2** Summary of Recorded Invasive Vertebrates in Brazil (*Continued*)

| Species/family | NT/ENT | Common name in Brazil | Originated from | Where it is invasive in Brazil | References |
|---|---|---|---|---|---|
| ***Family Characidae*** | | | | | |
| *Aphyocharax anisitsi* (Eigenmann and Kennedy, 1903) | NT | Tetra | Paraná-Uruguay basin | Upper Paraná River in marginal areas of the Itaipú reservoir and confluence areas of small tributaries in the reservoir; Mato Grosso in the municipality of Caceres; Rio Grande do Sul in the municipalities of Livramento, at Ibicuí River; Uruguaiana at Guarupa River; Rio de Janeiro in the municipalities of Macaé, Itacoatiara, and Petropolis. Normally occurring in streams and/or marginal areas of main rivers channel. Hydrographic regions L-M-I-H. | 30 |
| *Brycon amazonicus* (Spix and Agassiz, 1829) | NT | Jurturna | South America: Amazon River and its main tributaries in Brazil; Orinoco and Essequibo River basins | Paraíba do Sul River. Hydrographic region H. | Guilherme Sousa and Erica Caramaschi, personal communication. |
| *Brycon hilarii* (Valenciennes, 1850) | NT | Piraputanga | South America: Paraguay River basin | Upper Paraná River in marginal areas of the Itaipú reservoir and confluence areas of small tributaries in the reservoir. Hydrographic regions J-L. | 30 |

| | | | | | |
|---|---|---|---|---|---|
| *Bryconamericus exodon* (Eigenmann, 1907) | NT | Lambari | South America: Paraguay River basin | Upper Paraná River in marginal areas of the Itaipú reservoir and confluence areas of small tributaries in the reservoir. Hydrographic regions J-L. | 27, 30 |
| *Colossoma macropomum* (Cuvier, 1816) | NT | Tambaqui | South America: Amazon and Orinoco basins | Upper Paraná River in marginal areas of the Itaipú reservoir and confluence areas of small tributaries in the reservoir, rivers, dams and reservoirs in Paraíba e Pernambuco; Parnaíba basin; Grande and Doce rivers, Paraíba do Sul basin. Hydrographic regions J-M-L-E-H. | 30, http://exoticfish.bio.br/lista.htm |
| *Cynopotamus kincaidi* (Schultz, 1950) | NT | Dentudo | South America: Paraguay River basin; Uruguay River basin | Upper Paraná River in marginal areas of the Itaipú reservoir and confluence areas of small tributaries in the reservoir. Hydrographic region L. | 30 |
| *Gymnocorymbus ternetzi* (Boulenger, 1895) | NT | Tetra preto | South America: Paraguay and Guaporé River basins to Argentina | Upper Paraná River in marginal areas of the Itaipú reservoir and confluence areas of small tributaries in the reservoir; Paraíba do Sul basin. Hydrographic regions J-L-H. | 30, http://exoticfish.bio.br/lista.htm |
| *Hyphessobrycon eques* (Steindachner, 1882) | NT | Tetra or Mato Grosso | South America: Amazon, Guaporé, and Paraguay River basins | Paraíba do Sul River. Hydrographic region H. | Erica Caramaschi, personal communication |

*(Continued)*

**Table 4.2** Summary of Recorded Invasive Vertebrates in Brazil (*Continued*)

| Species/family | NT/ENT | Common name in Brazil | Originated from | Where it is invasive in Brazil | References |
|---|---|---|---|---|---|
| *Hyphessobrycon flammeus* (Myers, 1924) | NT | Engraçadinho | South America: Coastal rivers of Rio de Janeiro, Brazil | Upper Paraná River in marginal areas of the Itaipú reservoir and confluence areas of small tributaries in the reservoir. Hydrographic regions J-M-L. | 30 |
| *Metynnis maculatus* (Kner, 1858) | NT | | South America: Amazon and Paraguay River basins | Upper Paraná River in marginal areas of the Itaipú reservoir and confluence areas of small tributaries in the reservoir; Paraíba do Sul, Grande and Paranaíba basins. Hydrographic regions J-L-H-G. | 30 |
| *Metynnis mola* (Eigenmann and Kennedy, 1903) | NT | Pacu | South America: Paraguay-Paraná River basin | Upper Paraná River in marginal areas of the Itaipú reservoir and confluence areas of small tributaries in the reservoir. Hydrographic regions J-L. | 30 |
| *Knodus moenkhausii* (Eigenmann and Kennedy, 1903) | NT | Tetra | South America: Paraguay River basin | Upper Paraná River in marginal areas of the Itaipú reservoir and confluence areas of small tributaries in the reservoir. Hydrographic regions J-L. | 30 |
| *Piaractus brachypomus* (Cuvier, 1818) | NT | Pirapitinga | South America: Amazon and Orinoco River basins | Main channel of the São Francisco River; Paraíba do Sul basin. Hydrographic regions F-H. | Erica Caramaschi, personal communication. |

| | | | | | |
|---|---|---|---|---|---|
| *Piaractus mesopotamicus* (Holmberg, 1887) | NT | Pacú-caranha | Paraná-Uruguai basin | Grande and São Francisco rivers basins. Hydrographic region F. | http://exoticfish.bio.br/lista.htm |
| *Pygocentrus nattereri* (Kner, 1858) | NT | Piranha-vermelha | Amazonas, Paraná-Paraguai, Essequibo basins | Doce River. Hydrographic region H. | http://exoticfish.bio.br/lista.htm |
| *Roeboides descalvadensis* (Fowler, 1932) | NT | Saicanga | South America: Upper Paraguay River basin | Upper Paraná River in marginal areas of the Itaipú reservoir and confluence areas of small tributaries in the reservoir. Hydrographic regions M-L. | 27, 30 |
| *Salminus brasiliensis* (Cuvier, 1816) | NT | Dourado | South America: Paraná, Paraguay, and Uruguay River basins; Laguna dos Patos drainage, upper Chaparé and Mamoré River basin in Bolivia; Amazon River | Paraíba do Sul and Doce rivers. Hydrographic region H. | Erica Caramaschi, personal communication |
| *Triportheus nematurus* (Kner, 1858) | NT | Sardinha | South America: Paraná-Paraguay River basin | Upper Paraná River in marginal areas of the Itaipú reservoir and confluence areas of small tributaries in the reservoir. Hydrographic regions J-L. | 27 |
| **Family Crenuchidae** *Characidium laterale* (Boulenger, 1895) | NT | | Paraná-Uruguay River basins | Upper Paraná River in streams and marginal areas of main rivers. Hydrographic regions J-L. | 30 |

*(Continued)*

*Table 4.2* Summary of Recorded Invasive Vertebrates in Brazil (*Continued*)

| Species/family | NT/ENT | Common name in Brazil | Originated from | Where it is invasive in Brazil | References |
|---|---|---|---|---|---|
| **Family Curimatidae** | | | | | |
| *Cyphocharax gillii* (Eigenmann and Kennedy, 1903) | NT | Curimbatazinho | South America: Paraguay River basin in Brazil and Paraguay | Upper Paraná River in marginal areas of the Itaipú reservoir and confluence areas of small tributaries in the reservoir. Hydrographic regions J-L. | 30 |
| **Family Prochilodontidae** | | | | | |
| *Prochilodus argenteus* (Agassiz, 1829) | NT | Curimbatá-pacu | São Francisco basin | Doce River. Hydrographic region H. | http://exoticfish.bio.br/lista.htm |
| *Prochilodus costatus* (Valenciennes, 1850) | NT | Curimbatá-piôa | São Francisco basin | Main channel of Paraíba do Sul River and Jequitinhonha River basin. Hydrographic regions G-H. | Erica Caramaschi, personal communication; http://exoticfish.bio.br/lista.htm |
| *Prochilodus lineatus* (Valenciennes, 1836) | NT | Curimbatá | Paraná-Uruguai basin | São Francisco and Paraíba do Sul basins. Hydrographic regions F-H. | http://exoticfish.bio.br/lista.htm |
| *Prochilodus vimboides* (Kner, 1859) | NT | | Paraná and São Francisco basins and coastal dranaiges from Jequitinhonha and Paraíba do Sul Rivers | Doce River. Hydrographic region H. | http://exoticfish.bio.br/lista.htm |
| **Family Erythrinidae** | | | | | |
| *Erythrinus erythrinus* (Bloch and Schneider, 1801) | NT | Jejú | Central and South America: Amazon and Orinoco River basins and coastal rivers of the Guianas | Upper Paraná River in marginal areas of the Itaipú reservoir and confluence areas of small tributaries in the reservoir. Hydrographic regions M-L. | 27, 30 |

| Species | | Common name | Distribution | Location | Reference |
|---|---|---|---|---|---|
| *Hoplerythrinus unitaeniatus* (Spix and Agassiz, 1829) | NT | Traira, Jeju | Central and South America: São Francisco, Amazon, Paraná-Uruguay, Orinoco, and Magdalena River basins, and coastal rivers in Guyana, Suriname, and French Guiana | Upper Paraná River in marginal areas of the Itaipú reservoir and confluence areas of small tributaries in the reservoir. Hydrographic regions M-L. | 27, 30 |
| *Hoplias lacerdae* (Miranda-Ribeiro, 1908) | NT | Trairão | Ribeira do Iguape basin | Paraíba do Sul, Grande, Doce, São Francisco, Jequitinhonha, Mucuri, and Paranaíba basins. Hydrographic regions H-G-F. | http://exoticfish.bio.br/lista.htm |
| **Family Hemiodontidae** | | | | | |
| *Hemiodus orthonops* (Eigenmann and Kennedy, 1903) | NT | Bananinha | Paraná-Uruguay basin | Upper Paraná River in marginal areas of the Itaipú reservoir and confluence areas of small tributaries in the reservoir. Hydrographic region L. | 27 |
| **Family Apteronotidae** | | | | | |
| *Apteronotus albifrons* (Linnaeus, 1766) | NT | Sarapó, ituí-cavalo | Venezuela to Paraguay and Paraná rivers; also in the Amazon basin of Peru | Upper Paraná River in marginal areas of the Itaipú reservoir and confluence areas of small tributaries in the reservoir. Hydrographic regions J-L. | 30 |
| *Apteronotus brasiliensis* (Reinhardt, 1852) | NT | Sarapó, Ituí-cavalo | São Francisco, Paraná-Uruguai, and Parnaíba River basins | Upper Paraná River in marginal areas of the Itaipú reservoir and confluence areas of small tributaries in the reservoir. Hydrographic regions M-L. | 30 |

*(Continued)*

*Table 4.2* Summary of Recorded Invasive Vertebrates in Brazil (*Continued*)

| Species/family | NT/ENT | Common name in Brazil | Originated from | Where it is invasive in Brazil | References |
|---|---|---|---|---|---|
| *Apteronotus caudimaculosus* (de Santana, 2003) | NT | Sarapó, Ituí-cavalo | Brazil | Upper Paraná River in marginal areas of the Itaipú reservoir and confluence areas of small tributaries in the reservoir. Hydrographic regions J-M-L. | 30 |
| *Apteronotus ellisi* (Alonso de Arámburu, 1957) | NT | Sarapó, Ituí-cavalo | Paraná-Uruguay and Paraguai River basins | Upper Paraná River in marginal areas of the Itaipú reservoir and confluence areas of small tributaries in the reservoir. Hydrographic region L. | 30 |
| **Family Atherinopsidae** | | | | | |
| *Odontesthes bonariensis* (Valenciennes, 1835) | NT | Peixe-rei | Uruguai basin | Grande basin. Hydrographic region F. | http://exoticfish.bio.br/lista.htm |
| **Family Gymnotidae** | | | | | |
| *Gymnotus inaequilabiatus* (Valencienne, 1839) | NT | Tuvira | Paraná-Uruguay basin | Upper Paraná River in marginal areas of the Itaipú reservoir and confluence areas of small tributaries in the reservoir. Hydrographic regions M-L. | 30 |
| *Gymnotus paraguensis* (Albert and Crampton, 2003) | NT | Tuvira | Paraná-Uruguay basin | Upper Paraná River in marginal areas of the Itaipú reservoir and confluence areas of small tributaries in the reservoir. Hydrographic regions M-L. | 30 |

*(Continued)*

| | | | | | |
|---|---|---|---|---|---|
| **Family Hypopomidae** | | | | | |
| *Brachyhypopomus gauderio* (Giora and Malabarba, 2009) | NT | | Brazil: Rio Grande do Sul in Laguna dos Patos, Uruguay and Tramandaí River drainages; Paraná-Uruguai basin | Upper Paraná River in marginal areas of the Itaipú reservoir and confluence areas of small tributaries in the reservoir. Hydrographic region L. | 27 |
| *Brachyhypopomus pinnicaudatus* (Hopkins, 1991) | NT | | South America: eastern South America from the Catatumbo River basin, Orinoco and the Guianas to La Plata River basin; Amazon River basin in Peru | Upper Paraná River in rivers and streams and Rio Grande do Sul at Lagoa Mirim and Lagoa dos Patos. Hydrographic regions I-L. | 30 |
| **Family Rhamphichthyidae** | | | | | |
| *Rhamphichthys hahni* (Meinken, 1937) | NT | Tuvira | Paraná-Uruguay basin | Upper Paraná River in marginal areas of the Itaipú reservoir and confluence areas of small tributaries in the reservoir. Hydrographic region L. | 30 |
| **Family Auchenipteridae** | | | | | |
| *Ageneiosus inermi* (Linnaeus, 1766) | NT | | Paraná-Uruguay basin | Upper Paraná River in marginal areas of Itaipú reservoir. Hydrographic region L. | 30 |
| *Ageneiosus militaris* (Valenciennes, 1835) | NT | | Paraná-Uruguay basin | Upper Paraná River in marginal areas of Itaipú reservoir. Hydrographic region L. | 30 |

*Table 4.2* Summary of Recorded Invasive Vertebrates in Brazil (*Continued*)

| Species/family | NT/ENT | Common name in Brazil | Originated from | Where it is invasive in Brazil | References |
|---|---|---|---|---|---|
| **Family Callichthyidae** | | | | | |
| *Hoplosternum littorale* (Hancock, 1828) | NT | Tamboatá | Cisandine drainages from América do Sul to north of Buenos Aires | Paraíba do Sul, Grande, and São Francisco. Hydrographic regions F-H. | http://exoticfish.bio.br/lista.htm |
| *Lepthoplosternum pectorale* (Boulenger, 1895) | NT | Tamboatá, Tamoatá | Paraná-Uruguay basin | Upper Paraná River in marginal areas of the Itaipú reservoir and confluence areas of small tributaries in the reservoir. Hydrographic region L. | 27 |
| *Megalechis personata* (Valenciennes, 1840) | NT | Tamboatá, Tamoatá | South America: Amazon, Orinoco, and upper Paraguay River basins, as well as coastal rivers of the Guianas and northern Brazil | Upper Paraná River in marginal areas of the Itaipú reservoir and confluence areas of small tributaries in the reservoir. Hydrographic regions J-L. | 30 |
| **Family Clariidae** | | | | | |
| *Clarias gariepinus* (Scopoli, 1777) | ENT | Bagre-africano | Africa: almost Pan-Africa; Asia: Jordan, Israel, Lebanon, Syria, and southern Turkey | Paraná state in the main channel of the upper Paraná River; Rio Grande do Sul state at Lagoa dos Patos; Paraíba do Sul basin; São Francisco, Grande, Doce, and Mucuri basins. Hydrographic regions J-I-L-M-H-G-F. | 30, 87, Erica Caramaschi, personal communication; http://exoticfish.bio.br/lista.htm |

| | | | | | |
|---|---|---|---|---|---|
| **Family Doradidae** | | | | | |
| *Oxydoras eigenmanni* (Boulenger, 1895) | NT | Armado; Rique-rique | Paraná-Uruguay basin | Upper Paraná River in the Itaipú reservoir and confluence areas of small tributaries in the reservoir. Hydrographic region L. | 27, 30 |
| *Platydoras armatulus* (Valenciennes, 1840) | NT | Armado | Paraná-Uruguay basin | Upper Paraná River in the Itaipú reservoir and confluence areas of small tributaries in the reservoir. Hydrographic region L. | 30 |
| *Pterodoras granulosus* (Valenciennes, 1821) | NT | Armado | Paraná-Uruguay basin | Upper Paraná River in the Itaipú reservoir and confluence areas of small tributaries in the reservoir. Hydrographic region L. | 27, 30 |
| *Trachydoras paraguayensis* (Eigenmann and Ward, 1907) | NT | Armado | Paraná-Uruguay basin | Upper Paraná River in the Itaipú reservoir and confluence areas of small tributaries in the reservoir. Hydrographic region L. | 27, 30 |
| **Family Pseudopimelodiade** | | | | | |
| *Lophiosilurus alexandri* (Steindachner, 1877) | NT | Pacamã | São Francisco basin | Doce basin. Hydrographic regions H-G. | http://exoticfish.bio.br/lista.htm |
| **Family Heptapteridae** | | | | | |
| *Heptapterus mustelinus* (Valenciennes, 1835) | NT | | South America: La Plata and Uruguay River basins and coastal drainages of southern Brazil | Upper Paraná River in marginal areas of the Itaipú reservoir and confluence areas of small tributaries in the reservoir. Hydrographic region L. | 30 |

*(Continued)*

**Table 4.2** Summary of Recorded Invasive Vertebrates in Brazil (*Continued*)

| Species/family | NT/ENT | Common name in Brazil | Originated from | Where it is invasive in Brazil | References |
|---|---|---|---|---|---|
| *Hypophthalmus edentatus* (Spix and Agassiz, 1829) | NT | | Paraná-Uruguay basin | Upper Paraná River in marginal areas of the Itaipú reservoir and confluence areas of small tributaries in the reservoir. Hydrographic regions M-L. | 30 |
| *Megalonema platinum* (Günther, 1880) | NT | Jundiá-branco | South America: Paraná River basin | Upper Paraná River in marginal areas of the Itaipú reservoir and confluence areas of small tributaries in the reservoir. Hydrographic regions J-M-L. | 30 |
| *Pimelodella taenioptera* (Miranda Ribeiro, 1914) | NT | Mandi-chorão | South America: upper Paraguay River basin in Brazil | Upper Paraná River in marginal areas of the Itaipú reservoir and confluence areas of small tributaries in the reservoir. Hydrographic regions J-I-L. | 27, 30 |
| *Pimelodus fur* (Lütken, 1874) | NT | | South America: Das Velhas River basin in São Francisco River drainage, Brazil | Upper Paraná River in marginal areas of the Itaipú reservoir and confluence areas of small tributaries in the reservoir. Hydrographic regions J-L. | 30 |
| *Pimelodus ornatus* (Kner, 1858) | NT | Barbudo | South America: Amazon, Corantijn, Essequibo, Orinoco, and Paraná-Uruguay River basins; also in major rivers of the Guianas basins | Upper Paraná River in marginal areas of the Itaipú reservoir and confluence areas of small tributaries in the reservoir. Hydrographic region L. | 27, 30 |

*(Continued)*

| Species | Status | Common name | Native distribution | Invasive distribution | References |
|---|---|---|---|---|---|
| *Sorubim lima* (Bloch and Schneider, 1801) | NT | | South America: Amazon, Orinoco, Paraná-Uruguay and Parnaíba River basins | Upper Paraná River in marginal areas of the Itaipú reservoir and confluence areas of small tributaries in the reservoir. Hydrographic regions I-L. | 27, 30 |
| **Family Pimelodidae** *Pseudoplatystoma reticulatum* (Linnaeus, 1766) | NT | Bagre | Paraná-Uruguay basin | Upper Paraná River in marginal areas of the Itaipú reservoir and confluence areas of small tributaries in the reservoir. Hydrographic region L. | 27 |
| **Family Trichomycteridae** *Trichomycterus brasiliensis* (Lütken, 1874) | NT | | South America: upper São Francisco River in Minas Gerais and in smaller adjoining basins in southeastern Brazil. | Upper Paraná River in small tributaries adjacent to the reservoir. Hydrographic region L. | 30 |
| **Family Poeciliidae** *Poecilia reticulata* (Peters, 1859) | NT | Guppy | South America: Venezuela, Barbados, Trinidad, northern Brazil, and the Guyanas | Coastal streams from Rio de Janeiro (municipalities of Saquarema, Maricá, Angra dos Reis, Ilha Grande, Paraíba do Sul basin), Bahia; rivers and streams from Paraná (Tibagi and upper Paraná rivers), Goiás (Paranaíba basin and Ouvidor river); Paraíba do Sul, São Francisco, Doce, Grande, and Mucuri basins. Hydrographic regions J-I-L-F-H-G. | 30, 87, 89, 90; Rosana Mazzoni, personal communication |

*Table 4.2* Summary of Recorded Invasive Vertebrates in Brazil (*Continued*)

| Species/family | Common name in Brazil | NT/ENT | Originated from | Where it is invasive in Brazil | References |
|---|---|---|---|---|---|
| *Poecilia sphenops* (Valenciennes, 1846) | Moliésia | NT | Central, South, and North America | Paraíba do Sul basin. Hydrographic region H. | http://exoticfish.bio.br/lista.htm |
| *Poecilia vivipara* (Bloch & Schneider, 1801) | Barrigudinho, Guarú | NT | South America: Venezuela all along the coast to Rio de la Plata in Argentina | Upper Paraná River in marginal areas of the Itaipú reservoir and confluence areas of small tributaries in the reservoir. Hydrographic regions J-I-L. | 30 |
| *Xiphophorus hellerii* (Heckel, 1848) | Espadinha | NT | North and Central America: Rio Nantla, Veracruz in Mexico to northwestern Honduras | Upper Paraná River in marginal areas of the Itaipú reservoir and confluence areas of small tributaries in the reservoir; Paraíba do Sul and Doce basins. Hydrographic regions J-I-L-G-H. | 30, http://exoticfish.bio.br/lista.htm |
| *Xiphophorus maculatus* (Günther, 1866) | Plati | NT | North and Central America: Ciudad Veracruz, Mexico to northern Belize | Upper Paraná River in marginal areas of the Itaipú reservoir and confluence areas of small tributaries in the reservoir; Paraíba do Sul basin. Hydrographic regions J-I-L-H. | 30, http://exoticfish.bio.br/lista.htm |
| *Xiphophorus variatus* (Meek, 1904) | Plati | ENT | North America | Paraíba do Sul and São Francisco basins. Hydrographic regions F-H. | http://exoticfish.bio.br/lista.htm |
| ***Family Centrarchidae*** | | | | | |
| *Lepomis gibbosus* (Linnaeus, 1758) | Perca-sol | NT | North America: New Brunswick in Canada to South Carolina in the United States | Lavras Novas in Minas Gerais state at Custódio reservoir in the upper Doce River. Hydrographic regions F-G-H. | www.institutohorus.org.br |

| | | | | | |
|---|---|---|---|---|---|
| *Micropterus salmoides* (Lacépède, 1802) | NT | Achigã | North America: St. Lawrence—Great Lakes, Hudson Bay (Red River), and Mississippi River basins; Atlantic drainages from North Carolina to Florida and to northern Mexico | Upper Paraná River in the Itaipú reservoir; Doce, Grande, and Paranaíba basins. Hydrographic regions J-M-L-H-G-F. | 30, www.institutohorus.org.br, http://exoticfish.bio.br/lista.htm |
| **Family Scianidae** | | | | | |
| *Plagioscion squamosissimus* (Heckel, 1840) | NT | Corvina | South America: Amazon, Orinoco, Paraná-Paraguay, and São Francisco River basins, and rivers of Guianas | Lakes, dams, and reservoirs from Pernambuco, Paraná (Itaipú reservoir), Minas Gerais (Volta Grande reservoir), São Paulo (Barra Bonita reservoir); Parnaíba basin, Grande River basin. Hydrographic regions F-E-C-D-G-J-H-L. | 92, www.institutohorus.org.br, http://exoticfish.bio.br/lista.htm |
| **Family Cichlidae** | | | | | |
| *Astronotus crassipinnis* (Heckel, 1840) | NT | Acará-açú | South America: Amazon River basin, in the Bolivian Amazon and Madre de Dios River drainage in Peru; Paraná River basin in the Paraguay drainage in Paraguay and Brazil | Upper Paraná River in marginal areas of the Itaipú reservoir and confluence areas of small tributaries in the reservoir. Hydrographic regions J-L. | 27, 30 |
| *Astronotus ocellatus* (Agassiz, 1831) | NT | Apaiari, Oscar | South America: Amazon River basin in Peru, Colombia, and Brazil; French Guiana | Rivers and streams in Pernambuco; Paraíba do Sul basin. Hydrographic regions E-G. | www.institutohorus.org.br |

*(Continued)*

*Table 4.2* Summary of Recorded Invasive Vertebrates in Brazil (*Continued*)

| Species/family | NT/ENT | Common name in Brazil | Originated from | Where it is invasive in Brazil | References |
|---|---|---|---|---|---|
| *Cichla kelberi* (Kullander and Ferreira, 2006) | NT | Tucunaré amarela | Brazil | Upper Paraná River in marginal areas of the Itaipú reservoir and confluence areas of small tributaries in the reservoir. Hydrographic regions J-M-L. | 30 |
| *Cichla monoculus* (Spix and Agassiz, 1832) | NT | Tucunaré-açú | South America: Rio Solimões-Amazonas along the main channel and lower courses of tributaries; Peru, Colombia, and Brazil; including Araguari and lower Oyapock rivers north of the Amazon | Mato Grosso in rivers and floodplain areas in the Pantanal Matogrossense; Rio Grande do Norte in Campo Grande reservoir, Paraíba do Sul, Grande and Paranaíba rivers. Hydrographic regions J-I-L-M-E-F-H. | Rosana Mazzoni, personal communication |
| *Cichla ocellaris* (Bloch and Schneider, 1801) | NT | Tucunaré | Amazonas and Araguaia-Tocantins drainages | Rio de Janeiro in Rio Paraíba do Sul basin, living in rivers, artificial lakes, and reservoirs; Minas Gerais at Três Marias reservoir; Paraná in the upper Paraná River in marginal areas of the Itaipú reservoir and confluence areas of small tributaries in the reservoir, main channel of São Francisco River basin, Parnaíba basin; Doce, Grande, São Francisco and Mucuri rivers. Hydrographic regions H-F-G-L-D-M-C-J. | 30 |

| Species | Status | Common name | Native distribution | Brazilian distribution | References |
|---|---|---|---|---|---|
| *Cichla temensis* (Humboldt, 1821) | NT | Tucunaré-paca | Amazonia basin | São Francisco basin. Hydrographic region F. | http://exoticfish.bio.br/lista.htm |
| *Crenicichla niederleinii* (Holmberg, 1891) | NT | Piquí | South America: Paraná River basin in Argentina, Brazil and Paraguay | Upper Paraná River in marginal areas of the Itaipú reservoir and confluence areas of small tributaries in the reservoir. Hydrographic region L. | www.institutohorus.org.br |
| *Geophagus proximus* (Castelnau, 1855) | NT | | South America: Amazon River basin, in the Ucayali River drainage of Peru, and along the Solimoes-Amazon River to the Trombetas River | Upper Paraná River in marginal areas of the Itaipú reservoir and confluence areas of small tributaries in the reservoir. Hydrographic regions J-L. | www.institutohorus.org.br |
| *Hemichromis bimaculatus* (Gill, 1862) | ENT | Peixe-jóia | Asia | Paraíba do Sul basin. Hydrographic region H. | http://exoticfish.bio.br/lista.htm |
| *Laetacara curviceps* (Ahl, 1923) | NT | Curviceps, Acará-azul | Amazonian basin | Paraíba do Sul basin. Hydrographic region H. | http://exoticfish.bio.br/lista.htm |
| *Mikrogeophagus ramirezi* (Myers and Harry, 1948) | NT | Ramirezi | South America: Orinoco River basin, in the llanos of Venezuela and Colombia | Coastal streams from Rio de Janeiro (municipalities of Saquarema and Maricá; Paraíba doSul basin. Hydrographic region H. | Rosna Mazzoni, personal communication; http://exoticfish.bio.br/lista.htm |
| *Oreochromis niloticus* (Linnaeus, 1758) | ENT | Tilápia, Tilápia do nilo | Africa: coastal rivers of Israel; Nile from below Albert Nile to the delta; Jebel Marra; in West Africa natural distribution covers the basins of the Niger, Benue, Volta, Gambia, Senegal and Chad | Lakes, dams and reservoirs from Pernambuco and Ceará (Orós reservoir); coastal streams from Bahia; main channel and tributaries of Paraíba do Sul River basin; Grande, São Francisco, Mucuri and Paranaíba basins. Hydrographic regions E-F-G-H. | 30, 91, www.institutohorus.org.br, Erica Caramaschi, personal communication; http://exoticfish.bio.br/lista.htm |

*(Continued)*

*Table 4.2* Summary of Recorded Invasive Vertebrates in Brazil (*Continued*)

| Species/family | NT/ENT | Common name in Brazil | Originated from | Where it is invasive in Brazil | References |
|---|---|---|---|---|---|
| *Pterophyllum scalare* (Schultze, 1823) | NT | Acará-bandeira | Amazonas, Oyapock (Guiana Francesa) and Essequibo (Guiana) basins | Paraíba do Sul basin. Hydrographic region H. | http://exoticfish.bio.br/lista.htm |
| *Satanoperca pappaterra* (Heckel, 1840) | NT | Cará | South America: Amazon River basin, in the Guaporé River in Brazil and Bolivia; Paraná River basin, in the Paraguay River drainage in Brazil and northern Paraguay | Upper Paraná River in marginal areas of the Itaipú reservoir and confluence areas of small tributaries in the reservoir. Hydrographic regions J–L. | 30 |
| *Tilapia rendalli* (Boulenger, 1897) | ENT | Tilápia | Africa: Kasai drainage (middle Congo River basin), Lakes Tanganyika, Malawi, Zambesi, coastal areas from Zambesi Delta to Natal, Okavango and Cunene | Paraná state in the main channel of the upper Paraná and Tibagi rivers; Rio de Janeiro at Paraíba do Sul basin; Paraíba do Sul, Grande, Doce, São Francisco, Paranaíba, upper Tocantins River. Hydrographic regions B–L–J–M–H–I–G–F. | 30, 87, Rosana Mazzoni and Erica Caramaschi, personal communication; http://exoticfish.bio.br/lista.htm |
| *Family Achiridae* | | | | | |
| *Catathyridium jenynsii* (Günther, 1862) | NT | Linguado de água doce | Paraná-Uruguay basin | Upper Paraná River in marginal areas of the Itaipú reservoir and confluence areas of small tributaries in the reservoir. Hydrographic region L. | 27, 30 |

| | | | | |
|---|---|---|---|---|
| **Family Salmonidae** | | | | |
| *Oncorhynchus mykiss* (Walbaum, 1792) | ENT | Truta arco-íris | Southwest Atlantic: Argentina; eastern Pacific: Kamchatkan Peninsula and have been recorded from the Commander Islands east of Kamchatka and sporadically in the Sea of Okhotsk as far south as the mouth of the Amur River along the mainland | Minas Gerais and Rio de Janeiro states at Serra da Bocaina and Serra da Mantiqueira in mountain rivers and streams; river Grande basin. Hydrographic regions H-F. | 93, 94 |
| **Family Centrarchidae** | | | | |
| *Lepomis gibbosus* (Linnaeus, 1758) | ENT | Perca-sol | North America | Doce basin. Hydrographic regions F-G. | http://exoticfish.bio.br/lista.htm |
| **Family Cobitidae** | | | | |
| *Misgurnus anguillicaudatus* (Cantor, 1842) | ENT | Dojô | Asia | Paraíba do Sul River. Hydrographic region H. | http://exoticfish.bio.br/lista.htm |
| **Family Osphronemidae** | | | | |
| *Colisa lalia* (Hamilton, 1822) | ENT | Gourami-anão, Colisa | Asia | Paraíba do Sul River. Hydrographic region H. | http://exoticfish.bio.br/lista.htm |
| *Macropodus opercularis* (Linnaeus, 1758) | ENT | Peixe do paraiso | Asia | Paraíba do Sul River. Hydrographic region H. | http://exoticfish.bio.br/lista.htm |
| *Trichogaster chuna* (Hamilton, 1822) | ENT | Colisa-chuna | Asia | Paraíba do Sul River. Hydrographic region H. | http://exoticfish.bio.br/lista.htm |

(Continued)

*Table 4.2* Summary of Recorded Invasive Vertebrates in Brazil (*Continued*)

| Species/family | NT/ENT | Common name in Brazil | Originated from | Where it is invasive in Brazil | References |
|---|---|---|---|---|---|
| *Tricogaster trichopterus* (Pallas, 1770) | ENT | Tricogaster | Asia | Paraíba do Sul River. Hydrographic region H. | http://exoticfish.bio.br/lista.htm |
| **Family Polycentridae** | | | | | |
| *Polycentrus schomburgkii* (Müller and Troschel, 1849) | NT | Peixe-folha | Trinidad and Atlantic coastal rivers from Venezuela, Guiana, Suriname, French Guiana Francesa, and Brasil (Amapá) | Paraíba do Sul basin; Hydrographic region H. | http://exoticfish.bio.br/lista.htm |
| **Amphibians** | | | | | |
| **Family Leptodactylidae** | | | | | |
| *Leptodactylus labyrinthicus* (Spix, 1824) | NT | Rã-pimenta | Southeastern Brazil | Manaus in Central Amazonia. | 36 |
| **Family Ranidae** | | | | | |
| *Lithobates catesbeianus* (Shaw, 1802) | ENT | Rã-touro | Eastern and Central United States of America | In many municipalities of at least 11 Brazilian states (Piauí, Rio Grande do Norte, Pernambuco, Alagoas, Bahia, Espírito Santo, Rio de Janeiro, Minas Gerais, São Paulo, Santa Catarina, and Rio Grande do Sul). | 37, 38 |

| | | | | | |
|---|---|---|---|---|---|
| **Family Pipidae** *Xenopus laevis* (Daudin, 1802) | ENT | Rã-Africana | Southern Angola south to Cape Region of South Africa and then eastwards and northwards in savanna habitats to southern Sudan and then west to Nigeria | Although there is consistent evidence of the species introduction in Brazil (in Goiânia, Goiás state) and some suggestions of its presence in natural environments, there is a lack of its effective invasion of natural environments in Brazil. | 17, www.institutohorus.org.br |

**Reptiles**

| | | | | | |
|---|---|---|---|---|---|
| **Family Gekkonidae** *Hemidactylus mabouia* (Moreau de Jonnès, 1818) | ENT | Lagartixa-de-casa | Native from Africa | It is a synanthropic lizard in most environments in Brazil but it is already invasive in some habitats of native forests and in restingas (coastal sand-dune habitats). | 41 |
| **Family Liolaemidae** *Liolaemus lutzae* (Mertens, 1938) | | Lagarto-branco-da-praia; lagartixa-da-areia | Sand-dune habitats (restingas) of Rio de Janeiro state in Brazil | In the beach habitat of Praia das Neves restinga, in Presidente Kennedy Municipality, Espírito Santo state, southeastern Brazil, where a population was introduced. | 44 |
| **Family Teiidae** *Tupinambis merianae* (Linnaeus, 1758) | | Teiú, teju | Atlantic Forest biome and associated habitats of southeastern Brazil | In Fernando de Noronha Archipelago, Pernambuco state, northeastern Brazil. | 43, 46 |

(Continued)

*Table 4.2* Summary of Recorded Invasive Vertebrates in Brazil (*Continued*)

| Species/family | NT/ENT | Common name in Brazil | Originated from | Where it is invasive in Brazil | References |
|---|---|---|---|---|---|
| **Family Emididae** | | | | | |
| *Trachemys dorbigni* (Duméril and Bibron, 1835) | NT | Tigre-d'água | Uruguay and Argentina and Rio Grande do Sul state in Brazil | Aquatic environments of Santa Catarina Island (Florianópolis) and Palmas in Tocantins state. | 17, www.institutohorus .org.br |
| *Trachemys scripta* (Schoepff, 1792) | ENT | Tartaruga-da-orelha-vermelha | Mississippi Valley, United States of America | Some areas of the atates of Rio Grande do Sul, Santa Catarina, Paraná, São Paulo, Rio de Janeiro, Espírito Santo, Mato Grasso do Sul, Goiás, Tocantins, Piauí, and Paraíba. | 17, www.institutohorus .org.br |
| **Birds** | | | | | |
| **Family Psittacidae** | | | | | |
| *Amazona aestiva* (Linnaeus, 1758) | NT | Amazona-de-fronte-azul | Southwestern Brazil; Paraguay; Bolivia and northern Argentina | Established population at the forests of Ilha Santa Catarina Island (Florianópolis) in Santa Catarina state, Paraná; southern Brazil. | 95, 96, 98, www .institutohorus.org.br |
| **Family Columbidae** | | | | | |
| *Columba livia* (J. F. Gmelin) | ENT | Pombo-doméstico | Europe | Rio de Janeiro, Ceará, Acre, Goiás, Mato Grosso, Pernambuco, Piauí, Tocantins, São Paulo, Minas Gerais, Espírito Santo. | www.institutohorus.org.br |

| | | | | | |
|---|---|---|---|---|---|
| **Family Estrildidae** *Estrilda astrild* (Linnaeus) | Bico-de-lacre | ENT | Native of tropical and southern Africa, most areas south of 10° N | Established populations in the states of Rio de Janeiro, Rio Grande do Sul, Santa Catarina, Paraná, São Paulo, Minas Gerais, Pernambuco, Ceará, and Pará. | 95, 96, 97, 98, 100, www .institutohorus.org.br |
| **Family Passeridae** *Passer domesticus* (Linnaeus, 1758) | Pardal | ENT | Eurasia and North Africa | Well-established populations in the states of Rio de Janeiro, Rio Grande do Sul, Santa Catarina, Paraná, São Paulo, Minas Gerais, Espírito Santo, Bahia, Pernambuco, Rio Grande do Norte, Ceará, Pará, Goiás, Mato Grosso, and Mato Grosso do Sul. | 88, 95, 96, 97–100 |

**Mammals**

| | | | | | |
|---|---|---|---|---|---|
| **Family Callitrichidae** *Leontopithecus chrysomelas* (Kuhl, 1820) | Mico-leão-de-cara-dourada | NT | Northeastern and southeastern Brazil in Bahia and Minas Gerais states | Established population in Serra da Tiririca State Park and Municipal Ecological Reserve Darcy Ribeiro in Niteroi, Rio de Janeiro state. | Helena de Godoy Bergallo, personnal communication |
| **Family Cebidae** *Callithrix jacchus* (Linnaeus, 1758) | Sagui-do-nordeste, sagui-comum, mico | NT | Northeastern Brazil | Established populations in south and southeastern Brazil in urban and natural areas. | 101–103, www .institutohorus.org.br |

*(Continued)*

*Table 4.2* Summary of Recorded Invasive Vertebrates in Brazil (*Continued*)

| Species/family | NT/ENT | Common name in Brazil | Originated from | Where it is invasive in Brazil | References |
|---|---|---|---|---|---|
| *Callithrix penicillata* É. (Geoffroy, 1812) | NT | Sagui-do-cerrado, sagüi-de-tufo-, preto | Midwest Brazil, in Caatinga and Cerrado biome | Established populations in south and southeastern Brazil in urban and natural areas | 102, 103, ww.institutohorus.org.br |
| *Saimiri sciureus* (Linnaeus, 1758) | NT | Mico-de-cheiro | Amazon forest in Peru, Brazil, Colombia, Paraguay, Equador, Venezuela, Guiana and Suriname. | Established populations in Saltinho Biological Reserve, in Pernambuco state and Tijuca Forest National Park, in Rio de Janeiro state. | 101, www.institutohorus .org.br |
| **Family Canidae** | | | | | |
| *Canis familiaris* (Linnaeus, 1758) | ENT | Cachorro, cão-doméstico | Domestication of the gray wolf (*Canis lupus*) from Europe, Asia, and North America | Everywhere in Brazil and, in many areas, dogs invaded natural areas and conservation units. | www.institutohorus.org.br |
| **Family Felidae** | | | | | |
| *Felis catus* (Linnaeus, 1775) | ENT | Gato | Derivated from at least five founders wildcats from Europe, Near East, Asia, southern Africa, and China | Everywhere in Brazil and, in many areas, cats invaded natural areas and conservation units. | 55, www.institutohorus .org.br |
| **Family Bovidae** | | | | | |
| *Bubalus bubalis* (Linnaeus, 1758) | ENT | Búfalo | Southeastern Asia | In many areas in at least 10 states in Brazil (Pará, Rondônia, Amapá, Maranhão, Mato Grosso, Mato Grosso do Sul, Minas Gerais, São Paulo, Paraná, Rio Grande do Sul). | 68, www.institutohorus .org.br |

| Species | Status | Common name | Native range | Occurrence in Brazil | Reference |
|---|---|---|---|---|---|
| *Capra hircus* (Linnaeus, 1758) | ENT | Cabra | Southwest Asia and eastern Europe | Widespread in the semiarid region of northeastern Brazil, and established populations were recorded in Abrolhos Archipelago, Santa Catarina Island, and in a conservation unit in Rio Grande do Sul state. | 69, www.institutohorus.org.br |
| **Family Cervidae** *Cervus unicolor* (Kerr, 1792) | ENT | Veado-sambar | Southern Asia, Southeast Asia, southern China, Taiwan, and the islands of Sumatra and Borneo in Indonesia | In an environmental protection area in São Paulo state. | www.institutohorus.org.br |
| **Family Suidae** *Sus scrofa* (Linnaeus, 1758) | ENT | Javali, porco-monteiro | Europe, Asia, north of Africa | In natural areas in at least 11 Brazilian states (Acre, Maranhão, Mato Grosso, Mato Grosso do Sul, Bahia, Espírito Santo, Rio de Janeiro, São Paulo, Paraná, Santa Catarina, and Rio Grande do Sul). | 68, www.institutohorus.org.br |
| **Family Equidae** *Equus caballus* (Linnaeus, 1758) | ENT | Cavalo, cavalo-lavradeiro | | In savanna regions of Roraima state, in the indigenous lands Raposa Serra do Sol and São Marcos. | www.institutohorus.org.br |

*(Continued)*

*Table 4.2* Summary of Recorded Invasive Vertebrates in Brazil (*Continued*)

| Species/family | NT/ENT | Common name in Brazil | Originated from | Where it is invasive in Brazil | References |
|---|---|---|---|---|---|
| **Family Caviidae** | | | | | |
| *Kerodon rupestris* (Wied-Neuwied, 1820) | NT | Mocó | In Caatinga biome occuring from Piauí to north of Minas Gerais state | Fernando de Noronha Archipelago, Pernambuco state. | 104 |
| **Family Muridae** | | | | | |
| *Mus musculus* (Linnaeus, 1758) | ENT | Camundongo | Europe and Asia | Everywhere in Brazil and, in many areas, house mouse invaded natural areas and conservation units. | 17, 57, www .institutohorus.org.br |
| *Rattus rattus* (Linnaeus, 1758) | ENT | Rato | Indian Subcontinent | Everywhere in Brazil and, in many areas, black-rat invaded natural areas and conservation units. | 17, 51, 24, www .institutohorus.org.br |
| *Rattus norvegicus* (Berkenhout, 1827) | ENT | Ratazana | Northern China | Everywhere in Brazil in rural and urban areas. | 24, www.institutohorus .org.br |
| **Family Leporidae** | | | | | |
| *Lepus europaeus* (Pallas, 1778) | ENT | Lebre | Great Britain and western Europe, east to through the Middle East to Central Asia | Mainly in south Brazil (Rio Grande do Sul, Santa Catarina, and Paraná states), but also in Goiás and Minas Gerais states. | 17, 24, 67 |

activities followed by escapes from the aquaculture ponds and/or their use as game fish and for aquariology. Two Neotropical, but non-Brazilian species, were introduced for aquarists and presently are widely distributed in many different hydrographic and/or ecoregions in Brazil. *Poecilia reticulata* and *Mikrogeophagus ramirezi* (called guppy and ramirezi, respectively) are very widespread in aquatic environments in Brazil with recorded occurrences in six different hydrographic regions (Atlântico Sudeste, Atlântico Leste, São Francisco, Atlântico Sul, Uruguay, and Paraguay). *Poecilia reticulata*, the guppy, was introduced in Brazil as a biological "mosquito" controller and then became an important and widespread freshwater species used in aquaria. All the other aquarium fish species are restricted to one or even two hydrographic regions, mainly in the Atlântico Sudeste hydrographic region. Nowadays, many small-sized fish species of aquarist interest are introduced into freshwater systems, mainly in southeast basins of Brazil, although stocking with nonnative species in public waters is forbidden.

Concerning fish species used in fish cultures, *Salmo gairdneri*, *Oreochromis mossambicus* (Mozambique tilapia), and *Cyprinus carpio* (common carp) have been mentioned as the most widespread exotic species around the world.[22] Nonetheless, *Tilapia rendalli* and *Clarias gariepinus* (African sharptooth catfish) are presently registered as the most widespread exotic species in Brazil, which should be addressed due to their importance in fish cultures because of their adaptability to tropical climate and economical interests.[34]

Despite the large number of ENT species introduced in Brazilian freshwater systems, many native species have been transposed since the 1960s.[35] The movement occurred from the Amazonian basin into northern, southeastern, and southern regions of Brazil and followed the interests for fish culture. Following this, many cases of introductions have been reported for the Paraíba do Sul basin (Atlântico Sudeste hydrographic region) with drastic consequences for native species displacement as well. The introduction of *Salminus maxillosus* (golden dorado) is one of the most emblematic cases in Paraíba do Sul, due to its importance for fishing and also due to its voracious behavior, which perhaps accounts for the reduction of the stocks of two native *Brycon* species (Erica Caramaschi, pers. comm.).

More recently, the construction of hydroelectric power plants interrupted some river drainages and/or eliminated natural barriers, which has resulted in severe changes in natural ecosystems with important negative effects on freshwater fish diversity. Two historical cases were unequivocally responsible for massive invasions of freshwater fish fauna in Brazil. The first case was registered during the 1960s and was related to the Piumhi River drainage (Paraná hydrographic region) that had its outflow diverted into the headwaters of the São Francisco hydrographic region causing a devastating mixing of fish fauna from the upper Paraná and the upper São Francisco hydrographic region.[36] The second important example occurred during the 1980s when Sete Quedas Falls was inundated due to the construction of the Itaipú hydroelectric power plant and allowed a massive species invasion (upstream colonization) by endemic species from the lower into the upper Paraná River ecoregion.[27]

Although the relationship between hydrologic alterations and biological invasions is still not well understood, hydropower impoundments are frequently associated with these events. For example, we found that the larger number of cases of invasive fish species have been registered in the Paraná, Paraguay, and Uruguay hydrographic regions (Figure 4.1), being largely caused by the Sete Quedas case.

From the data presented in Figure 4.1, we observed that south and southeast regions of Brazil are highly affected by fish introduction, due to fish cultures or hydropower impoundments. In fact, these are the most populous regions in Brazil, but these are also the most studied and explored.

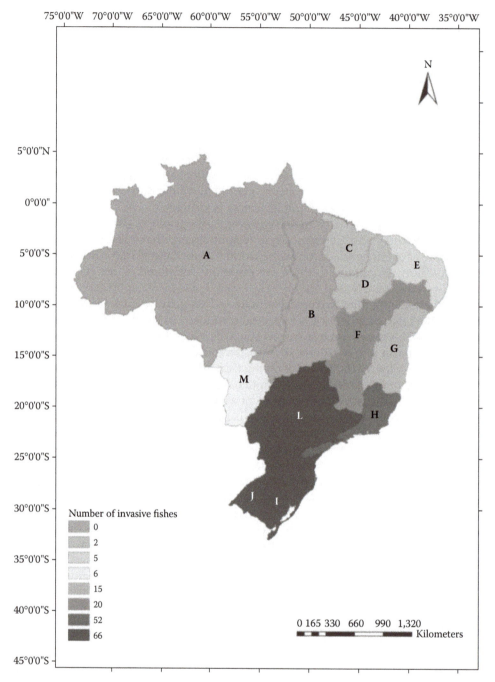

***Figure 4.1*** Number of invasive freshwater fish species in each Brazilian hydrographic region. A-Amazônica, B-Tocantins/Araguaia, C-Atlântico Nordeste Ocidental, D-Parnaíba, E-Atlântico Nordeste Oriental, F-São Francisco, G-Atlântico Leste, H-Atlântico Sudeste, L-Paraná, M-Paraguai, I-Atlântico Sul, and J-Uruguai.

## 4.2.2   Amphibians

Among amphibians, a total of three species have been recorded as being invasive exotic species in Brazil: the leptodactylid *Leptodactylus labyrinthicus* (Spix, 1824), the ranid *Lithobates catesbeianus* (Shaw, 1802), and the pipid *Xenopus laevis* (Daudin, 1802), although for the last species, their exotic status is well recorded, its invader status is still controversial. If *X. laevis* is considered to be an invader, presently the three invader species represent 0.35% of living amphibians known to occur in Brazil, *L. labyrinthicus* being of Neotropical origin, whereas the two other species are of ENT origin (Table 4.1). These species have already invaded different localities in different Brazilian states, with the records of amphibians as invasive in natural environments coming from the states along the eastern coast of Brazil and usually corresponding to only one case per state (Figure 4.2). Compared to the number of recorded invasive species of fishes and mammals, the number of invasive amphibians can be considered relatively low.

The labyrinth frog *L. labyrinthicus*, which is called Rã-pimenta in Brazil, occurs in southern Brazil and has become invasive in the Amazon region, presently having well-established populations in Manaus in Central Amazonia[37] and corresponds to the only case of invasion by amphibians presently known to the Amazon region (Figure 4.2). Although the Adolpho Ducke Forest Reserve is surrounded by the city of Manaus, *L. labyrinthicus* has as yet not been recorded in this conservation unit.[37]

The common bullfrog, *L. catesbeianus* (called Rã-touro in Brazil), is presently very widespread in natural environments in Brazil with recorded occurrences in different municipalities of 11 Brazilian states (Piauí, Rio Grande do Norte, Pernambuco, Alagoas, Bahia, Espírito Santo, Rio de Janeiro, Minas Gerais, São Paulo, Santa Catarina, and Rio Grande do Sul; Table 4.2).[38] This frog is one of the main species used in raniculture in Brazil.[39] *Lithobates catesbeianus* has a high capacity to adapt to many different environments and climates, which has facilitated its invasion of natural environments (especially lentic habitats such as rivulets, lakes, and lagoons) where it competes directly with native species.[40] This species usually invades natural environments of Brazil after escaping from raniculture farms, which results in its distribution usually occurring in the municipalities and states having raniculture farms. Most records of this species as an invader occur in Rio Grande do Sul state (southern Brazil) where available records indicate its presence as an invader in 21 municipalities. These invasion figures show how easy (in some cases) the escape of an exotic species (introduced for economic purposes) from farm cultures to nature can be if no rigorous controls exist; this indicates the need of rigorous controls to be devised and implemented. Presently, *Lithobates catesbeianus* is recorded as an introduced species in other countries on different continents, including Mexico, Cuba, Puerto Rico, Hispaniola, Jamaica, Spain, Crete, Malaysia, Java, Indonesia (Bali), Japan, Thailand, Korea, and Taiwan (China).[41]

The pipid frog *Xenopus laevis* (common platanna) is first recorded to have occurred recently in Goiás state (Goiânia municipality) in central Brazil, where it was introduced as a pet (specimens are found in pet shops). Although locals report that some specimens have been already found in water bodies of the region, the lack of more precise records leave the species status as controversial. In this case, the monitoring of pet shops and the water bodies of the region of Goiânia is urgently needed to prevent its possible establishment as an invasive species.

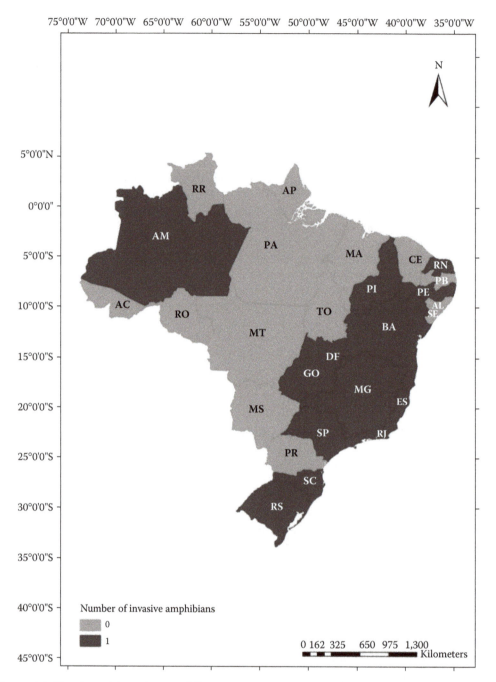

*Figure 4.2* Number of invasive amphibian species in each Brazilian state. Acronyms of states: AC-Acre, AL-Alagoas, AM-Amazonas, AP-Amapá, BA-Bahia, CE-Ceará, DF-Federal District, ES-Espírito Santo, GO-Goiás, MA-Maranhão, MG-Minas Gerais, MS-Mato Grosso do Sul, MT-Mato Grosso, PA-Pará, PB-Paraíba, PE-Pernambuco, PI-Piauí, PR-Paraná, RJ-Rio de Janeiro, RN-Rio Grande do Norte, RO-Rondônia, RR-Roraima, RS-Rio Grande do Sul, SC-Santa Catarina, SE-Sergipe, SP-São Paulo, and TO-Tocantins.

## 4.2.3   Reptiles

Among reptiles, five species have become invasive in some areas in Brazil: the gekkonid lizard *Hemidactylus mabouia* (Moreau de Jonnès, 1818), the sand lizard *Liolaemus lutzae* (Mertens, 1938), the teiid lizard *Tupinambis merianae* (Linnaeus, 1758), and the water tortoises *Trachemys dorbigni* (Duméril & Bibron, 1835) and *Trachemys scripta* (Schoepff, 1792). Of these species, *L. lutzae*, *T. merianae*, and *T. dorbigni* are native in some areas of Brazil being of Neotropical origin, whereas the African *H. mabouia* and the North American *T. scripta* are of ENT origin. These five invasive species represent approximately 0.71% of the living reptile species in Brazil (Table 4.1).

The most widespread exotic invasive reptile species in Brazil is the gekkonid *H. mabouia*, known to occur as an exotic species in virtually all urban environments in Brazil.[42] Available records of *H. mabouia* living in natural conditions,[43] based on data obtained from literature, supplemented with original field records of authors, showed a total of 36 records of occurrence of *H. mabouia* in natural habitats in 36 different localities in 13 Brazilian states. The states presenting higher instances of cases were Rio de Janeiro (seven), Bahia and São Paulo (six), and Espírito Santo (five different areas). Based on the data obtained by Rocha,[43] the invasion of Brazilian natural habitats by *H. mabouia* has already taken place for some decades, and presently there are consistent records of its invasion for nearly half the Brazilian states. An analysis of endoparasites of this gecko showed that *H. mabouia* already interacts directly with the local natural fauna.[44] *Hemidactylus mabouia* shared most of its helminth fauna with two other sympatric native lizard hosts, *Mabuya frenata* and *Tropidurus itambere*.[44] The helminth assemblage of this exotic gecko was entirely acquired from the local helminth species pool, not possessing any parasitic faunal species of the original African populations.[44] In some areas of Brazil, this gecko also became a host for native Pentastomida parasites of the lungs of lizards.[45] Although there is no study evaluating the resource sharing by this gecko with native lizards, its presence in nature suggests that a portion of available resources in each invaded habitat has already been taken by this invasive lizard.

The sand lizard *Liolaemus lutzae* is endemic to the sand-dune habitats (restingas) of Rio de Janeiro state in Brazil and is presently listed as critically endangered in the Brazilian Official Checklist of Endangered Fauna (Ministerio do Meio Ambiente of Brazil, list published online in 2003). This lizard had one population introduced outside its original distribution in a restinga area of Espírito Santo state,[46] a state just north of Rio de Janeiro state. The introduced population has been monitored for about two decades, and this monitoring has indicated that the population is presently well established.[46]

The teiid *Tupinambis merianae* found in eastern and southeastern Brazil as well as in Uruguay and Argentina was introduced in Fernando de Noronha Archipelago (about 300 km off the coast of Brazil) in 1950 in order to control established populations of mice, Norway rats, and black rats on the island.[17] From two originally introduced lizard couples, the population increased dramatically to between 2000 and 8000 on the main island as a result of abundance of spatial and food resources,[47] and presently this lizard, a serious egg predator,[48] has become a considerable problem negatively affecting different species of the fauna and flora.[47] The main impact of *Tupinambis merianae* is a decline in some species of terrestrial birds on the island due to its consumption of bird eggs.[47]

The two invasive water tortoise species of the genus *Trachemys* are good examples of how species formerly used as pets can become invasive as they are released into natural environments. *Trachemys dorbigni* occurs as a native species in the Rio Grande do Sul state in Brazil and in Uruguay and Argentina and became invasive in aquatic environments of Santa Catarina Island (Florianópolis) and Palmas in Tocantins state (central Brazil). *T. scripta*, native to the Mississippi River drainage system in the United States, is presently recorded as invasive in Rio Grande do Sul, Santa Catarina, Paraná, São Paulo, Rio de Janeiro, Espírito Santo, Mato Grosso do Sul, Goiás, Tocantins, Piauí, and Paraíba. These species are frequently used as pets and in aquariology, and the invasions reflect escape from aquariums and/or their intentional release.

The number of cases of invasive reptile and amphibian species can be considered low compared with fishes and mammals with the records coming predominantly from the eastern portion of Brazil, with no cases presently recorded for the states of the Amazon region and for Minas Gerais, Alagoas, and Sergipe (Figure 4.3).

## 4.2.4   Birds

According to data available from Instituto Horus, presently, four bird species are known to be invasive in Brazil, accounting for about 0.22% of the living species in Brazil (Table 4.1): the psittacid *Amazona aestiva* (Linnaeus, 1758), the pigeon *Columba livia* (Gmelin, 1789), the estrildid *Estrilda astrild* (Linnaeus, 1758), and the worldwide invasive passerid *Passer domesticus* (Linnaeus, 1758).

The psittacid *A. aestiva*, originally found in southwestern Brazil and in Paraguay, Bolivia, and northern Argentina, was introduced to the forests of Ilha Santa Catarina Island (Florianópolis) in Santa Catarina state, southern Brazil, where a population became established. Apparently, this case of bird invasion still remains restricted to the Ilha de Santa Catarina Island.

The passerine *E. astrild* (Linnaeus, 1758) is a small native of tropical and southern Africa, residing in most areas south of 10° N in Africa.[48] The species was introduced into Brazil by sailors of merchant ships that used to cross the Atlantic from Africa to Brazil during the last few centuries. Presently, this bird species has established populations in areas of the states of Pará, Ceará, Pernambuco, Minas Gerais, Rio de Janeiro, São Paulo, Santa Catarina, and Rio Grande do Sul.

The passerid bird *P. domesticus* was introduced by man to all continents (except in the Antarctic) and presently is the bird species having the largest geographic range on Earth. First introduced in Brazil in 1906, as a strategy to cope with mosquito problems in Rio de Janeiro, the species quickly dispersed in Brazil[49] and by 1971 reached the Amazon region, specifically in Marabá, Tocantins River.[50]

All Brazilian states present at least one case of an invasive bird species (Figure 4.4). Compared to the number of invasive species of amphibians and reptiles, the number of invasive bird species in Brazil can still be considered relatively low (four species) with the occurrence of species per Brazilian state varying from one to three (Figure 4.4). Although the number of invasive species is low, the extensive invasion by at least two of the bird invasive species in Brazil, the sparrow *P. domesticus* and the pigeon *C. livia*, is having huge effects on ecosystems and sympatric species in many communities, although we still lack studies providing data in Brazil on such subjects.

## 4.2.5   Mammals

Of the 16 invasive species of mammals recorded in Brazil, 11 came from outside Brazil, mainly from Asia and Europe (Table 4.2). The rodents (*Rattus rattus*, *R. norvegicus*, and *Mus musculus*) were disseminated around the world by ships[51] and arrived in Brazil

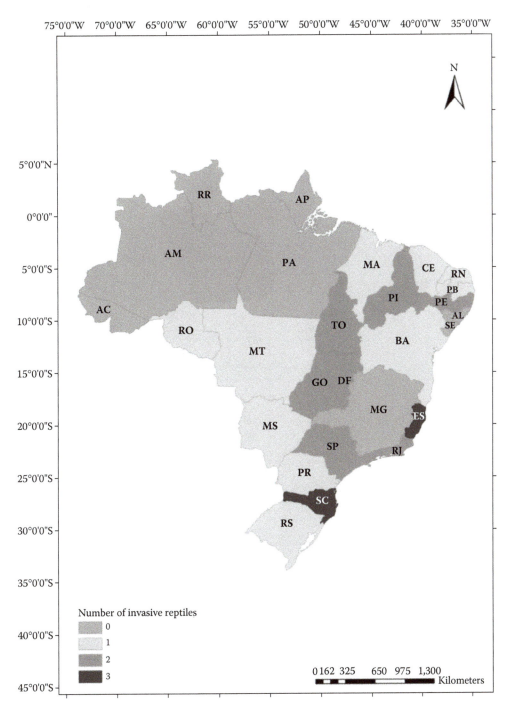

***Figure 4.3*** Number of invasive reptile species in each Brazilian state. Acronyms of states: AC-Acre, AL-Alagoas, AM-Amazonas, AP-Amapá, BA-Bahia, CE-Ceará, DF-Federal District, ES-Espírito Santo, GO-Goiás, MA-Maranhão, MG-Minas Gerais, MS-Mato Grosso do Sul, MT-Mato Grosso, PA-Pará, PB-Paraíba, PE-Pernambuco, PI-Piauí, PR-Paraná, RJ-Rio de Janeiro, RN-Rio Grande do Norte, RO-Rondônia, RR-Roraima, RS-Rio Grande do Sul, SC-Santa Catarina, SE-Sergipe, SP-São Paulo, and TO-Tocantins.

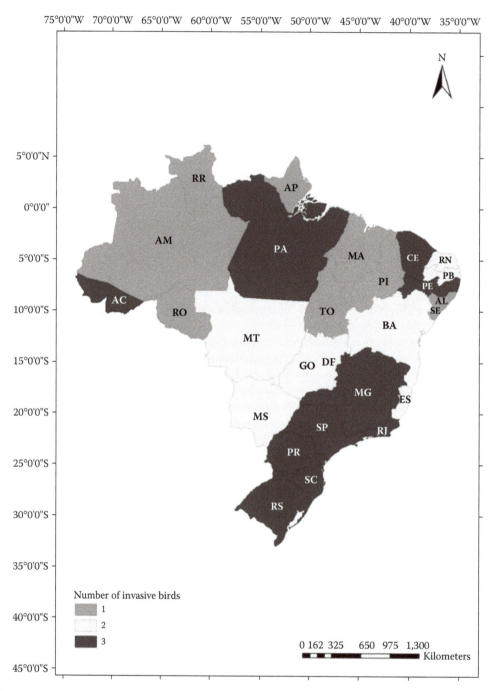

*Figure 4.4* Number of invasive bird species in each Brazilian state. Acronyms of states: AC-Acre, AL-Alagoas, AM-Amazonas, AP-Amapá, BA-Bahia, CE-Ceará, DF-Federal District, ES-Espírito Santo, GO-Goiás, MA-Maranhão, MG-Minas Gerais, MS-Mato Grosso do Sul, MT-Mato Grosso, PA-Pará, PB-Paraíba, PE-Pernambuco, PI-Piauí, PR-Paraná, RJ-Rio de Janeiro, RN-Rio Grande do Norte, RO-Rondônia, RR-Roraima, RS-Rio Grande do Sul, SC-Santa Catarina, SE-Sergipe, SP-São Paulo, and TO-Tocantins.

during the discovery and colonization period. These species as well as *Canis familiaris* and *Felis catus* spread throughout most of Brazil, in urban and rural areas, and in the present study, we consider that these species occur in all states (Figure 4.5). Yet many studies have shown that these species have established populations in natural environments as feral stocks.[52–59]

While studies on domestic cats (*Felis catus*) including population estimations, eradication projects (mainly on islands), and measuring the impacts of their predation are common in other parts of the world,[60–63] in Brazil there are few studies to estimate population size for this species and the other impacts that *F. catus* may cause. At a small village on Ilha Grande, an island in Rio de Janeiro state, cat density estimates were relatively high, reaching 6.62 cats per hectare using census and 3.9 cats per hectare using transects.[57] The same study showed that the population would only start decreasing 10 years from now with the current reproductive rate (only 10% of males did not breed). However, other forms of control used, such as prey poisoning and introduction of viral disease,[64–66] are unfeasible in the studied area for two reasons: (1) the presence of other animals including wild cats[67] and (2) the rejection of these methods by residents. According to Gonçalves da Silva et al.,[68] castration constitutes an efficient method for the eradication of exotic vertebrates, but only if applied together with some lethal method.

The European hare, *Lepus europaeus*, was introduced in South America, in 1888 in Argentina, and in 1896 in Chile. By 1983, the European hare invaded Uruguay, Paraguay, southern Bolivia, and Brazil and reached Peru in the second half of the 1990s.[17,24] However, residents of Londrina municipality in Paraná state say that the European hares appeared in the region in the late 1960s.[69] It is suspected that the European hare can compete for food resources with the native species of rabbit, *Sylvilagus brasiliensis*.[17]

Apparently, *Sus scrofa* was introduced in Brazil by settlers in the sixteenth century, but residents of the wetlands in Pantanal associated the release of feral pigs to the Paraguayan War (1864–1870).[70] According to these authors, the density of *S. scrofa* in Pantanal was about 9800 (SE = 1400) groups. Feral pigs are recognized as pests in many parts of the world, transmitting diseases to humans and other animals and negatively impacting the ecosystems colonized. Although there are no studies, feral pigs may be competing with Brazilian peccaries (*Tayassu pecari* and *Pecari tajacu*).[70] Despite the impacts that it may cause, feral pigs are even protected by law in Brazil because they are considered wildlife.[14] But, in 2005 the state of Rio Grande do Sul, through a normative instruction (Instrução Normativa N° 71, August 4, 2005), authorized the population control of feral pigs by capturing and killing (www.institutohorus.org.br).

It is believed that the goats (*Capra hircus*) should have come with European settlers. But their arrival in Brazil occurred in two steps: (1) from the sixteenth century to eighteenth century when goats were introduced as undefined breeds and (2) from the late nineteenth century on when modern breeds arrived.[71] The largest population of goats is found in northeastern Brazil, where the pastures maintain goats better than other animals.[72]

Water buffalo (*Bubalus bubalis*) were first introduced to Marajo Island in Pará state in 1895, and it was then introduced into other regions of Brazil (www.institutohorus.org.br). The northern region holds 62% of the buffalo population in Brazil, with a growth of 10% per year, whereas Pará state holds 41%.[73] Wild water buffalo in the Pantanal occur at few sites with an overall population of 5100 buffalos (SE = 600).[70] In other parts of the world, efforts to control and exterminate the water buffalo cost millions of dollars annually[74] because the water buffalo causes environmental damage in wetlands.[75,76] Despite the damage to wetlands, the Brazilian government has ignored the risks associated with the presence of water buffalo in the Pantanal.[14,70] In some areas, where the buffalo from livestock

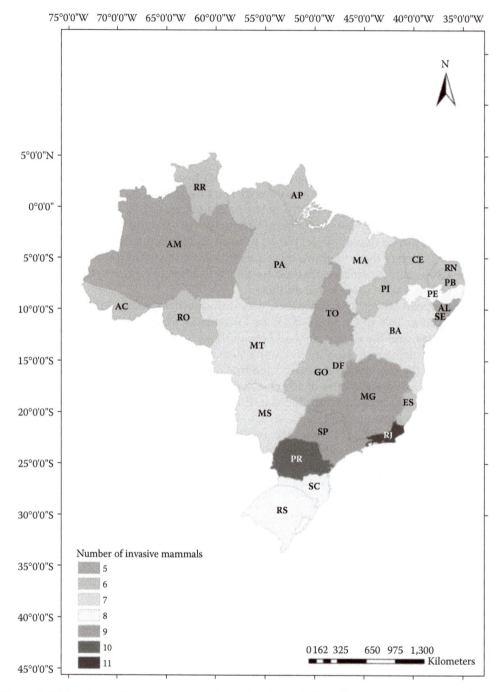

*Figure 4.5* Number of invasive mammal species in each Brazilian state. Acronyms of states: AC-Acre, AL-Alagoas, AM-Amazonas, AP-Amapá, BA-Bahia, CE-Ceará, DF-Federal District, ES-Espírito Santo, GO-Goiás, MA-Maranhão, MG-Minas Gerais, MS-Mato Grosso do Sul, MT-Mato Grosso, PA-Pará, PB-Paraíba, PE-Pernambuco, PI-Piauí, PR-Paraná, RJ-Rio de Janeiro, RN-Rio Grande do Norte, RO-Rondônia, RR-Roraima, RS-Rio Grande do Sul, SC-Santa Catarina, SE-Sergipe, SP-São Paulo, and TO-Tocantins.

farming invade protected areas, such as the Guaporé Biological Reserve, in Rondonia state, the impacts caused by buffaloes are already evident (www.institutohorus.org.br).

Horses (*Equus caballus*) were introduced during the Portuguese colonization, initially in Pernambuco and Bahia states. Nowadays, Brazil has the third largest horse population in the world, with 5.9 million head.[77] About 5000 wild horses live in indigenous land in Roraima state.[78]

There is no information as to when *Cervus unicolor* (Sambar deer) was introduced in Brazil, but there is a report of a population of this species already established with 30–40 animals in an Area of Environmental Protection in Lins and Cafelândia municipalities in São Paulo state (www.institutohorus.org.br).

The other five invasive mammal species have their origin in Brazil (Table 4.2). The majority of them are primates of relatively small size. The *Callithrix jacchus* (common marmoset) and the *C. penicillata* (black-tufted marmoset) have been introduced into new areas since the beginning of the twentieth century, being captured in their natural area and sold along Brazilian roads or in local markets as pets to increase the income of the local poor.[79,80] However, as they tend to be aggressive and cannot be tamed, the owners end up releasing them into both the wild and urban areas. Nowadays, *C. jacchus* and *C. penicillata* can be found in many areas in south and southeastern Brazil. These species present considerable negative impact on birds because they may feed on native species, especially bird eggs and nestlings, as well as lizards, tree frogs, and infant mammals.[81] Another major impact these species may cause is the hybridization with congeneric species. Six species of the genus *Callithrix* occurs in Brazil, four of them (*C. aurita, C. flaviceps, C. geoffroyi,* and *C. kuhlii*) being endemic to the Atlantic Forest, with *C. aurita* and *C. flaviceps* presently threatened with extinction.[82,83] Some hybrids between *C. penicillata* and *C. aurita* can already be seen in some conservation units, such as Serra dos Órgãos National Park and Bocaina State Park. Beyond the destruction of native forests,[82,83] *Callithrix aurita* is threatened with extinction by the loss of its gene pool due to hybridization with exotic congeneric species.

The story of the golden-headed lion tamarin, *Leontopithecus chrysomelas*, is different from other invasive mammals in Brazil. This species is on the International Union for Conservation of Nature (IUCN) list of endangered species mainly due to severe population reduction due to deforestation in the Atlantic Forest.[84] Its distribution is restricted to some fragments in Bahia state and formerly in northeastern Minas Gerais state. To ensure their survival, animals confiscated from illegal trade in the 1980s have been used for a captive breeding program.[85] A veterinarian had requested permission from the Brazilian Institute of Environment and Renewable Natural Resources (IBAMA) to establish in captivity some individuals of the species in the municipality of Niteroi, in Rio de Janeiro state. After his death, about 10 years ago, the animals were released in a municipal park. Today, many groups can be seen in the neighborhood of the park, as well as in Serra da Tiririca State Park, near the area where they were released. Although it is an endangered species, the presence and spread of *L. chrysomelas* in Rio de Janeiro state can be a threat to the golden lion tamarin, *L. rosalia*, which is also endangered and is endemic to the state.[86] If individuals of *L. chrysomelas* reach the areas where *L. rosalia* occurs, these congeneric species may compete for resources and hybridize, weakening the gene pool of the native species.[87]

The rodent *Kerodon rupestris* was introduced in the main island of Fernando de Noronha Archipelago in the middle of 1960s to serve as wild game for the military. This species feeds on fruits, seeds, and roots, biting the base of the bush or tree until it can be overthrown. The soil is then exposed to erosion or the establishment of invasive plants (www.institutohorus.org.br).

As observed for other vertebrate groups, the largest number of invasive species of mammals is found in south and southeastern states (RS = Rio Grande do Sul, SC = Santa Caterina, PR = Paraná, SP = São Paulo, RJ = Rio de Janeiro, MG = Minas Gerais and ES = Espirito Santo; Figure 4.5). This may be due to greater population density of people in the region and the largest number of researchers. The composition of invasive species in these states is mainly those common to all other states (*M. musculus, R. rattus, R. norvegicus, F. catus*, and *C. familiaris*) and primate invaders who came from other regions of Brazil.

## 4.3   Pooled vertebrate species

Considering the high biodiversity of vertebrates in Brazil (about 6623 species), the proportion of invasive species (2.06% or 137 species) cannot be considered negligible but is of special concern, especially considering that for the great majority of the vertebrate invasive species (or for vertebrates as a group), we still lack consistent data on population sizes, on monitoring the geographic expansion of the invasive species, and on invasive species' effects on indigenous organisms; in the face of this absence of data, the number of invasive vertebrate species in Brazil is quite significant. When analyzing the pooled vertebrate species in Brazilian states, we can see that all Brazilian states presently have instances of invasive vertebrates (Figure 4.6), although the frequency varies considerably. This variation in the number of cases of invasive vertebrates may result from historical processes that may have facilitated invasions in some Brazilian regions, from the level (or lack) of present knowledge regarding invasive species, and from the proximity to areas more densely populated with humans, which in turn facilitated exotic species introductions. In fact, the number of invasive vertebrate species is higher in states of eastern Brazil (Figure 4.6), where most of the Brazilian human population is concentrated.

Most cases of vertebrate invasions in Brazil are a result of intentional introductions due to economic needs such as farming or pet shops and, in the case of fishes, especially as a result of the alteration of the hydrographic basins due to the construction of dams for hydropower. In these cases, the main environmental changes are related to river transpositions and the elimination of natural barriers between isolated freshwater systems.[36,27] Other introductions and subsequent invasions are nonintentional; the species were introduced as a result of transport and trade of products among areas in Brazil or between continents.

## 4.4   Economic issues

Besides the ecological implications, the introduction of exotic species has important economic implications and, in general, may result in high costs to solve the prejudices caused by their negative effects.[10] In economic terms, there are still no comprehensive estimates on the negative effects of invasive exotic species in Brazil, although some consequences of their impact on habitat structure and integrity, the loss or decline of sympatric native species, and the dissemination of parasites and diseases are obvious. Although in many cases the invasive vertebrate species were introduced to produce economical improvements (e.g., farming, cultures, pet shops), some escaped to nature and presently constitute problems for ecosystems, potentially causing damage or prejudice often exceeding their economic

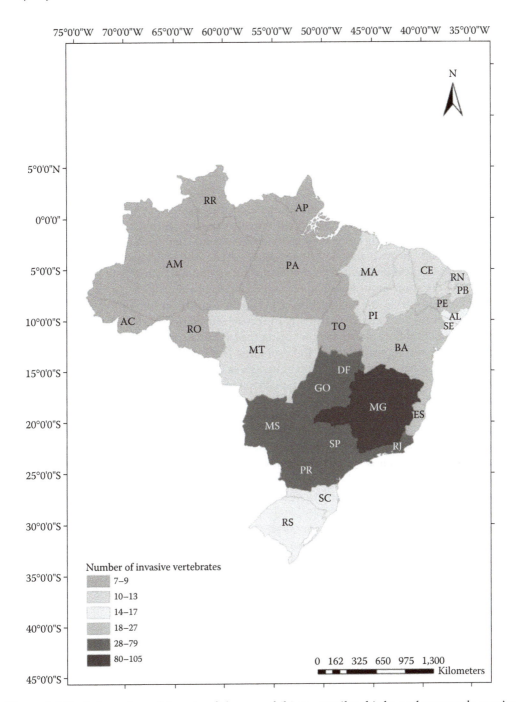

*Figure 4.6* Pooled number of invasive fishes, amphibians, reptiles, birds, and mammals species in each Brazilian state. Acronyms of states: AC-Acre, AL-Alagoas, AM-Amazonas, AP-Amapá, BA-Bahia, CE-Ceará, DF-Federal District, ES-Espírito Santo, GO-Goiás, MA-Maranhão, MG-Minas Gerais, MS-Mato Grosso do Sul, MT-Mato Grosso, PA-Pará, PB-Paraíba, PE-Pernambuco, PI-Piauí, PR-Paraná, RJ-Rio de Janeiro, RN-Rio Grande do Norte, RO-Rondônia, RR-Roraima, RS-Rio Grande do Sul, SC-Santa Catarina, SE-Sergipe, SP-São Paulo, and TO-Tocantins.

benefits. Also, some invasive species act as reservoirs for some parasites (e.g., rabies viruses in some mammals and lice in pigeons) causing diseases in humans and animals, which can result in higher costs for health care.

## 4.5   From here to where?

At this time, we consider the knowledge of invasive vertebrates and their effects in Brazil to be still germinal, but it is important to establish appropriate strategies to cope with a problem that can erode a considerable portion of the biological diversity of a megadiversified country such as Brazil. Some actions that can be taken are the development of a plan for the country regarding the identification (through the elaboration of national, state, and local lists of exotic invasive species), monitoring, control, and eradication of exotic invasives; a more strict and rigorous control of alien species arriving in the country as pets and for farming and cultures; a rigorous control by governmental agencies of culturing and farming exotic species; the investment of financial resources for the study of the distribution and current status of different invasive species (the governmental agencies of science and technology and education should open specific bidding to grant projects devoted for the study and monitoring of IAS); the elaboration of protocols for eradication of a defined portion of invasives (starting with invasives causing the most negative effects to natural environments and their fauna and flora and those that bring the highest economic costs); the development of studies determining estimates of the economic costs caused by invasives; and the mapping of the occurrence and distribution of invasives in all conservation units of the country, identifying the problems to present perspectives of control and eradication. Although the proportion of invasives among living vertebrates we found cannot be considered negligible, we still have time to start planning and to create consistent strategies and actions.

This chapter, despite covering a wide range of primary and secondary databases, supplemented with extensive inventories made by the authors and personal communications, is far from being conclusive. The dynamic aspect of invasive vertebrate distribution in natural systems, together with the high number of new species being described each year in many different regions of Brazil as well as the concentration of knowledge expertise and surveys developed for some geographic regions of the country is detrimental to obtaining complete and conclusive results. Nonetheless, the data presented is an exhaustive compilation of the available information for invasive vertebrate distribution in Brazilian systems and can be used as an important starting point for mitigating actions in the different biomes of the country.

## Acknowledgments

This chapter is part of the results of the "Exotic and Invader Species" (Project No. E-26.110.430/2007) supported by Fundação Carlos Chagas Filho de Amparo à Pesquisa do Estado do Rio de Janeiro (FAPERJ). It is also partially supported by grants of the FAPERJ to Carlos Freder D. Rocha (CFDR) through the "Programa Cientistas do Nosso Estado" (Processes E-26/102.404.2009 to CFDR and E-26/102.799.2008 to Helena Go Bergallo (HGB) and by the Conselho Nacional do Desenvolvimento Científico e Tecnológico (CNPq; Processes No. 476684/2008-0 and 307653/2003-0 to CFDR, Process No. 309527/2006-6 to HGB and Processes No. 301433/2007-0 and 470286/2008-3 to Rosana Mazzoni (RM). We thank the Programa Prociência of the Universidade do Estado do Rio de Janeiro, whose support allowed us our dedication to studies like these. The Instituto Biomas provided some logistic support for this study.

# References

1. MMA/SBF (Ministério do Meio Ambiente). 2002. *Avaliação e identificação de áreas e ações prioritárias para a conservação, utilização sustentável e repartição dos benefícios da biodiversidade nos Biomas Brasileiros.* Brasília: Ministério do Meio Ambiente.
2. Galindo-Leal, C., and I. G. Câmara. 2003. *The Atlantic Forest of South America: Biodiversity Status, Threats and Outlook.* Washington, DC: Island Press.
3. Leal, I., M. Tabarelli, and J. M. C. Silva. 2003. *Ecologia e Conservação da Caatinga.* Recife: Ed. Universitária da Universidade Federal de Pernambuco.
4. Rocha, C. F. D., H. G. Bergallo, M. A. S. Alves, and M. Van Sluys. 2003. *A biodiversidade nos grandes remanescentes florestais do Rio de Janeiro e nas restingas da Mata Atlântica.* São Carlos: Rima Ed.
5. Rocha, C. F. D., H. G. Bergallo, M. Van Sluys, M. A. S. Alves, and C. Jamel. 2007. The remnants of restinga habitats in the Brazilian Atlantic Forest of Rio de Janeiro State, Brazil: Habitat loss and risk of disappearance. *Braz J Biol* 67:263.
6. Bergallo, H. G., E. C. Fidalgo, C. F. D. Rocha, M. C. Uzêda, M. B. Costa, M. A. S. Alves, M. Van Sluys, M. A. Santos, T. C. C. Costa, A. C. R. Cozzolino (Orgs.). 2009. *Estratégias e ações para a conservação da biodiversidade no Estado do Rio de Janeiro.* 1st. ed., vol. 1. Rio de Janeiro: Instituto Biomas.
7. Abrams, P. A. 1996. Evolution and the consequences of species introductions and deletions. *Ecology* 77:1321.
8. Rodríguez, J. P. 2001. Exotic species introductions into South America: An underestimated threat? *Biodivers Conserv* 10:1983.
9. Rahel, F. J. 2000. Homogenization of fish faunas across the United States. *Science* 288:854.
10. Rappoport, E. H. 1992. Las implicaciones ecológicas y econômicas de La introducción de espécies. *Ciencia e Ambiente* 3:69.
11. Carlton, J. T. 1996. Pattern, process, and prediction in marine invasion ecology. *Biol Conserv* 78:97.
12. Elton, C. 2000. *The Ecology of Invasions by Animals and Plants.* Chicago: University of Chicago Press.
13. IUCN. 2000. IUCN guidelines for the prevention of biodiversity loss due to biological invasion. *Species* 31–32:28.
14. Magnusson, W. E. 2006. Homogenização biotic. In *Biologia da Conservação: Essências*, ed. C. F. D. Rocha, H. G. Bergallo, M. Van Sluys, and M. A. S. Alves, (Orgs.), 211. São Carlos: RIMA Editora.
15. Santos, A. R., C. F. D. Rocha, and H. G. Bergallo. 2010. Native and exotic species in the urban landscape of the city of Rio de Janeiro, Brazil: Density, richness, and arboreal deficit. *Urban Ecosyst* 13:209–222.
16. Santos, A. R., H. G. Bergallo, and C. F. D. Rocha. 2008. Paisagem urbana Alienígena. *Ciência Hoje* 41:68.
17. Matthews, S. 2005. *A América do Sul invadida: a crescente ameaça das espécies exóticas invasoras.* GISP, Programa Global de Espécies Invasoras.
18. Shannon, R. C. 1932. *Anopheles gambiae in Brazil. Am J Epidemiol* 15:634.
19. Causey, O., L. M. Deane, and M. P. Deane. 1943. Ecology of Anopheles gambiae in Brazil. *Am J Trop Med* 23:73.
20. Soper, F. L., and D. B. Wilson. 1943. *Anopheles gambiae in Brazil 1930 to 1940: Semination and Eradication.* New York: The Rockefeller Foundation.
21. Myers, G. S. 1945. A natural habitat of the house gecko (*Hemidactylus mabouia*) in Brazil. *Copeia* 1945(2):120.
22. Delariva, R. L. & A. A. Agostinho. 1999. Introdução de espécies: uma síntese comentada. *Acta Sci* 21(2):255–262.
23. Buckup, P. A., N. A. Menezes, and M. S. Ghazzi. 2007. *Catálogo das espécies de peixes de água doce do Brasil.* Série Livros 23. Rio de Janeiro: Museu Nacional.
24. Reis, N. R., A. L. Peracchi, W. A. Pedro, and I. P. Lima. 2006. *Mamíferos do Brasil.* Londrina: Ed. UEL.
25. Conselho Nacional de Recursos Hídricos (CNRH). 2003. *Resolução 32 de 15 de Outubro de 2003* – Institui a Divisão Hidrográfica Nacional. Published in Diário Oficial da União in 17th December 2003.

26. Abell, R., M. L. Thieme, C. Revenga, M. Bryer, M. Kottelat, N. Bogutskaya, B. Coad et al. 2008. Freshwater ecoregions of the world: A new map of biogeographic units for freshwater biodiversity conservation. *Bioscience* 58:403.

27. Julio-Júnior, H. F., C. D. Tós, A. A. Agostinho, and C. S. Pavanelli. 2009. A massive invasion of fish species after eliminating a natural barrier in the upper rio Paraná basin. *Neotrop Ichthyol* 7:709.

28. Vari, R. P., and L. R. Malabarba. 1998. Neotropical Ichthyology: An overview. In *Phylogeny and Classification of Neotropical Fishes*, ed. L. R. Malabarba, R. E. Reis, R. P. Vari, Z. M. S. Lucena, and C. A. S. Lucema, 1–11, vol. 1. Porto Alegre: Edipucrs.

29. Lima, F. C. T., L. R. Malabarba, P. A. Buckup, J. F. Pezzi da Silva, R. P. Vari, A. Harold, R. Benine et al. 2003. Genera Incertae Sedis in Characidae. In *Check List of the Freshwater Fishes of South and Central America*, ed. R. E. Reis, S. O. Kullander, and C. Ferraris Jr., 106. Porto Alegre: Edipucrs.

30. Langeani, F., R. M. C. Castro, O. T. Oyakawa, O. A. Shibatta, C. S. Pavanelli, and L. Casatti. 2007. Diversidade da ictiofauna do alto rio Paraná: composição atual e perspectivas futuras. *Biota Neotropica* 7:1.

31. Mazzoni, R., R. Iglesias-Rios, and S. A. Schubart. 2004. Movement patterns of *Astyanax janeiroensis* along a small stream in southeastern Brazil. *Ecol Freshw Fish* 13:231.

32. Kullander, S. O. 2003. Family Cichlidae (Cichlids). In *Check List of the Freshwater Fishes of South and Central America*, ed. R. E. Reis, S. O. Kullander, and C. Ferraris Jr., 605. Porto Alegre: Edipucrs.

33. Howes, G. J. 1991. Systematics and biogeography: An overview. In *Cyprinid Fishes, Systematic, Biology and Exploitation*, ed. I. J. Winfield and J. S. Nelson, 1. London: Chapman & Hall.

34. Braun, A. S., P. C. C. Milani, and N. F. Fontoura. 2003. Registro da Introdução de *Clarias gariepinus* (Siluriformes, Clariidae) na Laguna dos Patos, Rio Grande do Sul, Brasil. *Revista Biociências* 11:101.

35. Agostinho, A. A., and H. F. Júlio-Júnior. 1996. Ameaça ecológica: peixes de outras águas. *Ciência Hoje* 21:36.

36. Moreira-Filho, O., and P. A. Buckup. 2005. A poorly known case of watershed transposition between the São Francisco and upper Paraná River basins. *Neotrop Ichthyol* 3:449.

37. Lima, A. P., W. E. Magnusson, M. Meni, J. K. Erdtmann, D. J. Rodrigues, C. Keller, and W. Hödl. 2006. *Guia de sapos da Reserva Adolpho Ducke, Amazônia Central*. Fundação Banco Bilbao Vizcaya Argentaria (BBVA), Instituto Nacional de Pesquisas da Amazônia (INPA), Conselho Nacional do Desenvolvimento Científico e Tecnológico (CNPq) and Programa de Pesquisas em Biodiversidade/Ministério do Meio Ambiente.

38. Borges-Martins, M., M. Di-Bernardo, G. Vinciprova, and J. Measey. 2002. *Rana catesbeiana* (American Bullfrog). Brazil: Rio Grande do Sul. *Herpetol Rev* 33:319.

39. Fontanello, D., and C. M. Ferreira. 2004. Histórico da Ranicultura Nacional, 2004. http://www.aquicultura.br/ (accessed February 1, 2011).

40. Mattews, S. 2005. *América do Sul invadida. Programa Global de Espécies Invasoras*. Instituto Hórus. GISP, Programa Global de Espécies Invasoras.

41. Frost, D. R. 2009. Amphibian species of the world: An online reference. Version 5.3, American Museum of Natural History, New York. http://research.amnh.org/herpetology/amphibia/ (accessed May 27, 2010).

42. Anjos, L. A., and C. F. D. Rocha. 2008. The *Hemidactylus mabouia* Moreau de Jonnes, 1818 (Gekkonidae) lizard: An invasive alien species broadly distributed in Brazil. *Natureza & Conservação* 6:196.

43. Rocha, C. F. D., L. A. Anjos, and H. G. Bergallo. Conquering Brazil: The invasion by the exotic gekkonid lizard *Hemidactylus mabouia* in Brazilian natural environments. *Biol Invasions* Submitted.

44. Anjos, L. A., C. F. D. Rocha, D. Vrcibradic, and J. J. Vicente. 2005. Helmints associated with the exotic lizard *Hemidactylus mabouia* in an area of rock outcrops in southeastern Brazil. *J Helminthol* 79:307.

45. Anjos, L. A., W. O. Almeida, A. Vasconcellos, E. M. X. Freire, and C. F. D. Rocha. 2008. Pentastomids infecting an invader lizard, *Hemidactylus mabouia* (Gekkonidae) in northeastern Brazil. *Braz J Biol* 68:611.

46. Soares, A. H. B., and A. F. B. Araújo. 2008. Experimental introduction of *Liolaemus lutzae* (Squamata, Iguanidae) in Praia das Neves, State of Espírito Santo, Brazil: A descriptive study 18 years later. *Rev Bras Zool* 25:640.

47. Péres Jr., A. K. 2003. *Sistemática e conservação de lagartos do gênero Tupinambis (Squamata, Teiidae)*. PhD thesis. Brasília: Universidade de Brasília. Instituto de Biologia.

48. Cramp, S., and C. M. Perrins. 1994. *The Birds of the Western Palearctic*. Oxford, UK: Oxford University Press.

49. Sick, H. 1959. A invasão da América Latina pelo pardal, *Passer domesticus* Linnaeus 1758, com referência especial ao Brasil (Ploceidae, Aves). *Boletim do Museu Nacional, nova série, Zoologia* 207:1.

50. Borges, S. H., J. F. Pacheco, and A. Wittaker. 1996. New records of the house sparrow (*Passer domesticus*) in the Brazilian Amazon. *Ararajuba* 4:116.

51. Flannery, T. F. 1994. *The Future Eaters*. Chatswood: Reed Books.

52. Abreu Jr., E. F., and Köhler, A. 2009. Mastofauna de médio e grande porte na RPPN da UNISC, RS, Brasil. *Biota Neotropica* 9:1.

53. Bergallo, H. G., F. Martins-Hatano, D. S. Raíces, T. T. L. Ribeiro, A. G. Alves, J. L. Luz, R. Mangolin, and M. A. R. Mello. 2004. Os mamíferos da Restinga de Jurubatiba. In *Pesquisas de longa duração na Restinga de Jurubatiba. Ecologia, História Natural e Conservação*, ed. C. F. D. Rocha, F. A. Esteves, and F. R. Scarano, 215. São Carlos: Rima Editora.

54. Cerqueira, R., F. A. S. Fernandez, and M. F. Q. S. Nunes. 1990. Mamíferos da Restinga de Barra de Maricá. *Pap Avulsos Zool* 37:141.

55. Cherem, J. J., and D. M. Perez. 1996. Mamíferos terrestres de Floresta de Araucária, no município de Três Barras, Santa Catarina, Brasil. *Biotemas* 9:29.

56. Lacerda, A. C. R., W. M. Tomas, and J. Marinho-Filho. 2009. Domestic dogs as an edge effect in the Brasília National Park, Brazil: Interactions with native mammals. *Anim Conserv* 12:477.

57. Lessa, I. C. M., and H. G. Bergallo. Effects of castration on population density of domestic cats in an island of the Brazilian Atlantic Forest. Submitted.

58. Pessôa, F. S., T. C. Modesto, H. G. Albuquerque, N. Attias, and H. G. Bergallo. 2009. Non-volant mammals, Reserva Particular do Patrimônio Natural (RPPN) Rio das Pedras, municipality of Mangaratiba, State of Rio de Janeiro, Brazil. *Check List* 5:577.

59. Ribeiro, R., and J. Marinho-Filho. 2005. Estrutura da comunidade de pequenos mamíferos (Mammalia, Rodentia) da Estação Ecológica das Águas Emendadas, Planaltina, Distrito Federal, Brasil. *Rev Bras Zool* 22:898.

60. Baker, P. J., A. J. Bentley, R. J. Ansell, and S. Harris. 2005. Impact of predation by domestic cats *Felis catus* in an urban area. *Mammal Rev* 35:302.

61. Baker, P. J., S. E. Molony, E. Stone, I. E. Cuthill, and S. Harris. 2008. Cats about town: Is predation by free-ranging pet cats (*Felis catus*) likely to affect urban bird populations? *Ibis* 150 (Suppl. 1):86.

62. Nogales, M., A. Martíns, B. R. Tershy, C. J. Donlan, D. Veicth, N. Puerta, B. Wood, and J. Alonso. 2004. A review of feral cat eradication on islands. *Conserv Biol* 18:310.

63. Woods, M., R. A. McDonald, and S. Harris. 2003. Predation of wildlife by domestic cats *Felis catus* in Great Britain. *Mammal Rev* 33:174.

64. Brown, K. P. 1997. Impact of brodifacoum poisoning operations on South Islands Robins *Petroica australis australis* in a New Zealand Nothofagus forest. *Bird Conserv Int* 7:399.

65. Brown, K. P., N. Alterio, and H. Moller. 1998. Secondary poisoning of stoats (*Mustela erminea*) at low mouses (*Mus musculus*) abundance in a New Zealand Nothofagus forest. *Wildl Res* 25:419.

66. Van Resburg, P. J. J., J. D. Skinner, and R. J. Van Aarde. 1987. Effects of feline panleucopaenia on the population charecteries of feral cat on Marion Island. *J Appl Ecol* 24:63.

67. Pereira, L. G., S. E. M. Torres, H. S. Silva, and L. Geise. 2001. Non-volant mammals of Ilha Grande and adjacent areas in southern Rio de Janeiro state, Brazil. *Boletim do Museu Nacional* 459:1.

68. Gonçalves da Silva, A., S. O. Kolokotronis, and D. Wharton. 2010. Modeling the eradication of invasive mammals using the sterile male technique. *Biol Invasions* 12:751.

69. Ferracioli, P., M. B. F. Nascimento, H. Mori, and M. L. Orsi. 2009. *Ocorrência de Lepus europaeus Pallas, 1778 em trechos do Município de Londrina*. São Lourenço, MG: Anais do IX Congresso de Ecologia do Brasil.

70. Mourão, G. M., M. E. Coutinho, R. A. Mauro, W. M. Tomás, and W. E. Magnusson. 2002. Levantamentos aéreos de espécies introduzidas no Pantanal: porcos ferais (porco monteiro), gado bovino e búfalos. Boletim de Pesquisa. *Embrapa Pantanal, Corumbá* 28:7.

71. Oliveira, J. D. 2007. *Origem, distribuição e relação genética entre populações de Capra hircus do Nordeste do Brasil e sua relação com populações do Velho Mundo.* PhD thesis, Graduate Program in Genetics. São Paulo: Universidade de São Paulo.

72. Araújo, A. M., S. E. F. Guimarães, T. M. M. Machado, P. S. Lopes, C. S. Pereira, F. L. R. Silva, M. T. Rodrigues, V. S. Columbiano, and C. G. Fonseca. 2006. Genetic diversity between herds of Alpine and Saanen dairy goats and naturalized Brazilian Moxotó breed. *Genet Mol Biol* 29:67.

73. Marcondes, C. R., J. R. F. Marques, M. R. T. R. Costa, M. C. F. Damé, and L. G. Brito. 2007. *Programa de pesquisas da Embrapa Amazônia Oriental para o melhoramento genético de búfalos.* Documento 303. Belém, PA: Embrapa Amazônia Oriental.

74. Boulton, W. J., and W. J. Freeland. 1991. Models for the control of feral water buffalo (*Bubalus bubalis*) using constant levels of offtake and effort. *Aust Wildl Res, Victoria* 18:63.

75. Bayliss, P., and K. M. Yeomans. 1989. Distribution and abundance of feral livestock in the "top end" of the Northern Territory (1985–86), and their relation to population control. *Aust Wildl Res* 16:651.

76. Hill, R., and G. Webb. 1982. Floating grass mats of the Northern Territory floodplains, an endangered habitat. *Wetlands* 2:45.

77. Food and Agricultural Organization of the United Nations, 2002. Country Pasture/Forage Resource Profiles, Brazil. Food and Agricultural Organization of the United Nations, 2002. http://www.fao.org/ag/AGP/AGPC/doc/counprof/Brazil/brazil.htm (accessed February 1, 2011).

78. Braga, R. M. 2000. *Cavalo lavradeiro em Roraima: aspectos históricos, ecológicos e de conservação.* Brasília: Embrapa Comunicação para Transferência de Tecnologia.

79. Coimbra-Filho, A. F. 1984. A situação atual dos calitriquídeos que ocorrem no Brasil (Callitrichidae, Primates). In *A Primatologia no Brasil*, ed. M. T. Mello, 15. Brasília: Sociedade Brasileira de Primatologia.

80. Rylands, A. B., A. F. Coimbra-Filho, and R. A. Mittermeier. 1993. Systematics, geographic distribution, and some notes on the conservation status of the Callitrichidae. In *Marmosets and Tamarins: Systematics, Behaviour, and Ecology*, ed. A. B. Rylands, 95. Oxford, UK: Oxford Science Publications.

81. Digby, L., and C. E. Barreto. 1998. Vertebrate predation in common marmosets. *Neotrop Primates* 6:124.

82. Rylands, A. B., S. F. Ferrari, and S. L. Mendes. 2008. *Callithrix flaviceps.* IUCN Red List of Threatened Species. Version 2009.2. http://www.iucnredlist.org (accessed February 24, 2010).

83. Rylands, A. B., M. C. M. Kierulff, S. L. Mendes, and M. M. de Oliveira. 2008. *Callithrix aurita.* IUCN Red List of Threatened Species. Version 2009.2. http://www.iucnredlist.org (accessed February 24, 2010).

84. Kierulff, M. C. M., A. B. Rylands, S. L. Mendes, and M. M. de Oliveira. 2008. *Leontopithecus chrysomelas.* IUCN Red List of Threatened Species. Version 2009.2. http://www.iucnredlist.org (accessed February 22, 2010).

85. Konstant, W. R. 1986. Illegal trade in golden-headed lion tamarins. *Primate Conserv* 7:29–30.

86. Bergallo, H. G., C. F. D. Rocha, M. A. S. Alves, and M. Van-Sluys. 2000. *A fauna ameaçada de extinção do Estado do Rio de Janeiro.* 1st ed. Rio de Janeiro: EdUERJ.

87. Rhymer, J. M., and D. Simberloff. 1996. Extinction by hybridization and introgression. *Annu Rev Ecol Syst* 27:83.

88. Bovendorp, R. S., A. D. Alvarez, and M. Galetti. 2008. Density of the Tegu lizard (Tupinambis merianae) and its role as nest predator at Anchieta Island, Brazil. *Neotrop Biol Conserv* 3:9.

89. Oliveira, D. C., and S. T. Bennemann. 2005. Ictiofauna, recursos alimentares e relações com as interferências antrópicas em um riacho urbano no sul do Brasil. *Biota Neotropica* 5:1.

90. Lemes, E. M., and W. Garutti. 2002. Ecologia da ictiofauna de um córrego de cabeceira da bacia do Alto rio Paraná, Brasil. *Iheringia Ser Zool* 92:69.

91. Araujo, N. B., and F. L. Tejerina-Garro. 2007. Composição e diversidade da ictiofauna em riachos do Cerrado, bacia do ribeirão Ouvidor, alto rio Paraná, Goiás, Brasil. *Rev Bras Zool* 24:1.

92. Sarmento-Soares, L. M., R. Mazzoni, and R. F. Martins-Pinheiro. 2008. A fauna de peixes dos Rios dos Portos Seguros, extremo sul da Bahia, Brasil. *Boletim do Museu de Biologia Mello Leitão* 24:119.

93. Magalhães, A. L. B., and T. F. Ratton. 2005. Reproduction of a South American population of pumpkinseed sunfish *Lepomis gibbosus* (Linnaeus) (Osteichthyes, Centrarchidae): A comparison with the European and North American populations. *Rev Bras Zool* 22:477.

94. Lazzarotto, H., and E. P. Caramaschi. 2009. Introdução da truta no Brasil e na Bacia do rio Macaé, Estado do Rio de Janeiro: Histórico, Legislação e Perspectivas. *Oecologia Brasiliensis* 13:649.

95. Magalhães, A. L. B., R. F. Andrade, T. F. Ratton, and M. F. G. Brito. 2002. Ocorrência da truta arco-íris Oncorhynchus mykiss (Walbaum, 1792) (Pisces: Salmonidae) no alto rio Aiuruoca e tributários, bacia do rio Grande, Minas Gerais, Brasil. *Boletim do Museu de Biologia Mello Leitão (N. Ser.)* 14:33.

96. Azevedo, M. A. G. 2006. Contribuição de estudos para licenciamento ambiental ao conhecimento da avifauna de Santa Catarina, Sul do Brasil. *Biotemas* 19:93.

97. Mallet-Rodrigues, F., and M. L. M. Noronha. 2009. Birds in the Parque Estadual dos Três Picos, Rio de Janeiro state, southeast Brazil. *Cotinga* 31:96.

98. Telino-Junior, W. R., M. M. Dias, S. M. Azevedo-Junior, R. M. Lyra-Neves, and M. E. L. Larrazábal. 2005. Estrutura trófica da avifauna na reserva Estadual de Gurjaú, Zona da Mata Sul, Pernambuco, Brasil. *Rev Bras Zool* 22:962.

99. Anjos, L., K. L. Schuchmann, and R. Berndt. 1997. Avifaunal composition, species richness, and status in the Tibagi River basin, Paraná State, Southern Brazil. *Ornitol Neotrop* 8:145.

100. Tampson, V. E., and M. V. Petry. 2008. Nidificação e análise das guildas alimentares de aves no morro do Espelho, na zona urbana de São Leopoldo—RS. *Biodiversidade Pampeana* 6:63–9.

101. Cunha, A. A., and M. V. Vieira. 2004. Present and past primate community of the Tijuca Forest, Rio de Janeiro, Brazil. *Neotrop Primates* 12:153.

102. Eduardo, A. A., and M. Passamani. 2009. Mammals of medium and large size in Santa Rita do Sapucaí, Minas Gerais, southeastern Brazil. *Check List* 5:399.

103. Graipel, M. E., J. J. Cherem, and A. Ximenez. 2001. Mamíferos terrestres não voadores da Ilha de Santa Catarina, sul do Brasil. *Revista Biotemas* 14:109.

104. Oren, D. C. 1984. Resultados de uma nova expedição zoológica a Fernando de Noronha. *Boletim do Museu Paraense Emilio Goeldi, Zoologia* 1:19.

*section three*

---

*British Isles*

# chapter five

# Alien plants in Britain

*Mark Williamson*

## Contents

## 5.1 Introduction

There has only been one attempt[1] to estimate the cost, species by species, of a large set of native and introduced plants in the British Isles. That attempt was based primarily on the cost of herbicides and is very useful as far as it goes, but clearly does not estimate the other appreciable costs of some species. In this chapter, I examine various ways in which such costs might be estimated.

    The approach here is based on two programs of work with which I have been involved: The first is the study of the impacts of alien species and how to measure them[2,3] and the second is the economics section of the Global Invasive Species Programme (GISP).[4] It is important to note that economics is not accountancy, even though economic assessments will normally include a cost–benefit analysis. So, the GISP Economics Programme produced few cost estimates, and none that should be taken too seriously. The same applies to the cost figures in this chapter. Although I give some numbers, the importance and effect of alien invasive plants in the British Isles are given more reliably by an understanding of how costs arise and the policy options to contain them rather than by concentrating on narrowly based figures.

    This chapter deals with the British Isles, that is, the large islands of Britain and Ireland and numerous smaller associated islands. Politically it involves two sovereign states, the Republic of Ireland and the United Kingdom of Great Britain and Northern Ireland. For biological purposes, they are usually treated together. Britain and its associated islands are about 229,000 km², 131,000 of them in England. The population of Britain is about 54 million. (All the population figures here are based on the 1991 census.) Most of that population is in England, with a little less than 5 million in Scotland and about 2.8 million in Wales. Ireland is considerably smaller than Britain at about 84,000 km² and with a population of slightly more than 5 million. The total area considered here is about 313,000 km². In land area, England is only about 40%, but economically it is over three quarters of the total.

As a preliminary, it is desirable to know how many plant species of different status are thought to grow in the British Isles, which is more problematic than the invasion literature might lead you to expect. It is also necessary to clarify how impact may be measured and its relationship to cost. I will deal with those two points first and then consider 30 particular invasive alien plant species. Only after that will I consider the distribution of impact and cost over the British flora, with a view to getting an overall understanding of the impact of alien plants in the British Isles.

## 5.2   The number of British alien plant taxa

Both the number of British native plants and alien plants are uncertain. There are further doubts about the ecological status of some taxa. I will describe these uncertainties and show the effect they have on numbers.

With native plants the troubles come mostly from microspecies and hybrids. Hybrids are perfectly satisfactory taxa, recognizable and nameable, like the cordgrass hybrid *Spartina* × *townsendii*. The × indicates it is a known hybrid, in this case between the native *S. maritima* and the alien, American, *S. alterniflora*. *S.* × *townsendii,* like many hybrids, can only reproduce vegetatively, but it is the parent of *S. anglica* (discussed in Section 5.4 as one of 30 interesting species), which is, again like many hybrids, fertile. But most hybrids fail to form populations, fail to establish, and occur only near their parents. Counts of British species usually omit hybrids, but as there are around 400 of them listed in the floras, including them makes a large difference to the taxa counts. There is also the question of whether to count crosses between natives and aliens, like *S.* × *townsendii,* as native or alien; most floras, oddly in my view, call them native if they have arisen in the British Isles. They are nonindigenous species in the sense of not having been in Britain before agriculture.

Microspecies and critical species are common in the British flora. Critical species are those whose identifications need to be verified by an expert in the group but may nevertheless be perfectly good species in all senses. They are just difficult to identify. All microspecies are critical but are in groups that are often apomictic, so the definition of a species is unclear. Stace[5] estimates that there are 400 microspecies in *Rubus fruticosus* agg. (blackberries) and 250 in *Hieracium* (hawkweeds), almost all native. There are ordinary, noncritical species in those genera too. In *Taraxacum* (dandelions), 226 microspecies are recognized, 39 endemic, 76 are described as other natives, and 111 are considered aliens. With modern genetic techniques many more could be distinguished. Generally, none of these are included in counts comparing the British flora with others. With around 900 native microspecies, counting them in the total of native species would make a huge difference to comparisons.

Even so, there is doubt about the number of what I will call in this context native macrospecies. There have been three authoritative floras in the last 15 years, and counts from them produce 1311,[6–8] 1255,[9,10] and 1552 macrospecies.[5] Taking the highest of those macrospecies and the counts of hybrids and microspecies gives 2852 native species. But you could argue that the figure should be as low as 1255. I would suggest saying "about 1500 macrospecies" is a sensible basis for comparisons. It is not far from the 1407 natives picked out by the Ecological Flora Database.[11,12]

The next uncertainty is whether all those species are in fact native. Some of them may well be aliens. Almost all native species had to invade the British Isles after the last glaciation, so those known to be growing in the forested landscape of the Mesolithic, before agriculture, very roughly 5,000–10,000 years ago, are called native. It is customary to call native those that are present in the late glacial, notably some species of disturbed ground, even though some may well have died out and been reintroduced with agriculture. But there are many species for which there is no fossil or historic record and which might be native or not. In the floras, roughly 10% of the species have labels of uncertainty such as "probably native" or "possibly introduced."

The pair of complementary catalogs of alien plants[13,14] list 49 species as "accepted with reservations as native." One standard flora[9] lists 7 of these as unqualified native, another[6] lists 11, but there is only one species common to both sets, *Centaurea cyanus*, the corn flower. This is native on the basis of only one well-stratified pollen grain, of more grains that could have been washed down, and from its occurrence in postglacial, preagricultural deposits on the mainland of Europe. Salisbury[15] was remarkably indignant about this sort of procedure: "Hence the presence of seeds, still less of pollen grains, of a species affords little if any evidence as to its status, whether casual or more or less naturalized. To assert, because of the presence of the pollen of a species in prehistoric deposits, that it is 'native' is at once misjudged, misleading and well-nigh meaningless." That is too strong, but caution is needed.

As the number of native species is uncertain so is the number of aliens that would be called archaeophytes on continental Europe, those introduced before ca. 1500 AD.[16] Yet it is essential to include archaeophytes when estimating the cost of aliens as their impact is much the same as neophytes (those introduced after ca. 1500). *Aegopodium podagraria*, ground elder, and *Avena fatua*, wild oats, are two notable archaeophytes in the list of 30 species described in Section 5.4.

Most neophytes have a date when they were first introduced into the British Isles or first found outside cultivation, or both. But species first found relatively recently may still be labeled native. An example is *Gladiolus illyricus*, wild gladiolus. It is found in a few places in Hampshire in the extreme south of England, but these are nevertheless about 400 km north of mainland records. For a species with a showy flower in a county full of naturalists, the first date of 1856[17] and its disjunctive distribution suggest to me that it may well be alien. It is said that it "has the look of a genuinely wild species,[17]" but Webb[18] showed how unreliable such a criterion is.

However, even when species are clearly introduced, difficulties with the terms casual, persistent, established, and others lead to very different counts of the number of alien species. Table 5.1 gives the counts I published some years ago[7], counts that underpin the tens rule[8,19,20] that 10% of plant taxa imported into the British Isles become at least casual, while 10% of the casuals become established. Of the established, about 10% become pests, that is, economically sufficient Table 5.1 shows some of the different usages of "established"; the tens rule works with "fully established" rather than "locally established." Local floras, perhaps not surprisingly, seem generally to follow "locally established" as can be seen in Table 5.2. But the proportions in the set of county floras from the north of England are highly significantly different, showing that different standards are being used.

Various counts of the numbers of aliens in the whole British Isles are given in Table 5.2. There are three counts around 200 for fully established, going up to 945 for established in the weakest sense. That sense is in the limit of a single plant thriving: "at least one colony

**Table 5.1** The Number of British Plant Aliens by Status

| | |
|---|---|
| Severe pests | 11 |
| All pests | 39 |
| Widely naturalized | 56 |
| Fully naturalized | 196 |
| Subtotal including pests (established,[8,20] sensu Williamson and Fitter) | 210 |
| Locally naturalized (established as used in some county floras) | 348 |
| Subtotal, all above | 558 |
| Garden outcasts | 223 |
| Casuals | 898 |
| Subtotal, all above (introduced,[8,20] sensu Williamson and Fitter) | 1,642 |
| Other imports | 10,821 |
| Grand total | 12,507 |

*Source:* Data from Williamson, M., *Experientia*, 49:219, 1993.

**Table 5.2** Counts of Plant Taxa in the British Isles.[a]

| Source | Natives[a] | Established aliens | Established as % native | All aliens | All aliens as % native |
|---|---|---|---|---|---|
| **British counts** | | | | | |
| Ecological Flora Database[11,12] | 1407 | 196 | 14 | – | – |
| Williamson[7] (Table 5.1) | – | 210 or 558 | – | 1642 | – |
| Vitousek et al.[10] | 1255 | 945 | 75 | – | – |
| Alien catalogs[13,14] | – | 945 | – | 3467 | – |
| Stace[5] (Weber/Pysek count) | 1552 | 725 | 47 | – | – |
| Clapham et al.[6,7] | 1311 | 193 | 15 | – | – |
| **County counts** | | | | | |
| Cumbria[74] | 951 | – | – | 469 | 33 |
| 14 counties etc., mean[21,22] | 878.8 | – | – | 449.5 | 34 |
| *5 northern vicecounties*[29] | | | | | |
| Cheshire | 868 | 225 | 26 | 363 | 42 |
| South Lancashire | 831 | 266 | 32 | 685 | 82 |
| West Lancashire | 922 | 229 | 25 | 657 | 72 |
| Durham | 1000 | 430 | 43 | 656 | 66 |
| Northumberland | 949 | 279 | 29 | 626 | 66 |

[a] "Natives" mostly exclude hybrids and microspecies, but the usage is not consistent.

either reproducing by seed or vigorously spreading vegetatively."[13] It is fairly certain that the 745 or so species only locally or weakly established have negligible costs of any sort.

The number of casuals is far higher than that of established species, whatever criterion is used. The set of casuals includes very many garden escapes and occasional planted specimens. Some of them nevertheless have important costs, namely those that are so-called volunteers in crops. Volunteers come from previous crops on the same site. Oilseed rape *Brassica napus* and potato *Solanum tuberosum* (both hybrids as crops) rank eighth and twelfth in the herbicide costs estimated by Prus,[1] ahead of all species in the 30 interesting species considered below except for *Avenas* (wild oats) and *Veronica persica* (common field speedwell). Although the database I used in elaborating the tens rule[7] had only 1642 casual and established alien species in total, by searching for every record on single plants and other extreme casuals, the alien catalogs[13,14] raised the number to 3467. The numbers established, using whatever number you take from the previous paragraph, need to be subtracted from the total number of aliens to give the number of casuals. But, with the exception of the volunteers, the cost of these casuals will be totally negligible.

So what proportion of the British flora is alien? Lonsdale,[21] using some significantly heterogeneous data brought together by Crawley,[22] believed it was 31%, and Vitousek et al.[10] made it 43%. Using traditional figures of about 1500 native good species and about 200 fully established aliens gives 12%. Using the highest totals above, 2900 natives, including hybrids, critical and microspecies, and 3500 aliens seen in the wild since 1930, gives 55%. Chacun à son goût (to each their own). My own view is that the lowest of those three figures, 12%, gives the best feel for the noticeable impact of aliens in British vegetation. It is also quite close to the 9% that Lonsdale[21] estimated for the rest of Europe.

## 5.3   From impact to cost

The possible types of impacts of aliens and the ways in which they might be measured are both large.[2] The Lonsdale equation

$$I = R \times A \times E$$

where I is the overall impact, R is the range size, A is the abundance, and E is the effect per unit, brings some order. R and A are fairly straightforward, but E is still fairly complex. Nevertheless, the Lonsdale equation is about as complicated as present measuring techniques usually allow. It would be desirable to add the extra dimensions of species interaction, community structure, and so on, but for the present they are normally measured as the effect E of the invasive species. It is rare for the data to be good enough to measure multivariate effects, but when it is the results are interesting.[23] There are no such data available for British alien terrestrial plants as a set.

In theory, each impact could be converted into an estimate of cost or, even better, a functional relationship could be found between variation in the impact and variation in cost. Again, this is not possible with present data for most British alien plants. The Lonsdale equation does, however, allow us to say that when one of its three components is negligible, then the total impact and so the total cost will also be negligible. This simple rule, as will be seen, applies to a surprisingly large number of alien plants in Britain.

For some British aliens, I was able[3] to find five quantifiable measures of which two were the first two components of the Lonsdale equation: range and abundance. The other three related to weediness: weediness as perceived by a panel of scientists, weediness as measured by the cost of herbicides, and weediness as measured by the incidence of weeds in an agricultural survey. The correlations between these were only moderate[3] showing they were indeed measuring different aspects of impact. Different aspects of cost should, similarly, be measured by different things. How this might be done is best treated by considering individual species.

## 5.4   Thirty interesting aliens

### 5.4.1   Generalities

In order to describe in general the cost and impact of British nonindigenous plants, I have picked 30 for more detailed discussion. These are the 20 listed[22] as "The 'top twenty' British alien plant species" with 10 others, which have a major impact on some measure. Coincidentally, they include 10 that are not regarded in the alien catalog[13] as "naturalized" and another 10 that are not spreading according to the data of the *Sample Survey*.[24] The catalog[13] definition of naturalized is "Established extensively among native vegetation so as to appear native." As a first approximation, only species naturalized in that sense will have an important environmental impact, even though those not so naturalized are often conspicuous. It is more common to use "naturalized" just to mean "established,"[25] and the two usages cause some confusion. Economic impact can be important whether a plant is naturalized in the alien catalog[13] sense or not; arable weeds such as *Avena sterilis* (wild oat) and *Veronica persica* (common field speedwell) are examples of the latter.

The major and consistent estimate of cost for these 30 species are what I call the Prus cost.[1] These are stated as both the cost in pounds sterling per year and its natural logarithm, for example, 13.816 or £1 million. Prus calculated his weed cost for each species from three main variables: the value of herbicide sales, the cost of application, and the cost of cultivation. He derived these for each species by an ingenious use of government statistics; manufacturers' information; and surveys of farmers, foresters, nurserymen, and head gardeners. The result is a cost of control not just of agricultural weeds, but of all species in the British flora. Nevertheless, it is fundamentally an estimate derived from herbicide costs.

Estimates of rates of spread in the accounts below, called "sample survey estimates" and explained more fully in Section 5.5.3, are derived from comparisons of surveys in 1956–60[38] and 1987–88,[24] and are shown graphically in Figure 5.1.

### 5.4.2   Species accounts

1. *Acer pseudoplatanus*, sycamore. This is a native European tree species, and it is surprising that it failed to reach England after the last glaciation. It is often said to be a Roman introduction, but that is probably wrong. Jones[26] found records for Scotland from the fifteenth century, possibly earlier, but from England only from the sixteenth century, and it seems not to have been established in the wild before the eighteenth century. That is consistent with its absence in the archaeological record (A. Hall, pers. comm.). Now, it behaves like a native and disperses readily, though it is not spreading, having filled its range. "Ubiquitous in mixed and deciduous woodland, parkland, as a planted street tree and in shelter belts and hedgerows,"[27] "a Johnny-come-lately out nativing the natives in almost any situation, shading out native species."[17]

   However, it is difficult to estimate its impact in ways other than range. It is commonly of concern in nature reserves[3] and forestry. It is probably the main source of the "wrong sort of leaves" that delay trains in the autumn. The Prus[1] cost estimate is 10.71 = £44,802, which is very low, showing that it is usually controlled mechanically. The considerable benefits of the species should be put against all that. It forms a straight, handsome tree in exposed and polluted sites. So, it provides shelter for upland farms, near the sea, and it adorns towns and other places. Entomologists have

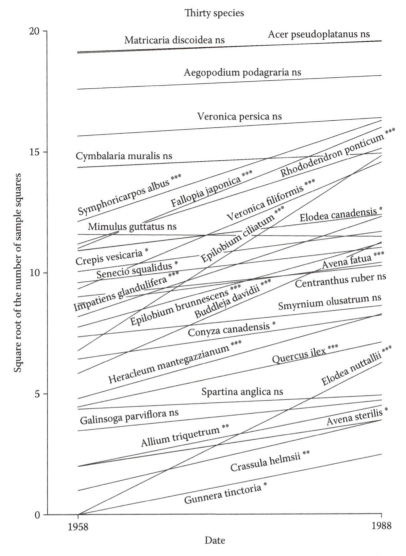

**Figure 5.1** The change in recorded number of sample hectads for the 30 species. The *Atlas* survey[38] was done between 1952 and 1960 but was more or less complete by 1958, and the *Sample Survey*[24] was done in 1987 and 1988. Note the square root scale of the ordinate.

mixed feelings about it. Any total cost figure, particularly one that allowed credit for the benefits, would, on my present information, just be a wild guess.

2. *Aegopodium podagraria*, ground elder. This perennial herb is an ancient introduction, apparently brought in by the Romans, possibly as a pot herb, possibly for medical reasons (it can be called goutweed). It is now a major garden weed in the British Isles and relatively rare away from gardens. Wilmore[27] describes those nongarden habitats: "widespread colonist of waste ground, disused gardens, roadside verges and other marginal land," not spreading, though far from ubiquitous in Scotland and Ireland. The total cost in the time and effort of gardeners must be considerable. It is often said that the only satisfactory way to get rid of bad infestations is to dig them

out completely. In practice, infestations in orchards and such places are usually left, or cut along with the grass. Small populations can be eliminated by painting glyphosate on the emerging leaves in spring; I have done this. Variegated forms are still sold to gardeners, but the benefits of this plant must be negligible in comparison. The Prus[1] cost estimate is 10.71 = £44,802 per year, which suggests a total cost of between £100,000 and £1 million per year.

3. *Allium triquetrum*, three-cornered leek or white bluebell, is a weed of rough, waste, and cultivated ground, copses, hedgerows, and waysides[5] and is a perennial herb that grows to about 45 cm. It was found almost entirely in southwest England until quite recently, but now seems to be spreading fast and diffusely to Ireland, Wales, the Isle of Man, and southern England. Although it can be quite abundant locally, the significant impact of this species is as a weed of bulb fields (of daffodils, etc.) in Cornwall and Isles of Scilly. There, it has been a serious weed since the nineteenth century. However, it is regarded as impossible to eradicate and "most islanders have abandoned any attempt at control."[28] So, its cost would have to be estimated from the loss through slower growth that it causes the growers. I have found no data on this. As a national cost, it would seem to be negligible. The Prus[1] cost estimate is 8.82 = £6768, which is small.

4. *Avena fatua*, wild oat, is an ancient invader, dating from the Bronze Age, 3000 years ago or so. Nevertheless, it is still largely a weed of lowland England, though it has been spreading for a surprisingly long time. In the northwest of England, the first historic record for Cheshire was 1805, for South Lancashire 1840, and for West Lancashire 1900,[29] despite there being Bronze Age records from Cheshire (A. Hall, pers. comm.). But it has also become more abundant in recent decades because of the difficulty of controlling a grass weed in a grass crop such as wheat with herbicides. It is a "weed of arable and waste ground and also a wool and bird-seed alien."[27] The Prus[1] cost estimate of 17.88 = £58,235,168 is by far the largest in the 30 species. But, it is not the most expensive in the list, being exceeded by *Alopecurus myosuroides*, blackgrass, and *Galium aparine*, cleavers or goosegrass, both of which are native, and *Matricaria recutita*, scented mayweed, which Prus[1] (following Webb[18]) regarded as a neophyte from the sixteenth century but which most floras call native.

5. *Avena sterilis*, winter wild oat, is a much more recent invader than *A. fatua*, introduced in the First World War, and has a much more restricted distribution in central England. It is found in similar places to *A. fatua* but usually on heavy soils and replacing *A. fatua* there.[5] Nevertheless, where it occurs, it is a major weed of cereals, giving a Prus[1] cost estimate of 16.36 = £12,736,724, which is a major national cost. It has not been found much outside crops and not in native vegetation.[14]

6. *Buddleja davidii* is usually known as buddleia though the Botanical Society[30] name is butterfly bush, a name that partly explains its popularity with gardeners. It is a shrub, to 2 m or more, with sprays (pyramidal panicles) of typically lilac flowers (also purple or white) in late summer. As a naturalized alien, it is mostly found in marginal and derelict habitats—waste ground, walls, banks, and scrub,[9] where costs and benefits may be evenly balanced. However, it is still spreading (the fourth fastest of the set of 30 in 1958–88; Figure 5.1) and may yet become an environmental threat. For instance, Everett[31] says "I am watching the march of Buddleja along the Kennet and Avon Canal, where it is ousting fen and water-margin natives such as Comfrey [*Symphytum officinale*] (foodplant of local Scarlet Tiger moths [*Callimorpha dominula*]), Meadowsweet [*Filipendula ulmaria*], and willows [*Salix* spp.] ... a stone's throw from the River Kennet proposed Special Area of Conservation" and goes on to point out

that it could fairly easily be eliminated now, but soon will not be. No action is being taken, and the species is being recommended by some conservationists for its value as a feeding source for adult butterflies—a familiar sort of story to invasion biologists, but it would be hard to claim there is an appreciable cost now. The Prus[1] cost estimate is 7.54 = £1881, which is negligible.

7. *Centranthus ruber*, or red valerian, is a garden plant, an erect perennial growing to 80 cm, which escapes to colonize walls, disused railway land, and other waste places. It is a seventeenth-century introduction, so it is not surprising that it is no longer spreading. It is unwanted in some places, leading to the Prus[1] cost estimate of 9.74 = £16,984, but its national impact and cost are trivial even though it can occur in native vegetation.[13]

8. *Conyza canadensis*, Canadian fleabane, is an annual herb introduced in the seventeenth century. It is a "quite widespread plant of urban derelict land, waste ground, disused railway land and marginal areas which seems to be increasing its range and abundance [in Yorkshire] in recent years"[27] and may possibly have been spreading nationally in 1958–88 (Figure 5.1). The Prus[1] cost estimate of 9.74 = £16,984 shows that it is sometimes unwanted. It is interesting as a plant that is more of a pest where native than where introduced.[32,33] In the British Isles, where it is not found among native vegetation[13] and is not a serious weed of cultivation, the total cost is trivial.

9. *Crassula helmsii*, New Zealand pigmyweed, is an herb grown by aquarists and discarded or planted in ponds, "well naturalized in many places in south England, rapidly spreading."[5] Clement and Foster[13] describe it as abundant and a threat (though they neglect to use the word naturalized). Its history in Britain and maps of its known distribution in 1969, 1979, 1989, and 1998 are given by Leach and Dawson.[34] Some ineffective attempts to control it by herbicide have been described.[35] The Prus[1] cost estimate is only 0.63 = £2, a derisory figure as a result of its very recent spread. It seems unlikely that control will be effective except very locally, and its cost, potentially large, should be estimated from the environmental cost of changed habitat and reductions in other species. I know of no way of doing this that I would believe.

10. *Crepis vesicaria*, beaked hawk's-beard, is an herb of grassy places, waysides, walls, and rough ground.[5] It is not in native vegetation[13] and probably no longer spreading, which is not surprising for an eighteenth-century introduction. The Prus[1] cost estimate of 9.19 = £9799 is trivial, and it is in the Crawley[22] top 20 merely because it is a commonly seen plant. There can be no appreciable national cost.

11. *Cymbalaria muralis*, ivy-leaved toadflax, is a common herb on English walls introduced in the seventeenth century. It is a "locally abundant plant of walls, disused quarry areas, builders' rubble, derelict sites, and marginal land."[27] Considering where it grows, the Prus[1] cost estimate of 9.74 = £16,984 is surprisingly high. It is not found in native vegetation and can be a pleasant adornment of walls, a minor benefit. Boyd Watt[36] gives the history of its introduction as a garden plant, noting that it is a prolific flowerer and deserves the name used in some parts of "mother of thousands." I would put its national net cost as zero.

12, 13. *Elodea canadensis*, or Canadian waterweed, and *Elodea nuttallii*, Nuttall's waterweed, are both found in streams, dykes, and canals, and other slow-moving or still water bodies.[27] The history of the spread of these two pond-weed species is given by Simpson.[37] Briefly, *E. canadensis* was first recorded in 1836, increased rapidly and often became a pest. But it declined in abundance, if not range, from the 1880s. It can still be locally abundant or dominant in some stretches[27] and is no longer spreading. *E. nuttallii* was only recorded in 1966 and is still spreading. It has often replaced *E. canadensis*, and although it can form large and extensive beds, it has rarely been regarded as a pest. The economic cost of these two species is essentially confined to the mid-nineteenth century; the present cost is negligible at a national scale. Environmentally, there may even be some benefit now from increased habitat heterogeneity and water oxygenation. The Prus[1] cost estimate for *E. canadensis* is 9.74 = £16,984, just about worth noting, but for *E. nuttallii* is only a derisory 3.09 = £22, reflecting its recent spread and confusion with *E. canadensis*. Together their total national cost must be less than £100,000.

14. *Epilobium brunnescens*, New Zealand willow herb, is a prostrate perennial herb, which (like number 29, *Veronica filiformis*) was introduced as a rock garden plant, first noted as a casual in 1908. There was confusion about its name for some time, there being many epilobia in New Zealand, and it was called *nerterioides*[15,38] and, earlier, *pedunculare*. New aliens are not infrequently difficult to identify. It is now found on "damp stony or marshy ground, often in upland terrain, as well as being a noticeable garden weed,"[27] still spreading (Figure 5.1) and occurring sometimes in natural vegetation. The Prus[1] cost estimate of 9.19 = £9799 presumably reflects its behavior in gardens. Outside, it just seems yet another minor if common addition to the flora of no consequence though of some interest. The Prus[1] cost is probably the right order of magnitude for the total cost.

15. *Epilobium ciliatum*, American willow herb, is another *Epilobium* with a changing name; it used to be called *E. adenocaulon*, and I would not be surprised if its name is changed again, as it is a member of a critical group. There seem to have been two important introductions, possibly of different genotypes (or even species). The introduction in Leicestershire before 1891 established but scarcely spread, and the introduction in Surrey was before 1930 and spread steadily[39] in all directions, including over Leicestershire. It was the fastest spreading alien in 1958–88 (Figure 5.1) and is very common in some areas. It is a perennial herb, a "weed species of disturbed ground, woodland edges, disused railway land, urban waste ground, and often frequent on damper stream or canal sides."[27] The Prus[1] cost estimate of 9.19 = £9799 is the same as that for *E. brunnescens*, but as it is a less serious garden weed and less in natural vegetation, I would put the total cost as less, even though it is the more abundant species.

16. *Fallopia japonica*, or Japanese knotweed, another perennial herb, was introduced as a garden flower and won prizes as such.[40] Nowadays, it is much disliked, even feared, particularly in cities as a "widespread aggressive colonist of waste ground, disused cemeteries, railway land, disturbed woodland herb layers and sometimes damper, rich organic soils."[27] It is also one of the two named in the Wildlife and Countryside Act of 1981; the other is *Heracleum mantegazzianum*, number 19 of this list. There is a Japanese Knotweed Alliance (JKA).[75] The problem with this plant comes from its rhizomes, which can grow to 2 m depth. They are difficult to kill by herbicide, and the plant can regenerate from small fragments (as little as 0.7 g), so digging may cause more harm than good. The Prus[1] cost estimate of 10.71 = £44,802 is not large. For the city of Swansea in Wales, JKA estimates that £1/m² for spraying glyphosate and £8/m² for landscaping would result in a bill of £9.5 million. But would any sensible authority pay that if it realized how ineffective glyphosate is with this plant? The Swansea planning department has actually spent is £140,000 over 6 years for treating established populations.[18] The Loughborough group[41,42,76] seems to me to show that endless sums can be spent on ineffective control. JKA would like to try classical biological control. This has never been used against a plant in the British Isles and would have to be extremely specific,[8] as there are closely related native species.

Although undoubtedly a major problem in some places, there are those who say the general problem with *Fallopia japonica* is exaggerated. Dickson[43] writes from personal knowledge that "Japanese Knotweed was already very common in the Glasgow area forty years ago … If it is a problem now it was a problem then," and argues against major attempts to control aliens in urban sites (and strongly for controlling aliens that may invade "vegetation of outstanding interest," cf. *Rhododendron ponticum*, number 24 of this list). Gilbert[44] finds merit in Japanese knotweed as a habitat for grass snakes (*Natrix natrix*) and otters (*Lutra lutra*) and as actually improving the habitat for spring woodland flowers in the Sheffield area.

Clearly any estimate of total cost is much affected by perception and whether the money is being well spent. My guess is that the cost of controlling it effectively, where it really needs to be controlled, could be as much as £1 million a year. It is doubtful if the cost of developing and testing biological control would be justified; better herbicide regimes[45] seem a more cost-effective and politically acceptable route.

17. *Galinsoga parviflora*, gallant soldier, is another noticeable perennial herb invader, which is a "well naturalized weed of cultivated and waste ground,"[5] a garden weed in some places. It is not a threat to seminatural vegetation and is no longer spreading. The Prus[1] cost estimate of 12.73 = £337,729 reflects its image with gardeners and seems high for another daisy-flowered weed with an amusing English name.

18. *Gunnera tinctoria*, giant rhubarb, is a spectacular herb with leaves almost 2 m across and 1.5 m high. It is "planted by lakes etc. and often self-sown where long-established; naturalized in scattered places through much of lowland British Isles."[5] The Prus[1] cost estimate of 0.63 = £2 reflects the smallness of the problem in general from this species, but it is spreading (Figure 5.1) and is a problem in some seminatural grassland, especially in the west of Ireland[46] where it can occur as stands suppressing all other plants. It is not known if control will be needed, how difficult it would be or what it would cost. I include it here as an example of the early stages of an invasive alien, which could conceivably become costly in the future.

19. *Heracleum mantegazzianum*, giant hogweed, is another impressive perennial herb with a reputation (not really deserved according to Dickson[43] but correct according to Wade et al.[47]) of causing serious dermatitis. The Prus[1] cost estimate of 9.74 = £16,984 is quite low, but this plant is one of the two named in the Wildlife and Countryside Act of 1981. It is "common along industrial river corridors and in wetland areas, also found locally along motorway verges and in waste ground and tall ruderal grassland."[27] The spread and management of this species and the next have been studied and modeled by the Durham group.[48,49] Although they conclude that successful management depends on understanding population structure and succeed in modeling such structure fairly successfully, they make no cost estimates.

20. *Impatiens glandulifera*, Himalayan balsam, is an annual herb, the tallest such in the British flora at 2 m. It is an "aggressive colonist of river and canal banks, sewage works, waste ground and damp carr woodland."[27] For its history and spread see Section 5.3. The Prus[1] cost estimate is only 9.74 = £16,984 as it is neither an agricultural weed nor a garden weed. As was noted in the previous species, management has been modeled,[48,49] but without estimating costs. As an annual species, it might be thought easy to either pull up or cut off the flowering stems, particularly as there is only a small seed bank. Most seeds germinate within a year. In practice, such measures usually only give temporary relief. Estimating the cost requires estimating the value of the biodiversity in the woodland. In some cases, it might be possible to put a value on the pheasant (*Phasanius colchicus*) shooting lost, but valuing the biodiversity

in a nature reserve such as Askham Bog near York is still an essentially subjective process. But, clearly, the cost must be several times the Prus cost, suggesting maybe £100,000–£500,000 a year, but all such figures are very foggy.

21. *Matricaria discoidea*, pineapple-weed, a small annual herb, is a "virtually ubiquitous species of waste ground, path edges, gardens, muddy gateways of arable and pasture fields, disused railway land, and marginal land and verges,"[27] but it does also occur a bit in the body of arable fields. The Prus[1] cost estimate is 13.59 = £798,108, which implies some farmers find it weedy. It is no longer spreading and is not found in native vegetation. Its characteristic habitat is bare ground unusable by other species, so to that extent it is a neutral addition to British biodiversity. It is difficult to see in what way this species can inflict a real cost of nearly a million pounds.

22. *Mimulus guttatus*, monkey flower, is a low-growing, but often prolifically flowering, perennial herb. It lives in "stream flush zones, pond edges, marshy grassland, and sometimes acidic wetland zones on moorland."[27] It seems not to threaten biodiversity or anything else, and its flowers can liven up otherwise rather drab habitats. It has completed its spread in Britain. The Prus cost estimate of 9.19 = £9799 is trivial, but there seems no reason to add to it.

23. *Quercus ilex*, evergreen oak or holm oak, is a fine tree. "Introduced; much planted for ornament, and often for shelter in east England; self-sown in south and central England, Wales, south Ireland and the Channel Islands."[5] The Prus cost estimate of 7.54 = £1881 shows that herbicide would not usually be used to control this species. As a fine tree, it brings many benefits, but it has costs too: "This species is locally becoming a threat to native vegetation"[13]; but then so are some native trees. The net cost is probably near zero, however, these effects are valued.

24. *Rhododendron ponticum*, rhododendron, is an "evergreen shrub in woodlands, ornamental parkland, and large gardens on acidic or semiacidic soils"[27] which can grow to 5 m. It has been much planted in woodland, particularly in Victorian times (nineteenth century) to give cover for pheasants (*Phasanius colchicus*) and for its profuse flowers. The British stock came primarily from southern Spain and much of it is hybrid, crossed particularly with *R. catawbiense* but also with *R. maxima* both from the Appalachians in the Unites States.[50] The hybrids may be important in allowing the taxon to thrive in harsher climates. In westerly parts of Britain, rhododendron can be a very serious problem, forming dense monocultures and shading out all other species. In the east, it is much more rarely a pest. It is still readily available from nurserymen, and there is often no reason why it should not be grown in gardens. Nevertheless, it is probably the major alien environmental weed in the British Isles.

The extent of the problem has been described[51-53] in many places. It is a problem in forestry, for national parks and conservation bodies, for the National Trust, which owns and manages buildings and land of historical and environmental importance, and for land owners in general. The Prus[1] cost estimate of 10.71 = £44,802 is a serious underestimate of the cost of rhododendron. This is because much of the control is mechanical, by either machines or hand cutting. Hand cutting may be necessary on difficult terrain and is often done by volunteers. Gritten[52] estimated the total cost of rhododendron control at £45 million in the Snowdonia National Park in North Wales. As there are less than 45,000 ha of woodland in Gwynedd[54] (the county containing the National Park), it would imply many thousand pounds per affected hectare, even allowing for some spread beyond woodlands, but it is not clear how the figure has been derived. On National Trust property, rhododendron bashing is second only to bracken bashing as hard labor by volunteers (W. Bundy, pers. comm.; bracken is the native fern *Pteridium aquilinum* and bashing means attacking in any physical way). Costing that is difficult.

One place where rhododendron is a threat to biodiversity is on the island of Lundy in the Bristol Channel. This is the only locality for the endemic Lundy cabbage, *Coincya wrightii* (Brassicaceae), whose closest relatives are in Spain.[55] Lundy cabbage is the only food plant for the flea beetle (*Psylliodes luridipennis*). *C. wrightii* is confined to 2500 m of the east coast of Lundy. Its range is restricted by grazing mammals and exposure to SW storms, so it occurs mostly on cliffs and in gullies. These are now being invaded by rhododendron, and the whole population of *C. wrightii* would probably eventually be shaded out[56] without control measures. Clearing rhododendron from cliffs is dangerous work, requiring skilled climbers and stringent safety controls. In 1997, it took 226 volunteer hours to clear 1 ha.[56] It may be possible to eliminate rhododendron from cliff sides and cliff tops with 5 m of the cliff edge by 2006 with 105 days work, or £26,880 overall[57] at commercial rates. That works out at almost £60,000 per hectare, reflecting both the difficulty of the terrain and the high cost of labor when paid for. Even so, it may be an underestimate as glyphosate, as applied, has not stopped regeneration and other herbicides have yet to be tried (S. Compton, pers. comm.).

The National Trust (for England, Wales, and Northern Ireland) and the National Trust for Scotland (NTS) have kindly provided me with some figures. In Scotland on the island of Arran, a 40-ha plot was managed at a cost, not counting volunteers, of £20,000. That is £500 per hectare with free labor. Although NTS is the largest charitable conservation organization in Scotland, it has only about 500 ha of rhododendron needing control, which leads to an estimate of £250,000 plus the value of volunteer labor, but it is total cost not annual cost.

The National Trust owns about 250,000 ha in all, of which about 25,000 ha is woodland managed by the trust. Rhododendron has been controlled on about 1000 ha in the last 10 years, though less than half of that involved dense rhododendron scrub. The cost averages around £2000–£2500 per hectare, with a range from £200 to £4000, again not including the value of volunteer labor but including both initial mechanical clearing and the labor and chemical costs of herbicide treatment of stumps and of regenerating leaves. That comes to at least £200,000/year in direct costs.

The extent of the rhododendron problem has not been quantified, so it is not possible to extrapolate from these figures to a total cost in the British Isles, either to what is or what should be spent. But clearly the figures would run into millions, if not tens of millions. Indeed, if Gritten[52] were to be believed, it would be hundreds of millions.

25. *Senecio squalidus*, Oxford ragwort, is well named as it seems it is a species that arose in Oxford Botanic Gardens. It is nonindigenous rather than an alien, as is *Spartina anglica*, number 27 of this list. Sometime in the seventeenth century, material from the hybrid swarm, on Mount Etna in Sicily, between *Senecio aethensis* and *S. chrysanthemifolius*,[58] was brought to Oxford and cultivated. By the 1790s, it was growing on walls in Oxford.[59] The current feral species "originated in cultivation"[58] and is fairly certainly the consequence of evolution and adaptation in the Botanic Gardens. Unusually for an invasive species, it is self-incompatible. Possibly the "genetic flexibility" in the system was crucial to its success[60]; only four S alleles have been found, S being the incompatibility locus.

Its spread has been rather irregular and not all that fast, partly along railways that give it suitable habitat. It is still spreading in Ireland and Scotland. Salisbury[15] claims that *squalidus* refers to the habitat, but that is not so. A common English name for it in the early nineteenth century was inelegant ragwort, which gives probably the best translation of the Latin and refers to the disposition of the ray florets: "one ca'n't help one's petals getting a little untidy."[61] The name Oxford ragwort seems to date from 1886.[62] Nowadays, it is found in "waste ground, disused railway land, canal towpaths, waysides and derelict land generally."[27] The Prus[1] cost estimate 9.19 = £9799 is surprisingly high for a species that is neither found in native vegetation nor a pest. I would be reluctant to put its cost at anything but zero. But it is a most interesting plant biologically.

26. *Smyrnium olusatrum*, alexanders, is a biennial herb, the only biennial in this list of 30 species. It is "fully naturalized on cliffs and banks, by roads and ditches and in waste places, mostly near the sea"[5] and is not spreading. Some of its habitats are natural, but it seems not to threaten biodiversity. The Prus[1] cost estimate is trivial at 7.54 = £1881 and seems a fair estimate of its cost.

27. *Spartina anglica*, common cord-grass, is a perennial grass of mud flats and is, like *Senecio squalidus* (number 25), nonindigenous but not alien. Most floras call it native, though the Joint Nature Conservation Committee editors[63] disagree, as do I. Its history is well known[8] and as stated above it is the fertile allotetraploid derived from the sterile diploid *S. townsendii* itself derived from the cross between the native *S. maritima* and the alien, American, *S. alterniflora* (which was the female[64] parent). All these are tidal mud species. *S. anglica* is useful for reclaiming mud flats and a serious problem in blocking channels. It is no longer spreading in the British Isles. Much of the present distribution comes from planting, and it is in fact declining in the south of England (A. Gray, pers. comm.). The Prus[1] cost estimate is 4.98 = £145, showing that herbicide is not used to control this species. Millions are spent controlling *S. anglica* overseas, for example in Tasmania and Washington state in the United States, but not in the British Isles. In view of its balance of costs and benefits, I would put the net cost in the British Isles as near zero.

28. *Symphoricarpos albus*, snowberry, is a low shrub, not infrequently planted for cover. It "occurs in woodland, scrub, thickets, ornamental parkland, churchyards, hedgerows, wasteland and large gardens"[27] and spreads vegetatively quite vigorously. The Prus[1] cost estimate of 4.98 = £145 shows that herbicide is not used to control this species. Indeed, it is only a problem when planted where its spread is unwanted. The cost of this species must be near zero.

29. *Veronica filiformis*, slender speedwell, was introduced as a rock garden plant and soon became invasive of lawns.[8] Some gardeners dislike it and try to control it. It was still spreading quite fast between 1958 and 1988 (Figure 5.1). In 1996, I said "Each April the lawns of the campus at the University of York turn blue with the flowers of *Veronica filiformis*,"[8] but this is no longer true. It has died back and now occurs in small patches, and that seems to be true elsewhere too. It has been found "in shorter mown grassland and verges, soft turf banks, and sometimes stream sides"[27] and, I would add, in longer grass as in my orchard. It is not found in native vegetation. The Prus[1] cost estimate is 9.74 = £16,984, which is nothing much but does show that some gardeners want pure grass lawns.

30. *Veronica persica*, common field speedwell, is a small annual herb, and this plant is a well-known and widespread agricultural weed. In Yorkshire, it is found in "arable land, waste ground, roadside verges, disused railway land and gardens."[27] It is not spreading, having reached its geographic limits, nor is it found naturalized among native vegetation.[13] The Prus[1] cost estimate is 17.41 = £36,397,112, a large figure reflecting its importance as an agricultural weed.

### 5.4.3   Roundup

Roundup is a trade name for an herbicide and so an appropriate title under which to summarize these 30 species. Remarkably, few of them are widespread and serious pests. *Avena fatua*, *A. sterilis*, and *Veronica persica* are important agricultural weeds. *Aegopodium podagraria* is a serious garden weed. *Acer pseudoplatanus*, *Impatiens glandulifera*, and *Rhododendron ponticum* can be major pests in woodland. *Fallopia japonica* and *Heracleum mantegazzianum* are the only two named in the Wildlife and Countryside Act of 1981 and are serious pests in some places, particularly riversides and urban areas. That completes the list of those that are, at present, of national importance in terms of impact and cost, just nine species.

With the Prus[1] costs, only nine again score more than 10, that is, have an estimated annual cost of more than £22,000. The bulk of the Prus costs, 98.6% of them, comes in *Avena* spp. and *Veronica persica*, the arable weeds. The total Prus cost for those three is £107 million. But, as noted under *Avena fatua*, some native species cost even more. The remaining species with a Prus cost of more than 10 are those listed in the previous paragraph less *Heracleum mantegazzianum* and *Impatiens glandulifera* but with the addition of *Matricaria discoidea* (occasionally a minor agricultural weed) and *Galinsoga parviflora* (a garden weed of restricted distribution).

Nevertheless, many are spreading quite fast. That will be quantified in Section 5.5.3. For instance *Crassula helmsii*, a weed of ponds and so on, is causing concern because of its unconstrained origin from aquarists and the difficulty of controlling it, whereas *Buddleja davidii*, at the moment one of the numerous aliens of waste and derelict land, may become an important environmental weed as it spreads. Many other species, both in the list of 30 species and in general, can be difficult pests in some circumstances. The importance of these species need to be looked at in the flora as a whole, and by comparing native and alien species, which is discussed in Section 5.5.

## 5.5   Overall estimates of impact and cost

On most measures, the cost or impact, species for species, is about the same for natives and aliens[3] in the British Isles. Here I consider in varying detail the distribution of such impacts over the British flora.

### 5.5.1   Abundance

Abundance is a basic element in the impact of any species. Unfortunately, abundance is difficult to measure with plants because of the variety of life forms and phenotypic plasticity, while vegetative reproduction can cause problems in deciding what unit to use. The use of biomass, which might seem to be the obvious common measure, has great difficulties because so much of it is underground.

The only extensive published survey that I have been able to find that relates to abundance is the one done by what was the Unit of Comparative Plant Ecology,[65] the Sheffield survey II. This survey recorded the presence and absence of each species in 1-m$^2$ quadrats. It also recorded presence in 10-cm$^2$ areas within the quadrats. That finer measure is called abundance[66] but is really gregariousness.[65,67] The 1-m$^2$ samples were taken in a way that can be "loosely described as a stratified random sampling scheme."[3]

It is well known that plotting the logarithm of abundance against the rank of the species gives a lightning strike curve. This is often called a diversity dominance curve. That such a curve is shown by the Sheffield survey (Figure 5.2) is consistent with my view

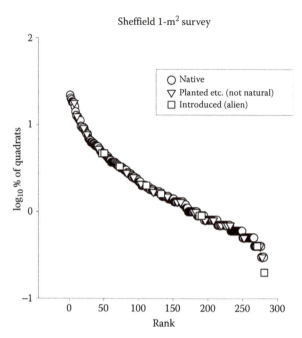

**Figure 5.2** Dominance diversity plot for 1-m² quadrat records of the Sheffield survey II,[65] showing native, planted, and introduced species simultaneously. Note the logarithmic scale of the ordinate.

that it primarily measures abundance. In Figure 5.2, I have distinguished the categories[65] native, planted, and introduced. The planted category includes both those not native to the Sheffield region but British natives and Sheffield region natives whose abundance has been increased by planting. It can be seen that all three categories follow the same distribution. There is no significant difference between them. Overall, aliens in Britain have the same abundance distribution as natives and so to that extent the same cost, species for species.

### 5.5.2  Range size

The range of aliens is the one collective character that distinguishes them from natives; aliens have, statistically, smaller range sizes.[3] This is so whether casuals are included or not, though casuals, as might be expected, have smaller range sizes than established species. When considering cost, casuals can be almost entirely disregarded.

The distribution of range sizes typically follows a logit-normal[12] distribution. When plotted as a diversity dominance curve, this usually gives a simple convex curve as can be seen for British natives in Figure 5.3. As far as I know, such a plot has not been published before and I call it an area dominance curve. It is the plot of the logarithm of the range of each species against its rank. The data in Figure 5.3 are the occurrence in hectads (10 km × 10 km grid squares) for the species in the Ecological Flora Database.[11]

It can be seen in Figure 5.3 that British aliens, nonnatives, are much less widely distributed and have a distinct turnup in the curve at the left-hand side. That is, they show a curve more like a typical abundance (diversity dominance) curve. The reason is probably that many of them are still spreading, as will be discussed in Section 5.5.3. But whatever the reason, Figure 5.3 shows that nonnative British plants have, on average, a much more

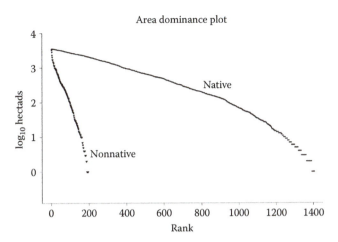

**Figure 5.3** The first published example of an area diversity plot. This is for hectad records from the Ecological Flora Database.[11] Note the logarithmic scale of the ordinate.

restricted range than natives, a lesser impact on this measure. This will tend to make their cost less than that of comparable native species.

### 5.5.3   Rate of spread

One reason why aliens are more narrowly distributed than natives could be that they are still spreading, that they have not reached their full range. This spread was shown for three *Impatiens* species by fitting logistic curves to vice-county records.[8,68] (Vice counties are subdivisions of civil counties to give roughly equal areas; they average 2200 km²). All three *Impatiens* spread from the first half of the nineteenth century and were expected to reach their full range early in the twenty-first century. If that were typical, and as most aliens were introduced in the nineteenth century or later, many aliens would still be spreading and so have misleadingly small recorded ranges. Only those introduced early or that had fast rates of spread, faster than the *Impatiens* spp., would be expected to have reached their maximum range.

A period of 100–200 years for an alien plant to reach its range limits in the British Isles is not surprising. Forest trees after the last glacial took 1000 or 2000 years.[8] The order of magnitude difference shows the magnitude and effectiveness of human dispersal, usually accidental, in spreading aliens. But the period is sufficiently long to make it difficult to compare rates of spread in different alien plant species.

There has been only one pair of extensive surveys that allow testing of whether aliens have been spreading. The pair were the distribution surveys of 1952–1960, *the Atlas*,[38] and 1987–1988 (*Sample Survey*).[24] The latter was intended partly to assess the changes in roughly 30 years and was deliberately a sample survey. Both surveys were based on the 10 km × 10 km squares or hectad of transverse Mercator grids. The first survey tried to be complete. The second was based on systematic sampling of one such square in every three in both dimensions, hence one in nine (except in coastal regions). The samples were taken systematically so were in a sense a sample survey of areas of 900 km², around 40% of the area of a vice county. Within each sampled hectad, only three tetrads (2 km × 2 km) were studied, but intensively. The effect was that the second survey was marginally the more efficient except for very rare or local species.

Unfortunately the organizers[33,69] of the *Sample Survey* and the editors of their report became overconcerned about the statistical validity of the comparison between the two surveys. There are, of course, sampling errors and biases in both, as there are in any large survey, and more effort was made to control them in the second survey than in the first. But records of readily found and recognizable taxa, the bulk of the flora, can certainly be compared. The statements "a statistical comparison is considered to be inappropriate" and "differences between the two data sets ... make valid comparisons extremely problematical"[24] seem quite unnecessarily cautious. This gloom seems to be the result of wishing both surveys to be comprehensive rather than samples. Treating both as sample surveys[70] allows many statistical comparisons.

For each of 1553 taxa (330 of them aliens), the *Sample Survey* maps show in which of the sampling survey hectads the taxon was recorded in the first (*Atlas*) survey, in the second (*Sample Survey*) survey, or in both. There are 318 such hectads in Britain, 110 in Ireland, 428 in total. From those maps, it is straightforward, if tedious, to count the records in each survey. For taxa that have not changed their distribution, if the surveys had been equally efficient, then the relationship of the two totals would not be significantly different from 1:1 and can be tested by $\chi^2$, a process familiar to all who have done some simple Mendelian breeding.

That was done for the 30 species discussed above, as was indicated in some of the accounts. The results for the set are shown in Figure 5.1, where I have dated them 1958 (when the field work was largely complete) and 1988 (when the sample survey finished). It can seen that 10 species have a nonsignificant spread, have probably not changed in range in this 30-year period, 6 are significant at the 5% level (*), 2 at the 1% (**), and 12 at the 0.1% (***). The main oddity is *Avena fatua*, an ancient invader, which apparently is spreading with *** significance. It is known to have become more common through the change of agricultural practices, and the map is perhaps better interpreted as meaning just that: a marked change in abundance leading to an increase in records.

The four that have spread fastest in this 30-year period are, in rank order, *Epilobium ciliatum*, *Heracleum mantegazzianum*, *Elodea nuttallii*, and *Buddleja davidii*, as can be seen, more or less, in Figure 5.1. But such figures cannot be used to put the species in rank order of their spreading potential. The ones that are not now spreading may include the fastest spreaders, ones that have reached their ecological limit relatively quickly. An arbitrary 30-year period for species that have been introduced at widely different times cannot be used to get comparative figures on dispersal ability. From the six most widespread but not statistically significant species, it is possible to examine the 1:1 assumption. That could be expressed as 50:50. Taking the apparent change in these six species gives 48.8 to 51.2, which is probably a measure of the efficiency of the two surveys and is so close to 50:50 as not to affect the significance levels importantly.

Thus, I would count 16 as definitely expanding their ranges between 1958 and 1988: the 11 *** species other than *Avena fatua*, both ** species and, in the * set, *Avena sterilis*, *Gunnera tinctoria* (both spreading from small ranges), and *Senecio squalidus* (clearly still spreading in Scotland and Ireland). That just leaves three that are significant at the 5% level (*), which may or may not really be spreading, as the statistical test is a crude one. They are *Crepis vesicaria*, *Conyza canadensis* (which may well be starting to invade Ireland), and *Elodea canadensis*. Out of the 26 for which I feel confident of their status, 16 (62%) are spreading. The range of the rest seems more or less static. Natives, in contrast, are both spreading and shrinking their ranges as different species respond individually to climatic and land-use changes.[71] If around three-fifths of British-established aliens are still spreading, it means that comparisons of geographic range are biased against them and that estimates of present costs underplay what the future cost will be.

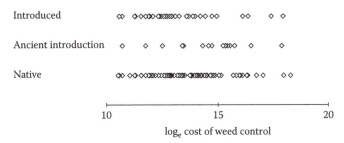

**Figure 5.4** The distribution of Prus[1] weed costs over a score of 10 for three categories of British plant. Note that the abscissa is in natural logarithms of pounds sterling.

### 5.5.4 Perceived weediness, abundance as weeds, and cost of control

I have used[3] three measures of the impact of plants as weeds: the perception of 49 species of annuals by a set of scientists[72]; the rank of incidence of dicotyledonous weeds[73] on English farms, which appears to be a measure of abundance; and an estimate of the economic cost, a weed cost, for all the British flora.[1] All three show the same major pattern when considering aliens: there are no important differences in the distribution of impact of natives and aliens. So, here I only want to present the third, the Prus[1] cost (Figure 5.4), showing the results for all species with a score more than 10, that is with a cost of more than £exp(10) = £22,026. Prus[1] modeled his full results with a three parameter function

$$p(x) = 1 - [\exp[-(x/b)^c]^d]$$

where $p(.)$ is the cumulative probability function; $x$ the weed cost of each species; and $b$, $c$, and $d$ the three parameters. This Prus function has the shape of a dominance diversity curve. The exponential nature of this function, or equivalently the logarithmic abscissa of Figure 5.4, explains why so few species contribute nearly all the cost, as was noted above in Sections 5.4.2 4 *Avena fatua* and 5.4.3.

## 5.6 Conclusion

The impact of British nonindigenous plants or aliens is, species for species, much the same as that of British natives when the impact is measured by abundance or weediness. The range of British aliens is, as a statistical distribution, less than that of natives, but this is partly because many aliens are still spreading and have yet to reach the limits of their distribution. Costing these impacts is difficult, but there seems little doubt that major costs come from nine or fewer species. The weed control costs of Prus come to more than £100 million, these being essentially agricultural costs. The environmental costs are very much more uncertain, but seem likely to be less. This suggests a total cost of aliens in the British isles of £200 million to £300 million. Other indications are that the adverse costs of native species are around twice that. Aliens are costly and natives are more so.

## Acknowledgments

I am very grateful to Richard Abbott (University of St Andrews), Humphry Bowen (Dorset), Wendy Bunny (National Trust), Steve Compton (Leeds University), Alastair Fitter (University of York), Allan Hall (University of York), John Harvey (National Trust),

Ray Hawes (National Trust), Stephen Jury (University of Reading), Duncan Stevenson (National Trust for Scotland), and Michael Usher (Scottish National Heritage) for much advice and information.

## References

1. Prus, J. L. 1996. *New Methods of Risk Assessment for the Release of Transgenic Plants.* PhD thesis, Cranfield University, Cranfield, Bedfordshire, UK.
2. Parker, I. M., D. Simberloff, W. M. Lonsdale, K. Goodell, M. Wonham, P. M. Kareiva, M. Williamson et al. 1999. Impact: Toward a framework for understanding the ecological effects of invaders. *Biol Invasions* 1:3.
3. Williamson, M. 1998. Measuring the impact of plant invaders in Britain. In *Plant Invasions: Ecological Mechanisms and Human Responses,* ed. U. Starfinger, K. Edwards, I. Kowarik, and M. Williamson, 57. Leiden: Backhuys.
4. Perrings, C., M. Williamson, and S. Dalmazzone, eds. 2000. *The Economics of Biological Invasions.* Cheltenham: Edward Elgar.
5. Stace, C. 1997. *New Flora of the British Isles.* 2nd ed. Cambridge: Cambridge University Press.
6. Clapham, A. R., T. G. Tutin, and D. M. Moore. 1987. *Flora of the British Isles.* 3rd ed. Cambridge: Cambridge University Press.
7. Williamson, M. 1993. Invaders, weeds and the risk from genetically manipulated organisms. *Experientia* 49:219.
8. Williamson, M. 1996. *Biological Invasions.* London: Chapman & Hall.
9. Stace, C. 1991. *New Flora of the British Isles.* Cambridge: Cambridge University Press.
10. Vitousek, P. M., C. M. D'Antonio, L. L. Loope, M. Rejmánek, and R. Westbrooks. 1997. Introduced species: A significant component of human-caused global change. *N Z J Ecol* 21:1.
11. Fitter, A. H., and H. J. Peat. 1994. The ecological flora database. *J Ecol* 82:415.
12. Williamson, M., and K. J. Gaston. 1999. A simple transformation for sets of range sizes. *Ecography* 22:674.
13. Clement, E. J., and M. C. Foster. 1994. *Alien Plants of the British Isles.* London: Botanical Society of the British Isles.
14. Ryves, T. B., E. J. Clement, and M. C. Foster. 1996. *Alien Grasses of the British Isles.* London: Botanical society of the British Isles.
15. Salisbury, E. 1961. *Weeds and Aliens.* New Naturalist Series Volume 43. London: Collins.
16. Pyšek, P. 1995. On the terminology used in plant invasion studies. In *Plant Invasions: General Aspects and Special Problems,* ed. P. Pyšek, K. Prach, M. Rejmánek, and M. Wade, 71. Amsterdam: SPB Academic Publishing.
17. Mabey, R. 1996. *Flora Britannica.* London: Sinclair-Stephenson.
18. Webb, D. A. 1985. What are the criteria for presuming native status? *Watsonia* 15:231.
19. Williamson, M., and K. C. Brown. 1986. The analysis and modelling of British invasions. *Philos Trans R Soc B* 314:505.
20. Williamson, M., and A. Fitter. 1996. The varying success of invaders. *Ecology* 77:1661.
21. Lonsdale, W. M. 1999. Global patterns of plant invasions and the concept of invisibility. *Ecology* 80:1522.
22. Crawley, M. J. 1987. What makes a community invasible? *Symp Br Ecol Soc* 26:429.
23. Williamson, M. 1987. Are communities ever stable? *Symp Br Ecol Soc* 26:353.
24. Palmer, M. A., and J. H. Bratton, eds. 1995. *A Sample Survey of the Flora of Britain and Ireland.* UK Nature Conservation 8. Based on a 1990 report for the Nature Conservancy Council by TCG Rich and ER Woodruff. Peterborough: Joint Nature Conservation Committee.
25. Richardson, D. M., P. Pyšek, M. Rejmánek, M. G. Barbour, F. D. Panetta, and C. J. West. 2000. Naturalization and invasion of alien plants: Concepts and definitions. *Divers Distrib* 6:93.
26. Jones, E. W. 1945. Biological flora of the British Isles. *J Ecol* 32:215.
27. Wilmore, G. T. D. 2000. *Alien Plants of Yorkshire.* Doncaster: Yorkshire Naturalists Union.
28. Lousley, J. E. 1971. *The Flora of the Isles of Scilly.* Newton Abbott: David & Charles.

29. Greenwood, E. F. 1999. Vascular plants: A game of chance? In *Ecology and Landscape Development: A History of the Mersey Basin*, ed. E. F. Greenwood, 195. Liverpool: Liverpool University Press.

30. Dony, J. G., S. L. Jury, and F. H. Perring. 1986. *English Names of Wild Flowers*. 2nd ed. London: Botanical Society of the British Isles.

31. Everett, S. 2000. Conservation news: Introductions and genetic conservation. *Br Wildl* 11:450.

32. Crompton, C. W., J. McNeill, A. E. Stahevitch, and W. A. Wojtas. 1988. Preliminary inventory of Canadian weeds, Agriculture Canada, Ottawa. *Tech Bull* 1988-9E.

33. Rich, T. C. G. 1998. Squaring the circles: Bias in distribution maps. *Br Wildl* 9:213.

34. Leach, J., and H. Dawson. 1999. *Crassula helmsii* in the British Isles: An unwelcome invader. *Br Wildl* 10:234.

35. Child, L. E., and D. Spencer-Jones. 1995. Treatment of *Crassula helmsii*: A case study. In *Plant Invasions: General Aspects and Special Problems*, ed. P. Pyšek, K. Prach, M. Rejmánek, and M. Wade, 195. Amsterdam: SPB Academic Publishing.

36. Boyd Watt, H. 1957. Notes on the introduction and distribution of Cymbalaria muralis Gaertn., Mey. & Scherb. in Scotland (written ca. 1932). *Proc Bot Soc Br Isles* 2:123.

37. Simpson, D. A. 1984. A short history of the introduction and spread of *Elodea* Michx in the British Isles. *Watsonia* 15:1.

38. Perring, F. H., and S. M. Walters. 1962. *Atlas of the British Flora*. London & Edinburgh: Thomas Nelson and Sons.

39. Preston, C. D. 1988. The spread of *Epilobium ciliatum* Raf. in the British Isles. *Watsonia* 17:279.

40. Bailey, J. P., and A. P. Conolly. 2000. Prize-winners to pariahs: A history of Japanese knotweed s. l. (Polygonaceae) in the British Isles. *Watsonia* 23:93.

41. Brock, J. H., L. E. Child, L. C. de Waal, and M. Wade. 1995. The invasive nature of *Fallopia japonica* is enhanced by vegetative regeneration from stem tissues. In *Plant Invasions: General Aspects and Special Problems*, ed. P. Pyšek, K. Prach, M. Rejmánek, and M. Wade, 131. Amsterdam: SPB Academic Publishing.

42. de Waal, L. C. 1995. Treatment of *Fallopia japonica* near water — a case study. In *Plant Invasions: General Aspects and Special Problems*, ed. P. Pyšek, K. Prach, M. Rejmánek, and M. Wade, 203. Amsterdam: SPB Academic Publishing.

43. Dickson, J. H. 1998. Plant introductions in Scotland. In *Species History in Scotland*, ed. R. A. Lambert, 38. Edinburgh: Scottish Cultural Press.

44. Gilbert, O. 1994. Japanese knotweed—what problem? *Urban Wildl News* 11(3):1.

45. Green, D. Japanese knotweed, Letter, *The Times*, July 1, 2000.

46. Hickey, B., and B. Osborne. 1998. Effect of *Gunnera tinctoria* (Molina) Mirbel on semi-natural grassland habitats in the west of Ireland. In *Plant Invasions: Ecological Mechanisms and Human Responses*, ed. U. Starfinger, K. Edwards, I. Kowarik, and M. Williamson, 195. Leiden: Backhuys.

47. Wade, M., E. J. Darby, A. D. Courtney, and J. M. Caffrey. 1997. *Heracleum mantegazzianum*: A problem for river managers in the Republic of Ireland and the United Kingdom. In *Plant Invasions: Studies from North America and Europe*, ed. J. H. Brock, M. Wade, P. Pyšek, and D. Green, 139. Leiden: Backhuys.

48. Collingham, Y. C., R. A. Wadsworth, B. Huntley, and P. E. Hulme. 2000. Predicting the spatial distribution of non-indigenous riparian weeds: Issues of spatial scale and extent. *J Appl Ecol* 37(Suppl. 1):13.

49. Wadsworth, R. A., Y. C. Collingham, S. G. Willis, B. Huntley, and P. E. Hulme. 2000. Simulating the spread and management of alien riparian weeds: Are they out of control? *J Appl Ecol* 37(Suppl. 1):28.

50. Milne, R. I., and R. J. Abbott. 2000. Origin and evolution of invasive naturalized material of *Rhododendron ponticum* L. in the British Isles. *Mol Ecol* 9:541.

51. Cross, J. R. 1975. Rhododendron ponticum L. (biological flora of the British Isles). *J Ecol* 63:345.

52. Gritten, R. H. 1995. *Rhododendron ponticum* and some other invasive plants in the Snowdonia National Park. In *Plant Invasions: General Aspects and Special Problems*, ed. P. Pyšek, K. Prach, M. Rejmánek, and M. Wade, 213. Amsterdam: SPB Academic Publishing.

53. Usher, M. B. 1986. Invasibility and wildlife conservation: Invasive species on nature reserves. *Philos Trans R Soc B* 314:695.

54. Locke, G. M. L. 1987. Census of woodlands and trees 1979–82, HMSO, London. *For Comm Bull* 63.

55. Compton, S. G., and R. S. Key. 2000. *Coincya wrightii* (O. E. Schultze), Stace (*Rhyncosinapis wrightii* (O. E. Schultze), Dandy ex A. R. Clapham), biological flora of the British Isles. *J Ecol* 88:535.

56. Compton, S. G., R. S. Key, R. J. D. Key, and E. Parkes. 1998. Control of *Rhododendron ponticum* on Lundy in relation to the conservation of the endemic plant Lundy cabbage *Coincya wrigtii*. *English Nat Res Rep* 263:1.

57. Compton, S. G., and R. S. Key. 1998. *Species Action Plan: Lundy Cabbage (Coincya wrightii) and Its Associated Insects*. Peterborough: English Nature.

58. Abbott, R. J., J. K. James, J. A. Irwin, and H. P. Comes. 2000. Hybrid origin of the Oxford Ragwort, *Senecio squalidus* L. *Watsonia* 23:123.

59. Kent, D. H. 1956. Senecio squalidus L. in the British Isles: 1, early records (to 1877). *Proc Bot Soc Br Isles* 2:115.

60. Hiscock, S. J. 2000. Genetic control of self-incompatibility in *Senecio squalidus* L. (Asteraceae): A successful colonizing species. *Heredity* 85:10.

61. Carroll, L. 1872. *Through the Looking-Glass and What Alice Found There*. London: Macmillan.

62. Druce, G. C. 1886. *The Flora of Oxfordshire*. Oxford: Parker.

63. Eno, N. C., R. A. Clark, and W. G. Sanderson, eds. 1997. *Non-Native Marine Species in British Waters: A Review and Directory*. Peterborough: Joint Nature Conservation Committee.

64. Ferris, C., R. A. King, and A. J. Gray. 1997. Molecular evidence for the maternal parentage in the hybrid origin of Spartina anglica. *Mol Ecol* 6:185.

65. Grime, J. P., J. G. Hodgson, and R. Hunt. 1988. *Comparative Plant Ecology*. London: Unwin Hyman.

66. Thompson, K., J. G. Hodgson, and K. J. Gaston. 1998. Abundance-range size relationships in the herbaceous flora of central England. *J Ecol* 86:439.

67. Williamson, M. 2001. Can the impacts of invasive species be predicted? In *Plant Invasions: Species Ecology and Ecosystem Management*, ed. G. Brundi, J. Brock, I. Camarda, L. Child, and M. Wade. Leiden: Backhuys.

68. Perrins, J., A. Fitter, and M. Williamson. 1993. Population biology and rates of invasion of three introduced Impatiens species in the British Isles. *J Biogeogr* 20:33.

69. Rich, T. C. G., and E. R. Woodruff. 1992. Recording bias in botanical surveys. *Watsonia* 19:73.

70. Le Duc, M. G., M. O. Hill, and T. H. Sparks. 1992. A method for predicting the probability of species occurrence using data from systematic surveys. *Watsonia* 19:97.

71. Thompson, K. 1994. Predicting the fate of temperate species in response to human disturbance and global change. In *Biodiversity, Temperate Ecosystems, and Global Change*, ed. T. J. B. Boyle, and C. E. B. Boyle, 61. Berlin: Springer-Verlag.

72. Perrins, J., M. Williamson, and A. Fitter. 1992. A survey of differing views of weed classification: Implications for regulation of introductions. *Biol Conserv* 60:47.

73. Schering Agriculture. 1986. *Weed Guide*. revised ed. Nottingham: Schering Agriculture.

74. Halliday, G. 1997. *A Flora of Cumbria*. Lancaster: Centre for North-West Regional Studies, University of Lancaster.

75. Cabi-bioscience.org. Sponsored listings for Japanese Knotweed Alliance http://www.cabi-bioscience.org/html/japanese_knotweed_alliance.htm (accessed May 13, 2010).

76. Child, L., M. Wade, and M. Wagner. 1998. Cost effective control of *Fallopia japonica* using combination treatments. In *Plant Invasions: Ecological Mechanisms and Human Responses*, ed. U. Starfinger, K. Edwards, I. Kowarik, and M. Williamson, 143. Leiden: Backhuys.

# chapter six

# Economic, environmental, and social dimensions of alien vertebrate species in Britain

*Piran C. L. White, Adriana E. S. Ford-Thompson,*
*Carolyn J. Snell, and Stephen Harris*

## Contents

## 6.1   Alien species, alien populations, and the process of invasions

Alien or introduced species are nonindigenous species that have been imported, bred, and become established in a particular region, either accidentally or deliberately. There are a variety of reasons for introduction of species to a new area, such as sport (shooting, fishing, and hunting), amenity or ornament, food, domestication as pets, or importation for utilitarian purposes (livestock, fur, and food). Manchester and Bullock provide examples of reasons for the introduction of selected alien species into the United Kingdom.[1] Whereas most invertebrate and microbe introductions worldwide have been accidental, most vertebrate and plant introductions have been intentional. For vertebrates, there are a few exceptions to this generalization, most notably for commensal rodents and where species have been imported initially to satisfy the demands for utilitarian or ornamental purposes.

Although some vertebrate species were introduced into Britain as early as the Iron or Bronze Ages, the majority of vertebrate introductions occurred during the late nineteenth and early twentieth centuries. At this time, there was a considerable interest in and fashion for "acclimatization," the history of which has been documented by Lever.[2]

There have been many attempts to understand the process of invasion by alien species, focusing on determinants of invasion success, the rate of spread of alien species, and the susceptibility of different environments to invasion. Williamson[3] summarized the various issues surrounding biological invasions as a conceptual framework, in which he separated the invasion process into four stages: (1) arrival and establishment, (2) spread, (3) equilibrium and effects, and (4) implications. In this chapter, we are concerned primarily with the third and fourth stages of the process, although, as will be illustrated in the examples later on, these are affected considerably by the earlier stages. The effect of invasion pressure (a product of the number of individuals being introduced and the number of introductions) on the likelihood of establishment is of particular relevance, as are population parameters such as the intrinsic rate of increase and dispersal ability, and in some cases, climatic or habitat matching.

Two general rules that have emerged from the empirical observations of invasions by alien vertebrates are as follows: (1) islands are more susceptible to successful invasion than continental regions; and (2) simple communities with fewer species are more susceptible than more diverse communities.[4] Britain is an island with a relatively low-diversity vertebrate fauna and should therefore be more susceptible to successful invasions by vertebrates than many other countries. Invasions may be the result of natural processes (e.g., range expansion, such as that by the collared dove, *Streptopelia decaocto*, across Europe from the 1930s), and such invasions may become more frequent as a result of climate change,[5] but it is deliberate or accidental introductions by humans that concern us here.

For some indigenous species, their numbers in a particular region may have been enhanced, or perhaps replaced altogether, by translocations of nonindigenous individuals or populations. This may be due to accidental escapes or deliberate releases from animals in captivity or it may be due to deliberate reinforcement for the purposes of nature conservation. For the red squirrel (*Sciurus vulgaris*), red kite (*Milvus milvus*), white-tailed eagle (*Haliaeetus albicilla*), goshawk (*Accipiter gentilis*), and capercaillie (*Tetrao urogallus*) in Britain, this has been done primarily for nature conservation purposes. For the goshawk and the capercaillie, additional reasons for enhancement were falconry and sport shooting, respectively. The capercaillie actually became extinct in the second half of the eighteenth century but was re-established successfully during the nineteenth century.[6] The goshawk and the white-tailed eagle also show the same pattern of extinction followed by reestablishment, and the populations of all three species now extant in Britain are therefore completely distinct from the original native wild stock. The same is true for the reindeer (*Rangifer tarandus*) population on the Cairngorm plateau in Scotland. Other examples of the influence of introduced animals on native stocks are the release of 400,000 mallard each year for shooting[7] and the reinforcement of Atlantic salmon (*Salmo salar*) and brown trout (*Salmo trutta*) for fisheries purposes.

Although the definition of an alien species per se is relatively clear-cut, the influence of alien animals may extend to populations of native species. Introductions of new genes are likely to have a significant impact on the genetic diversity of the native fauna, which is therefore a very important conservation issue in its own right. However, for the purposes of this chapter, we will confine ourselves to discussions of the impacts of alien species per se.

Most impacts of alien species are negative, and these can be grouped into five main categories: (1) consumption of other species via predation or herbivory; (2) competition with other species; (3) introduction or maintenance of disease; (4) interbreeding with native populations or species; and (5) disturbance of the environment (physical or chemical). However, in a few cases in Britain, alien species can also bring benefits. Impacts may have environmental, economic, or social dimensions, although the social dimensions of invasive species have received relatively less attention.[8]

In this chapter, rather than focusing on individual case studies, we will consider the environmental, economic, and social dimensions of invasive species in Britain in turn and illustrate each dimension with reference to particular introductions. In Section 6.2, we will first provide a species-based overview of the vertebrate introductions that have occurred in Britain. Where we use the term "introduced species," we take this to be the same as "alien species," and where appropriate we distinguish "introduced populations" in the same way, that is, populations of native species that are in fact made up of introduced individuals. When we refer to Britain, we mean the mainland and surrounding small islands, but we exclude the Isle of Man, the Channel Islands, and Ireland. We also restrict our account to "land" vertebrates, by which we mean those animals that spend at least part of their life on the land surface above mean sea level. This, therefore, excludes the cetaceans that inhabit British waters and the leatherback turtle (*Dermochelys coriacea*), which is a seasonal visitor to British waters in the North Atlantic, where it feeds almost exclusively on jellyfish.[9] It also excludes all entirely marine species of fish. However, it includes seals and fish that migrate between fresh and marine waters.

## 6.2 Overview of alien vertebrate introductions in Britain

### 6.2.1 Mammals

A total of 22 mammal species that have been introduced and bred in Britain are currently extant in the wild (Table 6.1), although one of these, the red-necked wallaby (*Macropus rufogriseus*), is on the verge of extinction. A further eight formerly introduced species are

Table 6.1 History, Population Status, and Significant Environmental and Economic Costs of Extant Introduced Vertebrates in Britain

| Common name | Scientific name | Date of introduction | Population estimate | Population change | Reference(s) for population data | Economic costs | Environmental costs |
|---|---|---|---|---|---|---|---|
| **Mammals** | | | | | | | |
| Red-necked wallaby | Macropus rufogriseus | 1850s on | 1 | −2 | 10 | | |
| Lesser white-toothed shrew | Crocidura suaveolens | Iron Age or earlier | 99,000 | 0 | 10 | | |
| Rabbit | Oryctolagus cuniculus | Norman (1066–1154) | 37,500,000 | 2 | 11 | **Fh E** | **Fh** |
| Brown hare | Lepus europaeus | Roman | 820,000–2,000,000 | 0 | 10 | | Fh |
| Gray squirrel | Sciurus carolinensis | 1876–1930 | 2,520,000 | 2 | 11 | **Fh** E | Fp C |
| Orkney and Guernsey voles | Microtus arvalis | Neolithic/Bronze Age | 4,000,000 | −2 | 10 | | |
| Harvest mouse | Micromys minutus | Postglacial | 1,425,000 | −2 | 11 | | |
| House mouse | Mus domesticus | Iron Age or earlier | 5,192,000 | −2 | 11 | **Fh D E** | |
| Common rat | Rattus norvegicus | 1728–29 | 6,790,000 | −2 | 11 | **Fh D** E | |
| Ship rat | Rattus rattus | Roman (3 AD) | 1,300 | −2 | 11 | **D** | |
| Edible dormouse | Glis glis | 1902 | 10,000 | 1 | 11 | Fh E | |
| Feral ferret | Mustela furo | Norman or fourteenth century | 2,500 | 0 | 11 | D | **H** |
| Mink | Mustela vison | 1930s | 110,000 | −2 | 11 | Fh D | **Fp** |
| Feral cat | Felis catus | Norman | 813,000 | 0 | 11 | Fh D | **Fp H** |
| Sika deer | Cervus nippon | 1860s on | 20,000 | 2 | 10 | **Fh** D | Fh H |

| Common name | Species | Date | Number | | Ref. | Fh | Fh D E | Fh |
|---|---|---|---|---|---|---|---|---|
| Fallow deer | *Dama dama* | Roman/Norman | 100,000 | 1 | 11 | | | Fh |
| Reeves' muntjac | *Muntiacus reevesi* | Early 1900s | 100,000 | 2 | 10 | | | Fh |
| Chinese water deer | *Hydropotes inermis* | 1915 | 1,500 | 1 | 10 | | | |
| Père David's deer | *Elaphurus davidianus* | 1963 on | 30 | 0 | 10 | | | Fh |
| Feral goat | *Capra hircus* | Neolithic | 3,565 | 0 | 11 | | | Fh |
| Feral sheep | *Ovis aries* | Neolithic | 5,000 | 0 | 10 | | | |
| Wild boar | *Sus scrofa* | 1800s on | 500 | 1 | 10 | | E | Fh E |
| **Birds** | | | | | | | | |
| Night heron | *Nycticorax nycticorax* | 1868 on | <50 | 0 | 12 | | | |
| Black swan | *Cygnus atratus* | | <30 | 0 | 12 | | | |
| Pink-footed goose | *Anser brachyrhynchus* | | <100 | 0 | 12 | | | |
| White-fronted goose | *Anser albifrons* | | <100 | 0 | 12 | | | |
| Bar-headed goose | *Anser indicus* | | 85 | 0 | 13 | | | |
| Snow goose | *Anser caerulescens* | | <200 | 0 | 12, 13 | | | |
| Canada goose | *Branta canadensis* | Early 1700s | 59,500 | 2 | 17 | | **Fh E** | Fh |
| Egyptian goose | *Alopochen aegyptiacus* | Late 1700s on | 906 | 2 | 13, 17 | | | |
| Muscovy duck | *Cairina moschata* | 1980s | <300 | 0 | 17 | | | |
| Wood duck | *Aix sponsa* | 1870s | <100 | 2 | 12, 17 | | | |
| Mandarin duck | *Aix galericulata* | 1745 | 7,000 | 2 | 17 | | | |
| Red-crested pochard | *Netta rufina* | 1937 on | <100 | 2 | 17 | | | |
| Ruddy duck | *Oxyura jamaicensis* | 1950s | 3,625 | 2 | 12, 17 | | | H |
| Red-legged partridge | *Alectoris rufa* | 1770 on | ca. 350,000a | 2 | 14, 17 | | | |

*(Continued)*

**Table 6.1** History, Population Status, and Significant Environmental and Economic Costs of Extant Introduced Vertebrates in Britain (*Continued*)

| Common name | Scientific name | Date of introduction | Population estimate | Population change | Reference(s) for population data | Economic costs | Environmental costs |
|---|---|---|---|---|---|---|---|
| Pheasant | *Phasianus colchicus* | 1040s on | 3,100,000[b] | 0 | 17 | | **H** |
| Golden pheasant | *Chrysolophus pictus* | 1890s on | 1,500 | 2 | 17 | | H |
| Lady Amherst's pheasant | *Chrysolophus amherstiae* | 1828 on | 150 | −2 | 17 | | |
| Ring-necked parakeet | *Psittacula krameri* | 1969 on | 20,000 | 2 | 15 | | |
| Monk parakeet | *Myiopsitta monachus* | 1899 on | 50 | 0 | 12 | | |
| Eagle owl | *Bubo bubo* | | <10 | 0 | 12 | | |
| Little owl | *Athene noctua* | 1842 and 1870s on | 18,000 | −1 | 17 | | |
| **Reptiles** | | | | | | | |
| Red-eared terrapin | *Trachemys scripta elegans* | 1990s | <100 | 0 | 9 | | |
| Wall lizard | *Podarcis muralis* | 1930s | 800 | 1 | 9 | | |
| Aesculapian snake | *Elaphe longissima* | 1970 | <100 | 0 | 9 | | |
| **Amphibians** | | | | | | | |
| Midwife toad | *Alytes obstetricans* | 1900s on | <3,000 | 0 | 9 | | |
| Edible frog | *Rana esculenta* | 1837 on | 15,000 | 1 | T. Langton, pers. comm. | | H |
| Marsh frog | *Rana ridibunda* | 1935 on | 15,000 | 1 | T. Langton, pers. comm. | | **Fp** H |
| Alpine newt | *Triturus alpestris* | 1920s | <1,000 | 0 | 9 | | |
| African clawed toad | *Xenopus laevis* | 1960 on | <500 | 0 | 9 | | |

| Common name | Scientific name | Introduced | Estimated number | Change | Ref. | | |
|---|---|---|---|---|---|---|---|
| Italian crested newt | *Triturus carnifex* | 1960s | <2,000 | 0 | 9 | | H |
| Yellow-bellied toad | *Bombina variegata* | 1954 | <200 | 0 | 9 | | |
| Bullfrog | *Rana catesbeiana* | 1980s | <200 | 1 | 9 | | |
| **Fish** | | | | | | | |
| Rainbow trout | *Oncorhynchus mykiss* | | <10 | 1 | 14 | | **Fp** |
| Pink salmon | *Oncorhynchus gorbuscha* | | V | −1 | 14 | | |
| Brook charr | *Salvelinus fontinalis* | | <10 | 1 | 14 | | H |
| Carp | *Cyprinus carpio* | | >30 | 1 | 14 | E | **Fp** H |
| Sunbleak | *Leucaspius delineatus* | 1990 | <10 | 1 | 15 | | |
| Goldfish | *Carassius auratus* | | >30 | 1 | 14 | | |
| Bitterling | *Rhodeus sericeus* | | <30 | 1 | 14 | | |
| Ike | *Leuciscus idus* | | <30 | 1 | 14 | | H |
| Wels | *Silurus glanis* | | <10 | 1 | 14 | Fp | Fp |
| Large-mouth bass | *Micropterus salmoides* | | <30 | −1 | 14 | | |
| Pumpkinseed | *Lepomis gibbosus* | | <30 | 1 | 14 | | Fp |
| Rock bass | *Ambloplites rupestris* | | <30 | −1 | 14 | | |
| Zander | *Stizostedion lucioperca* | 1878 | >30 | 1 | 14 | **Fp** | **Fp** |

*Note:* Environmental and economic costs represented as follows: Fh, feeding (herbivory); Fp, feeding (predation); C, competition; D, maintenance or introduction of disease; H, hybridization; E, environmental disturbance. Significant impacts are highlighted in bold. For mammals, birds, reptiles, and amphibians, population estimates are the estimated number of animals at the start of the breeding season. For reptiles and amphibians, figures are our estimates based on information provided in the references listed. For fish, population estimates are the number of stocks. V, vagrant for mammals, amphibians, and reptiles, population changes in terms of numbers and/or range represented as follows: 2, strong evidence of increase; 1, suggestions of increase; 0, probably stable; −1, suggestions of decrease; −2, strong evidence of decrease. For birds, population changes in terms of numbers and/or range represented as follows: 2, >25% increase; 1, 10%–25% increase; 0, increase or decrease of <10%; −1, 10%–25% decrease; −2, >25% decrease. For fish, population changes in terms of numbers and/or range represented as follows: 1, evidence of increase; 0, probably stable; −1, evidence of decrease.

a  2 million captive-bred birds released each year for shooting.[14]

b  25 million captive-bred birds released each year for shooting.[56]

**Table 6.2** Number of Extant Native and Introduced Species in Britain of the Different Vertebrate Groups

| Vertebrate group | Total number of species | Number of introduced species | Introduced species as percentage of total species |
|---|---|---|---|
| Mammals | | | |
| All species | 65 | 22 | 33.8 |
| Terrestrial species | 48 | 22 | 45.8 |
| Birds | 219 | 21 | 9.6 |
| Reptiles | 9 | 3 | 33.3 |
| Amphibians | 14 | 8 | 57.1 |
| Fish | 56 | 13 | 23.2 |

*Note:* Sources of reference are given in the text and Table 6.1. Terrestrial species of mammal exclude seals and bats.

now extinct. These figures do not include vagrant species. A total of 64 mammal species exist in breeding populations in Britain at the moment, and introduced species therefore represent 34% of the current mammal fauna in terms of species richness. If the terrestrial mammals only are included (i.e., excluding 2 species of seal and 15 species of bat), introduced species account for 46% of mammals currently extant in Britain (Table 6.2).

Some mammal species were introduced in the Iron and Bronze Ages and the Neolithic period, and a few species were introduced by the Romans and the Normans. Introductions of mammals at these times were either accidental, for example the house mouse (*Mus domesticus*) and common rat (*Rattus norvegicus*), or deliberate, for example, the introduced animals being used for wool and meat. This was the case for feral sheep (*Ovis aries*) and feral goats (*Capra hircus*) during the Neolithic period and rabbits (*Oryctolagus cuniculus*) in Norman times. However, rabbits were initially farmed in warrens, and substantial increases in wild populations did not occur until the mid-eighteenth century in England and not until the early nineteenth century in parts of Scotland and Wales.[11]

The majority of mammalian introductions occurred in the late nineteenth and early twentieth centuries, as part of the vogue for acclimatization. Many wild populations of alien species originated from escapes. In the majority of cases, these were escapes of species originally imported for ornamental purposes on private estates, for example Reeves' muntjac (*Muntiacus reevesi*), sika deer (*Cervus nippon*), fallow deer (*Dama dama*), Chinese water deer (*Hydropotes inermis*), and the red-necked wallaby (*Macropus rufogriseus*), which is now believed to be on the verge of extinction, if not extinct in the wild. In some cases, escapes were of species originally imported for fur, for example mink (*Mustela vison*) and muskrat (*Ondatra zibethica*). Finally, a few species were liberated deliberately into the wild, for example gray squirrel (*Sciurus carolinensis*) and edible dormouse (*Glis glis*).

## 6.2.2　Birds

The categorization of birds as native or introduced species is more difficult than that of mammals due to their greater mobility and therefore uncertainty about what constitutes their "native" range. The situation is complicated further because many native populations, especially waterfowl, have been reinforced by introduced individuals. The British Ornithologists' Union Records Committee's (BOURC) British List does not classify species as "native" or "introduced," instead referring to species occurring in an "apparently natural state" (category A) and species that have derived populations from introduced stock (category C).[16] Species can therefore occur in more than one category in the British List. For the purposes of this chapter, we have included those species for which the bulk of the population stems from introduced animals and excluded those species that are commonly regarded as native, although they may also have been reinforced by some introduced animals. We have also excluded any feral domesticated species that are originally descended from native wild stock (e.g., feral pigeon or rock dove, *Columba livia*) and those species for which there is no current evidence of breeding.

A total of 21 breeding bird species presently extant in Britain are introduced (Table 6.1), with at least another four previously introduced species now extinct. One other introduced species, Lady Amherst's pheasant (*Chrysolophus amherstiae*), is on the verge of extinction in the wild. A total of 219 bird species was recorded as breeding in Britain during 1988–91.[17] Research to update these figures is underway and due for completion in 2011. Based on the 1993 data, introduced species represent 10% of the British breeding bird fauna in terms of species richness (Table 6.2), although only eight of the extant species have populations in excess of 1000 individuals. In total, almost 300 nonnative bird species have been recorded in the wild in Britain and about 50 of these species have bred at some point (T. Langton, pers. comm.),[16,18] but in most cases, populations have not become established.

The earliest introduction to the British bird fauna, the pheasant (*Phasianus colchicus*), was introduced from the Norman times onward initially as a source of food and later fulfilled the additional role of a hunting quarry. The red-legged partridge (*Alectoris rufa*) was also introduced generally as a game bird from the 1820s on, although Lever records that the species was first brought to Britain in 1673.[6] At least part of the reason for the introduction of the little owl was for biological control of small mammals. However, such functional introductions are the exception rather the rule for birds, and all the other introduced species are the result of escapes from private collections or deliberate introductions for ornamental purposes.

### 6.2.3  Reptiles

Three reptile species known to be presently extant in Britain are introduced (Table 6.1), with at least a further three previously introduced now extinct. The British reptile fauna totals nine species,[9,19] so introduced species constitute 33% of the British reptile fauna in terms of species richness (Table 6.2).

Of the six recorded introductions of reptiles to the British mainland fauna, the wall lizard (*Podarcis muralis*) and the green lizard (*Lacerta viridis*) were introduced deliberately for ornamental purposes. Three of the other species—dice snake (*Natrix tessellata*), European pond tortoise (*Emys orbicularis*), and red-eared terrapin (*Trachemys scripta elegans*)—may also have been introduced deliberately on some occasions. However, since all three have been regularly kept as pets at various times, the main form of introduction of all three species, in common with the Aesculapian snake (*Elaphe longissima*), was probably escape.

### 6.2.4  Amphibians

Eight amphibian species known to be presently extant in Britain are introduced (Table 6.1), with at least five previously introduced species now extinct. The British amphibian fauna totals 14 species,[9,19] so introduced species constitute 57% of the British amphibian fauna in terms of species richness (Table 6.2).

Most introductions of amphibians to Britain have been deliberate. These include the Italian crested (*Triturus carnifex*) and Alpine (*Triturus alpestris*) newts, African clawed toad

(*Xenopus laevis*), edible frog (*Rana esculenta*), marsh frog (*Rana ridibunda*), yellow-bellied toad (*Bombina variegata*), fire-bellied toad (*Bombina bombina*), and painted frog (*Discoglossus pictus*), the latter two species now being extinct.[6,9] However, it is probable that some populations may also have originated from accidental releases.

## 6.2.5   Fish

Thirteen fish species currently found in Britain are introduced (Table 6.1), and at least three further introduced species recorded by Lever[6] are now extinct. A total of 55 fish species are listed by Maitland and Lyle as being found as wild populations in Britain at present.[20] To this list should be added *Leucaspius delineatus*, which was introduced in the 1990s as a result of importation for ornamental purposes and subsequent escapes.[21] Introduced species therefore represent 23% of the fish fauna in terms of species richness (Table 6.2). Of all the vertebrate groups, fish have been the most influenced by introductions. This has been due to a combination of introductions of native species outside their natural range and enhancement of populations of native species with captive-bred fish from a variety of, often unknown, origins. As a result, the native ranges of many fish species are now obscured and their genetic diversity irreparably compromised. The impact of introduced species of fish has therefore been greater than on any other vertebrate group and represents a major conservation issue.

The main reasons for the deliberate introduction of fish species are for either sport fisheries or ornamental purposes, or both in some cases. Examples of species introduced predominantly for sporting reasons include the rainbow trout (*Oncorhynchus mykiss*), brook charr (*Salvelinus fontinalis*), carp (*Cyprinus carpio*), ide (*Leuciscus idus*), wels (*Silurus glanis*), and zander (*Stizostedion lucioperca*). An example of an ornamental introduction is the goldfish (*Carassius auratus*), some wild populations of which have emanated from fish breeding in lakes on private estates.[6] A few species appear to have been introduced accidentally. These include bitterling (*Rhodeus sericeus*), which was introduced due to its use as live bait for perch (*Perca fluviatilis*), *Tilapia zillii*, a type of cichlid, and the guppy (*Poecilia reticulata*), which are both escapees from pet shops. Lever recorded that populations of *Tilapia zillii* and guppy were breeding in a stretch of the St. Helens Canal in Lancashire in the early 1960s, where the water was artificially heated by local industrial discharge.[6] Koi carp (*Cyprinus carpio*) are stocked extensively for anglers, with the present British rod-caught record standing at almost 20 kg.[22] The keeping of koi as pets is widespread as well, and it is likely that escapees from angling or pet stocks have also bred successfully in the wild, although there are no records of this at present.

# 6.3   Economic dimensions of introduced species

## 6.3.1   Consumption of other species or crops

Of the introduced mammal species, the rabbit, gray squirrel, fallow deer, sika deer, and Reeves' muntjac are those that currently have the most significant economic impact on forestry and/or agricultural crops in situ via their feeding behavior. Putman and Moore presented data on the number of enquiries concerning deer damage to agricultural interests in lowland Britain, received by the Wildlife and Storage Biology section of the Agricultural Development and Advisory Service (Ministry of Agriculture, Fisheries, and Food) over the period January 1987–March 1989.[23] These data are complicated by the fact that the native deer species, red (*Cervus elaphus*) and roe (*Capreolus capreolus*) may also cause significant damage in certain areas. Most inquiries came from east England, where fallow deer were the primary concern, and the second highest number came from southwest England, where enquiries were split fairly evenly between fallow, red, and roe deer. Overall, the number of inquiries relating to fallow deer was higher than for any other deer species, in particular in relation to damage to grass, cereals, and root crops. However, only 1% of all wildlife-related inquiries concerned deer of any species, which suggests that damage to lowland agriculture by deer is relatively insignificant compared with other wildlife species. In a separate study, Doney and Packer carried out a survey of deer damage to farms in four regions of England.[24] Damage to cereals was reported on 44% of farms where deer were present. However, 85% of respondents estimated the level of financial loss due to deer damage to be less than £500 per year. Langbein, in a survey of an area of southwest Britain, found a median annual deer damage level of £500, equivalent to £4.50 per hectare per year, over all properties, equivalent to £10.30 per hectare per year if only those properties which actually reported damage were included.[25]

Deer can cause damage to coniferous plantation forestry by browsing the leading shoots of young trees, stripping the bark from mature trees, or rubbing their antlers against trees (known as "thrashing" or "fraying").[26] The most common forms of financial loss are (1) the loss of incremental growth due to leader browsing and (2) the downgrading of wood produced, which can occur due to forking, marking, and/or staining as a consequence of fungal invasion, a problem which is exacerbated by deer damage to the bark.[27–29] Gill et al. calculated that the loss of value may be up to 8.4% in a worst-case scenario where 25% of all stems were forked.[29] Ward et al. calculated that the total yield was not significantly reduced until at least 40% of trees were forked.[30] However, their model also showed that loss of annual increment could not be tolerated if it exceeded 1 year. With 1 year of growth lost, the net present value of a stand was reduced from £512 per hectare to £112 per hectare, and a further year's growth loss resulted in a net loss of £288 per hectare. Deer damage therefore has the potential to be economically significant in forestry, and sika and fallow deer contribute to this damage. Reeves' muntjac is not a significant pest of commercial forestry, but may be a pest in short-rotation coppice. This is likely to become a more significant problem in the future, as the area of land used for biomass fuel production increases.

The relative impact of rabbit grazing has changed over time in correspondence with the impact of myxomatosis on the British rabbit population. Rabbit numbers in Britain were at their peak in the early 1950s. Following the introduction of myxomatosis in 1953, rabbit numbers declined by 99% within a few years. However, in less than two decades, evolutionary changes in the virus and the rabbits meant that an intermediate state developed whereby only around 50% of rabbits died due to the disease,[31,32] and by the 1980s, this mortality had fallen further to about 20%.[33] Surveys in Scotland showed 55.9% of farms had serious rabbit infestations prior to 1953, that this was reduced to 1.5% in 1969–1970 and then increased again to 26.5% in 1991.[34] It is estimated that there are currently 40 million rabbits in Britain, causing £115 million damage to agriculture each year, with an additional £5 million spent on population control.[35] Based on enclosure plots, the yield loss due to rabbit grazing on winter wheat and spring barley has been calculated as 1% and 0.5% per rabbit per hectare, respectively, which equates to £6.50 and £2 per rabbit.[36,37] Bell et al. quantified the effect of rabbit grazing on winter cereals under experimental conditions with varying rabbit densities up to 77 rabbits per hectare to represent natural variation in the population throughout the year.[38] They found yield losses could be as high as 35% and that some cultivars of wheat were more susceptible than others. This equates to a loss of approximately 0.5% per rabbit per hectare. Bell et al. also found the level of damage was not linearly related to rabbit density and suggested there may be a threshold damage level that determines the ultimate effect of grazing on crop yield.[38] These experiments are inevitably artificial in that rabbits in enclosures do not have access to any alternative food sources, which would probably reduce their overall impact on the crops, especially at high rabbit densities when they would switch to more profitable food items. Nevertheless, the figures do represent maximum likely impacts on yield.

Bark-stripping damage by gray squirrels, whereby squirrels peel off the outer bark from trees before scraping off and eating the sap-filled phloem tissue beneath,[39] is of major concern to foresters and woodland managers. Trees can be killed outright where stripping results in complete ring-barking of the stem. Stripping high up can also lead to the death of the crown and deformation of the tree. Less extensive stripping can assist fungal penetration, either possibly killing the tree or more commonly significantly downgrading any timber products. Damage usually takes place in May, June, and July, with less frequent occurrences in August.[40] Broad-leaved trees are most commonly affected, the most vulnerable being sycamore (*Acer pseudoplatanus*, itself an introduced species), followed by beech (*Fagus sylvatica*), oak (*Quercus robur*), and ash (*Fraxinus excelsior*).[40] There is much interannual variability in damage levels, and damage itself is usually concentrated in small areas within stands. In a study of beech stands in southern England, Rowe found that 87% of stands reported squirrel damage, but that in 52% of these, less than 20% of stems had been attacked.[41] In naturally regenerating oak, incidence of damage is related to size (height and girth) of the tree.[42] In a woodland in the Forest of Dean in southwest England, Mayle et al. found damage on 9%–38% of trees each year, with up to 17% of trees being ring-barked in any year.[42]

Damage also occurs on coniferous trees,[43] and this is of significant economic concern. Generally, less than 5% of damaged trees are killed.[44] Damage can occur on trees of all ages up to about 60 years old, but trees 10–40 years old are the most vulnerable. Squirrels also seem to concentrate on the most actively growing trees, such as those on the edges of a stand or the dominant trees within a stand.[40] However, despite a considerable investment in squirrel control in British forests, there are no published estimates of the economic importance of squirrel damage. When red squirrels were common, they were also extensively culled to limit damage to commercial timber crops. There are no data to show that current economic losses to gray squirrels are significantly higher than previous losses to red squirrels, although it is likely that they are, since gray squirrels live at higher densities.[45]

The major economic impact in Britain caused by the feeding behavior of common rats and house mice is on stored food. It has been recorded that 53% of farm grain stores[46] and 33% of commercial grain stores can be infested with common rats.[47] Even at relatively high densities, rats and mice do not consume large quantities of food. However, damage to wheat sacks by common rats is much more important economically than direct consumption of the stored wheat.[48] Both common rats and house mice can also contaminate significant amounts of stored food with their excretory products. For example, house mice produce 50 or more droppings per day, which are costly to remove from stored foods.[49] Despite the continuing problem of damage of stored foodstuffs caused by commensal rodents, there have been no detailed analyses of the economic costs and benefits of rodent control. However, there are increasing concerns about the negative welfare and environmental impacts of this level of rodenticide use, and recent research has shown that habitat management to reduce burrow availability and cover from predation may be effective as a complementary means of reducing rat populations on farms.[50]

Although ship rats (*Rattus rattus*) were previously important in Britain due to their consumption and contamination of stored foodstuffs, their very low population level now means that these impacts are economically insignificant. Similarly, from the 1950s until their extinction in 1989, coypu (*Myocastor coypus*) had a significant impact on root crops, in particular sugar beet.[51]

The only other introduced mammal that causes significant damage via consumption of crops is the edible dormouse. This species can cause significant visual damage to forestry crops locally in the Chilterns where it occurs. The favored species is larch (*Larix decidua*), but pine (*Pinus* spp.), spruce (*Picea* spp.), and beech may also be damaged. The edible dormouse may also cause localized damage to orchard crops.[52] However, there has been no economic assessment of this damage, and it is very localized.

Mink have often been cited as an agricultural pest, and predation on domestic poultry, game birds, and fish stocks has been widely reported.[6,53,54] However, the occurrence of prey items derived from domestic animals in the diet of mink is rare, even in studies that have involved mink populations living in the vicinity of farm buildings and game-rearing pens.[55] Tapper reported evidence of up to 180 killings in a single night in pens containing 400 game birds,[56] and such mass killings should have an economic impact on commercial shoots if carried out near to the time of release.[57] However, such incidents appear rare, and studies using scat analysis have shown that poultry and game birds generally make up less than 1% of the diet, with the highest recorded being 5.4%.[58,59] Since a typical domestic laying hen was worth about £2 in the mid-1980s, Harrison and Symes concluded that most incidents of mink predation did not represent a serious financial loss.[57]

Fish farms and fisheries provide a potentially abundant food source for mink, and Chanin and Linn found that salmonid fish accounted for 34.2% of the mink diet.[58] However, salmonids only feature significantly in the diet of mink feeding on rivers where salmon and trout make up a large proportion of the biomass. Mink damage was reported from 5% of fish farms in England and Wales during the mid-1980s,[57] but mink tend to take fish under 20 cm long, and overall their commercial impact on fish farming is negligible.[60] Declining populations of mink across much of Britain[60] mean that these figures are likely to represent overestimates of the current loss.

Of the introduced bird species, the Canada goose (*Branta canadensis*) is the only species that currently has a significant economic impact due to its consumption and trampling of crops and consumption, trampling, and fouling of amenity grassland. There are no data quantifying the level of this impact nationally, but localized damage can be significant. Simpson reported instances of crop damage in the United Kingdom costing £15,000, with 20% yield losses on winter cereals continuously grazed by Canada geese.[61] The population of Canada geese has increased dramatically in recent years, showing a proportional increase of 640% since the mid-1960s,[62] with populations in the United Kingdom currently growing at 8% per year.[63] This population increase has been accompanied by an increase in complaints of Canada goose damage to agricultural crops and grasslands.[63] The ring-necked parakeet (*Psittacula krameri*) has the potential to be a serious pest of orchards in southeastern England. At present, the population is at too low a level for this to be a widespread problem, but the population has increased tenfold in the last 10 years, and damage may already be locally significant.

The only introduced reptiles and amphibians or fish that have a potential economic impact through their consumption of other species or crops are the wels and the zander. These species can damage fisheries due to their predation on other smaller fish such as roach (*Rutilus rutilus*), bream (*Abramis brama*), and ruffe (*Gymnocephalus cernuus*).[64] However, this problem is localized, and there have been no attempts to assess it in financial terms.

### 6.3.2    Competition with other species

No introduced mammals, birds, reptiles, amphibians or fish in Britain currently have a significant economic cost due to their competition with native vertebrate species.

### 6.3.3    Introduction or maintenance of disease

The most important introduced mammals with significant potential for causing or maintaining disease with economic consequences in humans or other species are the commensal rodents—the ship rat, common rat, and house mouse. Although ship rats were historically very important due to their role as a vector in the transmission of bubonic plague (Black Death, *Pasteurella pestis*) in the Middle Ages, their low population densities mean that this is no longer a significant threat. The major disease problems from introduced mammals in Britain therefore now come from the common rat and the house mouse.

Common rats can act as hosts of a number of diseases that can affect humans and livestock.[65] In Britain, the major diseases are salmonellosis, leptospirosis, cryptosporidiosis, toxoplasmosis, and yersiniosis. *Salmonella* is commonly transmitted from pigs and poultry to humans, but house mice are often claimed to be the initial reservoir[66] and Davies and Wray reported *Salmonella* in 48.7% of house mice from poultry units.[67] However, Pocock et al. tested 341 samples (fecal and intestinal samples from mice and environmental samples) from four mixed agriculture farms in North Yorkshire and found none to be positive for *Salmonella*.[68] This study, along with that of Healing and Greenwood,[69] suggests that it is probably infected poultry that initially infect the mice with *Salmonella*, although the mice can then serve as a reservoir for reinfection of the poultry.[67,70]

Yersiniosis in humans is characterized by acute infection (pseudoappendicitis and enterocolitis) and immunologic complaints. *Yersinia* spp. are typically isolated from 3% to 10% of wild rodents tested.[68] Pocock et al. isolated *Yersinia* from 23 of 354 samples tested across four mixed agriculture farms and found prevalence of *Yersinia* in fecal and intestinal samples of 3.4% and 9.3%, respectively.[68] Cryptosporidiosis is another infectious gastroenteritis that affects humans and other mammals. Quy et al. found the average prevalence of *Cryptosporidium parvum* to be 24% in farmland rat populations and suggested that rats would be able to act as a potential source of infection for the disease on both livestock and arable farms.[71] Toxoplasmosis is a protozoan infection that can cause abortions and fetal abnormalities in pregnant women. Murphy et al. found *Toxoplasma* in 59% of 200 mice from properties in Manchester and also found evidence for vertical transmission of the infection from infected mothers to fetus.[72] In all these studies, while it can be demonstrated that commensal rodents are acting as hosts for disease, to a greater or lesser degree, the importance of their role in the transmission process relative to other modes of transmission, and the extent to which they are acting as true reservoirs for infection (i.e., able to sustain the infection in the absence of other host species), is difficult to quantify. Thus, it is currently impossible to quantify the relative economic impact of introduced rodents as agents of disease in Britain.

Wild deer can feature in the epidemiology of a wide range of diseases in livestock and humans in the United Kingdom by being a source of disease via various transmission routes.[73] Sika, fallow, and muntjac deer; feral ferrets (*Mustela furo*); mink; and common rats have been shown to carry bovine tuberculosis in Britain.[74,75] Of the deer, levels of infection and the consequent risk to cattle appear to be highest in muntjac and especially fallow deer.[74] However, the prevalence of bovine tuberculosis found in these species (with the exception of ferrets) is significantly lower than that found in badgers (*Meles meles*), a native species that is believed to represent the main wildlife reservoir for the disease.[76] The role of these introduced species in the maintenance of bovine tuberculosis, as well as other native species in which the disease has been found, remains unclear. Sika, fallow, and muntjac deer; feral sheep; feral goats; and common rats can also act as carriers of foot and mouth disease, which broke out in British livestock during 2001. However, wildlife did not appear to play a significant role in this outbreak, either as a reservoir or a vector of infection, and there is no evidence that the foot and mouth outbreak reached the reestablishing wild boar (*Sus scrofa*) populations in southern England.

Rabbits have been shown recently to act as a host for *Mycobacterium avium* subsp. *paratuberculosis*, which is the causative agent of paratuberculosis (Johne's disease) in livestock.[77,78] There is no evidence that any introduced species of bird, reptile, amphibian or fish in Britain is acting as a significant reservoir of an economically significant infectious disease.

## 6.3.4    Interbreeding with native species

Sika deer hybridizing with red deer may reduce income from stalking due to smaller trophy heads and carcasses. However, there appear to be no other cases of mammals, birds, reptiles, amphibians or fish where introduced species have led to significant negative economic impacts through their interbreeding with native species.

## 6.3.5    Disturbance of the environment

From the 1950s until their extinction in 1989, coypu caused local damage to agricultural interests in East Anglia due to their burrowing into the banks of watercourses, which could cause flooding.[51] Gray squirrels, house mice, and fat dormice can all cause economic damage to property by disturbance to their environment. Fat dormice chew through electric cables, roofing felt, and ceiling plaster.[52] Gray squirrels also enter the loft spaces of buildings[79] and will do similar damage. The actions of gray squirrels, house mice, and fat dormice can occasionally cause fires. These can be financially devastating to individual properties, although the overall economic impact of such damage is probably negligible.

The feeding activities on agricultural crops of introduced mammals such as rabbits, fallow deer, common rats, and, locally, wild boar and birds such as Canada geese can also cause physical damage to the crops, which will have economic implications. It is extremely difficult to isolate the effects of such disturbance from actual consumption of the crops, so the economic significance of this is unknown.

## 6.3.6    Economic benefits

It should be borne in mind that introduced species can bring benefits as well as costs. The relative importance of the benefits and costs varies according to the time span over which they are being assessed, and this can lead to varying conclusions regarding the desirability of different introduced species.

For example, coypu, mink, and muskrats brought economic benefits for the fur industry when they were first brought into the country, prior to escape and establishment in the wild. Sika deer, pheasants, and red-legged partridges generate significant revenues from game shooting,[80] and many alien fish, in particular species such as rainbow trout, carp, and zander, provide recreational and local economic benefits through angling. Recreational fishing is a very important activity in Britain, involving up to 2.2 million people spending a total of £3.15 billion each year.[81] Although much of this revenue will come from expenditure on fishing of native species such as salmon and brown and sea trout,[82] some will come from introduced species such as rainbow trout. Alien fish species can therefore bring considerable indirect benefits to the local economy in certain areas.

## 6.4   Environmental impacts of introduced species

### 6.4.1   Consumption of other species

The majority of introduced mammal species has significant environmental impacts through the effects of their consumption of other species. Fallow and sika deer, Reeves' muntjac, brown hares (*Lepus europaeus*), rabbits, feral goats, wild boar and, historically, coypu can all have significant environmental impacts on vegetation species through their herbivory. Whether grazing and browsing are viewed as beneficial or detrimental depends on the management objectives in place for a particular location at a particular time.

The rapid increase in numbers of both native and introduced deer in Britain[83] and in the rest of western Europe since the 1950s has led to increasing impacts of deer browsing and grazing on conservation. Light deer grazing can be beneficial in that it helps maintain early successional stages of grassland and hold back woodland encroachment.[84] Heavy browsing effectively reduces the structural diversity of woodland by removing the middle layer of regenerating trees. This can be detrimental to woodland bird species[85] and to small mammals and invertebrates, although it may beneficial to some birds of conservation importance with preference for more open habitats, such as wood warblers (*Phylloscopus sibilatrix*), pied flycatchers (*Ficedula hypoleuca*), and redstarts (*Phoenicurus phoenicurus*).[86]

These examples indicate that deer grazing can sometimes be beneficial to conservation. However, conservation objectives in Britain are generally geared toward woodland regeneration, when the effects of deer grazing and browsing will be detrimental. Heavy browsing of young saplings by deer at densities as low as five deer/km$^2$ can inhibit regeneration of natural woodland.[86–89] Coppice woodland is particularly vulnerable to deer damage. Coppice regrowth can be delayed or occasionally completely inhibited by deer browsing, even at relatively low deer densities,[90,91] and this can sometimes have knock-on consequences for invertebrate conservation.[91] New plantations may also suffer quite heavy damage levels. Key et al. found damage rates by fallow deer to terminal shoots in broad-leaved plantations of up to 93%, with considerable variation between tree species.[92] In addition to impacts on trees, introduced deer may also damage the ground flora. This is particularly the case of Reeves' muntjac, and Cooke has attributed declines in bluebell (*Endymion nonscripta*) and dog's mercury (*Mercurialis perennis*) in Monk's Wood National Nature Reserve to grazing from Reeves' muntjac.[93,94] Grazing by muntjac may also have adverse effects on plant communities since they provide much lower rates of endozoochorous seed dispersal compared with native roe deer.[95]

Rabbits can also have a significant impact on woodland regeneration,[96] but more importantly, rabbit grazing can cause shifts in entire plant communities in grasslands.[97–99] Because rabbits have such a dramatic effect on vegetation composition, failure to exclude rabbits can have considerable influence on the success of land restoration projects.[100]

Coypu had a dramatic impact on native vegetation in the Norfolk Broads in East Anglia when numbers were high during the late 1950s and early 1960s, and again in the late 1970s and early 1980s. They caused a massive decline in reed swamp in the Broads[101] and a big reduction in favored food plants, including cowbane (*Cicuta virosa*).[102] The effects of herbivory by coypu have contributed to very significant changes in the Norfolk Broads ecosystem. However, the precise contribution made by coypu is impossible to quantify, since their effects were also linked to changes in phosphate and nitrate pollution, boat activity, a greater frequency of avian and fish diseases, and changes in management of rivers and marshes.[103]

Brown hares may damage saplings or shrubs during hard winters when grazing is not available,[104] feral goats may damage young trees due to browsing and bark stripping,[105] and wild boar may adversely affect the ground flora of woodlands. However, because of the generally much lower densities of these species and the localized populations of feral goats and pigs, their overall conservation impacts are relatively insignificant compared with those of the introduced deer species and rabbits.

Other introduced mammal species have had major impacts on native species via predation. Mink have been implicated in the decline of water voles (*Arvicola terrestris*), coots (*Fulica atra*), moorhens (*Gallinula chloropus*), and various nesting seabirds. The decline of water voles was the most dramatic of any native British mammal during the twentieth century.[11] Many authors have cited mink predation as a cause of local reductions in water vole populations in various habitat types.[106–110] It has also been reported that mink predate waterfowl,[58] but populations of most duck species are increasing, and there is little evidence to link mink predation with any specific population declines. For example, Halliwell and Macdonald found no significant correlations between mink and moorhen abundance in the Upper Thames catchment.[109] However, Ferreras and Macdonald, studying the same geographic area, found that mink took between 16% and 27% of adult moorhens and 46% and 79% of moorhen broods, and 30% and 51% of adult coots and 50% and 85% of coot broods.[111] The presence of mink was also found to adversely affect the breeding success of coots, although for moorhens the evidence was less clear.

The impact of mink on coastal ground-nesting bird populations is especially severe on islands. Observations have shown that where mink gain access to smaller colonies (<200 pairs) of terns or gulls, it is unusual for any chicks to fledge.[112] For example, between 1989 and 1995, mink predation of nests and chicks caused widespread breeding failures of whole colonies of black-headed gulls (*Larus ridibundus*), common gulls (*Larus canus*), and common terns (*Sterna hirundo*) on small islands off the west coast of Scotland.[113] Mink predation has also had an adverse effect on populations of ground-nesting birds on the islands of Harris and Lewis and in the Sound of Harris.[114,115] In the latter instance, lapwing (*Vanellus vanellus*) and redshank (*Tringa totanus*) were in particular badly affected.

Other introduced mammals reported to prey on native species include feral cats (*Felis catus*) and gray squirrels. The major impact of feral cats is on birds. Predation by cats on birds is greatest in spring and summer, and predation rates can be high in relation to population productivity. For example, in a study in Bristol, the mean annual predation rate was 21 prey per cat.[116] Even low levels of predation by cats can have a significant impact on native birds at the population level.[117] Gray squirrels also predate bird eggs and nestlings,[85] although there is no quantitative evidence of the overall significance of this predation.

The other major invasive species predation impact comes from hedgehogs (*Erinaceus europaeus*), which although native to Britain, have been introduced to the Western Isles of Scotland. Nest predation by hedgehogs can reduce wader nest success by a factor of 2.4[118] and cause declines at the population level.[119]

Among the birds, the only introduced species that has any proven significant environmental impact via its consumption of native plants is the Canada goose. Overgrazing by Canada geese can have a significant adverse effect on reed beds, salt marshes, and other vegetation.[120]

The only introduced amphibian that could potentially have widespread impacts on native species via predation is the marsh frog. Observations in southeast England in the late 1970s and early 1980s suggested that there had been no detrimental impact on native frogs and toads.[121] However, uncertainty remains since the habitat where populations of marsh frogs first became established may not be that favorable to native amphibians. Moreover, the warm summers of the late 1980s encouraged populations to increase, and it is estimated that there are now about 30,000 adult green frogs (marsh frogs and edible frogs) in around 80 localized populations in southeast England.[122] This population increase may therefore be a more real threat to native amphibians if these populations continue to increase and spread.

Predation of native species is a cause for concern with some introduced fish species. Rainbow trout, carp, wels, pumpkinseed (*Lepomis gibbosus*), and zander all predate newt, frog, and toad larvae.[122] Wels and zander also feed on smaller fish such as bream and roach, and wels also feed on mammals and diving birds.[22] However, the overall impact of this predation on the populations of the native species is not known, and for fish such as bream and roach, the effects are more likely to be important from a fisheries rather than a conservation perspective.

## 6.4.2  Competition with other species

The best-documented significant competitive interaction between an introduced and a native mammal is that between the gray and the red squirrel. There have been disagreements in the scientific literature regarding the extent to which direct competition from gray squirrels has caused the decline of the native red squirrel. However, the weight of evidence suggests that competition is a major contributory factor in this decline. There is little evidence for direct interference competition between adult red and gray squirrels,[123] and the mechanism of exclusion seems to be feeding competition.[124] The red squirrel is a less efficient forager in deciduous woodland and in this habitat especially, gray squirrels can live at higher densities and achieve faster breeding rates.[45,124,125] Where gray squirrels are present, red squirrel breeding, fecundity, and recruitment rates also decline.[126,127] Removal of gray squirrels can lead to red squirrel recovery, and intensive trapping of gray squirrels since 1998 on the island of Anglesey off the coast of North Wales has allowed an expansion in both numbers and distribution of the red squirrel on the island.[128]

A competitive interaction between introduced mink and native otters, probably mediated through improved habitat (water) quality, has helped otter populations to recover in many catchments. Where otters, the dominant competitor in good-quality habitats, have increased, mink have frequently declined.[129–131] However, in some cases, both species can continue to coexist, because mink change their diet to consume fewer fish and also exhibit temporal shifts in activity patterns.[132]

Among fish communities, invasive salmonids such as rainbow trout can have direct adverse impacts on native species via direct competition, as well as indirect effects through fragmenting populations.[133] Loss in submerged vegetation of lakes due to heavy grazing pressure by introduced carp can lead to reduced trophic diversity in fish communities, especially where these communities may be less resilient due to other anthropogenic pressures on the ecosystem.[134]

## 6.4.3  Introduction or maintenance of disease

Introduced species frequently act as hosts for disease. However, apart from those diseases of economic importance discussed in Section 6.3.3, there is no evidence that introduced mammals, birds, reptiles, amphibians, or fish have brought in new diseases that are environmentally damaging. Little evidence exists that alien species are contributing significantly to problems of disease in native species, although there are some exceptions. Some alien fish species introduced to supplement native stocks for angling are leading to increased diseases,[1] and increased fish diseases due to artificial stocking may have had a role in changes to the ecosystem of the Norfolk Broads.

## 6.4.4   Interbreeding with native species

Three introduced mammal species have adverse environmental impacts due to their interbreeding with native species. These are the feral ferret, feral cat, and sika deer. Feral ferrets are fully interfertile with the native European polecat (*Mustela putorius*)[135] and are widely kept. Escapes are frequent, and populations have been established on the Scottish islands of Mull, Lewis, Bute, and Arran, and also on the Isle of Man and the mainland. Native polecats have increased in abundance and extended their range considerably since a low in the 1920s, spreading out from a stronghold in mid-Wales.[56,136] As native polecat populations continue to spread, they will undoubtedly encounter feral ferrets and interbreed with them. However, the extent to which this has happened already is unknown.

Feral cats are capable of interbreeding with native wildcats (*Felis silvestris*). Wildcats have shown a population recovery and spread in northern Scotland since the First World War,[11] but the population remains critically endangered.[137] Former problems of direct persecution and habitat loss have now been reversed or halted, but the major problem of hybridization with feral or domestic cats persists. Hybridization has probably been going on for several hundred years, and a small level of hybridization does not seem to have significant adverse effects on population persistence. Wildcats have traditionally been distinguished from feral cats on the basis of morphological characteristics.[138] However, recent work has suggested that rather than being a specific form, there may be a morphological cline among wild-living cats in Scotland, with the "wildcat" existing at one end of this cline, which is the furthest from the domestic cat under all criteria.[137,139] This in turn suggests that species-level approaches to this particular hybridization problem may be ineffective and that management is required at the population level instead.[140]

Sika deer are able to hybridize with native red deer, although where both species exist at good densities, or where they are reasonably well separated, hybridization seems rare.[141,142] Moreover, since red deer populations in southwest Scotland and most of England are not native, hybridization of these populations may not be a significant conservation issue.[11]

The only hybridization problem within Britain caused by an alien bird species, as opposed to introduced populations, concerns pheasants. However, in this case, the affected species is itself introduced. There has been a traditional emphasis on measuring the success of pheasant shoots according to their bag size, rather than the "quality" of the shoot, and this has led to up to 20 million birds being reared in captivity and released each year to supplement the "wild" stock.[56] Captive-bred birds have much lower survival rates over the winter and a significantly lower breeding success, and the continual dilution of the wild stock therefore represents a risk to the persistence of the genes in the wild stock.[143]

The other significant hybridization problem in birds associated with a species introduced to Britain concerns the ruddy duck (*Oxyura jamaicensis*). However, in this case, the actual location of the problem itself is elsewhere in Europe. Ruddy ducks originating from the Wildfowl Trust's collection in Gloucestershire have spread abroad and have now been reported from 13 other European countries.[144] In Spain and Turkey, they hybridize with native white-headed ducks (*Oxyura leucocephala*), resulting in fertile offspring, which may lead to genetic introgression and possibly even extinction of wild white-headed ducks in these countries.[144]

There are two potential hybridization problems between native and introduced species among the amphibians. The first is between the Italian crested newt and the native great crested newt (*Triturus cristatus*). Hybridization between the two species occurs freely around Newdigate in Surrey where the only well-established Italian crested newt population occurs in Britain.[9] However, although the hybrids are viable, they have a low fertility. Beebee and Griffiths consider that these hybrids therefore amount to a "dead end," and that this is one reason why the Italian crested newts have not spread significantly beyond the site occupied by this single population.[9]

The other amphibian hybridization problem is that between native pool frogs (*Rana lessonae*) and introduced marsh and edible frogs. It is by no means certain that the pool frog is in fact a native species,[145] but Beebee and Griffiths consider that there is sufficient evidence to indicate that it probably is.[9] Unfortunately, the last remaining population near Thetford in Norfolk declined to extinction in the mid-1990s.[9] Marsh, pool, and edible frogs are all capable of interbreeding.[9] Edible frogs are viable vertical hybrids of male pool and female marsh frogs. Matings between male and female edible frogs produce female marsh frogs, but few of these survive because they have low viability and the habitats occupied by edible–pool frog populations are not well suited to marsh frogs. For this latter reason, most matings of edible frogs that produce viable offspring are with pool frogs, and green frog populations in Britain, as elsewhere in western Europe, are therefore mixtures of edible and pool frogs.

A number of fish species hybridize with one another quite frequently in the wild,[22] and the extent of hybridization is increased considerably by movements of species, both native and introduced, by anglers. However, only 3 hybridizations out of 11 recorded to date involve introduced species, and one of these is between two introduced species rather than between a native and an introduced species. These hybridizations are native trout × brook charr, native bream × ide, and carp × native crucian carp (*Carassius carassius*). There is some disagreement about the status of crucian carp, but the most recent evidence suggests that it is a native species, having a natural range in the eastern and midland counties, although it is only present elsewhere in Britain as a result of introduction.[146] In addition to these interspecies hybridizations, hybridization between farmed and wild fish of the same species may dilute the wild genetic stock and lead to long-term introgression of gene pools.[147]

## 6.4.5   Disturbance of the environment

Physical disturbance of the environment that is of environmental, rather than direct economic, importance can be caused by the rooting behavior of wild boar. The effects of this seem to vary, and studies in different countries have shown both decreases and increases in ground flora diversity or regeneration as a result.[148,149] Where wild boar occur in parts of Kent and East Sussex, their rooting behavior can damage the ground flora in native woodlands. However, at present, this damage is extremely localized. In the past, both coypu and muskrat were responsible for significant disturbance to riparian bank habitats by their burrowing behavior. However, both these species are now extinct, muskrat being eradicated soon after their first escapes in the 1930s and coypu by 1989.[150]

Localized, high densities of pheasants, for example close to release sites or feeding sites, may cause extensive disturbance to the soil as well as increases in nutrient load, leading to more bare ground, more weeds, and fewer perennial plant species.[151] However, these localized adverse impacts are offset by the fact that woodland management practices for pheasants have wider benefits such as denser understory vegetation and richer bird communities.[152]

The action of common carp (and native crucian carp) feeding on invertebrates in bottom muds and silts often increases water turbidity, and this can contribute to changes in freshwater ecosystems by reducing light available for growing plants.[22] However, these effects are often interlinked with increased nutrient loading and algal growth, so the relative importance of the carp themselves is difficult to quantify. Moreover, in studies of the eutrophication of whole ecosystems, other factors such as phosphate and nitrate pollution and boating activity appear to be the primary causes of changes in ecosystem structure and function.[103]

### 6.4.6 Environmental benefits

Although the environmental impacts of invasive species are generally detrimental, the impacts of certain invasive species may be considered beneficial. Besides being a major agricultural pest, rabbits have a significant positive effect in increasing floral diversity on chalk downland and breckland.[153,154] Changes in rabbit grazing pressure have also been associated with changing status of some butterfly species. For example, reduced rabbit grazing caused by myxomatosis has been suggested as one of the reasons for the extinction of the British population of the large blue butterfly (*Maculinea arion*)[155] and increases in rabbit grazing have been linked to recovery of the Chalkhill blue butterfly (*Polyommatus coridon*)[156] and the silver-spotted skipper (*Hesperia comma*).[157] Rabbits also play an extremely important role in the food web, providing over 75% of the total energy available to mammalian predators.[158] The mandarin duck is subject to conservation measures in its native range, and therefore alien populations in Britain may represent important refuges for this species.[1] One alien vertebrate species, the brown hare, is the subject of the UK Biodiversity Action Plan to expand its numbers and range.[159]

## 6.5 Social dimensions of alien vertebrates in Britain

### 6.5.1 Cultural associations

Public bias toward particular taxonomic groups or species can potentially make invasive alien vertebrate management a source of contention if the alien species concerned falls into the "charismatic pest" category,[160] or if there are other deep-rooted cultural associations or attachments to the species. Although those that argue against a cull of an alien species frequently cite ethical arguments, cultural attachments nevertheless usually play an important role, making one species more desirable to "save" than another. This can have serious implications for how some alien species can be managed.

One of the best examples of this in Britain concerns the management of invasive hedgehogs in the Western Isles of Scotland, where they have an impact on wader populations through egg predation.[118] Hedgehogs are partially protected by law under the Wildlife and Countryside Act 1981, and their population may be in decline in some parts of Britain.[161] From 2003 to 2007, The Uist Wader Project led a hedgehog-culling campaign, which was met with significant resistance from animal rights groups and the general public.[162]

As a native species in mainland Britain, hedgehogs were voted favorite British animal in the charity-led project Wild about Gardens.[163] Furthermore, hedgehogs have appeared in popular culture many times, including children's stories such as *The Tale of Mrs. Tiggy-Winkle* by Beatrix Potter,[164] and the video game Sonic the Hedgehog (by SEGA, created in 1991). Hedgehogs also feature in philosophical literature, such as Isaiah Berlin's popular essay *The Hedgehog and the Fox*,[165] based on the ancient Greek poet Archilochus' much-cited quotation "The fox knows many things, but the hedgehog knows one great thing."[166] Repeated appearances in literature and popular culture are indicative of a significant cultural attachment to this species. This is likely to have played an important role in the public response to the hedgehog cull in the Uists, which subsequently contributed to the end of the culling project and a switch to a nonlethal management plan to trap and transport the hedgehogs to mainland Scotland.[167]

Wild boar provides another example of the social complexity arising when a species is, or has been, considered native in some parts of Britain. Wild boar became extinct in Britain around 700 years ago, due to overhunting and habitat loss.[168] Following failed reintroduction attempts in the subsequent centuries for hunting purposes, escapees from wild boar farms allowed populations to become reestablished in southern England in the early 1990s;[169] therefore, these populations do not originate from the native stock. Whereas some reestablished species have received a mostly positive response from the public, such as the threatened red kite (*Milvus milvus*), which became extinct in England and Scotland by the late nineteenth century,[170] wild boar have received a predominantly negative response, at least from the media.[171] Goulding and Roper found that the most frequently reported issue relating to wild boar was fear of attack on humans, despite the fact they were unable to find any confirmed reports of this occurring, and large populations exist in other parts of Europe without significant threat to human safety.[171] This negative response may seem even more surprising given the strong positive cultural associations of wild boar in Europe, with legendary status as a brave and noble animal, hunted and feasted on by royalty.[172] In modern-day Britain, the wild boar's fierce reputation appears to remain, but much of the more positive cultural associations have diminished over time, perhaps due to their absence for several centuries, the lack of reinforcement of the positive cultural associations, and the lessened status of hunting in Britain, which may reflect the decreased need for subsistence hunting and increased ethical considerations.[173]

Invasive alien species such as rats, on the other hand, are on the whole disliked in most countries.[174] This may stem from negative cultural associations, such as the role of the ship rat (*Rattus rattus*) in the spread of the bubonic plague (Black Death) in Europe, western Asia, the Middle East, and North Africa in the fourteenth century,[175] as well as their continued significant health impact as vectors of this disease in various parts of the world even today.[65,176] Therefore, lethal control of rats does not usually engage notable interest by the wider public.

Cultural associations can therefore have a significant impact on invasive alien species management, since they play an important role in forming public attitudes. Even without strong cultural ties, taxonomic bias still has to be contended with. While scientists may form strong ecological arguments for a particular management plan, there is a growing realization that cultural, societal, and public attitudes are not something that can be easily dismissed, and an understanding of them may be a crucial foundation for successful ecological management.

### 6.5.2 Public attitudes toward introduced species, impacts, and their management

The way that humans perceive the natural world is influenced by cultural background[177,178] and may change over time.[177] Therefore, attitudes toward biodiversity conservation or environmental management initiatives are not necessarily construed from a rational behavioral response. Studies have illustrated that certain taxonomic groups are favored over others, with some species attributed more value than others (in both economic and noneconomic terms). This bias occurs in conservation research,[179] in conservation projects,[180] and in the views of the general public.[181] Knegtering et al. compared different national studies on the attractiveness of species according to the public, finding that birds were ranked as most liked, followed by mammals, with invertebrates at the "least-liked" end of the scale.[182] These authors also argued that animal welfare or conservation-based nongovernmental organizations (NGOs) are likely to be influenced by species characteristics, taxon, and relative size in their support for, or lobby against, conservation policies.

Attitudes toward introduced species also vary significantly according to a number of factors including species type, the impact of the species (in social, economic, and environmental terms), and the proposed conservation method.[181] Additionally, policy makers' motivations for controlling the species are also said to influence public attitudes. For example, Bruskotter et al.[183] suggest that where livestock requires protection from excessive predation, lethal control methods are more likely to be acceptable. Indeed, in Scotland, Philip and Macmillan[184] found that respondents generally disagreed with sport shooting, although they were likely to agree with culling of predators to protect rare species. The type of conservation method may also influence attitudes; there is evidence to suggest that once the public understand the implications of nonlethal methods (such as high mortality rates among relocated species or the financial implications of trapping) they are more likely to agree with lethal methods.[184]

In addition, Bruskotter et al.[183] argued that the acceptability of lethal forms of control varies by contextual, social, and cognitive factors. Contextual factors refer to many of the issues discussed above, including the nature of the problem, proposed forms of control, and the species in question.[183] Social factors include socioeconomic or demographic groupings, the nature of contact with the species in question, and membership of particular stakeholder groups or organizations (e.g., animal rights or prohunting groups). Stakeholder or organization membership itself is likely to be influenced by socioeconomic and demographic characteristics,[185] while also reinforcing attitudes toward lethal methods of control. Cognitive factors encompass beliefs, values, and attitudes[183] and help to understand emotional responses to wildlife, domestic animals, and conservation methods.[186] There are a number of arguments put forward to explain emotional responses and beliefs, including the cultural factors described above. In addition to such factors, Hill et al.[187] suggested that the separation of an increasingly urbanized society from the processes of food production has led to a primarily emotionally driven response to wildlife. These authors also suggested that a lack of knowledge about the species in question, its impacts, and its control, is likely to affect attitudes. However, while there are thought to be emotive factors influencing public responses to conservation measures, Philip and Macmillan[184] disputed the view that the British public would be unlikely to support a cull of nonnative species solely on moral grounds.

## 6.5.3 Social benefits

Most research and policy on alien species has focused on economic and environmental impacts rather than social and cultural dimensions.[8] Where social research does exist, the main focus has been on public attitudes,[181,188] rather than social benefits per se. Although many alien species cause notable costs, nevertheless social benefits can arise, for instance through improving community awareness of wildlife and the environment, nonuse value due to their aesthetic characteristics, and recreational activities such as hunting and shooting.

Community awareness of wildlife and the environment is an important element of gaining support for conservation initiatives; however, evidence suggests that education relating to wildlife and nature is deficient. For example, Balmford et al.[189] revealed that British children at primary school level have an excellent capacity to learn about creatures, but they have a greater knowledge of Pokemón* "creatures" than of wildlife in Britain. They conclude that with an ever-rising urban population, reestablishment of children's connection with nature is necessary to enthuse future generations about the importance of wildlife conservation. While doing this through educating about native and threatened species may be both expected and desired, in some cases, particularly in urban areas where contact with nature can be particularly low, any positive promotion of environmental awareness may be warranted. Indeed, nature conservation in urban areas in Britain has placed more weight on the benefits and values of urban wildlife to local people, than on conservation of endangered species.[190]

With this in mind, there is potential for some of the more charismatic invasive species in urban areas, such as the gray squirrel, to be useful as a promotional tool for community wildlife awareness, despite their invasive status. This is likely to be particularly important for education of children from deprived urban areas, although research is currently lacking on this topic. For example, while promotion of the native red squirrel is a priority from a conservation perspective, in many deprived urban areas of Britain the gray squirrel may be one of the few wild mammals children see on a regular basis. As squirrels are one of the top 10 favorite animals in Britain,[163] gray squirrels therefore act as a potential source for gaining children's interest in the natural world around them, for example by watching squirrel behavior and learning about their ecology, either at home or at school. Although this is a role that could in theory be fulfilled just as adequately (or more so) by native species, in reality there is also an argument for making the best of what we do have, particularly when viewing from a social, rather than ecological, perspective.

For most ecologists or wildlife enthusiasts, whether an alien species is aesthetically pleasing or not, or provides some entertainment value, does not warrant an alien species the right of abode in Britain. Nevertheless, the enjoyment some people receive in seeing certain alien species, whether this be gray squirrels, muntjac deer, or ring-necked parakeets (*Psittacula krameri*), is an existent, although difficult to quantify, social benefit. The media play an important role in forming public attitudes,[171] and if the media and the public express particular admiration for an alien species, this may create a considerable force that wildlife managers must contend with. However, whereas the case for educational and aesthetic benefits of alien species may be seldom presented and may seem weak against most ecological arguments, the case for the social benefits and cultural dimensions of hunting alien animals is more familiar. It may also present a greater challenge for wildlife managers wishing to remove alien species, particularly when politics become involved, as can be seen in relation to the debate surrounding the management of introduced deer in Australia.[191]

In Britain, there are many species of mammals and birds, both native and introduced, which are actively hunted as a recreational activity (and for their meat). Hunted introduced species include birds such as the pheasant, red-legged partridge, the Canada goose, and pink-footed goose (*Anser brachyrhynchus*), and mammals such as muntjac, sika deer, fallow deer, Chinese water deer, rabbits, and more recently, wild boar. The act and cultural significance of recreational hunting in Britain stems back possibly as far as the Iron Age, becoming of greater significance in the Middle Ages,[192] and although it has altered

---

* Pokemón is a media franchise, created in 1996 by Satoshi Tajiri and owned by Nintendo, which includes a collectable card game with different Pokémon types or "creatures."

considerably over time, many rural citizens consider hunting and shooting a major part of their life and culture. A case can be made that hunting native species provides a superior cultural and recreational experience, and this is suggested by the relative importance of native species–based rod fisheries compared with those based on introduced species.[82] However, some field sports such as pheasant shooting, which is a significant part of a multibillion-pound shooting industry,[80] may not have a suitable native equivalent, or may be already so embedded in British rural culture and economy that it would be politically impossible to attempt to remove the species.

## 6.6    Analysis and conclusions

The proportion of alien species in the different vertebrate groups present in Britain varies quite widely, being highest for amphibians (57%), then mammals (34%), then reptiles (33%), and fish (24%), and by far the lowest for birds (6%; Table 6.2). If the nonterrestrial mammals are excluded, introduced species make up almost half (46%) of the terrestrial mammal fauna. The figures for terrestrial vertebrate groups are much higher than global averages and reflect both the changing attitudes of humans toward introductions over time and the factors affecting the richness of native species.

There are two main factors that have combined to cause the relative paucity of native terrestrial vertebrate species (and partly in consequence, the high proportion of introduced species) in Britain today. The first factor is the island nature of Britain. This has ensured its relative inaccessibility to terrestrial species since the disappearance of the land bridge to continental Europe following the last glacial retreat, a fact that was exacerbated by the relatively short time period between the retreat of the ice sheet and the flooding of the English Channel.[168] The second factor is the effect of humans in causing the extinction of native species. In the next few sections, we review the importance of these factors in shaping both the native and the introduced vertebrate fauna of Britain.

### 6.6.1    Impact of environmental change on the British native vertebrate fauna

The native species in the current British terrestrial (land or freshwater) vertebrate fauna date back to 15,000 years ago with the retreat of the last Ice Age in the Devensian period.[168] At the peak of glaciation, tundra-like conditions existed to the south of the ice sheet. It is extremely unlikely that any species of reptile or amphibian could have survived in such conditions.[9] However, it is possible that a few species of stenohaline fish (species that can survive only in fresh water) did survive in the ice-free conditions in the extreme south.[21] It is also possible that a few mammals—the mountain hare (*Lepus timidus*), stoat (*Mustela erminea*), and weasel (*Mustela nivalis*)—could have survived.[168] Nevertheless, it is clear that at this time, the vertebrate fauna of what was to become Britain was extremely species poor.

The glacial period ended abruptly about 13,000 years ago, and mean annual temperatures increased dramatically from –8°C to +8°C. Another cooling followed almost immediately, reverting to tundra-like conditions again 10,000 years ago, but this was followed by another very rapid warming up to 8°C again.[193] At this time, Britain was still connected to continental Europe by a land bridge, across which flowed rivers that were either connected to or shared a flood plain with the Rhine.[22] However, as the climate warmed, the ice melted and sea levels rose, and Britain became isolated. This isolation may have occurred as early as 9500 years ago, but certainly by 7000 years ago.[168] Thus, there was only between 500 and 3000 years for terrestrial vertebrates and fish from warmer climates to the south to colonize Britain. For the mammals, excluding bats and seals, only 33 species managed to cross into Britain before it was separated as an island. Recent genetic evidence suggests that some small mammal species colonized from the east (Scandinavia) rather than the south.[194] Britain's native mammal fauna is therefore probably a mixture of southern and eastern origins and even fewer species may have colonized across the land bridge than previously thought.

This shortage of time for colonization partly explains the relative paucity of Britain's native terrestrial vertebrate assemblage today, since it did not allow a full representation of the northwest European vertebrate fauna to reach Britain. Thus, the fish fauna of Britain (55 species) is much reduced compared with that of northwest Europe (ca. 80 species) and Europe as a whole (ca. 215 species).[22] Euryhaline fish (those species that can live in both freshwater and salt water) are relatively flexible in terms of their colonizing ability, and as the ice melted, these species were able to follow the coastline northwards and colonize any freshwaters that were accessible from the sea. However, stenohaline fish had to colonize using rivers across the land bridge. The tight environmental constraints on movements of stenohaline fish, combined with the fact that the number of euryhaline species is relatively small, also gave rise to a gradient of native fish species richness across Britain. Areas furthest away from the former land bridge, such as Scotland, northern England, and Wales, have fewer native species than the south and east of England.[22]

Other terrestrial vertebrates are similarly constrained by the environment in terms of their colonization movements, and the native fauna of these groups is also relatively species poor. For example, Britain's 23 reptiles and amphibians (which include 11 introduced species) compares with 22 species in the Netherlands, and more than 60 amphibians and twice as many reptiles in the whole of Europe.[9]

The mammal fauna in Britain was also formerly relatively species poor, but the difference has been reduced by introductions, especially of larger mammals. In fact, the number of introductions *pro rata* is greater for mammals than for any other taxon. There are currently 63 species of mammals in Britain, excluding pinnipeds, compared with 188 in Europe.[195,196] The comparative figures for the different orders (with European figures in brackets) are: insectivores, 8 (28); bats, 17 (35); primates, 0 (1); lagomorphs, 3 (8); rodents, 14 (68); carnivores, 11 (27), artiodactyls, 9 (20); and marsupials, 1 (1).

Due to their greater mobility and nonreliance on the land bridge for colonization, birds are significantly more abundant than other vertebrate groups in the British fauna, and there are proportionally more birds in the British fauna than is found globally.[197] The mobility of most bird species means that, given appropriate habitat and food resources, they are readily capable of extending their native ranges. For example, the crane (*Grus grus*) recolonized in 1981.[16] In the medium- to long-term future, accelerated climate change due to human activity is likely to result in some bird species expanding their breeding ranges into parts of southern Britain from continental Europe. Similar opportunities for natural colonization also exist in theory for bats. However, bats tend to be weaker flyers than birds, and populations of most species of bats are currently much reduced throughout northwestern Europe. Recolonization by species such as the extinct mouse-eared bat (*Myotis myotis*) will therefore remain very unlikely for the foreseeable future.

In the shorter term, habitat change is likely to be the most important driver of invasive species populations in Britain. One of the key habitat changes in the short to medium term will be an increase in broad-leaved scrub and woodland cover, encouraged by agri-environment schemes, management for ecosystem service provision, and a move toward biomass energy sources. The regeneration and growth of such woodlands will provide improved and more abundant habitats for deer, with the likelihood of increased conflicts between both introduced and native deer and the environmental and economic interests associated with the woodlands. This is already the case for native red and introduced sika deer in Scotland. Future climate change may be especially important for those species such as amphibians whose current range and/or breeding success is closely linked to prevailing environmental conditions.[9] However, it may also affect diversity and the balance between native and introduced species for groups less obviously affected by climate such as birds.[198]

## 6.6.2   Human-induced extinctions of native species

Although natural environmental conditions have resulted in a relatively species-poor terrestrial vertebrate fauna, their effects have been exacerbated by extinctions, many of them as a direct result of the actions of humans. Previous extinctions of vertebrates and the influence of humans are most easily traced from the archaeological record for mammals.

Of the 33 mammals extant in Britain around the time of Britain's separation from continental Europe, 10 are now extinct. The reindeer and tarpan (*Equus ferus*) became extinct at about the time the land bridge disappeared. The root vole (*Microtus oeconomus*) became extinct about 5–8000 years ago, elk (*Alces alces*) at 4000 years ago and the aurochs (*Bos primigenius*) about 3000 years ago.[168] These early extinctions were all probably due to a combination of habitat and climate change, although humans may also have contributed to the decline of the tarpan and aurochs.[168] Brown bears (*Ursus arctos*) became extinct about 2000 years ago,[168] and this was probably the first extinction of a native species in which human persecution played the most significant role. Since that time, there have been further extinctions in which human persecution has been the overriding factor. These have been the lynx (*Lynx lynx*) 2000 years ago (previously thought to be a much earlier extinction ca. 4000 years ago), wild boar 700 years ago (although it was maintained after that date as a park animal), beaver (*Castor fiber*) 400 years ago, and wolf (*Canis lupus*) 300 years ago.[168] The numbers of species extinctions are similar for mammals and birds at 10 and 9, respectively,[197] but the proportion of mammals driven to extinction is relatively higher.

## 6.6.3   Effects of competition on the success of introduced
##          terrestrial vertebrates

The relative species paucity of native terrestrial vertebrate fauna, combined with the effects of extinctions, implies that Britain offers a potentially large number of vacant niches for introduced species to occupy. However, the fauna is dynamic, remnant species may increase in abundance in response to extinctions, and the influence of humans on these processes may also be considerable.

Maroo and Yalden compared the current mammal fauna of Britain with a hypothesized Mesolithic mammal fauna, which was based on extrapolation from pristine habitats existing elsewhere in Europe that would have provided similar conditions.[199] They calculated that there has been a 175 million kg decrease in the biomass of wild mammals from 304 million kg in the Mesolithic to 129 million kg now. This is largely due to extinctions or large reductions in the abundance of large herbivores. Introduced species make up 68 million kg of the current fauna, and approximately 90% of this is due to just one species, the rabbit. Of the mainland native mammals, only the badger, fox (*Vulpes vulpes*), field vole (*Microtus agrestis*), mole (*Talpa europaea*), stoat, and water shrew (*Neomys fodiens*) have increased in abundance since the Mesolithic.

Therefore, although the success rates of alien species in Britain are not as high as those for Ireland, which was cut off by raising sea level while still under ice and tundra,[3] they are nevertheless higher than expected. Moreover, analysis of the data on population trends shown in Table 6.1 indicates that of 67 introduced vertebrate species for which data are available, 30 are currently increasing in abundance and/or range, 25 are stable, and 12 are declining. Bird, amphibian and fish invaders are doing especially well at present (23 increasing, 17 stable, and 5 decreasing), while mammalian invaders are doing less well on average (7 increasing, 8 stable, and 7 decreasing).

### 6.6.4   Costs of control and mitigation

The costs of damage per se, especially where the impacts are environmental rather than economic, are very difficult to quantify. Although there is some information on damage costs for a few introduced species, such information is missing for many more. However, the costs of control programs against the species can be used to give some idea of at least the minimum estimated present or potential future damage (Table 6.3).

In 1991–1992, Forest Enterprise (the organization that manages state-owned forests in Britain) spent £1.22 million on shooting deer.[29] There are no more recent figures. In Scotland, between April 1999 and March 2000, 4041 sika deer were culled (D. Goffin, pers. comm.), of which about 60% would have been culled by Forest Enterprise. The total cost of a Forest Enterprise ranger (including overheads, equipment, and capital costs such as vehicle maintenance and provision of larders for storing carcasses) is approximately £13 per hour. The effort required to cull each sika deer will vary considerably with location, population density, and the methods of shooting (whether by day or by night under license). It will also vary with age and sex of the deer. However, in high-density populations of between 20 and 70 sika deer per km$^2$, the average culling rate is approximately 1 sika culled per 15 man-hours. In densities from 2 to 20 per km$^2$, this rate would be considerably lower and in the range of 1 sika per 20–100 man-hours. However, since a smaller proportion of the total cull would be taken from lower density populations, the rate of 1 per 15 man-hours can be used. These effort-based culling estimates compare reasonably closely with independent estimates of 1 sika stag per 4–10 man-hours and 1 sika hind per 10–88 man-hours in a single very high-density population.[200] Based on the more general estimated figures, the cull of 4041 sika deer would have cost about £788,000. Although this is the total investment cost in control, about £200,000 of this was returned in the form of revenues from the sale of venison, so the net cost of sika control in Scotland in 1999–2000 would have been £588,000 (D. Goffin, pers. comm.).

To obtain data for control of the other introduced deer species, it is necessary to take an indirect approach. Toleman provided data on the total number of deer of different species culled in state-owned forests across Britain in 1994–1995.[201] A total of 26,499 deer were culled, composed of 9,589 red, 1,395 sika, 11,933 roe, 2,904 fallow, and 678 muntjac. Rangers from Forest Enterprise spend 70% of their time carrying out crop protection duties, which includes management of wildlife in the forests. Of this 70%, an average of 59% is spent on deer culling, although this figure varies considerably between different regions. Gill et al. stated that Forest Enterprise spent an average of £238 per km$^2$ per year on rangers.[29] Including additional 100% overheads as above, the average total cost of rangers is £476 per km$^2$ per year. Based on these figures, the total cost of controlling deer over the 8350 km$^2$ of Forest Enterprise forests throughout Britain in 1994–1995 was £1.64 million. It is likely that the true figure is at least £2.5 million now.

Detailed data are not available on the relative effort put into the culling of different deer species in forestry. However, an approximation can be made using the above figures based on estimates of the relative time input required to kill an individual deer of each species. This will vary with factors such as population density, habitat, and the behavior of

*Table 6.3* Costs of Damage and Control for the Introduced Vertebrate Species in Britain That Currently Cause the Most Significant Environmental or Economic Costs

| Common name | Scientific name | Summary of damage (econ./env.) | Costs of economic damage | Costs of control |
|---|---|---|---|---|
| | | **Mammals** | | |
| Rabbit | *Oryctolagus cuniculus* | **Econ., env.** | Agriculture: £115 million damage to agriculture each year in Britain[35] Yield loss 1% per rabbit per hectare on winter wheat (£6.50 per rabbit) and 0.5% per rabbit per hectare on spring barley (£2 per rabbit)[36,37] Winter cereal losses up to 35%[38] Forestry: £90 million damage costs (D. Goffin, pers. comm.) | Agriculture: £5 million spent on population control[35] Forestry: Ranger costs incurred in rabbit control by Forest Enterprise throughout Britain in 1994–1995[201] estimated at £529,000 |
| House mouse | *Mus domesticus* | **Econ.** | No estimates | Local authorities used 405,000 kg of rodenticide bait in 1997[209] (R. Quy, pers. comm.) at a total cost of ca. £10/kg = £4.1 million per year |
| Common rat | *Rattus norvegicus* | **Econ.** | No estimates | Arable agriculture: 1,169,103 kg of rodenticide bait used in 1996[208] (R. Quy, pers. comm.) at a total cost of £10/kg = £11.2 million per year |
| Gray squirrel | *Sciurus carolinensis* | **Econ., env.** | No estimates | Forestry: Ranger costs incurred in gray squirrel control by Forest Enterprise throughout Britain in 1994–1995[201] estimated at £334,000 |
| Feral ferret | *Mustela furo* | Econ., **env.** | No estimates | No available data |
| Mink | *Mustela vison* | Econ., **env.** | No estimates | Outer Hebrides control costs: £1.6 million for successful eradication mink from Uists, £2.2 million spent to date on eradication effort on Lewis and Harris (S. Roy, pers. comm.) |

*(Continued)*

Table 6.3 Costs of Damage and Control for the Introduced Vertebrate Species in Britain That Currently Cause the Most Significant Environmental or Economic Costs (*Continued*)

| Common name | Scientific name | Summary of damage (econ./env.) | Costs of economic damage | Costs of control |
|---|---|---|---|---|
| Feral cat | *Felis catus* | **Env.** | N/A | Cats Protection spend £2 million[211] and RSPCA £600,000 (C. Booty, pers. comm.) annually on feral cat neutering and rescue |
| Reeves' muntjac | *Muntiacus reevesi* | Econ., **env.** | No estimates | Forestry: 678 culled by Forest Enterprise throughout Britain in 1994–1995[201] at estimated cost of £14,000 |
| Fallow deer | *Dama dama* | **Econ., env.** | Agriculture: up to £10.30 per hectare per year in southwest England for all deer species[18] | Forestry: 2,904 culled by Forest Enterprise throughout Britain in 1994–1995[201] at estimated cost of £76,000 |
| Sika deer | *Cervus nippon* | **Econ., env.** | Forestry: £400 per hectare for 1-year growth delay due to deer browsing; a further year's growth loss results in a net loss of £288 per hectare[30] | 4,041 sika deer culled in Scotland April 1999–May 2000 at total annual cost of £788,000 (based on 15 ranger hours per sika culled in high-density populations at £13 per hour). Venison return from this is £200,000, so net cost of sika deer cull in Scotland is £588,000 per year (D. Goffin, pers. comm.) Forestry: 1,395 culled by Forest Enterprise throughout Britain in 1994–1995[201] at estimated cost of £220,000 |
| **Birds** | | | | |
| Ruddy duck | *Oxyura jamaicensis* | **Env.** | N/A | Eradication campaign ongoing at cost of £3.4 million (D. Cowan, pers. comm.) |
| Canada goose | *Branta canadensis* | **Econ.**, env. | Cereal agriculture: instances of damage up to £15,000 and 20% yield losses[61] | Amenity grassland: cost of ameliorating trampling and fouling damage up to £40 per bird[63]<br>27 licenses issued for control (19 for shooting, 8 for egg oiling) May 1999–April 2000, resulting in 276 birds and 793 eggs destroyed (R. Watkins, pers. comm.) |

|  |  |  |  |  |
|---|---|---|---|---|
| Pheasant | *Phasianus colchicus* | **Env.** | N/A | Shooting carried out for sport, which generates revenue, so no cost of control per se |
| **Amphibians** | | | | |
| Marsh frog | *Rana ribidunda* | **Env.** | N/A | No coordinated control |
| **Fish** | | | | |
| Rainbow trout | *Oncorhynchus mykiss* | **Env.** | N/A | No coordinated control |
| Carp | *Cyprinus carpio* | **Env.** | N/A | No coordinated control |
| Zander | *Stizostedion lucioperca* | Econ., env. | No estimates | No coordinated control |
| Annual totals | | | >£200 million | >£30 million |

*Note:* Summary of damage taken from details in Table 6.1, with nationally significant damage shown in bold type. Econ. = economic; env. = environmental.

individual rangers, but it is still possible to make generalizations based on species ecology and behavior. For sika deer, it has been estimated that it takes an average of 15 hours to cull one deer (D. Goffin, pers. comm.). Sika are particularly secretive deer with a preference for dense forest and a relatively high effort required per cull. Red deer prefer slightly more open habitats, are less secretive, and easier to cull. Youngson provides figures from which estimates of approximate mean rates of 7 hours per red deer (range 2.7–11.4) and 11 per sika (range 6.4–16.0) can be derived.[202,203] Fallow deer prefer slightly more open habitats than red and sika deer and often feed on rides within the forest. Roe deer are usually solitary or in small groups and show a preference for younger forest stands, which makes them slightly more difficult to cull than either fallow or red. Muntjac are quite secretive, preferring dense cover, but are also inquisitive, and this, combined with high densities, means they are easier to cull than roe deer. Generalizations of culling rates between species need to be treated with caution. Nevertheless, 5 hours per deer for roe, 2.5 for fallow, and 2 for muntjac in comparison with 15 for sika and 7 for red seem approximately realistic. Using these figures, together with data on regional variation in the number of deer culled and the area of Forest Enterprise-owned forests, the total amount spent on control of the different deer species in forestry in 1994/95 can be calculated. Averaging over the whole of Britain, in 1994–1995, £705,000 was spent on red deer control in forestry, £220,000 on sika, £626,000 on roe, £76,000 on fallow, and £14,000 on muntjac. These figures will have risen since 1994–1995, especially for sika (as confirmed by the more recent estimates above) and muntjac. Moreover, these estimates based on labor costs and culling rates gain support from the independently estimated data for sika culls above. In 1994–1995, 3.3 times fewer sika were culled than in 1999–2000, and the calculated cost for 1994–1995 was 3.5 times lower.

Deer damage to woodlands and forestry can also be prevented by fencing, but this is a costly option. It is also frequently ineffective in the long term since holes appear in the fence without continual maintenance and deer can get over the tops of fences during times of heavy snowfall. Fencing can provide protection for between 10 and 15 years depending on the level of maintenance.[204] Staines quoted a cost of around £3 million and a lifetime of around 20 years for 3500 km of deer fencing in Scotland, amounting to an annual outlay of around £500,000.[205] Four years later, Ratcliffe gave a cost of approximately £4 per hectare per year for deer fencing over 300,000 ha of Forestry Commission plantations in Scotland, giving an annual outlay of £1.2 million.[206] However, most of the large-scale deer fencing operations in forestry are primarily for protection against native red and roe deer, so it is impossible to determine the level of expenditure on fencing for introduced species from the data available. Deer fencing also carries indirect costs in that it results in increased mortality of capercaillie due to collisions with fences.[207] This represents a strong conservation reason, in addition to the financial ones, for fencing currently losing favor as a damage prevention strategy. Fences may also be used to exclude deer from nature reserves,[91] but the costs involved are minimal compared with those in forestry.

Total rabbit management costs in Britain in 2000–2001 amounted to approximately £30 million per year (R. Trout, pers. comm.). Total annual management costs within agriculture alone are estimated at £5 million.[35] The cost of rabbit and gray squirrel control in forestry in terms of ranger time can be calculated on the same basis as for deer control according to the proportion of ranger time spent on control of these species. In 1994–1995, this was 13.3% for rabbits and 8.4% for gray squirrels.[201] Over all Forest Enterprise forests in Britain, ranger costs in 1994–1995 were therefore £529,000 for rabbit control and £334,000 for gray squirrel control. These figures include overheads, capital, and general equipment costs associated with rangers, but exclude the cost of specific equipment such as fencing for rabbits. It should also be remembered that the costs of gray squirrel management are concentrated in the forests of England and, to a lesser extent, Wales.

Despite the huge industry associated with the production of rodenticides for the control of commensal rodents, no cost–benefit analysis of commensal rodent control in Britain has been published, and there are no data on the cost-effectiveness of rodenticide treatments. However, the amount of rodenticide used is considerable. In 1996, 1,169,103 kg of rodenticide bait was used by arable agriculture[208] and 405,000 kg by local authorities.[209] The cost of rodenticide application will vary considerably according to whether it is done under contract or by individual farmers, the type of rodenticide used, and the extent of the rodent infestation. Catalog prices for rodenticides range from £2 to £8 per kg, averaging £4.50, and commercial pest control contracts vary widely from £200 to £2500, including bait. The amount of rodenticide bait used per contract also varies, but the average is probably about 20 kg. The lower limit for an average total cost including bait and application would be about £10 per kg (R. Quy, pers. comm.). This gives a cost of commensal rodent control of at least £11.2 million for arable agriculture and £4.1 million for local authorities.

Most mink control is done on a relatively localized scale, so data on costs of control are not generally available. However, cost information is available for the large-scale mink eradication campaigns on islands in the Outer Hebrides, off the west coast of Scotland. Here, the successful campaign to eradicate mink from North and South Uist cost £1.6 million, and eradication campaigns on Lewis and Harris are ongoing with £2.2 million spent to date (S. Roy, pers. comm.).

The most widespread form of feral cat control in Britain is trapping and neutering.[210] The cost of the Royal Society for the Prevention of Cruelty to Animals (RSPCA) feral cat neutering effort is £600,000 each year (C. Booty, pers. comm.) In addition, various other organizations invest considerable resources in neutering unwanted feral cats. The largest of these is Cats Protection, which spends £2 million each year rescuing and neutering cats, work that is mostly done by volunteers.[211]

There are no data specifically on the cost of Canada goose control. However, 27 licenses were issued for control (19 for shooting and 8 for egg oiling) between May 1999 and April 2000, which resulted in 276 birds and 793 eggs being destroyed (R. Watkins, pers. comm.). There are also costs associated with amelioration of goose fouling of recreational areas, and one London park official estimated that the reinstatement of damaged grassland and cleaning of fouled paths cost £40 per bird.[63] For ruddy duck, a coordinated eradication campaign is underway, at a cost to date of around £3.4 million (D. Cowan, pers. comm.). For pheasants, shooting of birds is carried out for sport, which generates considerable revenue, so there are no costs of control per se. There are no relevant data for introduced amphibians or fish, due to the lack of coordinated control efforts against these species, which also reflects the lower level of perceived damage caused by them.

## 6.6.5   *British vertebrates and invasion theory*

Some introduced species that have become naturalized in the British fauna have only done so through persistent introductions or reinforcement of existing populations. This emphasizes the fact that invasion pressure is an important element determining the success of introductions.[3] For example, although mink started escaping from fur farms from 1929 on, the first record of mink breeding in the wild was not until 1956.[11] The same is also true for pheasants, which are enhanced each year throughout much of lowland Britain by the release of captive-reared birds, as referred to earlier. It is also the case for many fish populations. For example, carp, rainbow trout, and brook charr are among the many fisheries species for which populations are continually artificially enhanced.[22] The current British muntjac population stems from no more than eight different maternal lines, and in the two decades following the first releases in 1901, spread was slow and populations were small and very localized.[212] However, spread in the second part of the twentieth century has been much quicker due to further deliberate and accidental releases, frequently outside the existing main range.[212]

Of the extant introduced vertebrate species, 11 out of 22 mammals (50%), 3 out of 21 birds (14%), 1 out of 8 amphibians (25%), none out of 3 reptiles, and 3 out of 13 fish (31%) have had significant environmental or economic costs. Overall, 18 out of 67 extant introduced vertebrate species (27%) have had either significant environmental or economic costs or both, and 9 out of 67 (13%) have had significant economic costs alone.

The tens rule suggests that 10% of introduced species that become established should become pests,[213,214] defined by Williamson as species that have a negative economic effect.[3] Williamson set a standard roughness, or margin of error, for the rule, which has been put at between 5% and 20%.[215] According to the economic definition of pests, the percentage of introduced vertebrates to Britain becoming pests following establishment falls within the upper limits predicted by the tens rule. However, with a wider definition of pests as species that cause negative economic or environmental effects (or that have negative "total" economic effects, i.e., incorporating environmental values as well as financial ones), this proportion is well above that expected by the tens rule.

The proportion of alien vertebrates in Britain that have had significant negative environmental and economic impacts is high. This may reflect the low number of resident species and hence the ability of an introduced species to build up large numbers. Nevertheless, the scale of the impacts of introduced species in Britain is very much less than that experienced by countries with more typical insular biotas, such as New Zealand.[216] Of all the alien vertebrates in Britain, only the rabbit and possibly the coypu have had a sufficient environmental impact to significantly affect ecosystem structure and function, and only the rabbit, house mouse, common rat, and gray squirrel impose major economic costs on a national scale. Moreover, the economic impact of the gray squirrel is mainly confined to central and southern England and Wales. This picture is very different from that for countries with a more typical "island" fauna. However, some introduced species that are currently increasing in numbers and/or range, such as the Reeves' muntjac, sika deer, Canada goose, marsh frog, and zander, have the potential to impose much greater environmental and/or economic costs in the future if their populations continue to grow. Furthermore, natural increases could be exacerbated by habitat and landscape change and also by climate change.

## 6.6.6   *Changing attitudes toward introduced species*

The value of nature is often conceptualized in ecocentric or anthropocentric terms.[217] In this first conceptualization, nature is thought to be "sacred in its own right," (p. 442)[218] and animals are an "ends in themselves." (p. 30)[219] The second conceptualization is the presumption that humans are "dependent on and interconnected with the natural world" (p. 443)[218] for food, clothing, and other resources. This aligns closely with the current development of the environmental policy agenda toward ecosystem services and the ecosystem approach to environmental management.[220] These different understandings permeate the literature on public attitudes to nature, environmental movements, and wildlife management. This section considers how attitudes toward nature in the United Kingdom have changed significantly over the last century, with an increase in concern for animal welfare, and a backlash against the use of animals for the advancement of science or leisure purposes. As Sheail[221] argued, the status of a species is likely to be a function of changing human perceptions and behaviors, with the natural history of the species itself.

The environmental movement in the United Kingdom has developed in a different trajectory to mainland Europe, with a more notable focus on animal welfare and conservation. Serpell and Paul[222] noted that in the early nineteenth century, "England was regarded by its neighbors as one of the cruelest and least sentimental in Europe" (p. 127)[222] in respect to the treatment of animals. The modern environmental movement in the United Kingdom is said to be partly a result of the "crusade against cruelty to animals" that began during the 1880s,[223] and the resulting formation of numerous animal protection groups.[224] However, Garner[217] suggested that concern about cruelty to animals during this period was divided by class, with the greatest cause of concern being the treatment of animals by the urban working class (such as cruelty to horses and using animals for baiting), rather than cruelties inflicted by the aristocracy through hunting, or the scientific elite, through animal experiments.

Since then, the movement progressed from a focus on conservation and access to nature during the war years, the mass environmental movement characterized by moderate direct action and environmental politics of the 1960s and early 1970s, including a revived concern for animal welfare,[222] to a new generation of organizations, including radical direct action and anticapitalist groups in the 1980s.[225] In addition to this, Garner[224] identified this most recent period as one characterized by a steady increase of local campaigning organizations. The development of the environmental movement in this way has led to England being regarded as a world leader in the field of animal protection, with the RSPCA being one of the largest and wealthiest of its kind in Europe, if not the world.[222]

McCormick[223] argued that the development of the environmental movement described above has led to a long-standing conflict between the very different agendas of rural conservation groups and animal rights groups; perhaps this is unsurprising given the class divisions discussed above. There has been a reported shift in public opinion over the last three decades with tolerance for hunting and animal testing decreasing significantly in the 1980s, accounted for in part by the proliferation of animal rights campaigns.[224] The early 1990s saw an increase in disapproval of animal testing, killing animals for their fur, and a significant increase in and acceptance of vegetarianism.[224] In terms of the treatment of invasive species in the United Kingdom, when compared to other countries such as New Zealand, Australia, or Mexico, it appears that very few eradications have occurred.[181,226]

In his review of gray squirrel (*Sciurus carolinensis*) control, Sheail[227] discussed the pressure placed on the government by the agricultural lobby to deal with the threat posed to farming. Strong terms were used in the 1940s, with gray squirrels being described as "the most destructive animal in farming," with an educational campaign launched to ensure that the public were aware of the threat they posed. (p. 148)[227] In the early 1950s, a bounty scheme was introduced to encourage the culling of gray squirrels. However, while the gray squirrels were still regarded as a pest by the 1970s, it was thought that baiting through warfarin was a more effective form of control and reflected the shift toward the broader issues of nature conservation and animal welfare.[227] However, Sheail also contrasted the sympathetic public response toward gray squirrel and rabbit control with the much lower concern toward muskrat and coypu control.[221] For these two species, public and stakeholder concern appeared to be expressed in terms of negative effects on the local environment, outweighing wider arguments around animal rights. These eradications occurred before the animal rights movement had gathered momentum in the 1990s, leading Sheail[221] to question whether the public response would have been the same a decade later. These case studies highlight the complex, changing, and multifaceted nature of public attitudes toward conservation measures, reflecting the likely interplay of social, cognitive, and contextual factors.

### 6.6.7    The future of alien vertebrates in Britain

Alien species make up almost half the biomass of the current British mammal and bird faunas, although this proportion is considerably less for the other vertebrate groups. Eradicating all the alien species would not be practical and would have massive implications for the populations of remaining native species. Some authors argue that long-established alien species in Britain that cause no significant environmental or economic costs should not represent a major cause for concern,[1] although others disagree.[168] For some alien species, the ecological status of the species as a whole or the social benefits they bring to some sectors of society may amount to a positive desire to actually conserve them. Moreover, some alien species can actually make positive contributions to the environment or the economy.

Nevertheless, there are a number of alien species that do impose significant costs. The annual costs of economic damage due to alien vertebrates in Britain are in excess of £200 million, and the costs of control and mitigation are much smaller, in excess of £30 million (Table 6.3). The majority of the damage cost is due to rabbits, yet rabbit grazing can provide considerable biodiversity benefits in certain habitats. Although rabbits are by far the most economically damaging alien vertebrate species in Britain, their relative importance is exacerbated by the lack of data on the costs imposed by other alien vertebrates. The majority of mitigation costs are spent on general control of rabbits and commensal rodents, with some significant spending recently on island-based mink eradication campaigns.

In addition to the introduced vertebrate species that are currently acting as significant pests, there are also a number of species that could potentially cause environmental or economic damage in the future, even though they do no damage in their current state of abundance or distribution. Given that the majority of established alien vertebrates in Britain are currently increasing in range and/or abundance, known populations of these species should be subject to stringent monitoring, and a precautionary approach taken to control if there are signs that these populations are increasing.

## Acknowledgments

We are very grateful to Dave Cowan, Roger Quy, Graham Smith, and Richard Watkins (Central Science Laboratory), Dave Goffin and Dick Youngson (Deer Commission for Scotland), Tom Langton (Froglife), and Colin Booty (Royal Society for the Protection of Animals) for providing unpublished data and reports. We are also grateful to Charles Critchley (Forest Enterprise) for informative discussions. We thank Nick Arnold (Natural History Museum), Jeremy Greenwood (British Trust for Ornithology), Tom Langton (Froglife), Gordon McKillop (Sports Turf Research Institute), Mark Williamson and Dave Raffaelli (University of York), and Alwyne Wheeler for useful comments on the chapter.

## References

1. Manchester, S. J., and J. M. Bullock. 2000. The impacts of non-native species on UK biodiversity and the effectiveness of control. *J Appl Ecol* 37:845.
2. Lever, C. 1992. *They Dined on Eland: The Story of the Acclimatisation Societies.* London: Quiller Press.
3. Williamson, M. 1996. *Biological Invasions.* London: Chapman & Hall.
4. Brown, J. H. 1989. Patterns, modes and extents of invasions by vertebrates. In *Biological Invasions: A Global Perspective*, ed. J. Drake et al., 85. Chichester: John Wiley & Sons.
5. Mack, R. N., D. Simberloff, W. M. Lonsdale, H. Evans, M. Clout, and F. A. Bazzaz. 2000. Biotic invasions: Causes, epidemiology, global consequences, and control. *Ecol Appl* 10:689.

6. Lever, C. 1977. *The Naturalized Animals of the British Isle*. London: Hutchinson.

7. Harradine, J. 1983. Sporting shooting in the United Kingdom—some facts and figures. In *Proceedings of the Second Meeting of the IUGB Working Group on Game Statistics*, ed. F. J. Leeuwenberg and I. R. Hepburn, 63. Doorwerth, Netherlands: IUGB.

8. White, P. C. L., A. E. S. Ford., M. N. Clout., R. M. Engeman., S. Roy, and G. Saunders. 2008. Alien invasive vertebrates in ecosystems: Pattern, process and the social dimension. *Wildl Res* 35:171.

9. Beebee, T. J. C., and R. A. Griffiths. 2000. *Amphibians and Reptiles: A Natural History of the British Herpetofauna*. London: Harper Collins.

10. Harris, S., and D. W. Yalden. 2008. *Mammals of the British Isles: Handbook*. 4th ed. Southampton: The Mammal Society.

11. Harris, S. et al. 1995. *A Review of British Mammals: Population Estimates and Conservation Status of British Mammals Other Than Cetaceans*. Peterborough: Joint Nature Conservation Committee.

12. Ogilvie, M., and the Rare Breeding Birds Panel. 1999. Non-native birds breeding in the United Kingdom in 1996. *Br Birds* 2:176.

13. Delany, S. 1993. Introduced and escaped geese in Britain in summer 1991. *Br Birds* 86:591.

14. Tapper, S. ed. 1999. *A Question of Balance—Game Animals and their Role in the British Countryside*. Fordingbridge, Hampshire: The Game Conservancy Trust.

15. Pithon, J. A., and C. Dytham. 1999. Census of the British ring-necked parakeet Psittacula krameri population by simultaneous counts of roosts. *Bird Study* 46:112.

16. British Ornithologists' Union Records Committee. 2000. *The British List*. Tring, Hertforshire: BOU.

17. Gibbons, D. W., J. B. Reid, and R. A. Chapman. 1993. *The New Atlas of Breeding Birds in Britain and Ireland: 1988–1991*. London: Poyser.

18. Holmes, J. S. et al. 1998. Ducks breeding in the United Kingdom in 1994. *Br Birds* 91:336.

19. Langton, T. E. S., and C. L. Beckett. 1995. Reptiles and amphibians in the United Kingdom: A review of their status and international significance. Unpublished report to Joint Nature Conservation Committee.

20. Maitland, P. S., and A. A. Lyle. Conservation of freshwater fish in the British Isles: The current status and biology of threatened species. *Aquat Conserv* 1:25.

21. Farr-Cox, F., S. Leonard, and A. Wheeler. 1996. The status of the recently introduced fish *Leucaspius delineatus* (Cyprinidae) in Great Britain. *Fish Manag Ecol* 3:193.

22. Maitland, P. S., and R. N. Campbell. 1992. *Freshwater Fishes of the British Isles*. London: Harper Collins.

23. Putman, R. J., and N. P. Moore. 1998. Impact of deer in lowland Britain on agriculture, forestry and conservation habitats. *Mammal Rev* 28:141.

24. Doney, J., and J. Packer. 1998. An assessment of the impact of deer on agriculture. In *Population Ecology, Management and Welfare of Deer*, ed. C. R. Goldspink, S. King and R. J. Putman, 38. Manchester: Metropolitan University.

25. Langbein, J. 1998. Quantock landholders deer management questionnaire: Overview of main results. Unpublished document collated by Dr J. Langbein on behalf of the Quantock Deer Management and Conservation Group.

26. Gill, R. M. A. 1992a. A review of damage by mammals in north temperate forests. 1. Deer. *Forestry* 65:145.

27. Welch, D. et al. 1991. Leader browsing by red and roe deer on young sitka spruce trees in western Scotland. I. Damage rates and the incidence of habitat factors. *Forestry* 64:61.

28. Gill, R. M. A. 1992b. A review of damage by mammals in north temperate forests. 3. Impact on trees and forests. *Forestry* 65:363.

29. Gill, R. M. A., J. Webber, and A. Peace. 2000. *The Economic Implications of Deer Damage: A Review of Current Evidence*. Wrecclesham, Surrey: Forest Research Agency.

30. Ward, A. I., P. C. L. White, A. Smith, and C. H. Critchley. 2004. Modelling the cost of roe deer browsing damage to forestry. *Forest Ecol Manag* 191:301.

31. Fenner, F., and J. Ross. 1994. Myxomatosis. In *The European Rabbit: The History and Biology of a Successful Colonizer*, ed. H. V. Thompson and C. M. King, 205. Oxford: Oxford University Press.

32. Ross, J., and A. M. Tittensor. 1986. Influence of myxomatosis in regulating rabbit numbers. *Mammal Rev* 16:163.

33. Ross, J. et al. 1989. Myxomatosis in farmland rabbit populations in England and Wales. *Epidemiol Infect* 103:333.

34. Kolb, H. H. 1994. Rabbit *Oryctolagus cuniculus* populations in Scotland since the introduction of myxomatosis. *Mammal Rev* 24:41.

35. Smith, G. C., A. J. Prickett, and D. P. Cowan. 2005. Costs and benefits of rabbit control options at the local level. *Int J Pest Manage* 53:317.

36. McKillop, I. G. et al. 1996. Developing a model to predict yield loss of winter wheat due to grazing by European wild rabbits. In *Brighton Crop Protection Conference 1996: Pests and Diseases*, vol. 1, 145. Farnham: British Crop Protection Council.

37. Dendy, J., G. McKillop, S. Fox, G. Western, and S. Langton. 2004. A field trial to assess the effects of rabbit grazing on spring barley. *Ann Appl Biol* 145:77.

38. Bell, A. C., P. M. Byrne, and S. Watson. 1996. The effect of rabbit (*Oryctolagus cuniculus*) grazing damage on the growth and yield of winter cereals. *Ann Appl Biol* 133:431.

39. Kenward, R. E. 1989. Bark-stripping by grey squirrels in Britain and North America: Why does the damage differ? In *Mammals as Pests*, ed. R. J. Putman, 144. London: Chapman & Hall.

40. Gill, R. M. A. 1992. A review of damage by mammals in north temperate forests. 2. Small mammals. *Forestry* 65:281.

41. Rowe, J. 1984. Grey squirrel bark-stripping damage to broadleaved trees in southern Britain up to 1983. *Q J For* 78:231.

42. Mayle, B., J. Proudfoot, and J. Poole. 2009. Influence of tree size and dominance on incidence of bark stripping by grey squirrels to oak and impact on tree growth. *Forestry* 82:431.

43. Rowe, J., and R. Gill. 1985. The susceptibility of tree species to damage by grey squirrels in England and Wales. *Q J For* 79:183.

44. Pepper, H. W. 1990. *Grey Squirrel Damage Control with Warfarin, Forestry Commission Research Information Note 180*. London: HMSO

45. Kenward, R. E. et al. 1998. Comparative demography of red squirrels (*Sciurus vulgaris*) and grey squirrels (*Sciurus carolinensis*) in deciduous and coniferous woodland. *J Zool* 244:7.

46. Prickett, A. J. 1988. English Farm Grain Stores 1987 Part 1. Storage practice and pest incidence. ADAS Central Science Laboratory Research Report, Number 23. Slough: ADAS.

47. Prickett, A. J., and J. Muggleton. 1991. Commercial grain stores 1988/89 England and Wales part 2—pest incidence and storage practice. HGCA Research Report, Number 29. Kenilworth: HGCA.

48. Barnett, S. A. 1951. Damage to wheat by enclosed populations of *Rattus norvegicus*. *J Hyg Camb* 49:22.

49. Goldenberg, N., and C. Rand. 1971. Rodents and the food industry: An in-depth analysis for a large British food handler. *Pest Control* 39:24.

50. Lambert, M. S., R. J. Quy, R. H. Smith, and D. O. Cowan. 2008. The effect of habitat management on home-range size and survival of rural Norway rat populations. *J Appl Ecol* 45:1753.

51. Cotton, K. E. 1963. The coypu. *Riv Board Assoc Yearb* 11:31.

52. Morris, P. 2008. Edible dormouse *Glis glis*. In *Mammals of the British Isles: Handbook*, 4th ed., ed. S. Harris and D. W. Yalden, 82. Southampton: The Mammal Society.

53. Clarke, S. P. 1970. Feral mink in southwest England. *Mammal Rev* 1:92.

54. Chanin, P. R. F. 1981. The diet of the otter *Lutra lutra* and its relations with the feral mink *Mustela vison* in two areas of southwest England UK. *Acta Theriol* 26:83.

55. Dunstone, N., and J. D. S. Birks. 1987. The feeding ecology of mink (Mustela *vison*) in coastal habitat. *J Zool* 212:69.

56. Tapper, S. 1992. *Game Heritage—an Ecological Review from Shooting and Gamekeeping Records*. Fordingbridge, Hampshire: Game Conservancy Limited.

57. Harrison, M. D. K., and R. G. Symes. 1989. Economic damage by feral American mink (*Mustela vison*) in England and Wales. In *Mammals as Pests*, ed. R. J. Putman, 242. London: Chapman & Hall.

58. Chanin, P. R. F., and I. Linn. 1980. The diet of feral mink (*Mustela vison*) in southwest Britain. *J Zool* 192:205.

59. Birks, J. D. S. 1986. *Mink*. London: The Mammal Society.

60. Dunstone, N., and D. W. Macdonald. 2008. American mink *Mustela vison*. In *Mammals of the British Isles: Handbook*, 4th ed., ed. S. Harris and D. W. Yalden, 487. Southampton: The Mammal Society.

61. Simpson, W. 1991. Agricultural damage and its prevention. In *Canada Geese Problems and Management Needs*, ed. J. Harradine, 21. Rossett: BASC.

62. Kirby, J. S. et al. 1995. Index numbers for waterbird populations. III. Long-term trends in the abundance of wintering wildfowl in Great Britain, 1966/67–1991/92. *J Appl Ecol* 32, 536.

63. Allan, J. R., J. S. Kirby, and C. J. Feare. 1995. The biology of Canada geese *Branta canadensis* in relation to the management of feral populations. *Wildl Biol* 1:129.

64. Maitland, P. S., and A. A. Lyle. 1992. Conservation of freshwater fish in the British Isles: proposals for management. *Aquat Conserv* 2:165.

65. Gratz, N. G. 1994. Rodents as carriers of disease. In *Rodent Pests and Their Control*, ed. A. P. Buckle and R. H. Smith, 85. Oxford: CAB International.

66. Hart, C. A., A. J. Trees, and B. I. Duerden. 1997. Zoonoses. *J Med Microbiol* 46:4.

67. Davies, R. H., and C. Wray. 1995. Mice as carriers of *Salmonella enteritis* on persistently infected poultry units. *Vet Rec* 137:337.

68. Pocock, M. J. O. et al. 2001. Patterns of infection by Salmonella and Yersinia spp. in commensal house mouse (*Mus musculus domesticus*) populations. *J Appl Microbiol* 90:755.

69. Healing, T. D., and M. H. Greenwood. 1991. Frequency of isolation of *Campylobacter* spp., *Yersinia* spp. and *Salmonella* spp. from small mammals from two sites in southern Britain. *Int J Environ Health Res* 1:54.

70. Henzler, D. J., and H. M. Opitz. 1992. The role of mice in the epizootiology of *Salmonella enteritis* infection on chicken layer farms. *Avian Dis* 36:625.

71. Quy, R. J. et al. 1999. The Norway rat as a reservoir host of *Cryptosporidium parvum*. *J Wildl Dis* 35:660.

72. Murphy, R. G. et al. 2008. The urban house mouse (*Mus domesticus*) as a reservoir of infection for the human parasite *Toxoplasma gondii*: An unrecognised public health issue? *Int J Environ Health Res* 18:177.

73. Bohm, M., P. C. L. White, J. Chambers, L. Smith, and M. R. Hutchings. 2007. Wild deer as a source of infection for humans and livestock in the UK. *Vet J* 174:260.

74. Delahay, R. J. et al. 2007. Bovine tuberculosis infection in wild mammals in the South-West region of England: A survey of prevalence and a semi-quantitative assessment of the relative risks to cattle. *Vet J* 173:287.

75. Ward, A. I., G. C. Smith, T. R. Etherington, and R. J. Delahay. 2009. Estimating the risk of cattle exposure to tuberculosis posed by wild deer relative to badgers in England and Wales. *J Wildl Dis* 45:1104.

76. White, P. C. L., M. Böhm, G. Marion, and M. R. Hutchings. 2008. Control of bovine tuberculosis in British livestock: There is no 'silver bullet.' *Trends Microbiol* 16:420.

77. Daniels, M. J. et al. 2003. Do non-ruminant wildlife pose a risk of paratuberculosis to domestic livestock and vice versa in Scotland? *J Wildl Dis* 39:10.

78. Judge, J., R. Davidson, G. Marion, C. L. White, and M. R. Hutchings. 2007. Are rabbits *Oryctolagus cuniculus* a true wildlife reservoir for paratuberculosis? *J Appl Ecol* 44:302.

79. Gurnell, J., R. E. Kenward, H. Pepper, and P. W. W. Lurz. 2008. Grey squirrel *Sciurus carolinensis*. In *Mammals of the British Isles: Handbook*, 4th ed., ed. S. Harris and D. E. Yalden, 66. Southampton: The Mammal Society.

80. Public and Corporate Economic Consultants. 2006. *The Economic and Environmental Impact of Sporting Shooting*. Cambridge: PACEC.

81. Hughes, S., and S. Morley. 2000. Aspects of fisheries and water resources management in England and Wales. *Fish Manag Ecol* 7:75.

82. Butler, J. R. A., A. Radford, G. Riddington, and R. Laughton. 2009. Evaluating an ecosystem service provided by Atlantic salmon, sea trout and other fish species in the River Spey, Scotland: The economic impact of recreational rod fisheries. *Fish Res* 96:259.

83. Ward, A. I. 2005. Expanding ranges of wild and feral deer in Great Britain. *Mammal Rev* 35:165.

84. Ward, L. K., R. T. Clarke, and A. S. Cooke. 1994. Long term scrub succession deflected by fallow deer at Castor Hanglands National Nature Reserve. In *Annual Report of the Institute of Terrestrial Ecology (1993–4)*, 78. Swindon: National Environment Research Council.

85. Fuller, R. J., D. G. Noble, K. W. Smith, and D. Vanhinsbergh. 2005. Recent declines in populations of woodland birds in Britain: A review of possible causes. *Br Birds* 98:116.

86. Mitchell, F. J. G., and K. J. Kirby. 1990. The impact of large herbivores on the conservation of semi-natural woodland in the British uplands. *Forestry* 63:333.

87. Watson, A. 1983. Eighteenth century deer numbers and pine regeneration near Braemar, Scotland. *Biol Conserv* 25:289.

88. Scottish Natural Heritage. 1994. Red deer and the natural heritage. Scottish Natural Heritage Policy Paper, Scottish Natural Heritage Publications & Graphics, Perth.

89. Ramsay, P. 1997. *Revival of the Land—Creag Meagaigh National Nature Reserve*. Perth: Scottish Natural Heritage.

90. Kay, S. 1993. Factors affecting severity of deer browsing damage within coppiced woodland in the south of England. *Biol Conserv* 63:217.

91. Putman, R. 1996. Deer management of National Nature Reserves: Problems and practices. *English Nat Res Rep* 173:1.

92. Key, G., N. P. Moore, and J. Hart. 1997. Impact and management of deer in farm woodlands. In *Population Ecology, Management and Welfare of Deer*, ed. C. R. Goldspink, S. King and R. J. Putman, 44. Manchester: Manchester Metropolitan University.

93. Cooke, A. S. 1994a. Is the muntjac a pest in Monks Wood National Nature Reserve? *Deer* 9:243.

94. Cooke, A. S. 1994b. Colonisation by muntjac deer *Muntiacus reevesi* and their impact on vegetation. In *Monks Wood National Nature Reserve, the Experience of 40 Years 1953–1993*, ed. M. S. Massey and R. C. Welch, 45. Peterborough: English Nature.

95. Dolman, P. M., and K. Waber. 2008. Ecosystem and competition impacts of introduced deer. *Wildl Res* 35:202.

96. Watt, A. S. 1919. On the causes of failure of natural regeneration in British oakwoods. *J Ecol* 7:173.

97. Thomas, A. S. 1960. Changes in vegetation since the advent of myxomatosis. *J Ecol* 48:287.

98. Thomas, A. S. 1963. Further changes in vegetation since the advent of myxomatosis. *J Ecol* 51:151.

99. Watt, A. S. 1981. A comparison of grazed and ungrazed grassland A in East Anglian Breckland. *J Ecol* 69:499.

100. Cullen, W. R., C. P. Wheater, and P. J. Dunleavy. 1998. Establishment of species-rich vegetation on reclaimed limestone quarry faces in Derbyshire, UK. *Biol Conserv* 84:25.

101. Boorman, L. A., and R. M. Fuller. 1981. The changing status of reedswamp in the Norfolk Broads. *J Appl Ecol* 18:241.

102. Ellis, E. A. 1963. Some effects of selective feeding by the coypu (*Myocastor coypus*) on the vegetation of Broadland. *Trans Norfolk Norwich Nat Soc* 20:32.

103. Moss, B. 1983. The Norfolk broadland: Experiments in the restoration of a complex wetland. *Biol Rev* 58:521.

104. Jennings, N. 2008. Brown hare *Lepus europaeus*. In *Mammals of the British Isles: Handbook*, 4th ed., ed. S. Harris and D. W. Yalden, 210. Southampton: The Mammal Society.

105. Bullock, D. J. 2008. Feral goat *Capra hircus*. In *Mammals of the British Isles: Handbook*, 4th ed., ed. S. Harris and D. W. Yalden, 628. Southampton: The Mammal Society.

106. Jefferies, M. J., P. A. Morris, and J. E. Mulleneux. 1989. An enquiry into the changing status of the water vole *Arvicola terrestris* in Britain. *Mammal Rev* 19:111.

107. Woodroffe, G. L., J. H. Lawton, and W. L. Davidson. 1990. The impact of mink on water vole populations in the North York Moors National Park. *Biol Conserv* 51:49.

108. Strachan, C., and D. J. Jefferies. 1993. *The Water Vole Arvicola Terrestris in Britain 1989–1990: Its Distribution and Changing Status*. London: The Vincent Wildlife Trust.

109. Halliwell, E. C., and D. W. Macdonald. 1996. American mink *Mustela vison* in the Upper Thames catchment: Relationship with selected prey species and den availability. *Biol Conserv* 76:51.

110. Barreto, G. R. et al. 1998. The role of habitat and mink predation in determining the status and distribution of water voles in England. *Anim Conserv* 1:129.

111. Ferreras, P., and D. W. Macdonald. 1999. The impact of American mink *Mustela vison* on water birds in the Upper Thames. *J Appl Ecol* 36:701.

112. Craik, J. C. A. 1995. Effects of North American mink on the breeding success of terns and smaller gulls in west Scotland. *Seabird* 17:3.

113. Craik, J. C. A. 1997. Long-term effects of North American mink on seabirds in western Scotland. *Bird Study* 44:303.

114. Rae, S. 1999. The effects of predation on ground-nesting birds in the Outer Hebrides. Report to the Mink Eradication Scheme for the Hebrides.

115. Burton, N. H. K., and R. J. Fuller. 1999. A review of the status and population trends of ground-nesting birds vulnerable to mink predation on Harris and Lewis. British Trust for Ornithology Report No. 230, British Trust for Ornithology, Thetford, Norfolk.

116. Baker, P. J., A. J. Bentley, R. Ansell, and S. Harris. 2005. Impact of predation by domestic cats *Felis catus* in an urban area. *Mammal Rev* 35:302.

117. Beckerman, A. P., M. Boots, and K. J. Gaston. 2007. Urban bird declines and the fear of cats. *Anim Conserv* 10:320.

118. Jackson, D. B. 2001. Experimental removal of introduced hedgehogs improves wader nest success in the Western Isles, Scotland. *J Appl Ecol* 38:802.

119. Jackson, D. B., and R. E. Green. 2000. The importance of the introduced hedgehog (*Erinaceus europaeus*) as a predator of the eggs of waders (*Charadrii*) on machair in South Uist. *Biol Conserv* 93:333.

120. Wattaola, G., J. R. Allan, and C. J. Feare. 1996. Problems and management of naturalised introduced Canada geese *Branta canadensis* in Britain. In *The Introduction and Naturalisation of Birds*, ed. J. S. Holmes and J. R. Simons, 71. London: HMSO.

121. Beebee, T. J. C. 1981. Habitats of the British amphibians (3): River valley marshes. *Biol Conserv* 18:281.

122. Langton, T., and J. A. Burton. 1997. *Amphibians and Reptiles. Conservation Management of Species and Habitats.* Strasbourg: Council of Europe Publishing.

123. Wauters, L. A., and J. Gurnell. 1999. The mechanism of replacement of red squirrels by grey squirrels: A test of the interference competition hypothesis. *Ethology* 105:1053.

124. Kenward, R. E., and J. L. Holm. 1993. On the replacement of the red squirrel in Britain: A phytotoxic explanation. *Proc R Soc Lond B Biol Sci* 251:187.

125. Okubo, A. et al. 1989. On the spatial spread of the grey squirrel in Britain. *Proc R Soc Lond B Biol Sci* 238:113.

126. Wauters, L. A., P. W. W. Lurz, and J. Gurnell. 2000. Interspecific effects of grey squirrels (*Sciurus carolinensis*) on the space use and population demography of red squirrels (*Sciurus vulgaris*) in conifer plantations. *Ecol Res* 15:271.

127. Gurnell, J., L. A. Wauters, P. W. W. Lurz, and G. Tosi. 2004. Alien species and interspecific competition: Effects of introduced eastern grey squirrels on red squirrel population dynamics. *J Anim Ecol* 73:26.

128. Shuttleworh, C. 2001. No more grey days for Anglesey red squirrels. *Mammal News* 128:11.

129. Bonesi, L., and D. W. Macdonald. 2004. Impact of released Eurasian otters on a population of American mink: A test using an experimental approach. *Oikos* 106:9.

130. Bonesi, L., R. Strachan, and D. W. Macdonald. 2006. Why are there fewer signs of mink in England? Considering multiple hypotheses. *Biol Conserv* 130:268.

131. McDonald, R. A., K. O'Hara, and D. J. Morrish. 2007. Decline of invasive alien mink (*Mustela vison*) is concurrent with recovery of native otters (*Lutra lutra*). *Divers Distrib* 13:92.

132. Harrington, L. A., A. L. Harrington, N. Yamaguchi, M. D. Thom, P. Ferreras, T. R. Windham, and D. W. Macdonald. 2009. The impact of native competitors on an alien invasive: Temporal niche shifts to avoid interspecific aggression? *Ecology* 90:1207.

133. Fausch, K. D. 2007. Introduction, establishment and effects of non-native salmonids: Considering the risk of rainbow trout invasion in the United Kingdom. *J Fish Biol* 71:1.

134. Zambrano, L., M. R. Perrow, C. D. Sayer, M. L. Tomlinson, and T. A. Davidson. 2006. Relationships between fish feeding guild and trophic structure in English lowland shallow lakes subject to anthropogenic influence: Implications for lake restoration. *Aquat Ecol* 40:391.

135. Davison, A. et al. Hybridization and the phylogenetic relationship between polecats and domestic ferrets in Britain. *Biol Conserv* 87:155.

136. Blandford, P. R. S. 1987. Biology of the polecat *Mustela putorius*: A literature review. *Mammal Rev* 17:155.

137. Kitchener, A. C., N. Yamaguchi, J. M. Ward, and D. W. Macdonald. 2005. A diagnosis for the Scottish wildcat (*Felis silvestris*): A tool for conservation action for a critically-endangered felid. *Anim Conserv* 8:223.

138. French, D. D., L. K. Corbett, and N. Easterbee. 1988. Morphological discriminants of Scottish wildcats (*Felis silvestris*), domestic cats (*F. catus*) and their hybrids. *J Zool* 214:235.

139. Daniels, M. J. et al. 1998. Morphological and pelage characteristics of wild living cats in Scotland: Implications for defining the "wildcat." *J Zool* 244:231.

140. Daniels, M. J. et al. 2001. Ecology and genetics of wild-living cats in the north-east of Scotland and the implications for the conservation of the wildcat. *J Appl Ecol* 38:146.

141. Diaz, A., S. Hughes, R. Putman, R. Mogg, and J. M. Bond. 2006. A genetic study of sika (*Cervus nippon*) in the New Forest and in the Purbeck region, southern England: Is there evidence of recent or past hybridization with red deer (*Cervus elaphus*)? *J Zool* 270:227.

142. Senn, H. V., and J. M. Pemberton. 2009. Variable extent of hybridization between invasive sika (*Cervus nippon*) and native red deer (*C-elaphus*) in a small geographical area. *Mol Ecol* 18:862.

143. Robertson, P. A., and A. A. Rosenberg. 1988. Harvesting gamebirds. In *Ecology and Management of Gamebirds*, ed. P. J. Hudson and M. R. W. Rands, 177. Oxford: BSP Professional.

144. Hughes, B. 1996. The ruddy duck *Oxyura jamaicensis* in the Western Palearctic and the threat to the white-headed duck *Oxyura leucocephala*. In *The Introduction and Naturalisation of Birds*, ed. J. S. Holmes and J. R. Simons, 79. London: HMSO.

145. Snell, C. 1994. The pool frog: A neglected native? *Br Wildl* 6:1.

146. Wheeler, A. 2000. Status of the Crucian carp, *Carassius carassius* (L.), in the UK. *Fish Manag Ecol* 7:315.

147. Beveridge, M. C. M., L. G. Ross, and L. A. Kelly. 1994. Aquaculture and biodiversity. *Ambio* 23:497.

148. Bratton, S. P. 1975. The effect of the European wild boar (*Sus scrofa*) on Gray Beech Forest in the Great Smoky Mountains. *Ecology* 56:1356.

149. Welander, J. 1995. Are wild boar a future threat to the Swedish flora? *Ibex* 3:165.

150. Gosling, L. M., and S. J. Baker. 1989. The eradication of muskrats and coypus from Britain. *Biol J Linn Soc* 38:39.

151. Sage, R. B., M. I. A. Woodburn, R. A. H. Draycott, A. N. Hoodless, and S. Clarke. 2009. The flora and structure of farmland hedges and hedgebanks near to pheasant release pens compared with other hedges. *Biol Conserv* 142:1362.

152. Draycott, R. A. H., A. N. Hoodless, and R. B. Sage. Effects of pheasant management on vegetation and birds in lowland woodlands. *J Appl Ecol* 46:334.

153. Olofsson, J., C. de Mazancourt, and M. J. Crawley. 2008. Spatial heterogeneity and plant species richness at different spatial scales under rabbit grazing. *Oecologia* 156:825.

154. Lees, A. C., and D. J. Bell. 2008. A conservation paradox for the 21st century: The European wild rabbit *Oryctolagus cuniculus*, an invasive alien and an endangered native species. *Mammal Rev* 38:304.

155. Sumption, K. J., and J. R. Flowerdew. 1985. The ecological effects of the decline of rabbits (*Oryctolagus cuniculus* L.) due to myxomatosis. *Mammal Rev* 15:151.

156. Brereton, T. M., M. S. Warren, D. B. Roy, and K. Stewart. 2008. The changing status of the Chalkhill Blue butterfly *Polyommatus coridon* in the UK: The impacts of conservation policies and environmental factors. *J Insect Conserv* 12:629.

157. Davies, Z. G., R. J. Wilson, T. M. Brereton, and C. D. Thomas. 2005. The re-expansion and improving status of the silver-spotted skipper butterfly (*Hesperia comma*) in Britain: A meta-population success story. *Biol Conserv* 124:189.

158. Harris, S. et al. 2001. Abundance/mass relationships as a quantified basis for establishing mammal conservation priorities. In *Priorities for the Conservation of Mammalian Biodiversity: Has the Panda had its Day?*, ed. A. Entwistle and N. Dunstone, 101. Cambridge: Cambridge University Press.

159. Anon. 1995. *Biodiversity: The UK Steering Group Report*, vol. 2, *Action Plans*. London: HMSO.

160. Lockwood, J. A. 1997. Competing values and moral imperatives: An overview of ethical issues in biological control. *Agric Human Values* 14:205.

161. Battersby, J., ed., and Tracking Mammals Partnership. 2005. UK Mammals: Species Status and Population Trends. First Report by the Tracking Mammals Partnership, Joint Nature Conservation Committee/Tracking Mammals Partnership, Peterborough.. http://www.jncc.gov.uk/pdf/pub05_ukmammals_speciesstatusText_final.pdf (accessed June 18, 2010).

162. Webb, T. J., and D. Raffaelli. 2008. Conversations in conservation: Revealing and dealing with language differences in environmental conflicts. *J Appl Ecol* 45:1198.

163. Lincolnshire Wildlife Trust. 2005. Hedgehog rockets to UK's number one. http://www.lincstrust.org.uk/news/press-release.php?article=226 (accessed November 2005).

164. Potter, B. 1905. *The Tale of Mrs Tiggy-Winkle*. England: Frederick Warne & Co.

165. Berlin, I. 1953. *The Hedgehog and the Fox: An Essay on Tolstoy's View of History*. London: Weidenfeld & Nicolson.

166. Gould, S. J. 1997. Self-help for a hedgehog stuck on a molehill. *Evolution* 51:1010.

167. The Guardian. 2007. Uist hedgehog cull scrapped. http://www.guardian.co.uk/environment/2007/feb/20/animalrights.conservationandendangeredspecies (accessed February 2007).

168. Yalden, D. 1999. *The History of British Mammals*. London: Poyser.

169. Goulding, M. J., G. Smith, and S. J. Baker. 1998. Current status and potential impact of wild boar (*Sus scrofa*) in the English countryside: A risk assessment. Central Science Laboratory report to the Ministry of Agriculture, Fisheries and Food (MAFF). London: MAFF.

170. Evans, I. M., R. W. Summers, L. O. Toole, D. C. Orr-Ewing, R. Evans, N. Snell, and J. Smith. 1999. Evaluating the success of translocating Red Kites *Milvus milvus* to the UK. *Bird Study* 46:129.

171. Goulding, M. J., and T. J. Roper. 2002. Press response to the presence of free-living Wild Boar. *Mammal Rev* 32:272.

172. Rackham, O. 1986. *The History of the Countryside*. London: Phoenix Press.

173. Grandy, J. W., E. Stallman, and D. W. Macdonald. 2003. The Science and Sociology of Hunting: Shifting Practices and Perceptions in the United States and Great Britain. In *The State of the Animals II*, ed. D. J. Salem and A. N. Rowan, 107. Washington, DC: Humane Society Press.

174. Veitch, C. R., and M. N. Clout. 2001. Human dimensions in the management of invasive species in New Zealand. In *The Great Reshuffling: Human Dimensions of Invasive Alien Species*, ed. A. J. McNeely, 63. Gland, Switzerland and Cambridge, UK: IUCN.

175. Benedictow, O. J. 2004. *The Black Death 1346–1353. The Complete History*. Woodbridge: The Boydell Press.

176. Keeling, M. J., and C. A. Gilligan. 2000. Bubonic plague: A metapopulation model of a zoonosis. *Proc R Soc Lond B Bio Sci* 267:2219.

177. Kellert, S. R., M. Black, C. R. Rush, and A. J. Bath. 1996. Human culture and large carnivore conservation in North America. *Conserv Biol* 10:977.

178. Herzog, T. R., E. J. Herbert, R. Kaplan, and C. L. Crooks. 2000. Cultural and developmental comparisons of landscape perceptions and preferences. *Environ Behav* 32:323.

179. Clark, J. A., and R. May. 2002. Taxonomic bias in conservation research. *Science* 297:191.

180. Seddon, P. J., P. S. Soorae, and F. Launay. Taxonomic bias in reintroduction projects. *Anim Conserv* 8:51.

181. Bremner, A., and K. Park. 2007. Public attitudes to the management of invasive non-native species in Scotland. *Biol Conserv* 139:306.

182. Knegtering, E., L. Hendrickx, H. Van de Windt, and A. J. M. Schootuiterkamp. 2002. Effects of species' characteristics on nongovernmental organizations' attitudes towards species conservation policy. *Environ Behav* 34:378.

183. Bruskotter, J. T., J. J. Vaske, and R. H. Schmidt. 2009. Social and cognitive correlates of Utah residents' acceptance of the lethal control of wolves. *Hum Dimens Wildl* 14:119.

184. Philip, L. J., and D. Macmillan. 2003. *Public Perceptions of, and Attitudes Towards, the Control of Wild Animal Species in Scotland*. Stirling: University of Stirling.

185. Jordan, G., and W. A. Maloney. 2007. *Democracy and Interest Groups*. Basingstoke: Palgrave Macmillan.

186. Jacobs, M. H. 2009. Why do we like or dislike animals? *Hum Dimens Wildl* 14:1.

187. Hill, N. J., K. A. Carbery, and. E. M. Deane. 2007. Human-possum conflict in urban Sydney, Australia: Public perceptions and implications for species management. *Hum Dimens Wildl* 12:101.

188. Fitzgerald, G., N. Fitzgerald, and C. Davidson. 2007. Public attitudes towards invasive animals and their impacts. Canberra: Invasive Animals Cooperative Research Centre. http://www.invasiveanimals.com/downloads/12D5-Social-Literature-Review-Final-July-07.pdf (accessed June 18, 2010).

189. Balmford, A., L. Clegg, T. Coulson, and J. Taylor. 2002. Why conservationists should heed Pokémon. *Science* 295:2367b.

190. Goode, D. A. 1989. Urban nature conservation in Britain. *J Appl Ecol* 26:859.

191. Moriarty, A., A. English, and R. Mulley. 2009. Conservation Hunting and its role in game and feral animal management: A response to papers by the Invasive Species Council of Australia, Game Council, NSW.

192. Pounds, N. J. G. 1994. *The Culture of the English People: Iron Age to the Industrial Revolution.* Cambridge: Cambridge University Press.

193. Atkinson, T. C., K. R. Briffa, and G. R. Coope. 1987. Seasonal temperatures in Britain during the past 22,000 years, reconstructed using beetle remains. *Nature* 325:587.

194. Bilton, D. T. et al. 1998. The Mediterranean region of Europe as an area of endemism for small mammals rather than the major source for the postglacial colonisation of northern Europe. *Proc R Soc Lond B Bio Sci* 265:1219.

195. Corbet, G. B., and S. Harris. 1991. *The Handbook of British Mammals.* 3rd ed. Oxford: Blackwell Scientific Publications.

196. Mitchell-Jones, A. J. et al. 1999. *The Atlas of European Mammals.* London: Poyser.

197. Greenwood, J. J. D. et al. 1996. Relations between abundance, body size and species number in British birds and mammals. *Philos Trans R Soc Lond B Biol Sci* 351:265.

198. Lennon, J. J., J. J. D. Greenwood, and J. R. G. Turner. 2000. Bird diversity and environmental gradients in Britain: A test of the species-energy hypothesis. *J Anim Ecol* 69:581.

199. Maroo, S., and D. W. Yalden. 2000. The Mesolithic mammal fauna of Great Britain. *Mammal Rev* 30:243.

200. McLean, C. 2001. Costs of sika control in native woodland: Experience at Scaniport Estate. *Scott For* 55:109.

201. Toleman, R. D. L. 1995. Deer cull returns 1994–95. Unpublished report, Forestry Commission.

202. Youngson, R. W. 2000a. Red deer control in Fiunnary Forest, Argyll: Control effort. Unpublished summary of a joint project report between Forestry Commission and Red Deer Commission.

203. Youngson, R. W. 2000b. Costs of sika control in native woodland. Unpublished summary of a joint report between Forestry Commission and Red Deer Commission.

204. Mayle, B. 1999. Managing deer in the countryside, Forestry Commission Practice Note No. 6. Edinburgh: Forestry Commission.

205. Staines, B. W. 1985. Humane control of deer in rural areas. In *Humane Control of Land Mammals and Birds*, ed. D. P. Britt, 105. Potters Bar, Hertfordshire: Universities Federation for Animal Welfare.

206. Ratcliffe, P. R. 1989. The control of red and sika deer populations in commercial forests. In *Mammals as Pests*, ed. R. J. Putman, 98. London: Chapman & Hall.

207. Baines, D., and R. W. Summers. 1997. Assessments of bird collisions with deer fences in Scottish forests. *J Appl Ecol* 34:941.

208. De'Ath, A., D. G. Garthwaite, and M. R. Thomas. 1999. Rodenticide usage on farms in Great Britain growing arable crops 1996. PUSG Report 144, Ministry of Agriculture, Fisheries and Food Publications, London.

209. Bankes, J., and D. G. Garthwaite. 2000. Rodenticide usage by local authorities in Great Britain 1997, PUSG Report 155, Ministry of Agriculture, Fisheries and Food Publications, London. http://secure2.csl.gov.uk/plants/pesticideUsage/rodmunicip1998.pdf (accessed June 18, 2010).

210. Remfry, J. 1996. Feral cats in the United Kingdom. *J Am Vet Med Assoc* 208:520.

211. National Cat Centre. 2001. Cats protection. Haywards Heath, Sussex: National Cat Centre. http://www.cats.org.uk (accessed June 18, 2010).

212. Chapman, N., S. Harris, and A. Stanford. 1994. Reeves' muntjac *Muntiacus reevesi* in Britain: Their history, spread, habitat selection, and the role of human intervention in accelerating their dispersal. *Mammal Rev* 24:113.

213. Holdgate, M. W. 1986. Summary and conclusions: Characteristics and consequences of biological invasions. *Philos Trans R Soc Lond B Biol Sci* 314:733.

214. Williamson, M., and K. C. Brown, K. C. 1986. The analysis and modelling of British invasions. *Philos Trans R Soc Lond B Biol Sci* 314:505.

215. Williamson, M. 1992. Environmental risks from the release of genetically modified organisms (GMOs)—the need for a molecular ecology. *Mol Ecol* 1:3.

216. Lever, C. 1994. *Naturalized Animals: The Ecology of Successfully Introduced Species*. London: Poyser.

217. Garner, D. 1996. *Environmental Politics*. London: Harvester Wheatsheaf.

218. Ignatow, G. 2006. Cultural models of nature and society, reconsidering environmental attitudes and concern. *Environ Behav* 38:441.

219. Macnaughten, P. 2001. Animal futures: Public attitudes and sensibilities towards animals and biotechnology in contemporary Britain. A report for the Institute for Environment, Philosophy and Public Policy for the Agricultural and Environmental Biotechnology Commission.

220. Millennium Ecosystem Assessment. 2005. *Ecosystems and Human Well-Being: Biodiversity Synthesis*. Washington, DC: World Resources Institute.

221. Sheail, J. 2003. Government and the management of an alien pest species: A British perspective. *Landsc Res* 28:101.

222. Serpell, J., and E. Paul. 1994. Pets and the development of positive attitudes to animals. In *Animals and Human Society: Changing Perspectives*, ed. A. Manning and J. Serpell, 127. London & New York: Routledge.

223. McCormick, J. 1991. *Reclaiming Paradise: The Global Environmental Movement*. Bloomington: Indiana University Press.

224. Garner, D. 1993. The animal protection movement. *Parliamentt Aff* 46:333.

225. Rawcliffe, P.1998. *Environmental Pressure Groups in Transition*. Manchester: Manchester University Press.

226. Genovesi, P. 2005. Eradications of invasive alien species in Europe: A review. *Biol Invasions* 7:127.

227. Sheail, J. 1999. The grey squirrel (*Sciurus carolinensis*)—a UK historical perspective on vertebrate pest species. *J Environ Manage* 55:145.

*section four*

---

*Europe*

*chapter seven*

# Impacts of alien vertebrates in Europe

*Susan M. Shirley and Salit Kark*

## Contents

## 7.1  Introduction

Alien species are those species that have been introduced either intentionally or unintentionally into areas to which they are not native. There are a number of stages in the invasion process.[1] Species are imported for a variety of purposes, including as domestic pets or for hunting, livestock, and food, and then escape from captivity or are released into the wild. Once introduced, certain species are able to breed and establish self-sustaining populations. Some of these established populations spread to new areas through natural dispersal and expand their populations. In the absence of factors to regulate their population abundances, certain species may reach population abundances that impose negative impacts on the ecosystems and human activities in their introduced ranges. Our evolving understanding of the magnitudes of these impacts has led to growing concern.[2,3]

In this chapter, we provide an overview of the impacts of alien terrestrial vertebrates in Europe. We focus primarily on mammals and birds but also provide a brief overview of impacts of reptiles and amphibians. We begin with a general overview of introductions and impacts for each taxa group followed by a more detailed discussion of some of the species with the most important impacts on biodiversity and economic resources. We then highlight eradication efforts and future management of impacts. Finally, we propose a simple procedure to rank impacts of alien vertebrates, using alien birds as a case study.

## 7.2   Mammal introductions

The history of mammal introductions to Europe by humans goes back to the early Neolithic period as wild animals became commensal with humans (e.g., house mouse [*Mus musculus*]) or arose from escapes of deliberately domesticated species. In modern times (since 1500), at least 88 known species of alien mammals have been introduced into Europe.[4] Of these, approximately 20 species are native to some other part of Europe, whereas the majority of species are of a non-European origin, mainly Asia and North America. A high percentage (72%) of introduced mammal species have become established with self-sustaining populations, and the rate of successful introductions has increased exponentially with the largest number of introductions occurring since the mid-1800s. Many of these species were introduced for fur hunting and hunting as well as misguided attempts to "improve" the local fauna. Currently, about 59 invasive species are present in Europe.[4]

### 7.2.1   Alien mammal impacts in Europe

Many of the mammal species introduced into Europe have known negative ecological impacts and/or impacts on human activities. Approximately 40 mammal species are known to cause damage to crops, livestock, and other human activities, including human health, while 37 species have negative ecological impacts on native species, such as competition, hybridization, and predation.[4] Mammals in the Muridae (rats), Mustelidae (mink), and Echimyidae (coypus) families have the highest environmental impacts, while economic impacts were distributed relatively evenly across families (Figure 7.1). The

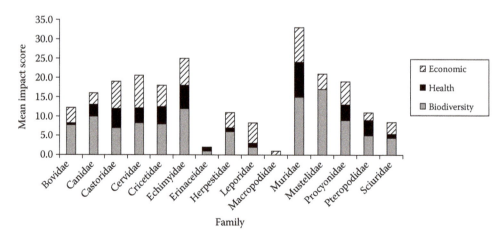

**Figure 7.1** Potential economic, biodiversity, and human health impact scores of alien mammals in Europe. Shown are average scores for all mammal families. (Data from Nentwig, W. et al., *Conserv Biol* 24:302, 2009.)

most prevalent types of impacts across all mammal orders are disease transmission to animals or humans (health impacts), herbivory (environmental impacts), and impacts on human health, livestock, and forestry (economic impacts). Several of the mammal species, including the sika deer, muskrat, and brown rat, have high impacts on both environmental and economic resources.[5]

### 7.2.2 Biodiversity impacts

A range of alien mammals have potential negative environmental impacts in Europe. Some are widely distributed, having been introduced many years ago, while others are recent introductions with limited ranges at present. One of the species with the most widespread distribution, *Rattus norvegicus*, lives in a variety of habitats in its introduced range, particularly human-associated habitats where it can reproduce year-round. As a predator unintentionally introduced through commercial trade, it is responsible for declines in several bird, small mammal, and reptile populations[4] and has one of the highest economic damage scores of any alien mammal in Europe.[5] Evidence of the detrimental effect on native mammal populations was shown by over sevenfold increases in the abundances of two shrew species on European islands when the brown rat was eradicated.[6] Although not easily quantified, the cost of *Rattus* spp. population control is very high. However, on an island in Corsica, costs of eradication of introduced *Rattus* spp. became less than the cost of control within 6 years and provided increased breeding success for seabirds and other ecological benefits.[7]

Several other widely distributed mammal species were originally introduced for fur farming, and populations became established following escapes. Muskrats (*Ondatra zibethicus*) and nutria or coypus (*Myocastor coypus*) are generalist herbivores that can cause severe damage to native wetland plants and reed beds.[8] Muskrat predation is also suspected of being responsible for declines of an endangered mussel *Unio crassus* in Germany.[8] Both are also implicated in *leptospirosis* disease transmission. Similarly, the carnivorous American mink (*Neovison vison*) and omnivorous raccoon dog (*Nyctereutes procyonoides*) prey on native birds, mammals, and amphibians. Both species also compete with native mammal species for food and den sites, causing displacement and native mammal population declines.[9,10] The American mink alone has been linked to declines in density of over 30 native species in Europe.

A few species such as the sika deer (*Cervus nippon*), raccoon (*Procyon lotor*), gray squirrel (*Sciurus carolinensis*), and Siberian chipmunk (*Tamias sibiricus*), currently with limited distributions in Europe, have serious environmental impacts and pose a concern because of their potential for rapid spread. The sika deer and raccoon were introduced for hunting and ornamental purposes and either escaped or were deliberately released into wild areas. In natural woodland habitats, heavy foliage browsing by sika deer can reduce the understory structural diversity and composition producing negative impacts for some animal species, although the creation of more open habitats may benefit other species.[11] Perhaps a more serious environment impact results from their ability to hybridize with the native red deer (*Cervus elaphus*) and create fertile hybrids. Hybridization has occurred in many areas of the United Kingdom, potentially threatening to replace this species in some habitats.[11] Sika deer are also implicated in the transfer of nematode diseases such as *Asworthius sidemi* to other wild deer populations.[12]

The gray squirrel has its largest populations in the United Kingdom and also a rapidly expanding population in Italy.[13] Its competition with the native red squirrel (*Sciurus vulgaris*) is one of the best examples of a negative competitive effect of an alien vertebrate species on another species. In many areas of the United Kingdom, red squirrel populations have declined in the presence of the gray squirrel apparently from reduced juvenile recruitment allowing gray squirrels to reach higher densities especially in deciduous forests.[14] A similar pattern is also found in Italy where the range of the red squirrel inhabited by the gray squirrel has contracted by over 70%.[15]

### 7.2.3   *Economic impacts*

In addition to environmental impacts, many mammal species introduced into Europe also have negative economic impacts. Species such as the muskrat, coypu, American mink, and raccoon dog that live in or along streams, lakes, and marshes disrupt a number of human activities. Muskrats and coypus engage in burrowing activity that causes significant damage to riverbanks, and feeding on agricultural crops such as sugar beets causes further damage. Annual damage costs are estimated at over €12 million in Germany[16] for muskrats. In Italy, despite control costs of €2.6 million from 1995 to 2000, costs for coypu damage to riverbanks and agriculture exceeded €11 million. As well as preying on native species, the American mink also negatively impacts organic chicken farming and fisheries.[17] Assessments of the extent of economic cost appear to vary by country; in Britain, costs of damage to livestock and fisheries are thought to be low overall,[11] whereas in Scandinavia, losses of game and fish are significant.[18] The cost of preventing damage to livestock and ecotourism by the American mink in Germany exceeds €4 million per year.[19] The raccoon dog has a similar reputation as a pest of game and fish in European Russia[20] and is also responsible for damage to vineyards in the Ukraine.[18]

Sika deer can cause considerable economic damage in areas where they are numerous (such as the United Kingdom and Switzerland) by feeding on both agricultural crops such as grass, oats, soybean, and corn as well as browsing young broadleaved and conifer forest plantations.[21,22] In the United Kingdom, losses in forestry plantations due to browsing of leading shoots of young trees were estimated at £400 per hectare for loss of growth for 1 year.[23] Likewise, the fallow deer (*Dama dama*) in high densities is a pest to agriculture and forestry causing damage to vegetable, fruit, and grain crops, and young tree plantations.[18] In the United Kingdom, they are the deer species most responsible for damage to agricultural crops and forestry.[24] Also in the United Kingdom, gray squirrels cause significant damage to forest trees by stripping the bark off young plantation trees, resulting in tree death or inferior wood products[25] although a comprehensive assessment of their economic damage has not been undertaken.[11] Broadleaved trees such as sycamore (*Acer pseudoplantanus*), beech (*Fagus sylvatica*), and oak (*Quercus* spp.) are more affected than conifer species, with some reports of up to 87% of stands reporting damage.[25] These broadleaved species are also common in the Alps of Italy where the gray squirrel's range is currently expanding.[13] Less serious damage has been reported to grain and nut crops as well as fruit trees.[26]

The European rabbit (*Oryctolagus cuniculus*), introduced as early as the eleventh century by the Romans and Normans, has become naturalized throughout western Europe[18] and causes extensive damage to a broad range of agricultural crops, including cereal and vegetable crops, forest plantations, and vineyards.[8] As burrowers, they also cause damage to levees, dykes, and other structures. In Germany, the total amount of damage is €5.62 million per year.[8] In the United Kingdom, the amount of damage has varied over time with rabbit population cycles induced by the introduction of myxomatosis disease.[11] An estimate of total damage for the 1980s was £100 million with an additional £30 million for control costs.[11]

## 7.2.4 Eradications of alien mammals in Europe

Mammals are generally believed to have more severe impacts than other vertebrates and, as a result, most control and eradication efforts have been directed toward mammal species. A review of introduced mammals in Europe,[27] suggests that a number of successful eradications have been carried out, most on islands off the coasts of several European countries such as Britain, Spain, Portugal, France, and Italy. In general, mammal species targeted for eradication on these islands have been those with the most severe impacts, for example *Rattus* spp., American mink. The majority of eradication attempts followed a five-point global strategy.[28] This strategy included providing an initial description of the problem within the context of the ecosystem and its other components, performing and monitoring the outcome of the eradication, assessing impacts of the eradication on native species, and establishing a protocol to prevent new invasions. All subsequent effects of the eradication of rats from the islands on native species were beneficial, resulting in increases in abundance, number of breeding pairs, or breeding success. In the last decade, American mink eradications have been successful on a number of islands off the coasts of Finland, Estonia, and the United Kingdom, also resulting in positive effects on native fauna.[9] Successful rabbit eradication campaigns were also conducted on several European islands off Spain and Portugal in the 1990s.[27]

There have been far fewer eradication efforts on the mainland areas of Europe due to inadequate legal support, public awareness, and funding.[27] A campaign to eradicate the gray squirrel from northern Italy failed due to opposition from radical animal rights groups, highlighting the need for a strong political commitment and legal authority in these situations.[13] Other eradications such as the gray squirrel in the United Kingdom also failed,[26] and control efforts are now undertaken on a local basis. Despite these challenges, a few eradications have been highly successful and illustrate the potential for these to be effective even in widespread populations. The nutria eradication in Britain is one example of a successful eradication campaign that was initially unsuccessful, but eventually succeeded through improvements to the overall strategy.[29] Coypus, native to South America and originally introduced to Britain for fur farming in 1929, escaped from farms and established a population covering the whole of East Anglia, reaching numbers in the tens of thousands by the 1950s. Nutria feed on native plants of reed beds as well as crops such as sugar beets. In addition, as burrowers, they cause extensive damage to drainage systems.[30] After the first eradication, numbers declined, but this period also coincided with an unusually cold winter in 1962. Subsequent milder winters resulted in an eruption in numbers revealing that trapping had been insufficient. A second eradication attempt in 1981 that made use of a long-term population study and improved trapping methods was ultimately successful. However, this species remains widely distributed across mainland Europe with the widest coverage in France.[31]

## 7.2.5 Future predicted impacts and management of alien mammals

In the last 10 years, several species have expanded their ranges in Europe, the American mink, nutria, sika deer, and raccoon considerably so by over 35% of grid cells measured.[5] Species in mammal groups such as cervids and sciurids have high potential impacts, but limited distributions at present. Genovesi et al.[4] report that 17 mammals are established in only one country, whereas others have localized distributions in a few countries. These species represent good candidates for eradication efforts as their local distributions increase the chance of success. For example, the Siberian chipmunk currently has local distributions in several countries including Italy, the United Kingdom, Germany, the Netherlands, and France. In its native range, this species has considerable impacts on both grain and nut crops and may compete with native forest rodents.[32] Well-organized eradication campaigns could eradicate or at least largely limit the invasive extent of these species. Where potentially invasive species are sold in pet shops, prevention of this practice for certain species could limit the potential for new introductions. Although there are no specific life-history characteristics that can predict the successful establishment and impacts of mammal species, the impacts of a species in its native range or other introduced areas can provide an early warning of possible impacts in Europe. In addition, those species with the most severe impacts appear to be habitat generalists, occurring in a wide variety of habitats.[5] This knowledge should inform efforts to prioritize species for eradication and management and to prevent their introduction in the first place.

## 7.3   Introductions of birds, reptiles, and amphibians

Since 1850, 197 bird species belonging to 37 different families are known to have been introduced into Europe.[33] The number of species introduced increased slowly from 1850 up until the 1940s and since then has increased at a much faster rate (Figure 7.2). Of these introductions, there are 77 bird species that have established populations in Europe as of 2000. There are 55 reptile and amphibian species belonging to 16 families that have been successfully introduced into Europe.[33] Although the numbers of alien reptile and amphibians have also increased in a similar pattern to birds over time, they have done so at a much slower rate (Figure 7.2). The highest numbers of alien bird, reptile, and amphibian species are found in the western European countries of the United Kingdom, France,

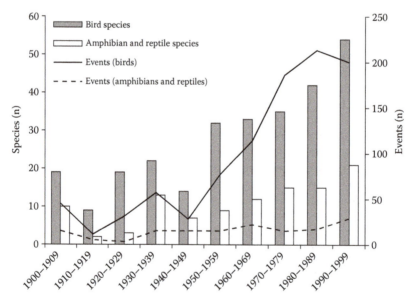

*Figure 7.2* Temporal trends in the number of introduced amphibian and reptile species (bars) and the number of introduction events (line) in Europe in the twentieth century. (From Kark, S., Solarz, W., Chiron, F., Clergeau, P., and Shirley, S. 2008. Alien birds, amphibians, and reptiles of Europe. In *Handbook of Alien Species in Europe*, ed Delivering Alien Invasive Species Inventories for Europe (DAISIE), 105. Dordrecht: Springer Science & Business Media. With permission.)

Spain, and Italy, with numbers decreasing as one moves eastward. Many of the introduced species were intentionally released for hunting and for improving the local fauna or were unintentionally introduced as escaped pets and zoo animals.

### 7.3.1 Impacts of birds

Among 140 birds introduced in Europe that are breeding, there are 64 species with known impacts either in Europe or known from their native range or other areas of introduction.[34] Most of the species that have impacts originated from the Afro-tropical, Indo-Malayan, and Palearctic regions. Economic impacts were reported for 56 species, whereas biodiversity and human health impacts were reported for 27 and 10 species, respectively (Table 7.1).[34] The most common type of economic impact reported was agricultural damage to grain or fruit crops. Biodiversity impacts were mainly negative competitive interactions with native species leading to their declines, predation, or hybridization. Economic impacts were widely distributed across several families, including Columbidae (doves), Estrildidae (weaver-finches), Ploceidae (weavers), and Psittacidae (parrots; Table 7.1). Several families including the Anatidae (ducks and geese), Corvidae (crows), Passeridae (sparrows), Phasianidae (pheasants), and Sturnidae (starlings) had both reported economic and biodiversity impacts.

*Table 7.1* The Number of European Alien Bird Species within Families with Reported Biodiversity, Health, and Economic Impacts

| Family | Biodiversity | Health | Economic | Total species with impacts | Percentage of species with impacts |
|---|---|---|---|---|---|
| Anatidae | 7 | 2 | 8 | 12 | 41 |
| Threskiornithidae | 1 | | | 1 | 100 |
| Numididae | | | 1 | 1 | 100 |
| Phasianidae | 3 | 1 | 7 | 8 | 50 |
| Meleagrididae | | | 1 | 1 | 100 |
| Columbidae | 2 | 2 | 3 | 3 | 60 |
| Psittacidae | 4 | 3 | 14 | 14 | 78 |
| Cacatuidae | | | 2 | 2 | 100 |
| Strigidae | 1 | | | 1 | 50 |
| Pycnonotidae | 1 | | 1 | 1 | 100 |
| Timaliidae | 1 | | | 1 | 25 |
| Corvidae | 3 | | 3 | 3 | 100 |
| Sturnidae | 2 | 1 | 2 | 2 | 22 |
| Passeridae | 2 | 1 | 2 | 2 | 100 |
| Ploceidae | | | 3 | 3 | 23 |
| Estrildidae | | | 5 | 5 | 43 |
| Fringillidae | | | 3 | 3 | 100 |
| Odontophoridae | | | 1 | 1 | 100 |
| Total | 27 | 10 | 56 | 64 | |

*Source:* Modified from Shirley, S. M., and S. Kark, *Glob Ecol Biogeogr* 18:450, 2009.

*Note:* The total number of species with reported impacts in each family includes species with more than one type of impact.

### 7.3.1.1 Biodiversity impacts

Specific types of biodiversity impacts of alien birds include competition and/or predation on native species as well as hybridization. The Anatidae (ducks, geese, and swans) family has the greatest number of species with environmental impacts (Table 7.1). More than one type of impact has been described for several species in this group including the Canada goose (*Branta Canadensis*), barnacle goose (*Branta leucopsis*), Egyptian goose (*Alopochen aegyptiacus*), and graylag goose (*Anser anser*). The Canada goose, originally released primarily for hunting, is well established in northern Europe, and populations are increasing in many countries. At high population numbers, it damages shoreline habitats by grazing on aquatic plants and trampling habitat.[8] In addition to removing vegetation, contamination of soil and water by goose feces can lead to the eutrophication of water bodies and degraded habitat for aquatic species. The Canada goose is also known to compete for feeding, nesting, and roosting sites with native waterfowl such as the graylag goose.[35] Predation on both young and adults of several smaller waterfowl has been observed in Sweden.[36] The Canada goose is reported to hybridize with native species such as the graylag goose and the barnacle goose across its European range. Other geese species such as the barnacle goose, Egyptian goose, and graylag goose have similar impacts including herbivory, competition with native species, hybridization, and pollution of water bodies with their feces, although their overall impacts are less widespread because current populations are much smaller.[37,38]

A nonnative water bird species currently causing much concern in Europe is the ruddy duck (*Oxyura jamaicensis*). It has the ability to hybridize with the white-headed duck (*Oxyura leucocephala*), one of the rarest birds in the world and already endangered due to habitat loss and overhunting.[39] Hybridization between these two species is known to produce fertile offspring and will contribute to genetic introgression and possibly the extinction of the European white-headed duck population in the longer term. The largest populations of the ruddy duck are found in the United Kingdom, due to escapes and intentional releases from waterfowl collections, but individuals have naturally dispersed to the European mainland. Ongoing eradications are now underway in more than 15 countries. Another water bird, the sacred ibis (*Threskiornis aethiopicus*), is known to prey upon the nests of water birds and amphibians. In France, predation has been observed on the nests of at least 11 other water bird species, particularly colonies of terns and herons, as well as several threatened amphibian and insect species.[40] Although this species currently occurs in only a few countries, it feeds at nearby dumps and as a generalist predator poses a threat to a number of species. The ability for year-round feeding has contributed to its fast population expansion.

The parrot family has a number of species with negative biodiversity impacts in their native ranges. Although several species are currently at very low population levels in Europe, two species, the rose-ringed parakeet (*Psittacula krameri*) and the monk parakeet (*Myiopsitta monachus*), are now established and increasing in population size in several European countries. A cavity-nesting species, the rose-ringed parakeet nests in old nests of woodpeckers. Because they nest earlier than other native cavity nesters, in Belgium this species has been associated with local declines in the population sizes of native nesters such as the European nuthatch (*Sitta europaea*).[41] Although little information is available, the monk parakeet is also known to compete aggressively with native species for territory.[42]

Several other alien bird species currently have localized distributions but are known to be generalist predators in their native and other introduced ranges and cause declines in native bird species. These include the house crow (*Corvus splendens*) and the common myna (*Acridotheres tristis*), both of which are on the list of the "100 of the World's Worst Invasive Alien

Species" (http://www.issg.org/worst100_species.html). The house crow is a highly invasive bird and occurs locally in the Netherlands and has been responsible for harassment and predation of native species in several localities leading to population declines.[43] The common myna has recently been introduced and is currently still localized to several areas of southwestern Europe and adjacent islands. It has been reported to cause declines in native bird species through predation and competition.[44,45] These species are also known carriers of a number of diseases that can be transmitted to humans and wildlife. Although populations in Europe are now low, both these species take advantage of human feeding and refuse areas to form large roosting flocks and rapidly increase their populations, making eradication an urgent issue.

### 7.3.1.2   Economic and human health impacts

In addition to biodiversity impacts, a number of species also have economic and human health impacts. For example, Canada geese can cause damage to agricultural crops including rye, oats, and wheat, as well as root crops. Local damage of £15,000 and yield losses of 20% have been reported in one area of the United Kingdom.[46] However, the amount of damage reported varies widely among sites depending on the numbers present, and large-scale assessment of damage has not yet been undertaken. Because Canada geese form large flocks, they have been involved in occasional collisions with airplanes, especially at airports bordering wetlands.[35,47] While goose control measures entail some costs, the potential costs from actual collisions and death or injury to passengers is likely to be much greater. An airstrike in Reno, Nevada cost US$250,000 in aircraft repairs.[47]

Economic damages by parrots have not yet been well studied in Europe although several species are serious pests in their native and introduced ranges.[48] The rose-ringed parakeet is responsible for substantial crop losses, estimated in one account at US$15 million to cereal and fruit crops in India and Pakistan.[42,48] In Europe, there has been some reported damage to fruit crops by parakeets although the economic impacts have not been thoroughly assessed. However, there are concerns about future damage to maize and sunflower crops increasingly grown in Britain.[48] Similarly, monk parakeet populations are expanding rapidly and pose a serious concern if they expand into agricultural areas. Their large bulky communal nests built on electrical utility poles also present a fire hazard and cause damage to the electrical infrastructure.

Game birds such as pheasants, often introduced for hunting, can have considerable impacts through their feeding on seeds and seedlings of grain crops such as corn, wheat, oats, and vegetable crops. Nine pheasant species (56% of the total) introduced into Europe have reported impacts in agriculture in their native and introduced ranges.[34] For example, the pheasant (*Phasianus colchicus*) is a pest of vegetable and fruit crops in Germany, causing total damages of €1.3 million per year.[8] A number of other alien bird species from tropical Asia and Africa, especially in the Ploceidae (weavers) and Estrildidae (weaver-finches) families, are currently breeding and in the process of establishing populations in Europe.[34] Several of these species have significant agricultural impacts in their home ranges and other introduced areas.[49] These populations will require careful monitoring and early control action to prevent future impacts in Europe. Warmer countries of the Mediterranean region are likely to be especially vulnerable.

### 7.3.1.3   Eradications of alien birds in Europe

The most well-known widespread eradication campaigns have been undertaken against the ruddy duck due to its great threat to the native white-headed duck described in Section 7.3.1.1. In several countries, eradication attempts are underway and have been successful in eradicating ruddy duck from Spain and Iceland and reducing populations substantially in the United Kingdom and France.[38] The eradication in Spain is particularly important since this is an area of overlap between the two species. Despite these successes, the ruddy duck is increasing in some countries such as the Netherlands and has been reported breeding in Germany and Denmark. Estimates to control the ruddy duck in the United Kingdom over a 4- to 6-year period were €4.4 million.[50] The problem of obtaining public support in the United Kingdom was made more complicated because the benefits would not be seen in that country even though it was assumed to be the source of most of the individuals arriving in continental Europe.

Some local efforts have been successful in eradicating populations of the common myna in several of the Canary Islands of Spain (Susana Saavedra, pers. comm.). However, these efforts are local and lacking a coordinated national effort. Similarly, despite their potential ability to cause damage to agricultural crops, there has been no real effort to eradicate parrot populations. However, in the United Kingdom, licenses to kill parrots will be available in special cases where there is damage to crops or native wildlife.[51]

### 7.3.1.4  Future predicted impacts and management of alien birds

The number of alien birds introduced into Europe has increased exponentially over the last century and particularly since the 1950s (Figure 7.2).[33] Amphibian and reptile introductions have also increased although less dramatically. With increasing trade, both internationally and within Europe, and expansion of urban areas and immigration of people from other countries, we suggest that introductions will continue on these pathways in the absence of strong regulation.

Several alien bird species in the Anatidae, Corvidae, Passeridae, Phasianidae, and Sturnidae families are associated with moderately or seriously negative impacts on both economic resources and biodiversity (Figure 7.3).[34] Certain species traits are also associated with considerable impacts. Species that occur in a variety of habitats, have more than one brood in a breeding season, and form large feeding or roosting flocks are of particular concern. Species in these families or possessing these traits that currently occur in Europe should be considered for eradication where populations are small and localized to avoid population spread and future impacts, and new introductions should be

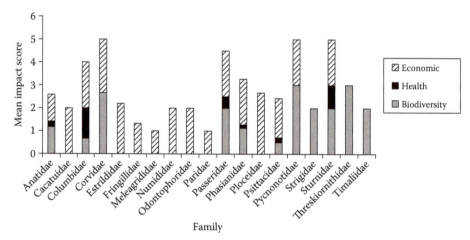

**Figure 7.3** Potential economic, biodiversity, and human health impact scores of alien birds in Europe. Shown are average scores for all bird families.

restricted. Most of the future alien bird, amphibian, and reptile introductions are likely to be a consequence of the pet trade. In 2005, the European Union temporally banned the import of wild-caught birds after some imported birds died from avian flu in the United Kingdom.[52] Subsequently, the European Commission is considering a permanent ban on the wild-bird trade[53] that would be expected to significantly reduce the number of new introductions. In the absence of an outright ban, risk assessments should be undertaken for all species imported into Europe.

To date, only 12% of alien bird species established in Europe have spread to regions outside the country to which they were introduced[54]; however, we expect natural dispersal to other European countries to increase as populations grow larger. For example, the Canada goose has doubled its population throughout Europe within the last decade and is continuing to expand into southern and eastern Europe.[38] Other geese species such as the barnacle, Egyptian, and graylag geese are showing similar rapid expansions in areas of introduction. The expansion of trade and integration of the countries of the former Soviet Union are also expected to lead to increased introductions into those countries that were previously isolated.[54] The impact of climate change on invasive species is rather difficult to predict. However, several alien bird species originate from warmer regions such as Asia and Africa, and their distributions in Europe appear to be limited by minimum temperatures (Shirley, unpublished data).[55] Warming trends in Europe may lead to the expansion of their distributions (Shirley, unpublished data). These ongoing drivers of introductions demonstrate the need for the European Union to develop rigorous policies preventing the import and controlling existing populations of alien bird species to prevent significant impacts in the future.

## 7.3.2 Impacts of reptiles and amphibians

Most amphibians and reptiles were introduced to Europe as pets or sources of food.[33] Many of these pets were intentionally released or escaped from captivity. The main impacts are to native biodiversity, largely predation, competition, and transmission of pathogens. Amphibians have twice the number of reported impacts as reptiles with impacts described for the Pipidae (tongueless frogs), Ranidae (frogs), and Salamandridae (salamanders) families.[33] The Ranidae family had 71% of established species with reported impacts. Three reptile families, the Emydidae (turtles), Lacertidae (lizards), and Colubridae (snakes), together representing 77% of established species, have reported impacts.

The most well-known example of negative impacts by amphibians is the American bullfrog (*Lithobates catesbianus*). The bullfrog occurs in a wide variety of habitats feeding on native frogs, snakes, and turtles.[33] It is also present as a vector for the fungus *Batrachochytrium dendrobatidis* that carries the chytridiomycosis disease responsible for worldwide amphibian declines.[56] Although established in several European countries, a local population was eradicated in the United Kingdom at a cost of £29,000.[57] Similar concerns about predation on native herpetofauna have been raised about the marsh frog (*Rana ridibunda*) in the United Kingdom although to date there has been no reported negative impacts.[11] The common slider (*Trachemys scripta*), a common pet species in Europe, is breeding in a few countries. Although now banned as an import in Europe, it continues to be introduced into wild areas by the release of pets. This reptile feeds on native reptiles, amphibians, small mammals, and birds and is reported to compete with native turtles for nesting and basking sites.[58]

There have been a few successful eradications of the American bullfrog in Europe, one in the United Kingdom and two in Germany[59] suggesting that eradication of amphibians is a viable strategy if populations are caught early enough.

## 7.4    Prioritizing alien invasive birds—how do we decide which are the worst of the worst?

Most conservation scientists and practitioners acknowledge that not all alien species have similar negative impacts in their introduced ranges[60] and that major problems are caused by a smaller subset of species. A major objective of invasion biology is to be able to predict beforehand those species that are likely to become a problem. Lists of "the worst aliens" are a widely used tool to focus awareness on the problem of invasive species and their impacts on native ecosystems, economic resources, and human health and well-being. Initiated originally by the Global Invasive Species Program[61] in 1989 as "100 of the World's Worst Invasive Alien Species" (http://www.issg.org/worst100_species.html), these lists have been effective in approaching the public who beforehand were largely unaware of a looming ecological problem. One major problem common to these sorts of "conservation lists" is that they are often based, at least initially, on "expert opinion"[62] rather than explicit scientifically objective criteria, and the alien species list is no exception (Piero Genovesi, pers. comm.). In particular, the presence on the list fails to consider specific criteria that might be important especially if the list is to be considered in a large regional or global context.

As part of a European Union consortium on alien/invasive species, we endeavored during a consortium meeting to produce a list of "Worst Alien Species in Europe."[63] The process resulted in much discussion relating to both the proportional representation of various taxonomic groups, for example, plants, invertebrates, and vertebrates, and also which species should be selected within the groups. During the discussion, it became clear that different experts had varying opinions on which species should be selected, based on different criteria, areas of interest, and points of view. For example, in some cases expert perspectives were influenced by the current situation in their home country rather than a consideration of the region as a whole, while in others, local priorities were less important. To address the subjectivity inherent in producing such lists, we believe there is a need for a more scientific, transparent analysis of species to focus effort and funding. The goals of a list should be clearly defined in advance, and it should be simple to understand, repeatable (i.e., based on published literature), and easy to update.

We developed a procedure to provide an objective and relatively straightforward method for ranking species that can be used both for awareness building and risk assessment. We ranked species according to three major categories that represent the currently described range of impacts for alien birds in Europe. Major impacts on native biodiversity include predation on eggs, nestlings and adults of native bird species, competition for food, competition for nesting cavities, and hybridization with native species. Impacts on economic resources generally comprise grazing or feeding on grain, fruit, and other agricultural resources. Described impacts on human health and well-being include disease transmission, contamination from droppings, threats to air safety, and excessive noise. Here, we limit our analyses to alien/invasive birds, but the approach can easily be extended to consider species across taxonomic groups.

### 7.4.1    Quantifying impacts

We classified impacts for alien bird species in continental Europe and its associated islands.[34] We considered alien species to be those species that are introduced outside their natural geographic range through human activities, either intentionally or accidentally. This definition also includes cases where a species is native to some regions of Europe and subsequently introduced to new areas, for example from mainland Europe to the

Canary Islands or the United Kingdom. Quantitative information on alien bird impacts is generally unavailable or difficult to obtain, especially at continental scales. Much of the information is anecdotal or restricted to local areas. In particular, information on impacts in introduced ranges is poorly documented for many species. For this reason, most of the information used for our analyses was collected from impacts documented in the species' native ranges or other introduced ranges. While we recognize that impacts may differ in Europe, this approach represents the best approximation to date and species impacts elsewhere are a reliable predictor of impacts in a novel area.[5,42]

Given a paucity of quantitative data, we classified known impacts reported in the literature from 1 to 3, with 1 representing a minor impact and 3 being the most serious impact. The decision to use three rankings allows for adequate sample sizes for rankings, while maintaining enough variation in the data to infer meaningful biological patterns. For each of the categories described above, we generated an ordinal scale associated with specific criteria (Table 7.1). Data on impacts was taken from a variety of published and unpublished sources.[42,64,65] We then ranked the species for each of the categories using the criteria. Finally, to compare the species' rankings across the categories, we summed the ranks across the three criteria and ranked the totals. Rankings for each category and the total are shown in descending order.

Because the overall impact of a species is likely related to population sizes, we also considered three categories that describe the spatial and population extent of species' invasibility and potential impacts: distribution, abundance, and rate of spread. Distribution rankings were based on numbers of 50 × 50 km grid cells in Europe. Abundances were obtained from BirdLife International, Banks et al., and Strubbe.[38,66,67] We recognize that distribution and abundance are likely to be positively correlated for many species producing similar rankings; however, departures from this general pattern may also be instructive. For example, the introduced ruddy duck (*Oxyura jamaicensis*) has its highest population in the United Kingdom, and it is widely believed that any eradication efforts by countries in continental Europe are useless without strong measures in the source country.[39,68] We derived an index of population impact by taking the scores for each category and dividing the total by the total maximum possible score. We then multiplied the total impact score by the index to arrive at a "realized" impact representing the current impact based on population. We did not weight the impacts although this could be added if policy makers were more concerned about a specific category, for example, economic damage.

## 7.4.2 Impact scores

The criteria used for ranking in each category are shown in Table 7.2. Species are listed according to impact rank in descending order (Table 7.3). Economic impacts were the most frequently documented category of impact with a total of 56 species, with 11 and 27 species with serious and moderate impacts, respectively (Table 7.3). Biodiversity impacts were documented for 18 species, 8 of which were considered serious (Table 7.3). Known impacts on human health and well-being are few, with only four species having moderate or serious impacts (Table 7.3).

The spatial and population extent of species impacts are described by an index that represents rankings for distribution, abundance, and spread according to the categories in Table 7.2. A total of 20 species had distributions in a variety of European countries, with 8 species recorded in at least 5 countries and the remaining 12 species widespread across Europe. A total of 14 species were moderately or highly abundant within Europe, while the majority of species (50/64) had low abundances (<5000) in any one country. A total of 17 species had moderate or high rates of spread.

*Table 7.2* Criteria for Ranking Impacts Reported for Alien Bird Species in Europe and Spatial Extent of Populations

| Ranking criteria for impacts | Spatial extent ranking criteria |
|---|---|
| **Biodiversity impacts** | **Distribution** |
| 1. Aggression or displacing at least one native bird species from feeding grounds | 1. Localized distributions; less than 50 grid cells |
| 2. Competition with at least one native bird species resulting in local population | 2. Present in several European countries; 100–400 grid cells |
| 3. Predation on >1 native bird species (generalist predator) or hybridization with at least one native bird species | 3. Widespread across Europe; over 400 grid cells |
| **Economic impacts** | **Abundance** |
| 1. Minor feeding on crops | 1. Low populations (<5,000) in every country |
| 2. Light damage to crops from feeding | 2. Medium or high populations (>5,000) in at least one country |
| 3. Considerable damage to crops through feeding and trampling | 3. High populations (>25,000) in several countries across Europe |
| **Human health impacts** | **Spread** |
| 1. Fecal droppings, low potential for disease transmission, noise | 1. Low rate of spread (<10%) in one country or no spread |
| 2. Fecal droppings, moderate potential for disease transmission, potential threats to air safety | 2. Low rate of spread (<10%) in more than one country |
| 3. Potential for serious disease transmission to humans, avian flu | 3. High rate of spread (average > 10% per year) in one or more countries |

*Source:* Modified from Shirley, S. M., and S. Kark, *Glob Ecol Biogeogr* 18:450, 2009.

*Note:* Impacts are ranked according to a general ordinal scale with the highest ranking representing the greatest impacts.

The alien monk parakeet and common myna had the top total impact scores of 7 (out of a possible maximum of 9) followed closely by the Canada goose, house crow, and rose-ringed parakeet. Considering the index of population expansion changes the ranking order; Canada goose, monk parakeet, and rose-ringed parakeet all still rank highly, but two dove species, rock dove (*Columba livia*) and collared turtledove (*Streptopelia decaocto*), rank much higher for realized impacts. Although their individual impacts scores are moderate, the wide distributions and large populations of the rock dove and collared turtledove increase their overall impact. Conversely, the house crow and common myna have high total impact scores, but their realized impacts are lower as their populations are currently very small and localized.

### 7.4.3 Interpreting the ranking procedure and further issues

A review of the rankings among categories reveals that the composition of species ranked with moderate or serious impacts, as expected, varies substantially among the categories of impacts. Species with negative impacts on economic resources, such as by feeding on fruit and grain crops, are not necessarily the same species with strongly negative biodiversity impacts. In all, 5 out of 10 species with the worst economic impacts (Alexandrine parakeet [*Psittacula eupatria*], chestnut munia [*Lonchura Malacca*], common peafowl [*Pavo cristatus*], Java sparrow [*Padda oryzivora*], and village weaver [*Ploceus cucullatus*]) had minor

*Table 7.3* Economic, Biodiversity, Human Health, and Total Impact Scores for Alien Birds in Europe

| Family | Common name | Species name | Biodiversity impact | Human health impact | Economic impact | Potential impact | Realized impact |
|---|---|---|---|---|---|---|---|
| Psittacidae | Monk parakeet | *Myiopsitta monachus* | 3 | 1 | 3 | 7 | 3.9 |
| Sturnidae | Common myna | *Acridotheres tristis* | 3 | 2 | 2 | 7 | 2.3 |
| Anatidae | Canada goose | *Branta canadensis* | 2 | 2 | 2 | 6 | 5.3 |
| Corvidae | House crow | *Corvus splendens* | 3 | 0 | 3 | 6 | 2.0 |
| Psittacidae | Rose-ringed parakeet | *Psittacula krameri* | 2 | 1 | 3 | 6 | 4.0 |
| Anatidae | Egyptian goose | *Alopochen aegyptiacus* | 2 | 0 | 3 | 5 | 3.3 |
| Corvidae | Common magpie | *Pica pica* | 3 | 0 | 2 | 5 | 1.7 |
| Passeridae | House sparrow | *Passer domesticus* | 2 | 1 | 2 | 5 | 1.7 |
| Pycnonotidae | Red-vented bulbul | *Pycnonotus cafer* | 3 | 0 | 2 | 5 | 1.7 |
| Anatidae | Mute swan | *Cygnus olor* | 2 | 0 | 2 | 4 | 3.1 |
| Columbidae | Rock dove | *Columba livia* | 1 | 2 | 1 | 4 | 3.6 |
| Corvidae | Blue magpie | *Urocissa erythrorhyncha* | 2 | 0 | 2 | 4 | 1.3 |
| Passeridae | Spanish sparrow | *Passer hispaniolensis* | 2 | 0 | 2 | 4 | 1.8 |
| Phasianidae | Common peafowl | *Pavo cristatus* | 0 | 1 | 3 | 4 | 1.3 |
| Psittacidae | Nanday parakeet | *Nandayus nenday* | 1 | 1 | 2 | 4 | 1.8 |
| Psittacidae | Alexandrine parakeet | *Psittacula eupatria* | 1 | 0 | 3 | 4 | 1.8 |
| Anatidae | Ruddy duck | *Oxyura jamaicensis* | 3 | 0 | 0 | 3 | 1.3 |
| Columbidae | Collared turtledove | *Streptopelia decaocto* | 0 | 2 | 1 | 3 | 3.0 |
| Estrildidae | Chestnut munia | *Lonchura malacca* | 0 | 0 | 3 | 3 | 1.0 |
| Estrildidae | Java sparrow | *Padda oryzivora* | 0 | 0 | 3 | 3 | 1.0 |
| Phasianidae | Ring-necked pheasant | *Phasianus colchicus* | 0 | 0 | 3 | 3 | 2.7 |
| Ploceidae | Village weaver | *Ploceus cucullatus* | 0 | 0 | 3 | 3 | 1.0 |
| Sturnidae | Crested myna | *Acridotheres cristatellus* | 1 | 0 | 2 | 3 | 1.0 |
| Threskiornithidae | Sacred ibis | *Threskiornis aethiopicus* | 3 | 0 | 0 | 3 | 2.0 |
| Anatidae | Common pintail | *Anas acuta* | 0 | 0 | 2 | 2 | 0.7 |
| Anatidae | Muscovy duck | *Cairina moschata* | 2 | 0 | 0 | 2 | 0.7 |

*(Continued)*

*Table 7.3* Economic, Biodiversity, Human Health, and Total Impact Scores for Alien Birds in Europe (*Continued*)

| Family | Common name | Species name | Biodiversity impact | Human health impact | Economic impact | Potential impact | Realized impact |
|---|---|---|---|---|---|---|---|
| Anatidae | Magellan goose | *Chloephaga picta* | 0 | 0 | 2 | 2 | 1.1 |
| Anatidae | Black swan | *Cygnus atratus* | 1 | 1 | 0 | 2 | 0.9 |
| Anatidae | Ruddy shelduck | *Tadorna ferruginea* | 2 | 0 | 0 | 2 | 1.1 |
| Cacatuidae | Sulfur-crested cockatoo | *Cacatua galerita* | 0 | 0 | 2 | 2 | 0.7 |
| Columbidae | Laughing turtledove | *Streptopelia senegalensis* | 0 | 0 | 2 | 2 | 0.7 |
| Estrildidae | Red avadavat | *Amandava amandava* | 0 | 0 | 2 | 2 | 0.7 |
| Estrildidae | Orange-breasted waxbill | *Amandava subflava* | 0 | 0 | 2 | 2 | 0.7 |
| Fringillidae | Greenfinch | *Carduelis chloris* | 0 | 0 | 2 | 2 | 0.7 |
| Numididae | Helmeted guinea fowl | *Numida meleagris* | 0 | 0 | 2 | 2 | 0.7 |
| Odontophoridae | California quail | *Callipepla californica* | 0 | 0 | 2 | 2 | 0.7 |
| Phasianidae | Chukar partridge | *Alectoris graeca* | 0 | 0 | 2 | 2 | 0.7 |
| Phasianidae | Red-legged partridge | *Alectoris rufa* | 0 | 0 | 2 | 2 | 1.1 |
| Phasianidae | Lady Amherst's pheasant | *Chrysolophus amherstiae* | 0 | 0 | 2 | 2 | 0.7 |
| Ploceidae | Red bishop | *Euplectes orix* | 0 | 0 | 3 | 2 | 0.7 |
| Ploceidae | Southern masked-weaver | *Ploceus velatus* | 0 | 0 | 2 | 2 | 0.7 |
| Psittacidae | Fischer's lovebird | *Agapornis fischeri* | 0 | 0 | 2 | 2 | 0.7 |
| Psittacidae | Rosy-faced lovebird | *Agapornis roseicollis* | 0 | 0 | 2 | 2 | 0.7 |
| Psittacidae | Orange-winged amazon | *Amazona amazonica* | 0 | 0 | 2 | 2 | 0.7 |
| Timaliidae | Red-billed leiothrix | *Leiothrix lutea* | 2 | 0 | 0 | 2 | 0.7 |
| Anatidae | European wigeon | *Anas penelope* | 0 | 0 | 1 | 1 | 0.6 |
| Anatidae | Bean goose | *Anser fabalis* | 0 | 0 | 1 | 1 | 0.4 |
| Anatidae | Barnacle goose | *Branta leucopsis* | 0 | 0 | 1 | 1 | 0.6 |

| Family | Common name | Scientific name | | | | | |
|---|---|---|---|---|---|---|---|
| Estrildidae | Black-rumped waxbill | *Estrilda troglodytes* | 0 | 0 | 1 | 1 | 0.4 |
| Fringillidae | European goldfinch | *Carduelis carduelis* | 0 | 0 | 1 | 1 | 0.4 |
| Fringillidae | Canary | *Serinus canaria* | 0 | 0 | 1 | 1 | 0.4 |
| Meleagrididae | Turkey | *Meleagris gallopavo* | 0 | 0 | 1 | 1 | 0.3 |
| Paridae | Great tit | *Parus major* | 0 | 0 | 1 | 1 | 0.3 |
| Phasianidae | Black francolin | *Francolinus francolinus* | 0 | 0 | 1 | 1 | 0.3 |
| Phasianidae | European partridge | *Perdix perdix* | 0 | 0 | 1 | 1 | 0.4 |
| Psittacidae | Yellow-collared lovebird | *Agapornis personatus* | 0 | 0 | 1 | 1 | 0.3 |
| Psittacidae | Yellow-crowned parrot | *Amazona ochrocephala* | 0 | 0 | 1 | 1 | 0.3 |
| Psittacidae | Yellow-headed amazon | *Amazona oratrix* | 0 | 0 | 1 | 1 | 0.3 |
| Psittacidae | Blue-crowned parakeet | *Aratinga acuticaudata* | 0 | 0 | 1 | 1 | 0.3 |
| Psittacidae | Red-masked parakeet | *Aratinga erythrogenys* | 0 | 0 | 1 | 1 | 0.3 |
| Psittacidae | Mitred parakeet | *Aratinga mitrata* | 0 | 0 | 1 | 1 | 0.3 |
| Psittacidae | Budgerigar | *Melopsittacus undulatus* | 0 | 0 | 1 | 1 | 0.3 |

or no known biodiversity impacts. Surprisingly, however, the top 13 species had moderate or high rankings across two out of three categories. These species had both considerable impacts on economic resources such as agricultural crops and the potential to negatively affect native bird biodiversity in an area by causing declines in native populations by predation, hybridization, or competition. In addition, many of the species with multiple impacts are present in several European countries where their populations are currently expanding.

Other species with high overall rankings highlight several issues that should be considered for developing an integrated policy to deal with alien birds. One of the most unclear issues is whether certain categories of impacts should be given priority. For example, should species with negative impacts on biodiversity trump those with negative economic impacts? Several species such as the ruddy duck and sacred ibis (*Threskiornis aethiopicus*) have serious biodiversity impacts, but populations are currently relatively restricted. The ruddy duck represents a serious threat to the continued existence of the native white-headed duck through hybridization and is currently being controlled by several countries.[39,68] The sacred ibis, a generalist predator with high potential to cause declines in native nesting species, is currently at low population levels, but increasing in several countries.[69] Alternatively, as a precautionary approach in the absence of empirical studies, the use of impacts in the ranking criteria could be bypassed altogether, and the rankings could be based solely on measures of population growth and spread.[70] Our proposed ranking approach allows for impacts to be weighted according to the specific goal of the assessment, while retaining all information thereby insuring flexibility for future conservation policy decisions.

Another important issue relates to the time frame of impacts, in particular whether we prioritize species that are already a problem or emphasize those with large potential to become a serious problem. In contrast to those species with established populations such as the Canada goose or the rock dove, certain established bird species have potential for rapid growth after reaching thresholds, for example common myna, bar-headed goose (*Anser indicus*), and house crow. Two of these species, the common myna and house crow, both generalist nest predators with the potential to spread rapidly, can potentially cause serious population declines in native species if they become established.[45] An assessment of the rankings could then focus efforts toward alien populations in the early stages of invasion when they can be more easily eradicated[71,72] rather than widespread populations that may be controlled, but with no real hope of eradication.

Like other global lists developed for the International Union for Conservation of Nature (IUCN), initial efforts based on expert opinion eventually require an objective framework to be effective over different scales and across global regions.[62] An objective ranking methodology such as that presented here offers conservation biologists, policymakers, and those involved in risk assessment a way to prioritize species for action with limited funds. It is possible that our rankings reflect a bias toward species with a longer history of introductions and establishment and therefore better information on impacts. However, we believe the likelihood of this bias is small because the majority of impacts described refer to species' native ranges. Rankings can be updated as new information becomes available enhancing the quality and usefulness of the approach and eventually facilitating geographic comparisons. In addition, the analysis can be extended with species-specific information such as life-history characteristics for potential high-impact aliens that can be further analyzed to infer more general patterns of invasibility and potential for negative impacts.[34] Increasing concern about the impacts and associated costs of alien species suggests that the time has come for an objective framework such as that proposed that can be compared with expert opinion and incorporated into protocols for prevention and control.

## 7.5 Concluding remarks

The number and diversity of alien vertebrate species are continuing to increase in Europe due to increasing globalization of trade and the resulting import of species for activities such as the pet trade, fur farming, and hunting. In this chapter, we have highlighted several of the species that have serious impacts on both native biodiversity and economic resources. Much of the information on impacts in the past has come from localized anecdotal accounts and research studies. The recent completion of large synthetic studies in Europe has yielded new insights, and we are now beginning to make generalizations about impacts as well as recommendations for management of groups of taxa. For species with relatively limited distributions and especially those isolated to islands, eradication appears to be the most cost-effective option in the long term for preventing future problems. Species with widespread distributions will require coordinated efforts between countries and public education in order for eradication campaigns to be successful. Prioritizing species for management using ranking systems is a flexible tool for those involved in managing alien species and can be updated as new information becomes available. A ranking of potential impacts can be useful in determining which species should be banned from import, whereas a ranking of the realized impacts that considers population distributions and sizes may be informative for developing policies to deal with those species that are already established. This type of systematic approach will also improve our ability to predict how alien species will respond in the future to changes in land use and climate.

## Acknowledgments

We thank O. Hatzofe for his thoughtful comments on earlier drafts of this chapter and members of the DAISIE (Delivering Alien Invasive Species Inventories for Europe) Vertebrate Consortium for providing valuable discussion. Funding was provided by the FP6 European Union DAISIE Project to Salit Kark.

## References

1. Duncan, R. P., T. M. Blackburn, and D. Sol. 2003. The ecology of bird introductions. *Annu Rev Ecol Syst* 34:71.
2. Pimentel, D., L. Lach, R. Zuniga, and D. Morrison. 2000. Environmental and economic costs of nonindigenous species in the United States. *Bioscience* 50:53.
3. Simberloff, D. 2004. A rising tide of species and literature: A review of some recent books on biological invasions. *Bioscience* 54:247.
4. Genovesi, P., S. Bacher, M. Kobelt, M. Pascal, and R. Scalera. 2008. Alien mammals of Europe. In *Handbook of Alien Species in Europe*, ed. Delivering Alien Invasive Species Inventories for Europe (DAISIE), 397. Dordrecht: Springer.
5. Nentwig, W., E. Kuhnel, and S. Bacher. 2009. A generic impact-scoring system applied to alien mammals in Europe. *Conserv Biol* 24:302.
6. Pascal, M., F. Siorat, O. Lorvelec, P. Yesou, and D. Simberloff. 2005. A pleasing consequence of Norway rat eradication: Two shrew species recover. *Divers Distrib* 11:193.
7. Pascal, M., O. Lorvelec, V. Bretagnolle, and J. M. Culioli. 2008. Improving the breeding success of a colonial seabird: A cost-benefit comparison of the eradication and control of its rat predator. *Endanger Species Res* 4:267.
8. Gebhardt, H. 1996. Ecological and economic consequences of introductions of exotic wildlife (birds and mammals) in Germany. *Wildl Biol* 2:205.
9. Bonesi, L., and S. Palazon. 2007. The American mink in Europe: Status, impacts, and control. *Biol Conserv* 134:470.

10. Kauhala, K., and M. Winter. 2008. *Nyctereutes procyonoides* (Gray), raccoon dog, (Canidae, Mammalia). In *Handbook of Alien Species in Europe*, ed. Delivering Alien Invasive Species Inventories for Europe (DAISIE), 365. Dordrecht: Springer.

11. White, P. C. L., and S. Harris. 2002. Economic and environmental costs of alien vertebrate species in Britain. In *Biological Invasions: Economic and Environmental Costs of Alien Plant, Animal, and Microbe Species*, ed. D. Pimentel, 113. Ithaca, NY: CRC Press.

12. Genovesi, P., and R. Putman. 2008. *Cervus Nippon* (Temminck), sika deer, (Cervidae, Mammalia). In *Handbook of Alien Species in Europe*, ed. Delivering Alien Invasive Species Inventories for Europe (DAISIE), 361. Dordrecht: Springer.

13. Bertolino, S., and P. Genovesi. 2003. Spread and attempted eradication of the grey squirrel (*Sciurus carolinensis*) in Italy, and consequences for the red squirrel (*Sciurus vulgaris*) in Eurasia. *Biol Conserv* 109:351.

14. Wauters, L. A., P. W. W. Lurz, and J. Gurnell. 2000. Interspecific effects of grey squirrels (*Sciurus carolinensis*) on the space use and population demography of red squirrels (*Sciurus vulgaris*) in conifer plantations. *Ecol Res* 15:271.

15. Wauters, L. A., I. Currado, P. J. Mazzoglio, and J. Gurnell. 1997. Replacement of red squirrels by introduced grey squirrels in Italy. In *The Conservation of Red Squirrels, Sciurus vulgaris L.*, ed. J. Gurnell and P. Lurz, 5. London: People Trust for Endangered Species.

16. Genovesi, P. 2008. *Ondatra zibethicus* (Linnaeus), muskrat, (Muridae, Mammalia). In *Handbook of Alien Species in Europe*, ed. Delivering Alien Invasive Species Inventories for Europe (DAISIE), 366. Dordrecht: Springer.

17. Pedersen, J. A. 1964. Villiminken i Norge. *Tidsskr norske landbruk* 71:41.

18. Long, J. L. 2003. *Introduced Mammals of the World: Their History, Distribution and Influence.* Collingwood, Australia: CSIRO Publishing.

19. Bonesi, L. 2008. *Mustela vison* (Schreber), American mink, (Mustelidae, Mammalia). In *Handbook of Alien Species in Europe*, ed. Delivering Alien Invasive Species Inventories for Europe (DAISIE), 363. Dordrecht: Springer.

20. Yanushevich, A. I., ed. 1966. *Acclimitization of Animals in the USSR.* Jerusalem: Israel Program for Scientific Translations.

21. Willett, J. A. 1970. Wild deer: Their status and distribution, 1970. (i) Deer in England and Wales. *J Br Deer Soc* 2:498.

22. Bartŏs, L. 2009. Sika deer in continental Europe. In *Sika Deer: Biology and Management of Native and Introduced Populations*, ed. D. R. McCullough, S. Takatsuki, and K. Kaji, 573. Tokyo: Springer.

23. Ward, A. I. 2001. *The Ecology and Sustainable Management of Roe Deer in Multiple Use Forestry.* York: University of York.

24. Putman, R. J., and N. P. Moore. 1998. Impact of deer in lowland Britain on agriculture, forestry and conservation habitats. *Mammal Rev* 28:141.

25. Rowe, J. 1984. Grey squirrel bark-stripping damage to broadleaved trees in southern Britain up to 1983. *Q J For* 78:231.

26. Thompson, H. V., and T. R. Peace. 1962. The grey squirrel problem. *Q J For* 56:33.

27. Genovesi, P. 2005. Eradications of invasive alien species in Europe: A review. *Biol Invasions* 7:127.

28. Lorvelec, O., and M. Pascal. 2005. French attempts to eradicate non-indigenous mammals and their consequences for native biota. *Biol Invasions* 7:135.

29. Baker, S. 2006. The eradication of coypus (*Myocastor coypus*) from Britain: The elements required for a successful campaign. In *Assessment and Control of Biological Invasion Risks*, ed. F. Koike, M. N. Clout, M. Kawamichi, M. De Poorter, and K. Iwatsuki, 142. Kyoto and Gland: Shoukadoh Book Sellers and IUCN.

30. Gosling, L. M., and S. J. Baker. 1989. The eradication of muskrats and coypus from Britain. *Biol J Linn Soc* 38:39.

31. Bertolino, S. 2008. *Myocastor coypus* (Molina), coypu, nutria (Myocastoridae, Mammalia). In *Handbook of Alien Species in Europe*, ed. Delivering Alien Invasive Species Inventories for Europe (DAISIE), 364. Dordrecht: Springer.

32. Chapuis, J. L. 2008. *Tamias sibiricus* (Laxmann), Siberian chipmunk (Sciuridae, Mammalia). In *Handbook of Alien Species in Europe*, ed. Delivering Alien Invasive Species Inventories for Europe (DAISIE), 372. Dordrecht: Springer.

33. Kark, S., W. Solarz, F. Chiron, P. Clergeau, and S. Shirley. 2008. Alien birds, amphibians, and reptiles of Europe. In *Handbook of Alien Species in Europe*, ed. Delivering Alien Invasive Species Inventories for Europe (DAISIE), 105. Dordrecht: Springer.

34. Shirley, S. M., and S. Kark. 2009. The role of species traits and taxonomic patterns in alien bird impacts. *Glob Ecol Biogeogr* 18:450.

35. Watola, G., J. Allan, and C. Feare. 1996. Problems and management of naturalised introduced Canada geese *Branta Canadensis*, in Britain. In *The Introduction and Naturalisation of Birds*, ed. J. S. Holmes and J. R. Simons, 71. London: HMSO.

36. Fabricius, E. 1974. Intra- and interspecific territorialism in mixed colonies of the Canada goose *Branta canadensis* and the greylag goose. *Anser anser. Ornis Scand* 5:25.

37. Percival, S. M., and D. C. Houston. 1992. The effect of winter grazing by barnacle geese on grassland yields on Islay. *J Appl Ecol* 29:35.

38. Banks, A. N., L. J. Wright, I. M. D. MacLean, C. Hann, and M. M. Rehfisch. 2008. Review of the status of introduced non-native waterbird species in the area of the African-Eurasian water-bird agreement: 2007 update. British Trust for Ornithology, Thetford, Norfolk. http://www.unep-aewa.org/meetings/en/mop/mop4_docs/meeting_docs_pdf/mop4_12_non_native_species_corr1.pdf (accessed June 10, 2010).

39. Hughes, B., J. A. Robinson, A. J. Green, Z. W. D. Li, and T. Mundkur. 2006. International single species action plan for the conservation of the white-headed duck *Oxyura leucocephala*. CMS Technical Series No. 13 and AEWA Technical Series No. 8, Bonn: BirdLife International, The Wildfowl and Wetlands Trust, and Wetlands International. http://ec.europa.eu/environment/nature/conservation/wildbirds/action_plans/docs/intl_white_headed_duck.pdf   (accessed June 10, 2010).

40. Yésou, P., and P. Clergeau. 2005. Sacred ibis: A new invasive species in Europe. *Birding World* 18:517.

41. Strubbe, D., and E. Matthysen. 2007. Invasive ring-necked parakeets *Psittacula krameri* in Belgium: Habitat selection and impact on native birds. *Ecography* 30:578.

42. Long, J. L. 1981. *Introduced Birds of the World*. New York: Universe Books.

43. Ryall, C. 1992. Predation and harassment of native bird species by the Indian house crow *Corvus splendens* in Mombasa, Kenya. *Scopus* 16:1.

44. Ali, S., and S. D. Ripley. 1964. *Handbook of the Birds of India and Pakistan*. Bombay: Oxford University Press.

45. Pell, A. S., and C. R. Tidemann. 1997. The impact of two exotic hollow-nesting birds on two native parrots in savannah and woodland in eastern Australia. *Biol Conserv* 79:145.

46. Simpson, W. 1991. Agricultural damage and its prevention. In *Canada Geese Problems and Management Needs*, ed. J. Harradine, 21. Rossett: British Association for Shooting and Conservation.

47. Allan, J. R., J. S. Kirby, and C. J. Feare. 1995. The biology of Canada geese *Branta canadensis* in relation to the management of feral populations. *Wildl Biol* 1:129.

48. Feare, C. J. 1996. Rose-ringed parakeet *Psittacula krameri*: A love-hate relationship in the making? In *The Introduction and Naturalisation of Birds*, ed. J. S. Holmes and J. R. Simons, 107. London: HMSO.

49. Manikowski, S. 1984. Birds injurious to crops in west Africa. *Trop Pest Manage* 30:379.

50. Owen, M., D. Callaghan, and J. Kirby. 2006. Guidelines on avoidance of introductions of non-native waterbird species. AEWA Technical Series No. 12. Bonn: The United Nations Environment Program (UNEP) and AEWA (Agreement on the Conservation of African-Eurasian Waterbirds). http://www.unep-aewa.org/publications/technical_series.htm (accessed June 10, 2010).

51. Natural England. 2009. Statement on monk and ring-necked parakeets. Press release issued 2009. http://www.naturalengland.org.uk/about_us/news/2009/031009.aspx (accessed February 1, 2010).

52. Senior, K. 2006. Battle over wild bird trade ban. *Front Ecol Environ* 10:509.

53. Carrete, M., and J. L. Tella. 2008. Wild-bird trade and exotic invasions: A new link of conservation concern? *Front Ecol Environ* 6:207.

54. Chiron, F., S. M. Shirley, and S. Kark. 2010. Behind the iron curtain: Socio-economic and political factors shaped exotic bird introductions into Europe. *Biol Conserv* 143:351.

55. Shwartz, A., D. Strubbe, C. J. Butler, E. Matthysen, and S. Kark. 2009. The effect of enemy-release and climate conditions on invasive birds: A regional test using the rose-ringed parakeet (*Psittacula krameri*) as a case study. *Divers Distrib* 15:310.

56. Garner, T. W. J., S. Walker, J. Bosch, A. D. Hyatt, A. A. Cunningham, and M. C. Fisher. 2005. Chytrid fungus in Europe. *Emerg Infect Dis* 11:1639.

57. Lorvelec, O., and M. Détaint. 2008. *Lithobates catesbeianus* (Shaw), American bullfrog (Ranidae, Amphibia). In *Handbook of Alien Species in Europe*, ed. Delivering Alien Invasive Species Inventories for Europe (DAISIE), 362. Dordrecht: Springer.

58. Scalera, R. 2008. *Trachemys scripta* (Schoepff), Common slider (Emydidae, Reptilia). In *Handbook of Alien Species in Europe*, ed. Delivering Alien Invasive Species Inventories for Europe (DAISIE), 374. Dordrecht: Springer.

59. Ficetola, G. F., C. Coic, M. Detaint, M. Berroneau, O. Lorvelec, and C. Miaud. 2007. Pattern of distribution of the American bullfrog *Rana catesbeiana* in Europe. *Biol Invasions* 9:767.

60. Simberloff, D. 2006. Invasional meltdown 6 years later: Important phenomenon, unfortunate metaphor, or both? *Ecol Lett* 9:912.

61. Mooney, H. A. 1999. The global invasive species program (GISP). *Biol Invasions* 1:97.

62. Keller, V., N. Zbinden, H. Schmid, and B. Volet. 2005. A case study in applying the IUCN regional guidelines for national red lists and justifications for their modification. *Conserv Biol* 19:1827.

63. Shirley, S. M., and Kark, S. 2006. Amassing efforts against alien invasive species in Europe. *PLoS Biol* 4:1311.

64. Cramp, S., and K. E. L. Simmons. 1977. *Handbook of the Birds of Europe, the Middle East and North Africa*. Oxford: Oxford University Press.

65. del Hoyo, J. et al., eds. 1992–2009. *Handbook of the Birds of the World*. 14 volumes. Barcelona: Lynx Editions.

66. BirdLife International. 2004. *Birds in Europe: Population Estimates, Trends and Conservation Status*. Cambridge: BirdLife International.

67. Strubbe, D. 2009. *Invasive Ring-Necked Parakeets Psittacula Krameri in Europe: Invasion Success, Habitat Selection and Impact on Native Bird Species*. Antwerp: University of Antwerp.

68. Blair, M. J., McKay, H., A. J. Musgrove, and M. M. Rehfisch. 2000. Review of the status of introduced non-native waterbird species in the agreement area of the African-Eurasian waterbird agreement research contract CR0219. British Trust for Ornithology, Thetford, Norfolk.

69. Clergeau, P., and P. Yesou. 2006. Behavioural flexibility and numerous potential sources of introduction for the sacred ibis: Causes of concern in western Europe. *Biol Invasions* 8:1381.

70. Pysek, P., and D. M. Richardson. 2006. The biogeography of naturalization in alien plants. *J Biogeogr* 33:2040.

71. Simberloff, D. 1997. Eradication. In *Strangers in Paradise. Impact and Management of Nonindigenous Species in Florida*, ed. D. Simberloff, D. C. Schmitz, and T. C. Brown, 221. Washington, DC: Island Press.

72. Clout, M. N., and C. R. Veitch. 2002. Turning the tide of biological invasion: The potential for eradicating invasive species (proceedings of the international conference on eradication of island invasives). Occasional paper of the IUCN Species Survival Commission No. 27, IUCN, Gland.

# chapter eight

# Invasive patterns of alien terrestrial invertebrates in Europe

*Alain Roques*

## Contents

## 8.1   Introduction

Invertebrates represent the majority of living organisms and, hence, form a large part of the alien species problem. Although alien invertebrates, and especially insects, have been considered problematic and sometimes even more problematic than native pests in many regions such as Australasia, South Africa, and North America,[1] Europe traditionally has been much less concerned.[2,3] Unlike alien animals and plants, no checklist of alien terrestrial invertebrates had been available in any of the European countries until the beginning of the twenty-first century. However, several invertebrate pests of economic importance have invaded the continent in recent years, raising interest in the issue of alien invertebrates. Among many others, the arrival in Europe of the American western corn rootworm, *Diabrotica virgifera virgifera*,[4] the Asian long-horned beetles, *Anoplophora* spp.,[5] the Asian tiger mosquito, *Aedes albopictus*,[6] the pine wood nematode, *Bursaphelenchus xylophilus*,[7] or the spectacular spread all over Europe of the horse-chestnut leaf miner, *Cameraria ohridella*, originating from the southern Balkans,[8] caused much public concern. Thus, checklists of alien invertebrates began to be compiled from 2002 onwards, successively covering a number of European countries: Austria,[9] Germany,[10] the Netherlands,[11] the Czech Republic,[12] Scandinavia,[13] the United Kingdom,[14,15] Italy,[16] Serbia and Montenegro,[17] Switzerland,[18] Albania, Bulgaria, and Macedonia,[19] and Hungary.[20] However, aliens are not respecting political barriers all the more, as the weakening customs and quarantine controls within the European Union (EU) may allow species to spread easily between EU countries once they are successfully established.[21] This situation has triggered the need for a continental survey of the already established alien invertebrate species.

The Delivering Alien Invasive Species Inventories for Europe (DAISIE) project, which was initiated in 2005, achieved a significant milestone in the knowledge of alien species in Europe. Gathering the research data of 19 scientific institutions across 15 countries, DAISIE compiled an inventory of all alien species known to have established before 2008 within the geographic limits of the European continent. DAISIE also supplied national and regional lists of alien fungi, bryophytes, vascular plants, invertebrates, fishes, amphibians, reptiles, birds, and mammals. The inventory revealed a dominance of vascular alien plants, accounting for 55% of all alien taxa, but terrestrial invertebrates taxa (23%) far outnumbered vertebrate (6%) and fungi taxa (5%).[22] By November 2008, a total of 1517 alien terrestrial invertebrates, including 1429 arthropods, had been identified as being established in Europe.[23] A recent update increased the number of alien arthropods to 1590 in 2010.[24]

This substantial work allowed researchers to figure out the relative importance of the different taxa of alien invertebrates comparatively to those of other plant, animal, and fungal groups, as well as to compare their respective habitats,[25] pathways,[26] and environmental and economic impacts.[27] This chapter presents the most important patterns exhibited by the terrestrial invertebrates alien to Europe.

## 8.2  To colonize Europe, it is better to be an insect herbivore from Asia!

By June 2010, a total of 1693 nonnative invertebrate species had established since 1500 CE in at least some part of Europe, including 1460 species from other continents and 233 cosmopolitan species of uncertain origin (defined as "cryptogenic"). Arthropods, mostly insects, predominate, accounting for 93.9% of the alien terrestrial invertebrate fauna (Table 8.1). However, limited taxonomic knowledge of several nonarthropod groups have likely led to underestimating their importance unless such species constitute phytosanitary threats or are vectors of disease.

Unlike the other groups of animals or plants, invertebrate species are mostly introduced unintentionally.[22] Rabistch[28] estimated that only 14% of alien arthropods are deliberately introduced, the most important purpose being biological control, essentially the release of hymenopteran parasitoids (180 species according to Rasplus et al.)[29] and ladybeetles such as the Harlequin ladybeetle, *Harmonia axyridis*.[30] Other intentional introductions include a few saturniid moths for silk production, "pet" arthropods such as walking sticks, insects reared for food for vertebrate pets, and edible snails.[31] Correcting for the total invertebrates, the percentage of deliberate introductions is estimated to be 12.9%. Accidental introductions of alien invertebrates predominantly occurred through the horticultural and ornamental trade (38%), including ornamental plants for planting, cut flowers, bonsais, seeds, and aquarium plants. Otherwise, invertebrate aliens arrived with contaminated stored products (18%),

*Table 8.1* Composition of the Alien Invertebrate Fauna Established in Europe by June 2010
(Species Introduced after 1500)

| Phylum | Exotic species | Cryptogenic species | Total species | Percentage |
|---|---|---|---|---|
| Platyhelminthes | 19 | 1 | 20 | 1.2 |
| Nematoda | 48 | 5 | 53 | 3.1 |
| Nemertea | 3 | 0 | 3 | 0.2 |
| Annelida | 12 | 1 | 13 | 0.8 |
| Mollusca | 14 | 0 | 14 | 0.8 |
| Chilopoda | 10 | 6 | 16 | 0.9 |
| Diplopoda | 12 | 2 | 14 | 0.8 |
| Pauropoda | 1 | 0 | 1 | 0.1 |
| Symphila | 2 | 1 | 3 | 0.2 |
| Acari | 99 | 3 | 102 | 6.0 |
| Aranea | 47 | 0 | 47 | 2.8 |
| Crustacea | 17 | 0 | 17 | 1.0 |
| Insecta | 1176 | 214 | 1390 | 82.1 |
| Total | 1460 | 233 | 1693 | |

*Source:* Updated from DAISIE, ed., *The Handbook of Alien Species in Europe*, Heidelberg: Springer, 2009; Roques, A., M. Kenis, D. Lees, C. Lopez-Vaamonde, W. Rabitsch, J. Y. Rasplus, and D. B. Roy, eds., *Alien Terrestrial Arthropods of Europe*, Two volumes, Sofia: Pensoft, 2010.

*Table 8.2* Feeding Regime of Alien Invertebrate Fauna Established in Europe

| Regime | Alien | Cryptogenic | Total | Percentage |
|---|---|---|---|---|
| Phytophagous | 729 | 61 | 790 | 45.9 |
| Parasitic or predator | 495 | 68 | 563 | 32.7 |
| Detritivorous | 241 | 104 | 345 | 20.0 |
| Unknown | 21 | 2 | 23 | 1.3 |

*Source:* Updated from DAISIE, ed., *The Handbook of Alien Species in Europe*, Heidelberg: Springer, 2009; Roques, A., M. Kenis, D. Lees, C. Lopez-Vaamonde, W. Rabitsch, J. Y. Rasplus, and D. B. Roy, eds., *Alien Terrestrial Arthropods of Europe*, Two volumes, Sofia: Pensoft, 2010.

vegetables and fruits (12%), fresh and manufactured wood material (10%), and in association with animal husbandry (3%).[23] In the specific case of the United Kingdom, Smith et al.[15] showed that the plant trade, particularly in ornamental plants, accounted for nearly 90% of human-assisted introductions of phytophagous insects. Even the introduction of alien forest insects in Europe appeared to be more a consequence of the trade of plants for planting rather than of forest products.[21] A few other alien species (95 species according to Rabitsch)[28] arrived in Europe as stowaways with used tires, containers, cars, and other vehicles. These introductions include mosquitoes, such as the Asian tiger mosquito, *Aedes albopictus*, and the Asian rock pool mosquito, *A. japonicus*, imported as eggs or larvae through the trade of secondhand tires[6,32]; hornets such as the Asian hornet, *Vespa velutina nigrithorax*, introduced in France with bonsai potteries[33]; ants such as the Argentine ant, *Linepithema humile*, and the garden ant, *Lasius neglectus*, translocated with soil and litter accompanying ornamental plants, with logs, or with other commodities offering shelter[28]; and cockroaches.

As a result of the predominance of the plant trade in the movement of alien species, the phytophagous regime largely dominates among alien invertebrates, accounting for nearly half the species (Table 8.2). The volume of international trade, including ornamental and horticultural products, has continuously increased during the last centuries and still grew from the early 1950s on at an average annual rate of 7.5% for manufactured goods and 3.5% for agricultural products.[34] Not surprisingly, the establishment of alien invertebrates has increased exponentially since the sixteenth century (Figure 8.1a). However, considering the first record in Europe as a proxy for the date of arrival of alien invertebrates, a significant acceleration has been observed since the last part of the twentieth century as a probable result of globalization. Thus, an average of 19.6 alien species has been newly reported annually in Europe between 2000 and 2008, nearly double the 12.4 species reported annually from 1950 to 1974. This trend is clearly related to a steep increase in the relative importance of phytophagous species in the newly reported alien invertebrate fauna (Figure 8.1a). Whereas parasitoids/predators and detritivores contributed quite equally as phytophages to enrich the alien fauna until the 1950s, parasitoids and predators markedly increased during the period 1950–2000 probably due to the infatuation for biological control agents. Since the 1990s, the explosion of the ornamental plant trade has likely triggered the arrival of alien phytophagous invertebrates, as has been shown for the insects that feed on woody plants.[21]

The lack of global inventories of alien invertebrate faunas makes it difficult to compare the European situation with that of other continents. However, such comparisons are possible at guild level and suggest some differences in the invasion processes among continents. Whereas the alien arrival rate was accelerating in both North America and Europe as shown by the incidence of interceptions at border checkpoints (see McCullough et al.[35] and Roques and Auger-Rozenberg,[36] respectively), the temporal patterns of establishment of alien forest insects largely differed between Europe, Canada, and the whole of North America during the twentieth century (Figure 8.1b).[21] In contrast to Europe, the

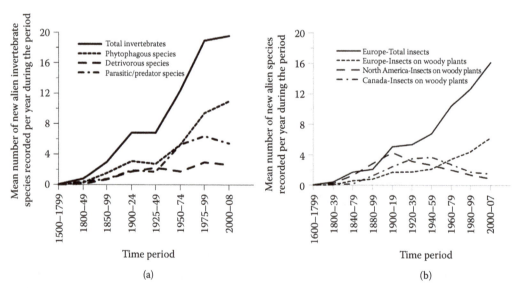

*Figure 8.1* Temporal changes in the mean number of new records per year in Europe of alien invertebrates. (a) Total number of alien invertebrates and detail per feeding regime. (Updated from Roques, A. 2010. Taxonomy, time and geographic patterns. In *Alien Terrestrial Arthropods of Europe*, ed. A. Roques et al., Chapter 2. *BioRisk* 4:11. http://pensoftonline.net/biorisk/index.php/journal [accessed August 5, 2010].) (b) Total number of alien insects and detail of alien insects associated with woody plants in Europe compared to North America. (Modified from Roques, A., *N Z J For Sc* 40 (suppl):77, 2010.)

establishment rate of alien species in North America increased from the late nineteenth century until about the 1920s, then declined rapidly thereafter. When considered separately, the same trend was observed for Canada, but the decline occurred later (1960s–1970s). Langor et al.[37] attributed these two declines to the quarantine legislations enacted in the United States in 1912 and in Canada in 1976, respectively. The disestablishment of internal customs processes (such as quarantine procedures) as a result of the formation of the EU in 1993 has probably made it easier for alien species to be introduced and spread further once established. Also, the quarantine legislation of the EU is less stringent than that in North America for some groups of organisms. A comparison of the quarantine interceptions during the period 1995–2005 with the establishments of alien insects and mites during the same period in Europe revealed large discrepancies[21]; the major groups of invaders (e.g., aphids, midges, scales, leafhoppers, and psyllids) remain largely undetected, whereas the groups that were predominantly intercepted (e.g., long-horned and bark beetles) made little contribution to the established alien entomofauna (Figure 8.2). Similar results were obtained at the country level for Austria, the Czech Republic, and Switzerland.[38]

Asia appears to be the dominant supplier of alien invertebrates since the beginning of the twentieth century, far beyond North America (Table 8.3). Surprisingly, this contribution from Asia to the alien invertebrate fauna in Europe appears relatively stable over the last century, fluctuating around 30% of the introduced species, whereas the commercial relationships between Europe and Asia, especially China, largely increased during the same period, and the major trade routes nowadays involve Asian harbors and airports.[21] The alien invertebrate

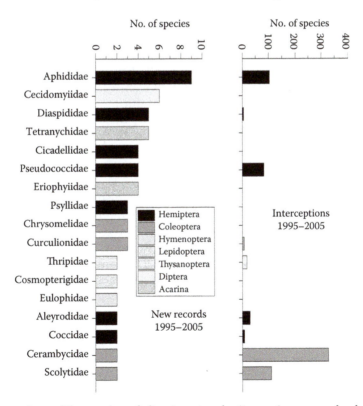

*Figure 8.2* Comparison of the number of alien insect and mite species on woody plant hosts established in Europe during 1995–2005, with the number of interceptions during the same period by family. (Modified from Roques, A., *N Z J For Sc* 40 (suppl):77, 2010.)

*Table 8.3* Temporal Variation in the Origin of the Alien Invertebrates Established in Europe. Values Are Expressed as the Relative Contribution of Each Origin to the Total Number of Species Having Arrived during the Considered Period (Right Column)

| Period | Africa | Asia | Australasia | North America | South America | Tropical | Cryptogenic | Total No. Species |
|---|---|---|---|---|---|---|---|---|
| <1900 | 11.7 | 20.0 | 3.4 | 11.2 | 13.2 | 8.8 | 31.7 | 205 |
| 1900–1949 | 12.8 | 27.5 | 9.0 | 17.1 | 12.5 | 4.9 | 16.2 | 345 |
| 1950–1974 | 13.5 | 31.2 | 4.5 | 29.9 | 7.4 | 4.5 | 9.0 | 311 |
| 1975–1999 | 12.7 | 29.2 | 6.7 | 27.8 | 10.4 | 4.4 | 8.8 | 479 |
| 2000–2008 | 12.0 | 31.4 | 10.3 | 20.6 | 9.7 | 4.0 | 12.0 | 175 |
| Overall | 12.7 | 28.3 | 6.7 | 22.7 | 10.6 | 5.1 | 14.0 | 1515 |

*Note:* Calculated on a subset of 1515 species, 168 species without reliable data.

contribution from North America has slightly decreased since 1975, whereas the contribution from Australasia has slightly increased, which is probably related, at least partly, to the establishment of a fauna accompanying the planting of eucalyptus trees in southern Europe.[21] However, the native origin of alien species in Europe largely differs according to phylum. Most alien lepidopterans and hemipterans originate from Asia, but mites, hymenopterans, and dipterans arrive predominantly from North America (see Section 8.3).[24]

## 8.3   A lot become established, but few spread all over Europe

A major pattern of most alien invertebrates once arrived in Europe is their slow spread (Figure 8.3). A total of 776 species (i.e., 45.8% of all alien invertebrates) have only spread in one or two European countries until 2010. Obviously, they mostly include recently arrived species, but nearly half the species arrived during the period 1950–74 show little dispersal. On average, the range of dispersal linearly increased with the duration of the alien species' presence in Europe (Figure 8.3). Only 30 species (1.8%), insects in stored products, aphids, and the potato cyst nematode, *Globodera rostochiensis*, have succeeded in invading more than 35 European countries. Most of these widely dispersed species arrived during the eighteenth and nineteenth centuries (70%), only one, *Omonadus floralis* (Coleoptera Anthicidae), was newly recorded just after 1950. Generally, detritivorous species appeared to have dispersed significantly more than phytophagous species and parasitoids or predators.[24] However, a few species show a large capability to spread during a limited time; the North American black locust gall midge, *Obolodiplosis robiniae*, initially found in Italy in 2003, colonized most of Europe in a few years,[39] as did the horse-chestnut leaf miner, *Cameraria ohridella*, since its detection in Austria in 1989[40] and the western conifer seed bug *Leptoglossus occidentalis*, first detected in 1999 in Italy.[41] Similarly, Ciosi et al.[4] recently showed that all the populations of central and southeastern Europe of the North American western corn rootworm, *Diabrotica virgifera virgifera*, probably originated from a single expanding population initially introduced in Serbia in 1992. In contrast, a species of concern such as the subterranean termite, *Reticulitermes flavipes*, is only present in France several centuries after its initial introduction from North America.[42]

Large differences exist between European countries in their total number of recorded alien invertebrates. Italy and France (>700 species), followed by Great Britain (ca. 600 species), host many more species than other countries but are also where the majority of alien invertebrates are recorded for the first time.[24] Although sampling efforts and local

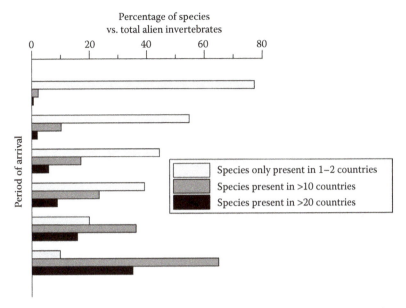

***Figure 8.3*** Geographic spread of the alien terrestrial invertebrates with regard to their date of first record in Europe.

taxonomic expertise may differ, the number of alien invertebrate species reported appeared positively correlated with the country surface area but even more strongly with macroeconomic variables reflecting the trading activity of the country, such as the size of the road network and the volume of recent merchandise imports and agriculture imports.[23] Pyšek et al.[26] found only two significant explanatory variables, national wealth and human population density, to explain the variation in alien richness of different animal and vegetal taxa of Europe.

Alien invertebrates show a strong affinity for the habitats intensively disturbed by human activities (Figure 8.4). The highest percentage of alien invertebrates species occur in parks and gardens (22.3%) and cultivated habitats (21.9%), while slightly less occur in human settlements (21.1%). Altogether, human-made habitats host 76.3% of the arthropods alien to Europe, most of these species are likely to occur in several different habitats. In contrast, less than 10% of the alien species have yet colonized natural and seminatural habitats, such as wetlands, riparian habitats, grasslands, and heathlands, and less than 15% occur in woodlands and forests. Pyšek et al.[26] stated that alien plants are also mostly found in human-made, urban, or cultivated habitats, unlike vertebrates that are more evenly distributed among habitats, the most invaded being aquatic and riparian habitats, woodland, and cultivated land.

Some habitats are differentially preferred by certain taxonomic groups, for instance, artificial (human settlements) habitats are invaded to a high degree by spiders. Indeed, more than 90% of alien spiders are found in buildings.[43] Psocids is another well-represented group, with 81.6% of its alien species found in buildings in Europe.[44] In contrast, alien hemipterans, especially aphids, and lepidopterans have predominantly colonized parks and gardens, 78.9% and 56.7% of their species being observed there, respectively.[45,46] Greenhouses constitute another important man-made habitat type, which hosts most of the alien myriapods (64.7%)[47] and thrips (55.8%).[48] Greenhouses are not escape-proof facilities, as confirmed by surveys in the areas surrounding such buildings, which recently

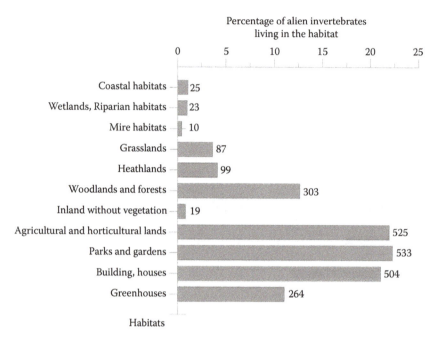

**Figure 8.4** Main European habitats colonized by the alien terrestrial invertebrates. The number over each bar indicates the absolute number of alien species recorded per habitat. Note that a species may have colonized several habitats. (Updated from Lopez-Vaamonde, C., M. Glavendekić, and M. R. Paiva. In *Alien Terrestrial Arthropods of Europe*, ed. A. Roques et al., Chapter 4. *BioRisk* 4:45, 2010a. http://pensoftonline.net/biorisk/index.php/journal [accessed August 5, 2010].)

revealed the escape of alien scale species affecting ornamental shrubs such as *Coccus hesperidum* and *Icerya purchasi* in Switzerland.[18] More than 200 alien species are listed as living in greenhouses in Europe.[28]

Why do most introduced terrestrial arthropods apparently stay confined to human-modified habitats in their new European distribution range? Lopez-Vaamonde et al.[49] proposed different hypotheses: (1) disturbed urban and semiurban areas may have a lower resistance to aliens, especially because of a lower pressure of potential natural enemies and, for phytophagous aliens, less vigorous host plants; (2) some species may prefer human-related habitats in their native range and are thus more likely to be carried into a new area by human transport than species living in natural environments.[38] For instance, exotic ornamental plants are generally used in man-made habitats such as nurseries, parks, gardens, roadside plantings, and shelter belts. Most alien phytophagous species introduced along with these ornamentals remain as yet strictly associated with their original, exotic host (46.4% in Europe);[50] they have not so far colonized native trees and thus they become established only in parks and gardens and in hedgerows where such exotic plants are planted. However, there might simply be a large time lag between the introduction to human habitats and adaptation and spread to natural habitats. Species such as the Asian longhorn beetles, *Anoplophora* spp. (Coleoptera, Cerambycidae), were restricted for a long time to urban areas in cultivated lanes (e.g., lanes planted with poplars), whereas they have the potential to live in natural forests where potential host plants occur. Twenty-two years after its arrival in North America, *A. glabripennis* was found in 2008 in natural forests dominated by *Acer* trees.[5]

## 8.4    Impact: Only a few are harmful… for the moment

The arrival of alien invertebrates already significantly modified the composition of the fauna presently observed in Europe. Roques[24] notes that a total of 30 families had no representatives in Europe before the arrival of aliens, including seven families of myriapods, four of mites, three of psocids and lice, two of scales, cockroaches, and beetles, and one of spiders, leafhoppers, lepidopterans, thrips, and stick insects. In some other families, alien species could be overrepresented (e.g., in scales) where aliens now represent nearly half the total Diaspididae fauna observed in Europe (60 out of 130 species [44.6%]), a third of the Coccidae fauna (23 out of 70 species [32.3%]), and a fourth of the Pseudococcidae fauna (37 out of 141 species [25.7%]). Similar high proportions of aliens are observed for psocids (Pachytroctidae, 66.7%; Ectopsocidae, 57%; and Liposcelidae, 26.4%), hemipterans (Aleyrodidae, 39.1% and Adelgidae, 36%), hymenopterans (Agaonidae, 40%; Aphelinidae, 24.2%; and Siricidae, 23.8%), and saturniid lepidopterans (30%). Even if the relative proportions are lower, the arrival of a large number of alien species has also largely modified the faunal taxonomic structure in dermestid beetles (21.9% aliens), tetranychid mites (15.1%), drosophilid flies (14.8%), and encyrtid chalcids (7.2%). In the chalcid genus *Megastigmus* (Hymenoptera), 10 out of 21 species specialized in the exploitation of tree seeds are not native to Europe.[51] Such changes in guild composition are likely to affect the functioning of ecosystems, but the precise effects are still little documented.

Vilà et al.[27] estimated that an ecological impact could be associated to 13.8% of the alien invertebrates in Europe, a value much smaller than that observed for alien terrestrial vertebrates and freshwater flora and fauna (>30%). However, experimental investigations to measure precisely the environmental impacts of alien invertebrates in Europe are still lacking.[52] Regarding herbivory, phytophagous insects largely dominate the alien fauna in Europe, but only a few examples can yet be cited about their impact on native plant biodiversity in Europe, contrasted to the North American situation where several alien forest pests of Eurasian origin dramatically affect forest ecosystems.[52] The most problematic alien species is probably the pine wood nematode, *Bursaphelencus xylophilus*, responsible for pine wilt disease, which arrived in Portugal in 1999 and affected maritime pine stands.[7] The immediate application of containment measures limited its impact to a small area until 2008, when it then spread all over Portugal[53]; this nematode is directly threatening pine forests of Spain and southwestern France. Another source of concern is the recent arrival in the Moscow region of the Asian emerald ash borer, *Agrilus planipennis*.[54] In North America, this woodborer already killed over 15 million ash trees, *Fraxinus* spp., in a few years.[55] The Asian emerald ash borer's expansion to the west of Russia is seriously threatening the European ash forests. In a similar way, the recent expansion throughout the Mediterranean Basin of two severe alien pests of palms, the red palm weevil (*Rhynchophorus ferrugineus*) and the palm moth (*Paysandisia archon*), is potentially threatening the survival of three endemic, endangered palm species in Europe: *Phoenix theophrasti* in Crete, *P. canariensis* in the Canary Islands, and *Chamaerops humilis* in the western Mediterranean region.[56,57] Alien seed insects have also been shown to reduce drastically the potential of natural regeneration of native plants. Two exotic species of *Megastigmus* are dramatically decreasing the seed yield of wild roses.[58] The rapid expansion throughout Europe of the western conifer seed bug, *Leptoglossus occidentalis*, since its introduction in Italy in 1999 is also a threat to conifer seeds.[41] The spread of the Asian chestnut gallmaker, *Dryocosmus kuriphilus*, in Italy, Slovenia, southeastern France, and, recently, Hungary[59] affect chestnut fruiting, causing yield reduction up to 70%.

Little information is available about niche displacements by aliens and intraguild competition for resources. Fabre et al.[60] showed that alien seed chalcids tend to displace the native ones for the exploitation of cedar and fir seeds. The parasitoid *Cales noacki* (Aphelinidae), deliberately introduced from South America, has replaced the native aphelinid *Encarsia margaritiventris* as the dominant parasitoid of the *Viburnum* whitefly, *Aleurotuba jelineki*, in Italy.[61] Another example of displacement of native parasitoids is the introduction of the North American aphid parasitoid *Lysiphlebus testaceipes* to control *Aphis spiraecola*, which affected the two congeneric parasitoids, *L. fabarum* and *L. confusus*, in Mediterranean countries.[62]

A few more data exist about the impact of alien predators and parasites on native prey. Among the 12 alien terrestrial planarians (predator flatworms) established in the United Kingdom, the New Zealand flatworm, *Arthurdendyus triangulatus*, has been shown to reduce lumbricid earthworm populations, with a possible major impact on soil ecosystems.[63] The aggressive garden ant, *Lasius neglectus*, and the Argentine ant, *Linepithema humile*, were observed to displace native ants as well as other invertebrates such as Lepidoptera in the Iberian Peninsula,[64] the impact of the latter species also resulting in a negative effect on seed dispersal of native plants. The ladybeetle *Harmonia axyridis*, introduced from eastern Asia, is also rapidly expanding outdoors in Europe.[65] Because of its tendency for intraguild predation, this large, aggressive, polyphagous coccinellid could have a dramatic impact on the abundance of native coccinellids and dramatically reduce their available niches.[66] Laboratory tests have already shown that European species are vulnerable to predation by *Harmonia axyridis* although this remains to be ascertained in field surveys.[67] A newly introduced hornet, *Vespa velutina*, is of increasing relevance for beekeepers. This predator of bees has invaded most of southern and western France[33] a few years after its accidental introduction from China in 2004. This introduced hornet's effective impact on the survival of bee populations and other hymenopteran competitors is currently under assessment. Another threat for bees, the parasitic varroa mite, *Varroa destructor*, has already spread to 24 European countries and is considered partly responsible for the decline of honeybees.[68]

Economic and sanitary impacts of alien invertebrates in Europe are much better documented. Many of these aliens affect European agriculture, horticulture, and forestry, mainly through yield losses and management costs but also through quarantine measures, market effects, and foreign trade impact. Vilà et al.[27] estimated that 24.2% of the alien invertebrates have such impacts. A review of the most problematic pests for outdoor agriculture and horticulture, greenhouses, stored products, forestry, and urban trees was recently supplied by Kenis and Branco.[69] However, only a few studies attempted to measure the induced costs. Pimentel[70] has calculated for the British Isles that, because arthropods annually damage or destroy approximately 10% of the crops and 30% of the pests are of exotic origin, alien arthropods cause yield losses of US$960 million per year. According to Kenis and Branco,[69] a similar calculation for the entire EU would lead to annual economic losses of approximately €10 billion caused by alien arthropods, not including control, eradication, or quarantine costs, nor costs linked to foreign trade impact or market effects. Only for the western corn rootworm, *D. virgifera virgifera*, Baufelt and Enzian[71] estimated the potential pecuniary losses in maize at €147 million per year due to the pest introduction in a number of European countries, based on a conservative average yield loss of 10%. The annual cost due to alien pests affecting stored products in Germany was estimated to reach €4.6–€12.3 million for the flour moth, *Ephestia kuehniella*, and €11.2–€35.3 million for the grain beetles *Rhyzopertha dominica* and *Oryzaephilus surinamensis*.[72] Consequently, most European countries apply costly regulatory control measures to prevent alien pest establishment in their countries. The costs of such measures to prevent the entry into Finland of the Colorado potato beetle, *Leptinotarsa decemlineata*, have been estimated to average €171,000 per year in the period 1999–2004.[73] An intensive eradication program carried out to control an outbreak of the melon thrips, *Thrips palmi*, in a United Kingdom greenhouse in 2000 cost £178,000.[74] Roosjen et al.[75] estimated that the annual cost of another thrips, *Frankliniella occidentalis*, to the Dutch greenhouses could be US$30 million, plus a further US$19 million from the effects of tomato spotted wilt tospovirus transmitted by the thrips.

Nearly 100 of the alien invertebrate species can affect, directly or indirectly, human and animal health in Europe.[23] Besides being a biting nuisance, six of the seven introduced species of mosquitoes in the family Culicidae are capable of transmitting diseases through bites of female mosquitoes.[76] The most important one, *Aedes albopictus*, is now established along the Mediterranean coast from southeastern France to northern Greece and is the vector of chikungunya disease, many arboviruses, avian plasmodia, and dog heartworm filariasis. The first outbreak of chikungunya observed in Europe occurred during the summer of 2007 in an area of central Italy infested by *A. albopictus*.[77] Other alien culicids may be vectors of the West Nile virus (*Aedes japonicus*,[32] *Culex tritaeniorhynchus*, *C. vishnui*, *O. atropalpus*), Japanese encephalitis (*A. japonicus*, *C. tritaeniorhynchus*), and Sindbis virus (*C. tritaeniorhynchus*).

Several alien ectoparasites can transmit viruses and pathogenic bacteria to humans and domestic animals. Three alien rat fleas are also able to feed on other mammals, including humans, to which they can transmit the bubonic plague when carrying the bacteria *Yersinia pestis*.[78] These fleas include a temperate species, *Nosopsyllus fasciatus*, from Asia, and the tropical rat flea, *Xenopsylla cheopis*, probably originating from the Nile area.[79] *X. cheopis* is also a vector of another human disease, the murine typhus fever caused by the bacteria *Rickettsia typhi*.[80] It became synanthropic in most of southern Europe where it could not survive as before because of large temperature variations between summer and winter within human habitats.[79] The third species, *Xenopsylla brasiliensis*, originates from tropical Africa, invaded the Canary Islands,[80] and occurs sporadically in port areas. Several chewing lice of cryptogenic origin are important pests of poultry farming, in particular *Menopon gallinae*, *Goniocotes gallinae*, and *Eomenacanthus stramineus*.[81] Chewing lice parasitizing mammals in Europe are listed in the study by Mey.[82] Some species are known to be of alien origin, such as the South American species *Gyropus ovalis*, *Gliricola porcelli*, and *Trimenopon hispidum*, arriving in Europe with guinea pigs and causing scratching, hair loss, and scabs to domestic and laboratory animals.[83] Other species worth mentioning are the cryptogenic dog louse, *Trichodectes canis*, and the sheep louse, *Bovicola ovis*, which cause pruritus and skin infections such as eczema. Several ticks in the Ixodidae family are of important sanitary concern, such as the North American *Dermacentor variabilis*, potential vector of lyme disease; *Rhipicephalus rossicus*, found in Romania, which may transmit the Crimean–Congo hemorrhagic fever; and three cryptogenic species in the genus *Hyalomma* (*H. anatolicum*, *A. excavatum*, and *A. truncatum*) affecting cattle.[68] A few alien spiders of medical importance to humans have also established in or around buildings in Europe, such as an Australian black widow, *Latrodectus hasselti*, from Australia, and two American *Loxosceles*, *L. laeta* and *L. rufescens*.[43]

Alien endoparasites of veterinary and medical concern include nematodes, flatworms, and mites. Thirteen alien species of nematodes may cause zoonosis in cattle, sheep (e.g., *Protostrongylus rufescens*, established in the Czech Republic),[12] game animals (e.g., *Ashworthius sidemi* in Poland),[84] poultry (*Ascaridia dissimilis*),[85] and humans (e.g., *Strongyloides stercoralis*, responsible for strongylosis).[12] Similarly, 7 of the 11 alien flatworms have a direct health impact on poultry and mammals including humans; for example, the American liver fluke, *Fascioloides magna*,[86] and the Asian liver fluke, *Clonorchis sinensis*, are found in Germany.[10] An Asian mite, *Epidermoptes bilobatus*, is a bird endoparasite causing avian scabies by burrowing into the skin.[68]

Finally, a number of alien invertebrates represent important sources of indoor allergens such as the tropical fowl mite, *Ornithonyssus bursa*,[87] the house dust mite, *Dermatophagoides evansi*, in Germany,[88] and the scuttle fly *Megaselia scalaris*.[89] They also include several species of cockroaches carrying allergens in their body, saliva, and fecal matter, and can cause asthmatic and skin reactions in humans. Because of their movement between waste and food materials, cockroaches can also acquire, carry, and directly transfer to food and eating utensils bacterial pathogens that cause food poisoning, diarrhea,[90] or typhoid. About 40 species of bacteria pathogenic to humans have been naturally found in or on cockroaches.[91]

## 8.5   Some specific patterns of the alien invertebrate groups

Invertebrate groups largely differ in the number of established alien species, colonized habitats and impacts.

### 8.5.1   Terrestrial flatworms (Platyhelminthes)

Alien flatworms include 12 predatory planarians mostly specialized on earthworms and other small invertebrates and 8 internal trematode parasitoids of veterinary (dog, sheep, poultry) and medical concern (see Section 8.4). The arrival of alien planarians, particularly marked in the United Kingdom, largely modified the faunal composition of this group. Thus, the three species native to the United Kingdom are largely outnumbered by introduced exotics, of which there are at least nine species.[92] Representative of this group are the New Zealand flatworm, *Arthurdendyus triangulatus*, the Australian flatworm, *Australoplana sanguinea*, and two *Kontikia* species (*K. andersonni* and *K. ventrolineata*), also from Australasia. The New Zealand flatworm was first recorded in Northern Ireland in 1963 where it supposedly arrived with a shipment of roses or bulbs from Christchurch, New Zealand.[93] A retrospective study suggested that it spread from botanical gardens to horticultural wholesalers, then to domestic gardens, and later invaded agricultural land.[63] At present, it is widespread in Great Britain and Ireland, particularly in Northern Ireland and central Scotland[92] and in the Faroe Islands.[94] It may have a large impact on the native lumbricid earthworm populations, with very high rates of flatworm population growth achieved when food is not limited and when horticultural practices favor population growth conditions. Earthworm species vary in their vulnerability to predation by *A. triangulatus* with surface-active and anecic species being considered the most at risk.[92] Climate models suggest that *A. triangulatus* may thrive in other parts of Europe, such as western Norway, southern Sweden, Denmark, Germany, and northern parts of Poland, at least in places where cool and damp conditions are found, with potential impact on soil ecosystems.[95] On the whole, most alien Platyhelminthes originate from Asia and Australasia, each of which supplied 40% of the species.

### 8.5.2   Nemerteans

The terrestrial species of this predominantly marine group are essentially native to Australia, New Zealand, and the Oceanic islands. Only three alien species have been found in Europe, at first in greenhouses and probably transported along with overseas plants imported for botanical gardens. *Antiponemertes pantini*, from New Zealand, was found outdoors in a garden in the Scilly Islands (the United Kingdom) in 1997.[96] The Australian *Geonemertes* (*Argonemertes*) *dendyi* is also present in the British Isles. It is likely that the increase in plant trade will allow the introduction of more nemerteans in the future as it has already been observed in the United States.[96,97]

## 8.5.3 Nematodes

A total of 53 alien nematodes have been recorded in Europe (Table 8.1). This number is an underestimation because studies of alien nematodes mostly focus on species of economic importance as phytosanitary or medical threats. Thus, a large number of alien nematodes recorded in Europe are agriculture pests such as *Globodera pallida, G. rostochiensis,*[98] and five species of *Meloidogyne* affecting crops, especially potato, in a large number of European countries. Eight alien species of *Xiphinema*, originating from the Americas, are also serious pests of *Vitis* and *Prunus.*[99] The American pine wood nematode, *Bursaphelenchus xylophilus,* introduced to Portugal in 1999, is a major threat for pine forests in southern Europe (see Section 8.4). Nearly half (24) of the alien species are internal parasites of cattle, poultry, game animals, rodents, and other animals. Thus, the Asian *Ashworthius sidemi* infests great numbers of wild ruminants (red deer, roe deer, and European bison) in Poland.[84] At least three alien species infest cockroaches (*Blatticola blattae, Hammerschmidtiella diesingi,* and *Leidynema appendiculatum*).[12]

Asia and North America contribute equally (16 species [30.1%]) to the nematode fauna alien to Europe. However, all Asian species are internal parasites, whereas 44.4% of the phytophagous species originated from North America (66.7% from the Americas). Unlike most groups of alien invertebrates, the nematodes show a rather large range expansion in Europe, with 24.5% of the species present in 10 countries and more. Nematodes have very limited potential for natural movement and need vectors such as long-horned beetles in the genus *Monochamus* for the pine wood nematode.[7] For phytophagous species, the most likely pathway of introduction in Europe is through the movement of infected or contaminated vegetal material. Infected host plants or host products such as bulbs, tubers, or wood for *B. xylophilus* can easily transport nematodes, as shown by interceptions.[36] For a number of parasitic species, nematodes have followed their exotic animal hosts introduced to Europe such as the raccoon nematode, *Baylisascaris procyonis.*[100]

## 8.5.4 Annelids (earthworms)

According to Hendrix,[101] at least 100 earthworm species have achieved distributions beyond their places of origin, but only 14 exotic annelids have yet been observed in Europe.[23] They essentially include species related to the degradation of organic wastes such as the Japanese red worm, *Eisenia japonica,* and the tropical *Eudrilus eugeniae.*[102] Intra-European invasion of earthworms has been detailed for Romania[103] and northeastern Europe.[104]

As with other invasive organisms, earthworm introductions appear to be facilitated by global commerce, both inadvertently with the importation of soil-containing materials (e.g., agricultural and horticultural products) and intentionally for use in commercial applications (e.g., waste management and land bioremediation). The characteristics of some earthworm species (e.g., parthenogenesis, environmental plasticity, ability to aestivate) appear to make them particularly successful as invaders.[101]

## 8.5.5 Mollusks

Unlike the marine and freshwater environments, terrestrial habitats have been little colonized by alien mollusks, with only 14 reported species. They include a predatory worm slug, *Boettgerilla pallens* (= *B. vermiformis*), originating from the Caucasus where it lives in submountainous and mountainous forests. It was introduced to Europe in the mid-twentieth century, probably with potted plants and vegetables. Since then it has quickly been spreading in synanthropic habitats of central, western, and northern Europe but it has also colonized natural habitats.[105] Several exotic species of snails first restricted to greenhouses have then been found outdoors in central Europe (e.g., the orchid snail, *Zonitoides arboreus*).[106] Other snails were deliberately introduced as edible species, such as *Helix nucula*, imported from Egypt and Libya to Crete.[31]

Besides these truly alien species, a number of other terrestrial mollusks have been introduced from southern and western Europe toward northern and eastern countries. Among them, the Iberian slug, *Arion vulgaris* (= *lusitanicus*), unintentionally introduced probably several times with plant material, package, and waste materials. As a defoliator, it has a large impact on plants in parks, gardens, and cultivated lands, and it outcompetes native slug species due to its large size and high population density peaks.[107] Other *Deroceras* slug species and snails such as *Milax gagates* and *Cryptomphalus aspersus* were unintentionally translocated within Europe.[108]

## 8.5.6   Crustaceans

A total of 17 exotic terrestrial crustacean species—13 isopods (woodlice), and 4 amphipods (lawn shrimps)—have established on the continent, mainly in the eastern and central countries.[109] In addition, 21 species native to Europe were introduced in a European region to which they are not native. The establishment of alien crustacean species in Europe slowly increased during the twentieth century without any marked changes during the recent decades. Almost all species alien to Europe originate from subtropical or tropical areas, except the woodlouse *Protracheoniscus major*, which is likely a native of central Asia. The most widely distributed alien woodlouse in Europe is the tropical American *Trichorhina tomentosa*, whereas the most widely distributed amphipod is *Talitroides alluaudi*, possibly originating from the Seychelles Islands. Most of the initial introductions are recorded in greenhouses, botanical gardens, and urban parks, probably associated with passive transport of soil, plants, or compost. All alien woodlice are still confined to synanthropic habitats (e.g., urban parks, villages, private gardens), but three lawn shrimp species in the family Talitridae have colonized natural habitats.[109] Among them, the Australian *Arcitalitrus dorrieni* may have some effects on the soil and leaf litter of deciduous and coniferous woodlands it has invaded in the Netherlands, Ireland, and western parts of Great Britain. O'Hanlon and Bolger[110] estimated that 24.7% of annual litter fall in a coniferous woodland is ingested by this species and suggested that this introduced species may play a more important role than native macrofaunal species in nutrient turnover.

## 8.5.7   Myriapods

The recently published first review of alien myriapods in Europe[47] can be summarized as follows. Currently, 40 species belonging to 23 families and 11 orders are considered alien to Europe, which accounts approximately for about 1.8% of all species known on the continent. They mostly include millipedes (Diplopoda) with 20 alien species and centipedes (Chilopoda) with 16. All of the alien myriapods have most probably been accidentally introduced to Europe with plant material involved with human activities and trade. Another possible new pathway is their intentional import as "pet animals" (large *Scolopendra* spp. as well as some large and colorful millipedes of the orders Spirobolida, Spirostreptida, and Sphaerotheriida), and their eventual escape from pet keepers. Introductions of alien myriapods into Europe probably began several centuries ago, even though a precise arrival date is hard to determine. Only 10 of 40 species were recorded for the first time in Europe in the nineteenth century, whereas most of the records date from the twentieth (26 species) and twenty-first centuries (4 records). They predominantly originate from tropical and subtropical regions (28 species out of 40 [70%]). The largest number of alien myriapods (25) has been recorded from Great Britain, followed by Germany with 12, France with 11, and Denmark with 10 species. In general, northern and economically more developed countries with high levels of imports and numerous busy sea ports are richer in alien species. Man-made artificial environments (pastures and cultivated lands, greenhouses, and urban and suburban areas) constitute the main habitat types hosting alien myriapods. Species of tropical and subtropical origin are likely to be restricted to greenhouses or equivalent artificially warmed habitats. Some of them in the summer season in the southern countries perhaps could survive outdoors in close proximity to the hothouses. Only 12 species (ca. 30%) have been reported from natural habitats in Europe, mostly centipedes. For instance, the Australasian *Lamyctes emarginatus* has been reported from gardens, roadsides, hedges, embankments, agricultural lands, and woodlands. It predominates in open and disturbed areas with sparse vegetation.[111] So far, the only alien millipede that has invaded some natural ecosystems is the east Asian species *Oxidus gracilis*, nowadays present in 33 European countries and found in forests close to suburban and urban areas woodlands.[112] At present, alien myriapods do not cause serious threats to the European economy, and there is insufficient data on their impact on native fauna and flora.

## 8.5.8   Spiders

Nentwig and Kobelt[43] reviewed a total of 47 spider species alien to Europe, which corresponds to 1.3% of the native spider fauna. Alien spiders essentially include species in the families Theridiidae (10 spp.) and Pholcidae (7 spp.), Sparassidae, Salticidae, Linyphiidae, Oonopidae (4–5 species each), and 11 additional families. The body sizes of alien Theridiidae, Pholcidae, and Salticidae imported to Europe were significantly larger than that of the native species, which may reflect a better capability of spider specimens with larger body sizes to survive the physical transport conditions, especially the temperature and humidity inside a standard ship container.[113] Even if the known number of alien spider introductions is an underestimation, a continuous increase in new records is observed: 12 first species records in the nineteenth century, 24 records for the twentieth century, and already 11 records for the first years of the twenty-first century.[43] Kobelt and Nentwig[113] predict that at least one additional alien spider species will annually arrive in Europe for the near future.

One-third of the alien spiders have an Asian origin, one-fifth comes from North America and Africa, and most species (55%) appear to originate from tropical habitats. Spiders are typically introduced as stowaways. They can survive shipment in or at containers or construction materials for periods long enough to reach most other continents. In the past, banana or other fruit shipments were an important pathway of introduction; today, potted plants and probably container shipments in general are more important. Consequently, many alien spiders are detected in a harbor, in buildings at or close to a harbor, and in or at warehouses.[114] France, Belgium, the Netherlands, Germany, and Switzerland possess the highest numbers of alien spider species. These countries are also the ones with the highest level of imports. Most alien spiders establish in and around man-made buildings, and only few species colonize natural habitats. However, the most frequently occurring alien spider in Europe, the North American linyphiid *Mermessus trilobatus*, has established in grassland and ruderal habitats.[115] First detected in southern Germany in the 1980s, *M. trilobatus* has spread since then and has become integrated into many natural spider communities. *M. trilobatus* belongs to the smaller linyphiids and is unlikely to outcompete a native species. Competition experiments indeed proved that the invasive success of *M. trilobatus* is not facilitated by strong competitiveness. More generally, no environmental impact of alien spider species is known so far, and Burger et al.[116] assume a high resilience of native spider communities. Some alien species are theoretically venomous to humans, such as the sicariids *Loxosceles laeta* and *L. rufescens* and the Australian black widow *Latrodectus hasselti*. However, there is no record in Europe reporting bites from these species.[43]

### 8.5.9   Mites and ticks

The inventory of the alien Acari of Europe includes 101 species, of which 87 mites and 9 ticks, belonging to 16 different families.[68] Exotic tick species, which are obligate ectoparasites, include tortoise tick, *Hyalomma aegyptium*, ticks found on snakes, such as *Amblyomma latum* and *A. exornatum*, and *Hyalomma dromedarii*, introduced with dromedaries into the Canary Islands. The brown dog tick, *Rhipicephalus sanguineus*, is spreading in parts of Europe beyond its current range because of the movement of domestic dogs. The alien mite species include 14 species that have a parasitic or predator regime. Among these species, eight invaded Europe with rodents such as muskrats and brown rats, whereas two others are parasites of birds (*Epidermoptes bilobatus* and *Ornithonyssus bursa*). Three phytoseiid species, *Phytoseiulus persimilis*, *Neoseiulus californicus*, and *Iphiseius degenerans*, were mainly introduced as predatory species against agricultural pests. The last mite, *Varroa destructor*, originated from southeast Asia, where it was confined to its original host, the Asian honeybee, *Apis cerana*. First reported in the 1970s in eastern Europe on *A. mellifera*, *V. destructor* spread rapidly all over Europe and is presently considered a major factor of bee decline in Europe.

The remaining 77 mite species are all phytophagous and mostly belong to two families, Eriophyidae (37 spp.) and Tetranychidae (spider mites, 27 spp.). These two families include the most significant agricultural pests. However, most alien Eriophyids have a very restricted distribution. More than 30% of the species have been observed in only one country (12 species), more than 50% (20 species) in 2–5 countries, but approximately 13% (5 species) in 6–11 countries. Only one cryptogenic species, the pear blister mite *Eriophyes pyri*, has been recorded in 32 European countries. Among the Tetranychidae, 19 alien species are found around the Mediterranean Basin and 12 in the rest of Europe. With relatively warm winters, the Mediterranean region provides suitable climatic living conditions for many species of temperate climates, but also for the establishment of many species of tropical or subtropical origin. Except for *Panonychus citri* and the cryptogenic species *Tetranychus ludeni*, which can be found in glasshouses in Europe, all tropical alien spider mites are restricted to the area around the Mediterranean Sea. In addition, three species of false spider mites (Tenuipalpidae) are major invaders in Europe, that is, *Brevipalpus californicus*, first recorded in 1960 and mainly observed on citrus around the Mediterranean Basin; the privet mite, *Brevipalpus obovatus*; and the phalaenopsis mite, *Tenuipalpus pacificus*.

The introduction rate of alien Acari has exponentially increased during the twentieth century with plant trade and the movement of agricultural commodities. Most of the alien mite species originated from North America (52%), and many from Asia (25%). In contrast, nearly half of the ticks (four spp.) alien to Europe originated from Africa.

### 8.5.10 Insects

#### 8.5.10.1 General patterns

A total of 1390 alien insect species had been recorded in Europe by June 2010.[117] Three orders namely Coleoptera, Hemiptera, and Hymenoptera (which has been previously underestimated by Roques et al.)[23] largely dominate, accounting for nearly 73% of total alien arthropods, representing 28.6%, 22.9%, and 21.4%, respectively (Figure 8.5a). Diptera, Lepidoptera, Thysanoptera, and Psocoptera have much less importance (<8%). The other orders are anecdotal. Some insect orders show no alien species in Europe, whereas others contribute important components of the native fauna such as Trichoptera. More generally, at the order level, the taxonomic composition of the alien fauna significantly differs from that of the native European arthropod fauna.[23] Hemiptera are nearly three times better represented in the alien fauna than in the native fauna (22.9% v. 8.0%). The alien entomofauna also includes proportionally more thrips (3.7% v. 0.6%), psocids (3.5% v. 0.3%), and cockroaches (1.3% v. 0.2%) than the native fauna, but far fewer dipterans (6.2% v. 21.1%) and hymenopterans (21.4% v. 25.3%). Differences are less pronounced for Coleoptera (28.6% v. 30.2%) and Lepidoptera (7.0% v. 10.2%).

The alien insect fauna is highly diverse with a total of 199 families involved. However, only 35 of these families contribute 10 or more alien species and 11 families more than 30 species (Figure 8.5b). These 11 families include mostly hemipterans, such as aphids

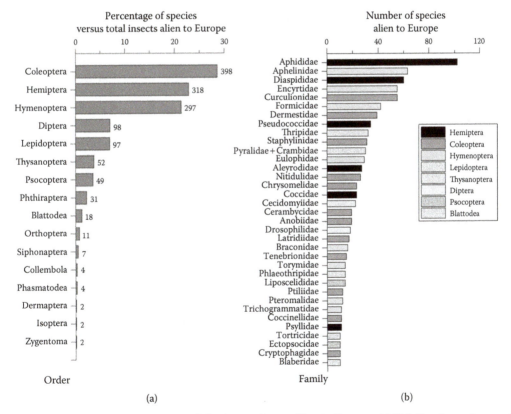

***Figure 8.5*** Taxonomic composition of the insect fauna alien to Europe. (a) Relative importance of the different taxonomic orders. The number over each bar indicates the total number of alien species observed per order. (b) Families showing more than 10 alien species.

(Aphididae) with the highest number of alien species (102 spp.) and scales (Diaspididae - 60 spp. and Pseudococcidae - 37 spp.), as well as hymenopteran chalcids used for biological control such as Aphelinidae (63 spp.), Encyrtidae (55 spp.), and thrips (Thripidae - 36 spp.). All of these except snout beetles (Curculionidae - 55 spp.) and ants (Formicidae - 43 spp.) are tiny arthropods.

### 8.5.10.2   *Coleoptera*

Twelve families of Coleoptera have more than 10 alien representatives in Europe (Figure 8.5b). The most numerous is Curculionidae (55 spp. including bark beetles),[118] followed by Dermestidae (39 spp.), and Staphylinidae (31 spp.). On the whole, alien detritivorous species (44.2%) slightly outnumber the phytophagous ones (40.2%). The number of new records of alien coleopteran species increased continuously during the twentieth century with a striking acceleration during the last decades: 1.8 new species per year during 1950–1974, 3.4 during 1975–2000, and 5.8 during 2000–2008. Alien coleopterans mostly originate from Asia (25%), followed by Africa (17.8%), but a large number of species are cosmopolitan with unknown origin (20.4%). These average first reported times and geographic origin patterns largely vary between and among taxa. For instance in Chrysomelidae, seed beetles dominated the introductions in Europe until 1950, but since then there has been an exponential increase in the rate of arrival of leaf beetles, reaching an average rate of 0.6 species per year during the period 2000–2009.[119] Most alien chrysomelids originated from Asia, but this pattern was mainly due to seed beetles, of which a half are of Asian origin, whereas leaf beetles predominantly originate from North America (36.4%).

Most of the coleopteran species have been introduced accidentally via international transport mechanisms except a few species used for biological control such as ladybeetles (11 spp.).[30] The most important pathway for unintentional introduction is the horticultural and ornamental trade, then stored products, crops, and wood and wood derivates. However, the pattern may differ according to the family. In long-horned beetles, wood-derived products such as wood packaging material and palettes, and bonsais constitute invasive pathways of increasing importance.[120]

The arrival of alien coleopterans largely changed the faunal composition in some families. Dermestidae has 39 species reported as aliens in addition to 139 species native to Europe.[121] In seed beetles, aliens presently account for 9.4% of the total European seed beetle fauna.[119] Most alien coleopterans are presently concentrated in man-made habitats, with 57.3% of the species observed in urban and semi-urban habitats, whereas only a few have colonized natural habitats such as forests (19.3%).[49]

### 8.5.10.3   *Hemiptera*

Only six families of Hemiptera have more than 10 alien representatives in Europe, but they include the most numerous ones: Aphididae (aphids) and Diaspididae (scales; Figure 8.5b). Most of the 102 alien aphid species originate from temperate regions of the world[45] with no significant variation in the geographic origin over time. The average introduction rate was 0.5 species per year since 1800; and unlike the other insect orders, the mean number of newly recorded species per year decreased since 2000.[45] Scale insects numerically represent a major group of aliens in Europe. Due to their small size and capacity for concealment, many species, mainly belonging to the families Diaspididae, Pseudococcidae, and Coccidae, have been accidentally introduced to Europe, mostly originating from tropical regions, mostly from Asia.[122] The fruit tree and ornamental trade appears to be their usual pathway of introduction. At present, alien scales represent an important component of the European entomofauna, accounting for about 30% of the total scale fauna.

Apart from aphids and scales, 52 additional Sternorrhyncha species alien to Europe have been identified to be Aleyrodidae (27 whitefly species including the highly crop-damaging species *Bemisia tabaci* and *Trialeurodes vaporariorum*), Adelgidae (9 adelgids), Phylloxeroidea (2 phylloxerans), and Psylloidea (14 species of jumping plant lice).[123] At present, alien species represent 39% of the total whitefly fauna and 36% of the total adelgid fauna occurring in Europe. The proportion is insignificant in other families. The arrival of alien phylloxerans and adelgids appeared to peak during the first part of the twentieth century. In contrast, the mean number of new records per year of alien aleyrodids and psylloids increased regularly after the 1950s. For these latter groups, an average of 0.5–0.6 new alien species has been recorded per year in Europe since 2000. Alien aleyrodids and psylloids mainly originated from tropical regions, while the adelgids and phylloxerans came equally from North America and Asia. Most of these alien species are presently observed in man-made habitats, especially in parks and gardens (78.9% of the species),[49] but alien adelgids are mainly observed in forests because of their association with conifer trees used for afforestation.[123]

In addition, 12 alien Auchenorrhyncha species have been recorded (mostly Cicadellidae-8 species), mainly originating from North America, such as the vine leafhopper *Scaphoideus titanus* first recorded in 1958 in France,[123] and 23 alien Heteroptera.[41] Most Heteropterans belong to the families Miridae (20 spp.), Tingidae (8 spp.), and Anthocoridae (7 spp.) and mostly originated from North America, such as the western conifer seed bug, *Leptoglossus occidentalis*, but some came from Asia, such as the brown marmorated stink bug, *Halyomorpha halys*, just detected in 2007 in Switzerland.[41] The rate of heteropteran introductions has exponentially increased during the twentieth century; and since 1990, an approximate arrival rate of seven species per decade has been observed. Most alien Heteroptera species are observed in central and western Europe (the Czech Republic, Germany, the Netherlands, and Great Britain). Ornamental trade and movement as stowaways with transport vehicles seem their major introductory pathways.[41]

### 8.5.10.4 Hymenoptera

Rasplus et al.[29] presented the first comprehensive review of the Hymenoptera alien to Europe. Their study revealed that nearly 300 species of Hymenoptera belonging to 30 families have been introduced to Europe. Two-thirds of these alien Hymenoptera are parasitoids or hyperparasitoids introduced mostly for biological control. Besides these parasitoids, 35 phytophagous species, 47 predator species, and 3 species of pollinators have been introduced. Seven families of wasps together represent about 80% of the alien Hymenoptera introduced to Europe (Aphelinidae, Encyrtidae, Formicidae, Eulophidae, Braconidae, Torymidae, and Pteromalidae). The three most diverse families are Aphelinidae (60 species representing 32% of the Aphelinid European fauna), Encyrtidae (55 species), and Formicidae (42 species), whereas the superfamily Chalcidoidea includes two-thirds of the total Hymenoptera species introduced to Europe. The first two families are associated with mealybugs, a group that also includes numerous aliens to Europe. The number of new records of alien Hymenoptera exponentially increased with time during the last 200 years, reaching a maximum of five new species per year between 1975 and 2000. North America provided the greatest part of the alien hymenopterans (96 species, 35.3%), followed by Asia (84, 30.9%) and Africa (49, 18%). Three Mediterranean countries (but only their continental parts) host the largest number of alien hymenopterans: Italy (144 spp.), France (111 spp.) and Spain (90 spp.). Intentional introduction, mostly for biological control, has been the main pathway of introduction for Hymenoptera. Consequently, the most invaded habitats are agricultural and horticultural as well as greenhouses. Ants have probably displaced native species, and this is also true for introduced parasitoids that are suspected to displace native parasitoids by competition, but reliable examples are still scarce (see Section 8.4).

### 8.5.10.5 Diptera

A total of 98 dipteran species distributed in 22 families have been listed as aliens to Europe,[124] which represent less than 0.5% of the native fauna (19,400 native species in 125 families). The suborder Brachycera dominates the alien fauna with 66 species (18 families), including several vegetable pests in the family Agromyzidae (*Liriomyza chinensis*, *L. huidobrensis*, and *L. trifolii*); major fruit pests in the family *Tephritidae* (*Ceratitis capitata*; *Rhagoletis cingulata*, *R. completa*, and *R. indifferens*); 18 species of Drosophilidae including the North American detritivorous *Chymomyza amoena* and fruit pests such as *Drosophila suzukii* discovered in 2009 in Italy; 9 Phoridae species that feed on mushrooms (e.g., the Australasian *Megaselia gregaria*) or are detritivores; and 8 species of Tachinidae predators. A total of 32 exotic species (four families) are observed in the suborder Nematocera, especially midges (Cecidomyiidae, 23 species) and mosquitoes (Culicidae, 7 species most being of high medical concern; see Section 8.4). Alien dipterans are predominantly phytophagous (35.6%), whereas a lesser portion are zoophagous (28.6%) or detritivorous/mycetophagous (29.6%).

The arrival of alien dipterans accelerated since the second half of the twentieth century. A total of 20 species were newly recorded from 2000 to 2009, that is, an average of 2.2 species per year. Unlike some other insect orders, North America appears to be the dominant contributor to the alien dipteran fauna, with almost one-third (30.6%) of the species originating from this continent, far greater than Asia (19.4%), while a significant percentage of species came from Africa (16.3%). Horticultural and ornamental trade is probably the most significant pathway of introduction, with a total of 30 species more or less closely associated with plants; *Horidiplosis ficifolii*, a midge causing leaf galls on *Ficus benjamina* was probably imported with infected fig plants in containers from southeastern Asia. The trade of vegetable and fruit crops likely resulted in the introduction of larvae of 10 other species such as *Liriomyza* spp. (Agromyzidae), tephritid fruit flies, and some midges. The movement of stored products seems responsible for the introduction of another 10 species, mostly drosophilids but also several species associated with the mushroom trade such as some phorids (*Megaselia tamilnaduensis*, *M. scalaris*) and mycetophilids (*Leia arsona*). Movement of compost is the probable pathway for two species of Stratiomyidae (*Exaireta spinigera* and *Hermetia illucens*), while three species are associated with animal husbandry such as *Crataerina melbae* (Hippoboscidae) and *Chonocephalus depressus* (Phoridae). A very limited proportion of alien Dipterans resulted from intentional introductions. Only three predator species were released for biological control and have subsequently become established. Two of them, *Hydrotaea aenescens* (Muscidae) and *Hermetia illucens* (Stratiomyidae), have been introduced from North America to control house flies in poultry farms and stables. The third one, *Feltiella acarisuga* (Cecidomyiidae), is a cryptogenic species of cosmopolitan distribution preying exclusively on tetranychid red spider mites, which has been intentionally released, mostly in glasshouses to protect crops.

Nearly 65% (64.1%) of the alien Dipteran species established in Europe are only present in man-made habitats, essentially around and in buildings, in agricultural lands, parks and gardens, and glasshouses. In addition, 16 of the 35 phytophagous aliens (45.7%) remain strictly related to their original, exotic plants used as ornamentals in the vicinity of human habitations such as bamboo, *Gleditsia*, *Buddleia*, and *Robinia pseudoacacia*. Woodlands and forests have been colonized by a few alien species (11.7%). Ecological impacts on native fauna and flora are not really documented, but large economic impacts on crops are reported for 14 alien dipterans.[124]

### 8.5.10.6 Lepidoptera

A total of 97 nonnative Lepidoptera species (about 1% of the known fauna) in 20 families and 11 superfamilies have established in Europe, of which 30 alone are Pyraloidea.[46] Although a number of species have been in Europe for hundreds of years, 74% have established during the twentieth century and arrivals are accelerating with an average of 3.2 alien Lepidoptera newly established per year between 2000 and 2007. Asia is the main contributor (37.2%), ahead of Africa (28.2 % including the Macaronesian Islands) and North America (20.5%).

Most alien Lepidoptera have been introduced accidentally to Europe (96.9%) except some saturniid species imported from Asia into Europe for silk production in the nineteenth century, which subsequently became naturalized, especially in urban areas. The import of ornamental plants (particularly palms, geraniums, and azaleas) is most likely responsible for the introduction of several species such as the palm moth, *Paysandisia archon*, the geranium bronze, *Cacyreus marshalli*, and *Caloptilia azaleella*. Transport also plays an important role in the dispersal of some species such as some tiny adult gracillariids (e.g., the horse-chestnut leaf miner, *Cameraria ohridella*). The escape or release for weddings from tropical butterfly houses may create a new pathway, and such exotic butterflies have been recorded flying freely in cities.[46]

About 50.5% of alien Lepidoptera live indoors in domestic, industrial, and other artificial habitats such as greenhouses. Parks and gardens host 52.6% of alien species, where they are frequently introduced along with their original, exotic host plant, while 25.8% have colonized agricultural land.[49] Only a few species have established in a more or less natural environment, mostly in woodlands. Examples of the latter include the arctiid *Hyphantria cunea*, the gracillariid *Phyllonorycter issikii* in central Europe, and the saturniid *Antheraea yamamai* in the Balkans. Most alien Lepidoptera are phytophagous (78.3%), whereas detritivores represent only 21.6%. A negative economic impact has been recorded for 16 alien species. The Indian meal moth *Plodia interpunctella* may severely affect stored grain and grain products, dried fruits, and seeds in households and warehouses. The Asian tortricid *Grapholita molesta* and yponomeutid *Prays citri* constitute serious pests in orchards in many parts of Europe, as well as the North American arctiid *Hyphantria cunea* on deciduous trees. Some species can also cause an aesthetic impact. *Cameraria ohridella* causes premature defoliation of the white-flowered horse-chestnut *Aesculus hippocastanum* in most towns of central and western Europe. The trees do not die, but the aesthetic impact is so severe that in some countries heavily infested trees have been felled and removed. Little is known about the ecological impact of alien lepidopterans, but Lopez-Vaamonde et al.[46] identified four species that may have potential impacts by threatening the native flora (*Diaphania perspectalis*, *Cacyreus marshalli*, and *Paysandisia archon*) or through apparent competition with native leaf miners (*Cameraria ohridella*).[125]

### 8.5.10.7 Other insects

A total of 52 alien species of thrips (Thysanoptera), belonging to four families, have been recorded in Europe.[48] Species introduced before 1950 mostly originated from the United States' tropical and subtropical areas, but the subsequent arrivals generally originated from Asia and from the United States to some extent. Five countries host more than 30% of the European alien thrips fauna and two alien thrips (*Heliothrips haemorrhoidalis* and *Frankliniella occidentalis*) occur in more than 50% of the European countries and islands. Most species (94%) have been accidentally introduced, probably with the trade of ornamental greenhouse plants since all widespread alien species in Europe are greenhouse pests or predators. Only three predator species (*Franklinothrips vespiformis*, *Franklinothrips megalops*, and *Karnyothrips melaleucus*) have been deliberately introduced for biological control. Alien thrips are mostly phytophagous (75%) and seldom predators (13.5%) or detritivores (11.5%). Cultivated habitats are preferentially (94.2%) invaded by exotic thrips, including greenhouses that provide suitable habitat for 55.8% of the invasive species in Europe. According to Reynaud,[48] less than 10 of the alien thrips can be considered to have an impact on human activities, such as the western flower thrips, *Frankliniella occidentalis*, which has become a key pest in a large range of agricultural and floricultural production areas in Europe and is vectoring a number of plant diseases. Two other North American *Frankliniella* species are known in Europe, but with a very limited distribution and without economic impact. The potential introduction of the melon thrips (*Thrips palmi*), a vector of six plant viruses at least, represents a continuous threat to glasshouse ornamental and vegetable crops in Europe. Numerous interceptions have been reported on cut flowers, fruits, and vegetables; and several outbreaks have been found in glasshouses in the Netherlands and the United Kingdom since 1988. *T. palmi* is considered to be absent in Europe; although it was detected outdoors within flowers of kiwi fruit in Portugal in 2004, it is no longer found there.

Among the 231 species of psocids (Psocoptera) present in Europe, 49 (21.2%) are considered to be of alien origin.[44] Most alien species originated from tropical and subtropical areas, especially from Africa. Many of them are food pests, moving along with stored products. A total of 39 of these species occur in buildings in Europe.

A total of 31 alien louse (Phthiraptera) species are listed by Kenis and Roques.[126] They include 24 chewing lice and 7 sucking lice of 12 different families. The families Gonionidae (Ischnocera) and Menoponidae (Amblycera) largely dominate the alien entomofauna of chewing lice. Asia is the major supplier of alien Phthiraptera, which are mostly associated with poultry farming, game birds, guinea pigs, and invasive alien mammals. The recent period did not show any acceleration in aliens' arrival in Europe. Alien fleas include six species in the families Pulicidae and Ceratophyllidae. Three of them are primarily associated with rats and are capable of transmitting major human diseases such as bubonic plague and murine typhus (see Section 8.4).

A total of 37 alien species have been recorded in the orders Blattodea, Isoptera, Orthoptera, Phasmatodea, and Dermaptera.[91] These belong to 14 different families. Most of these species show a detritivorous feeding regime (22 spp.), whereas 12 species are phytophagous and 2 are predators. A majority of these species were first observed between 1900 and 1975, and unlike most insect groups, the mean number of species newly recorded per year has not shown any acceleration since 1975. They mostly originated from Central or South America and Asia (10 species each, 27%), followed by Africa (7, 18.9%). More than 75% of the species are associated with artificial habitats (houses, buildings, and greenhouses) and cultivated areas. Blattodea and Isoptera have huge economic and/or medical importance (see Section 8.4).

## References

1. Pimentel, D., L. Lach, R. Zuniga, and D. Morrison. 2002. Environmental and economic costs associated with non-indigenous species in the United States. In *Biological Invasions: Economic and Environmental costs of Alien Plant, Animal and Microbe Species*, ed. D. Pimentel, 285. Boca Raton, FL: CRC Press.
2. Niemelä, P., and W. J. Mattson. 1996. Invasion of North American forests by European phytophagous insects: Legacy of the European crucible? *BioScience* 46:741.
3. Mattson, W. J., H. Vanhanen, T. Veteli, S. Sivonen, and P. Niemelä. 2007. Few immigrant phytophagous insects on woody plants in Europe: Legacy of the European crucible? *Biol Invasions* 9:957.
4. Ciosi, M., N. J. Miller, S. Toepfer, A. Estoup, and T. Guillemaud. 2011. Stratified dispersal and increasing genetic variation during the invasion of Central Europe by the western corn rootworm. *Diabrotica virgifera virgifera. Evolutionary Applications* 4:54–70.

5. Haack, R. A., F. Hérard, J. H. Sun, and J. J. Turgeon. 2010. Managing invasive populations of Asian longhorned beetle and Citrus longhorn beetle: A worldwide perspective. *Annu Rev Entomol* 55:521.

6. Eritja, R., R. Escosa, J. Lucientes, E. Marquès, D. Roiz, and S. Ruiz. 2005. Worldwide invasion of vector mosquitoes: Present European distribution and challenges for Spain. *Biol Invasions* 7:87.

7. Mota, M. M., H. Braasch, M. A. Bravo, A. C. Penas, W. Burgermeister, K. Metge, and E. Sousa. 1999. First report of *Bursaphelenchus xylophilus* in Portugal and in Europe. *Nematology* 1:727.

8. Valade, R., M. Kenis, A. Hernandez, S. Augustin, N. Mari Mena, E. Magnoux, R. Rougerie, F. Lakatos, A. Roques, and C. Lopez-Vaamonde. 2009. Mitochondrial and microsatellite DNA markers reveal a Balkan origin for the highly invasive horse-chestnut leaf miner *Cameraria ohridella* (Lep. Gracillariidae). *Mol Ecol* 18:3458.

9. Essl, F., and W. Rabitsch, eds. 2002. *Neobiota in Österreich [Alien in Austria]*. Wien: Umweltbundesamt.

10. Geiter, O., S. Homma, and R. Kinzelbach. 2002. Bestandsaufnahme und Bewertung von Neozoen in Deutschland: Untersuchung der Wirkung von Biologie und Genetik ausgewählter Neozoen auf Ökosysteme und Vergleich mit den potentiellen Effekten gentechnisch veränderter Organismen. [Inventory and evaluation of invasive species in Germany: study of the effect of biology and genetics of selected invasive species on ecosystems and comparison with the potential effects of genetically-modified organisms.] Berlin: Umweltbundesamt. http://www.umweltdaten.de/publikationen/fpdf-k/k2141.pdf (accessed August 5, 2010).

11. Reemer, M. 2003. *Invasieve Arthropoda in Nederland: Een eerste inventarisatie*. [Invasive arthropods in the Netherlands: a first inventory.] Leiden: European Invertebrate Survey—Nederland.

12. Šefrová, H., and Z. Laštůvka. 2005. Catalogue of alien animal species in the Czech Republic. *Acta Univ Agric et Silviculturae Mendelianae Brunensis* 53:151.

13. Nobanis. 2005. The North European and Baltic Network on Invasive Alien Species. http://www.nobanis.org (accessed August 5, 2010).

14. Hill, M., R. Baker, G. Broad, P. J. Chandler, G. H. Copp, J. Ellis, D. Jones et al. 2005. Audit of Nonnative Species in England, Research Report 662. English Nature, Peterborough.

15. Smith, R. M., R. H. A. Baker, C. P. Malumphy, S. Hockland, R. P. Hammon, J. C. Ostojá-Starzewski, and D. W. Collins. 2007. Recent non-native invertebrate plant pest establishments in Great Britain: Origins, pathways, and trends. *Agric For Entomol* 9:307.

16. Pelizzari, G., L. Dalla Monta, and V. Vacante. 2005. List of alien insect and mite pests introduced to Italy in sixty years (1945–2004). In *Plant Protection and Plant Health in Europe: Introduction and Spread of Invasive Species*, eds. D. V. Alford and G. F. Backhaus, 275. BCPC Conference, Humboldt University, Berlin, Germany, 9–11 June 2005. Alton: British Crop Protection Council.

17. Glavendekic´, M., L. Mihajlovic´, and R. Petanovic´. 2005. Introduction and spread of invasive mites and insects in Serbia and Montenegro. In *Plant Protection and Plant Health in Europe: Introduction and Spread of Invasive Species*, eds. D. V. Alford and G. F. Backhaus, 229. BCPC Conference, Humboldt University, Berlin, Germany, 9–11 June 2005. Alton: British Crop Protection Council.

18. Kenis, M. 2005. Insects-Insecta. In *An Inventory of Alien Species and Their Threat to Biodiversity and Economy in Switzerland*, ed. R. Wittenberg, 131. Délemont: CABI Bioscience.

19. Tomov, R., K. Trencheva, G. Trenchev, E. Cota, A. Ramadhi, B. Ivanov, S. Naceski, I. Papazova-Anakieva, and M. Kenis. 2009. *Non-Indigenous Insects and Their Threat to Biodiversity and Economy in Albania, Bulgaria and Republic of Macedonia*. Sofia-Moscow: Pensoft.

20. Ripka, G. 2010. An overview of the alien arthropod pest species in Hungary (I.) (in Hungarian). *Növényvédelem [Plant protection]* 46:45.

21. Roques, A. 2010. Alien forest insects in a warmer world and a globalized economy: Impacts of changes in trade, tourism and climate on forest biosecurity. *N Z J For Sc* 40 (suppl):77.

22. DAISIE, ed. 2009. *The Handbook of Alien Species in Europe*. Heidelberg: Springer.

23. Roques, A., W. Rabitsch, J. Y. Rasplus, C. Lopez-Vaamonde, W. Nentwig, and M. Kenis. 2009. Alien terrestrial invertebrates of Europe. In *The Handbook of Alien Species in Europe*, ed. DAISIE, 63. Heidelberg: Springer.

24. Roques, A. 2010. Taxonomy, time and geographic patterns. Chapter 2. In *Alien Terrestrial Arthropods of Europe*, ed. A. Roques et al., chapter 2. *BioRisk* 4:11. http://pensoftonline.net/biorisk/index.php/journal (accessed August 5, 2010).

25. Pyšek, P., S. Bacher, M. Chytry, V. Jarošik, J. Wild, L. Celesti-Grapow, N. Gasso et al. 2009. Contrasting patterns in the invasions of European terrestrial and freshwater habitats by alien plants, insects and vertebrates. *Glob Ecol Biogeogr* 19:317.

26. Pyšek, P., V. Jarošík, P. E. Hulme, I. Kühn, J. Wild, M. Arianoutsou, S. Bacher et al. 2010. Disentangling the role of environmental and human pressures on biological invasions across Europe. *Proc Natl Acad Sci* 107:12157.

27. Vilà, M., C. Basnou, P. Pyšek, M. Josefsson, P. Genovesi, S. Gollasch, W. Nentwig et al. and DAISIE Partners. 2009. How well do we understand the impacts of alien species on ecosystem services? A pan-European cross-taxa assessment. *Front Ecol Environ* 8:135.

28. Rabitsch, W. 2010. Pathways and vectors of alien arthropods in Europe. In *Alien Terrestrial Arthropods of Europe*, ed. A. Roques et al., chapter 3. *BioRisk* 4:27. http://pensoftonline.net/biorisk/index.php/journal (accessed August 5, 2010).

29. Rasplus, J. Y., C. Villemant, M. R. Paiva, G. Delvare, and A. Roques. 2010. Hymenoptera. In *Arthropod Invasions in Europe*, ed. A. Roques et al., chapter 12. *BioRisk* 4:669. http://pensoftonline.net/biorisk/index.php/journal (accessed August 5, 2010).

30. Roy, H., and A. Migeon. 2010. Ladybeetles (Coccinellidae). In *Alien Terrestrial Arthropods of Europe*, ed. A. Roques et al., chapter 8.4. *BioRisk* 4:293. http://pensoftonline.net/biorisk/index.php/journal (accessed August 5, 2010).

31. Welter-Schultes, F. W. 1998. Human-dispersed land snails in Crete, with special reference to *Albinaria* (Gastropoda: Clausiliidae). *Bio Gallo-Hellenica* 24:83.

32. Schaffner, F., C. Kaufmann, D. Hegglin, and A. Mathis. 2009. The invasive mosquito *Aedes japonicus* in Central Europe. *Med Vet Entomol* 23:448.

33. Villemant, C., J. Haxaire, and J. C. Streito. 2006. Premier bilan de l'invasion de *Vespa velutina* Lepeletier en France (Hymenoptera, Vespidae). *Bull de la Société entomologique de France* 111:535.

34. World Trade Organisation (WTO). 2007. *International Trade Statistics 2007*. Geneva: WTO Publications. http://www.wto.org/english/res_e/statis_e/statis_e.htm (accessed August 5, 2010).

35. McCullough, D. G., T. T. Work, J. F. Cavey, A. M. Liebhold, and D. Marshall. 2006. Interceptions of nonindigenous plant pests at U.S. ports of entry and border crossings over a 17 year period. *Biol Invasions* 8:611.

36. Roques, A., and M. A. Auger-Rozenberg. 2006. Tentative analysis of the interceptions of nonindigenous organisms in Europe during 1995–2004. *EPPO Bull* 36:490.

37. Langor, D. W., L. DeHaas, and R. G. Foottit. 2009. Diversity of non-native terrestrial arthropods on woody plants in Canada. *Biol Invasions* 11:5.

38. Kenis, M., W. Rabitsch, M. A. Auger-Rozenberg, and A. Roques. 2007. How can alien species inventories and interception data help us prevent insect invasions? *Bull Entomol Res* 97:489. http://pensoftonline.net/biorisk/index.php/journal (accessed August 5, 2010).

39. Glavendekić, M., A. Roques, and L. Mihajlović. 2010. An ALARM case study: The rapid colonization of an introduced tree, black locust by an invasive North American midge and its parasitoids. In *Atlas of Biodiversity Risks—from Europe to the globe, from stories to maps*, ed. J. Settele et al, 158. Sofia & Moscow: Pensoft.

40. Augustin, S., M. Kenis, R. Valade, M. Gilbert, A. Roques, and C. Lopez-Vaamonde. 2010. A stowaway species probably arriving from the Balkans, the horse chestnut leafminer, *Cameraria ohridella*. In *Atlas of Biodiversity Risks—From Europe to the Globe, From Stories to Maps*, ed. J. Settele et al, 160. Sofia & Moscow: Pensoft.

41. Rabitsch, W. 2008. Alien true bugs of Europe (Insecta: Hemiptera: Heteroptera). *Zootaxa* 1827:1.

42. Perdereau, E., A. G. Bagnères, S. Dupont, and F. Dedeine. in press. High occurrence of colony fusion in a European population of the American termite *Reticulitermes flavipe*, *Insectes Soc.*

43. Nentwig, W., and M. Kobelt. 2010. Spiders (Araneae). In *Alien Terrestrial Arthropods of Europe*, ed. A. Roques et al., chapter 7.3. *BioRisk* 4:131. http://pensoftonline.net/biorisk/index.php/journal (accessed August 5, 2010).

44. Schneider, N. 2010. Psocids (Psocoptera). In *Alien Terrestrial Arthropods of Europe*, ed. A. Roques et al., chapter 13.2. *BioRisk* 4:793. http://pensoftonline.net/biorisk/index.php/journal (accessed August 5, 2010).

45. Coeur d'acier, A., N. Pérez Hidalgo, and O. Petrović-Obradović. 2010. Aphids. In *Alien Terrestrial Arthropods of Europe*, ed. A. Roques et al., chapter 9.2. *BioRisk* 4:435. http://pensoftonline.net/biorisk/index.php/journal (accessed August 5, 2010).

46. Lopez-Vaamonde, C., D. Agassiz, S. Augustin, J. De Prins, W. De Prins, S. Gomboc, P. Ivinskis et al. 2010. Lepidoptera. In *Alien Terrestrial Arthropods of Europe*, ed. A. Roques et al., chapter 11. *BioRisk* 4:603. http://pensoftonline.net/biorisk/index.php/journal (accessed August 5, 2010).

47. Stoev, P., M. Zapparoli, S. Golovatch, H. Enghoff, N. Akkari, and A. Barber. 2010. Myriapods (Myriapoda). In *Alien Terrestrial Arthropods of Europe*, ed. A. Roques et al., chapter 7.2. *BioRisk* 4:97. http://pensoftonline.net/biorisk/index.php/journal (accessed August 5, 2010).

48. Reynaud, P. 2010. Thrips (Thysanoptera). In *Alien Terrestrial Arthropods of Europe*, ed. A. Roques et al., chapter 13.1. *BioRisk* 4:767. http://pensoftonline.net/biorisk/index.php/journal (accessed August 5, 2010).

49. Lopez-Vaamonde, C., M. Glavendekić, and M. R. Paiva. 2010a. Invaded habitats. In *Alien Terrestrial Arthropods of Europe*, ed. A. Roques et al., chapter 4. *BioRisk* 4:45. http://pensoftonline.net/biorisk/index.php/journal (accessed August 5, 2010).

50. Roques, A. 2008. The pan-European inventory of alien species established on trees on shrubs, a tool for predicting taxa and ecosystems at risk-final results of the DAISIE project. In *Alien Invasive Species and International Trade*, 2nd meeting of IUFRO Working Unit 7.03.12. Shepherdstown: National Conservation Training Center. http://www.forestry.gov.uk/pdf/IUFRO_Shepherdstown_Roques_Sheperdstown_end.pdf/$FILE/IUFRO_Shepherdstown_Roques_Sheperdstown_end.pdf (accessed August 5, 2010).

51. Roques, A., and M. Skrzypczynska. 2003. Seed-infesting chalcids of the genus *Megastigmus* Dalman (Hymenoptera: Torymidae) native and introduced to Europe: Taxonomy, host specificity and distribution. *J Nat Hist* 37:127.

52. Kenis, M., M. A. Auger-Rozenberg, A. Roques, L. Timms, C. Péré, M. J. W. Cock, J. Settele, S. Augustin, and C. Lopez-Vaamonde. 2009. Ecological effects of invasive alien insects. *Biol Invasions* 11:21.

53. Mota, M. M., K. Futai, and P. Vieira. 2009. Pine wilt disease and the pinewood nematode, *Bursaphelenchus xylophilus*. In *Integrated Management of Fruit Crops and Forest Nematodes*, ed. A. Ciancio, K. G. Mukerji, 253. New York: Springer Science.

54. Baranchikov, Y., E. Mozolevskaya, G. Yurchenko, and M. Kenis. 2008. Occurrence of the emerald ash borer, *Agrilus planipennis* in Russia and its potential impact on European forestry. *OEPP/EPPO Bull* 38:233–8.

55. Poland, T. M., and D. G. McCullough. 2006. Emerald ash borer: Invasion of the urban forest and the threat to North America's ash resource. *J For* 104:118.

56. Hollingsworth, T. 2004. Status of *Paysandisia archon* (Burmeister) (Lepidoptera: Castniidae) in southern Europe. *Br J Entomol Nat Hist* 17:33.

57. EPPO. 2008. Datasheets on quarantine pests. *Rhynchophorus ferrugineus. Bull OEPP/EPPO* 38:55.

58. Auger-Rozenberg, M. A., E. Budrys, T. Petanidou, M. Glavendekic, R. Bommarco, S. Bonzini, G. Kröel-Dulay et al. 2010. The ALARM field site network, an outstanding tool for the survey of invasive insects infesting seeds of wild roses in Europe. In *Atlas of Biodiversity Risks: From Europe to the Globe, From Stories to Maps*, ed. J. Settele et al., 156. Sofia & Moscow: Pensoft.

59. Csoka, G., F. Wittmann, and G. Melika. 2009. The oriental sweet chestnut gall wasp (*Dryocosmus kuriphilus* Yasumatsu, 1951) in Hungary. *Novenyvdelem* 45:359.

60. Fabre, J. P., M. A. Auger-Rozenberg, A. Chalon, S. Boivin, and A. Roques. 2004. Competition between exotic and native insects for seed resources in trees of a Mediterranean forest ecosystem. *Biol Invasions* 6:11.

61. Viggiani, G. 1994. Recent cases of interspecific competition between parasitoids of the family Aphelinidae (Hymenoptera: Chalcidoidea). *Norw J Agric Sci Suppl* 16:353.

62. Tremblay, E. 1984. The parasitoid complex (Hym.: Ichneumonoidea) of *Toxoptera aurantii* (Hom.: Aphidoidea) in the Mediterranean area. *Entomophaga* 29:2003.

63. Boag, B., and G. W. Yeates. 2001. The potential impact of the New Zealand flatworm, a predator of earthworms, in western Europe. *Ecol Appl* 11:1276.

64. Carpintero, S., J. Reyes-López, and L. Arias de Reyna. 2005. Impact of Argentine ants (*Linepithema humile*) on an arboreal ant community in Doñana National Park, Spain. *Biol Conserv* 14:151.

65. Brown, P. M. J., T. Adriaens, H. Bathon, J. Cuppen, A. Goldarazena, T. Hagg, M. Kenis et al. 2008. *Harmonia axyridis* in Europe: Spread and distribution of a non-native coccinellid. *Biocontrol* 53:5.

66. Legaspi, J. C., B. C. Legaspi, A. M. Simmons, and M. Soumare. 2008. Life table analysis for immatures and female adults of the predatory beetle, *Delphastus catalinae*, feeding on white-flies under three constant temperatures. *J Insect Sci* 8(7). http://www.insectscience.org/8.07/ (accessed August 5, 2010).

67. Ware, R. L., and M. E. N. Majerus. 2008. Intraguild predation of immature stages of British and Japanese coccinellids by the invasive ladybird *Harmonia axyridis*. *Biocontrol* 53:169.

68. Navajas, M., A. Migeon, A. Estrada-Pena, A. C. Mailleux, P. Servigne, and R. Petanovic. 2010. Mites and ticks (Acari). In *Arthropod Invasions in Europe*, ed. A. Roques et al., chapter 7.4. *BioRisk* 4:149. http://pensoftonline.net/biorisk/index.php/journal (accessed August 5, 2010).

69. Kenis, M., and M. Branco. 2010. Impact of alien terrestrial arthropods in Europe. In *Alien Terrestrial Arthropods of Europe*, ed. A. Roques et al., chapter 5. *BioRisk* 4:51. http://pensoftonline.net/biorisk/index.php/journal (accessed August 5, 2010).

70. Pimentel, D. 2002. Non-native invasive species of arthropods and plant pathogens in the British Isles. In *Biological Invasions: Economic and Environmental costs of Alien Plant, Animal and Microbe Species*, ed. D. Pimentel, 151. Boca Raton, FL: CRC Press.

71. Baufeld, P., and S. Enzian. 2005. Maize growing, maize high-risk areas and potential yield losses due to Western Corn Rootworm (*Diabrotica virgifera virgifera* LeConte) damage in selected European countries. In *Western Corn Rootworm: Ecology and Management*, S. Vidal, U. Kuhlmann, and C. R. Edwards, 285. Wallingford, CT: CABI Publishing.

72. Reinhardt, F., M. Herle, F. Bastiansen, and B. Streit. 2003. *Economic Impact of the Spread of Alien Species in Germany*. Berlin: Federal Environmental Agency (Umweltbundesamt).

73. Heikkilä, J., and J. Peltola. 2007(2006 by Roques). Phytosanitary measures under uncertainty: A cost benefit analysis of the Colorado potato beetle in Finland. In *New Approaches to the Economics of Plant Health*, ed. G. J. M. Alfons and O. Lansink, 147. Heidelberg: Springer-Verlag.

74. MacLeod, A., J. Head, and A. Gaunt. 2004. An assessment of the potential economic impact of *Thrips palmi* on horticulture in England and the significance of a successful eradication campaign. *Crop Prot* 23:601.

75. Roosjen, M., J. Buurma, and J. Barwegen. 1998. Verbetering schade-inschattingsmodel quarantaine-organismen glastuinbouw. *Verslagen en Mededelingen, Plantenziektenkundige Dienst. Wageningen* 197:1.

76. Taylor, M. A., V. Jackson, I. Zimmer, S. Huntley, A. Tomlinson, and R. Grant. 2006. *Qualitative Veterinary Risk Assessment: Introduction of Exotic Diseases (other than Rabies) in the UK*. Sand Hutton: Central Science Laboratory, Veterinary Surveillance Team.

77. Enserink, M. 2007. Tropical disease follows mosquitoes to Europe. *Science* 317:1485.

78. Audouin-Rouzeau, F. 2003. *Les Chemins de la Peste—Le Rat, la Puce et l'Homme*. Rennes: Presses universitaires de Rennes.

79. Beaucournu, J. C. 1999. Diversité des puces vectrices en fonction des foyers pesteux. *Bull de la Société de pathologie exotique* 92:419.

80. Beaucournu, J. C., and H. Launay. 1990. Les puces (Siphonaptera) de France et du Bassin méditerranéen occidental. *Faune de France* 76. Paris: Fédération Française des Sociétés de Sciences Naturelles.

81. Sychra, O., P. Harmat, and I. Literák. 2008. Chewing lice (Phthiraptera) on chickens (*Gallus gallus*) from small backyard flocks in the eastern part of the Czech Republic. *Vet Parasitol* 152:344.

82. Mey, E. 1988. Uebersicht über die Säugtiere-Mallophagen Europas. *Angew Parasitol* 29:113.

83. Stojcevic, D., Z. Mihaljevic, and A. Marinculic. 2004. Parasitological survey of rats in rural regions of Croatia. *Vet Med Czech* 49:70.

84. Drozdz, J., A. W. Demiaszkiewicz, and J. Lachowicz. 2003. Expansion of the Asiatic parasite *Ashworthius sidemi* (Nematoda, Trichostrongylidae) in wild ruminants in Polish territory. *Parasitol Res* 89:94.

85. Šnábel, V., A. Permin, H. B. Magwisha et al. 2001. On the species identity of *Ascaridia galli* (Schrank, 1788) and *Ascaridia dissimilis* (Perez Vigueras, 1931): A comparative genetic study. *Helminthology* 38:221.

86. Novobilský, A., E. Horáčková, L. Hirtová, D. Modrý, and B. Koudela. 2007. The giant liver fluke Fascioloides magna (Bassi 1875) in cervids in the Czech Republic and potential of its spreading to Germany. *Parasitol Res* 100:549.

87. Gjelstrup, P., and A. P. Møller. 1985. A tropical mite, *Ornithonyssus bursa* (Berlese, 1888) (Macronyssidae, Gamasida) in Danish swallow (*Hirundo rustica*) nests, with a review of mites and ticks from Danish birds. *Entomologiste Medd* 53:119.

88. Musken, H., J. T. Franz, R. Wahl, A. Paap, O. Cromwell, G. Masuch, and K. C. Bergmann. 2000. Sensitization to different mite species in German farmers: Clinical aspects. *J Investig Allergol Clin Immunol* 10:346–51.

89. Disney, R. H. L. 2008. Natural history of the scuttle fly, *Megaselia scalaris*. *Annu Rev Entomol* 53:39.

90. Burgess, N. R., and K. N. Chetwyn. 1981. Association of cockroaches with an outbreak of dysentery. *Trans R Soc Trop Med Hyg* 75:332.

91. Rasplus, J. Y., and A. Roques. 2010. Dictyoptera (Blattodea, Isoptera), Orthoptera, Phasmatodea and Dermaptera. In *Alien Terrestrial Arthropods of Europe*, ed. A. Roques et al., chapter 13.3. *BioRisk* 4:807. http://pensoftonline.net/biorisk/index.php/journal (accessed August 5, 2010).

92. Cannon, R. J., R. H. Baker, M. Taylor, and J. P. Moore. 1999. A review of the status of the New Zealand flatworm in the United Kingdom. *Ann Appl Biol* 135:597–614.

93. Christensen, O. M., and J. G. Mather. 1999. Den Newzealandske Fladorm—En eksotisk organisme der truer regnorme, 1–3. http://mit.biology.au.dk/zoology/zoodk/fladorm.html (accessed August 5, 2010).

94. Bloch, D. 1992. A note on the occurence of land planarians in the Faroe Islands. *Fródskaparrit* 38:63.

95. Boag, B., K. A. Evans, R. Neilson, G. W. Yeates, P. M. Johns, J. G. Mather, and O. M. Christensen. 1995. The potential spread of terrestrial planarians *Artioposthia triangulata* and *Australoplana sanguinea var. alba* to continental Europe. *Ann Appl Biol* 127:385.

96. Moore, J., R. Gibson, and H. D. Jones. 2001. Terrestrial nemerteans thirty years on. *Hydrobiologia* 456:1.

97. Moore, J. 1985. The distribution and evolution of terrestrial nemertines. *Am Zool* 25:15.

98. Grubini, T. D., L. Ontrec, T. G. Buljak, and S. Blümel. 2007. The occurrence and distribution of potato cyst nematodes in Croatia. *J Pest Sci* 80:21.

99. Lamberti, F., and A. Ciancio. 1993. The diversity of *Xiphinema americanum* and related species and the problems associated with taxonomic identification. *J Nematol* 25:332.

100. Küchle, M., H. L. J. Knorr, S. Medenblik-Frysch, A. Weber, C. Bauer, and G. O. H. Naumann. 1993. Diffuse unilateral subacute neuroretinitis syndrome in a German most likely caused by the raccoon roundworm, *Baylisascaris procyonis*. *Graefe's Arch Clin Exp Ophthalmol* 231:48.

101. Hendrix, P. F. 2006. Biological invasions belowground—earthworms as invasive species. *Biol Invasions* 8:1201.

102. Dominguez, J., C. A. Edwards, and J. Dominguez. 2001. The biology and population dynamics of *Eudrilus eugeniae* (Kinberg) (Oligochaeta) in cattle waste solids. *Pedobiologia* 45:341.

103. Pop, V. V., and A. A. Pop. 2006. Lumbricid earthworm invasion in the Carpathian Mountains and some other sites in Romania. *Biol Invasions* 8:1219.

104. Tiunov, A. V., C. M. Hale, A. R. Holdsworth, and T. S. Vsevolodova-Perel. 2006. Invasion patterns of Lumbricidae into the previously earthworm-free areas of northeastern Europe and the western Great Lakes region of North America. *Biol Invasions* 8:1223.

105. Reise, H., J. M. Hutchinson, R. G. Forsyth, and T. J. Forsyth. 2000. The ecology and rapid spread of the terrestrial slug *Boettgerilla pallens* in Europe with reference to its recent discovery in North America. *Veliger* 43:313.

106. Dvořák, L., and J. Kupka. 2007. The first outdoor find of an American snail *Zonitoides arboreus* (Say, 1816) from the Czech Republic. *Malacol Bohem* 6:1.

107. Kozlowski, J. 2005. Host plants and harmfulness of the *Arion lusitanicus* (Mabille, 1868) slug. *J Plant Prot Res* 45:221.

108. Wittenberg, R. 2006. *Invasive Alien Species in Switzerland. An Inventory of Alien Species and Their Threat to Biodiversity and Economy in Switzerland.* Environmental studies series, 29. Bern: Federal Office for the Environment. http://www.bafu.admin.ch/publikationen/publikation/00028/index.html?lang=en (accessed August 5, 2010).

109. Cochard, P. O., F. Vilisics, and E. Sechet. 2010. Alien terrestrial crustaceans (Isopods and Amphipods). In *Alien terrestrial arthropods of Europe*, ed. A. Roques et al., chapter 7.1. *BioRisk* 4:81. http://pensoftonline.net/biorisk/index.php/journal (accessed August 5, 2010).

110. O'Hanlon, R. P., and T. Bolger. 1999. The importance of *Arcitalitrus dorrieni* (Hunt) (Crustacea: Amphipoda: Talitridae) in coniferous litter breakdown. *Appl Soil Ecol* 11:29.

111. Andersson, G. 2006. Habitat preferences and seasonal distribution of developmental stadia in *Lamyctes emarginatus* (Newport, 1844) (*L. fulvicornis* Meinert, 1868) and comparison with some *Lithobius* species (Chilopoda, Lithobiomorpha). *Norw J Entomol* 53:311.

112. Arndt, E., H. Enghoff, and J. Spelda. 2008. Millipedes (Diplopoda) of the Canarian Islands: Checklist and key. *Biol Invasions* 8:893.

113. Kobelt, M., and W. Nentwig. 2008. Alien spider introductions to Europe supported by global trade. *Divers Distrib* 14:273.

114. Van Keer, K. 2007. Exotic spiders (Araneae): Verified reports from Belgium of exported species (1976–2006) and some notes on apparent neozoan invasive species. *Nieuwsbrief Belg Arachnologische Ver* 22:45.

115. Schmidt, M. H., S. Rocjer, J. Hanafi, and A. Gigon. 2008. Rotational fallows as overwintering habitat for grass-land arthropods: The case of spiders in fen meadows. *Biodivers Conserv* 17: 3003.

116. Burger, J. C., M. A. Patten, T. R. Prentice, and R. A. Redak. 2001. Evidence for spider community resilience to invasion by non-native spiders. *Biol Conserv* 98:241.

117. Roques, A., M. Kenis, D. Lees, C. Lopez-Vaamonde, W. Rabitsch, J. Y. Rasplus, and D. B. Roy, eds. 2010. *Alien Terrestrial Arthropods of Europe*. Two volumes. Sofia: Pensoft.

118. Sauvard, D., M. Branco, F. Lakatos, M. Faccoli, and L. R. Kirkendall. 2010. Weevils and Bark Beetles (Coleoptera, Curculionoidea). In *Alien Terrestrial Arthropods of Europe*, ed. A. Roques et al., chapter 8.2. *BioRisk* 4:219. http://pensoftonline.net/biorisk/index.php/journal (accessed August 5, 2010).

119. Beenen, R., and A. Roques. 2010. Leaf and seed beetles (Coleoptera, Chrysomelidae). In *Alien Terrestrial Arthropods of Europe*, ed. A. Roques et al., chapter 8.3. *BioRisk* 4:267. http://pensoftonline.net/biorisk/index.php/journal (accessed August 5, 2010).

120. Cocquempot, C., and A. Lindelöw. 2010. Longhorn beetles (Coleoptera, Cerambycidae). In *Alien Terrestrial Arthropods of Europe*, ed. A. Roques et al., chapter 8.1. *BioRisk* 4:193. http://pensoftonline.net/biorisk/index.php/journal (accessed August 5, 2010).

121. Denux, O., and P. Zagatti. 2010. Coleoptera families other than Cerambycidae, Curculionidae sensu lato, Chrysomelidae sensu lato and Coccinelidae. In *Alien Terrestrial Arthropods of Europe*, ed. A. Roques et al., chapter 8.5. *BioRisk* 4:315. http://pensoftonline.net/biorisk/index.php/journal (accessed August 5, 2010).

122. Pellizzari, G., and J. F. Germain. 2010. Scales (Hemiptera, Superfamily Coccoidea). In *Alien Terrestrial Arthropods of Europe*, ed. A. Roques et al., chapter 9.3. *BioRisk* 4:475. http://pensoftonline.net/biorisk/index.php/journal (accessed August 5, 2010).

123. Mifsud, D., C. Cocquempot, R. Mühlethaler, M. Wilson, and J. C. Streito. 2010. Other Hemiptera Sternorrhyncha (Aleyrodidae, Phylloxeroidea, and Psylloidea) and Hemiptera Auchenorrhyncha. In *Alien Terrestrial Arthropods of Europe*, ed. A. Roques et al., chapter 9.4. *BioRisk* 4:511. http://pensoftonline.net/biorisk/index.php/journal (accessed August 5, 2010).

124. Skuhravá, M., M. Martinez, and A. Roques. 2010. Diptera. In *Alien Terrestrial Arthropods of Europe*, ed. A. Roques et al., chapter 10. *BioRisk* 4:553. http://pensoftonline.net/biorisk/index.php/journal (accessed August 5, 2010).

125. Péré, C., S. Augustin, R. Tomov, L. H. Peng, T. C. J. Turlings, and M. Kenis. 2010. Species richness and abundance of native leaf miners are affected by the presence of the invasive horse-chestnut leaf miner. *Biol Invasions* 12:1011.

126. Kenis, M., and A. Roques. 2010. Lice and fleas (Phthiraptera and Siphonaptera). In *Alien Terrestrial Arthropods of Europe*, ed. A. Roques et al., chapter 13.4. *BioRisk* 4:833.

*chapter nine*

# Invasive plant pathogens in Europe

**Ivan Sache, Anne-Sophie Roy, Frédéric Suffert,
and Marie-Laure Desprez-Loustau**

## Contents

## 9.1  Introduction

In the second half of the nineteenth century, grapevine and potato crops in Europe were destroyed by diseases caused by invasive plant pathogens, such as the Oomycetes *Plasmopara viticola* and *Phytophthora infestans* (causing grapevine downy mildew and potato late blight, respectively) and the Ascomycete *Erysiphe necator* (causing grapevine powdery mildew). These "great invasions" by pathogens of non-European origin significantly contributed to the individualization of plant pathology as a science distinct from botany,[1] and the threat represented by the introduction of nonnative plant pathogens was quickly emphasized. In an international congress held at the Hague (the Netherlands) in 1891, the Danish plant pathologist Emil Rostrup advocated the setup of "measures for preventing the importation of living plants or seeds from contaminated areas. (p. 6)"[2] In a seminal book subsequently translated in several languages, the Swedish plant pathologist Jakob Eriksson stated that the increased prevalence and severity of plant diseases (in Europe) should be related to the recent emergence of new diseases and their spread worldwide.[2]

From the 1878 Phylloxera International Convention of Bern to the adoption in 1951 by the FAO Conference of the International Convention on Plant Protection (IPPC), a series of international conventions sought to relieve European agriculture from "foreign parasites. (p. 6)"[3] The purpose of IPPC, revised in 1979 and 1997, is to secure at a global level a common and effective action against the introduction and spread of plant pests. This treaty also supplies a framework for phytosanitary measures to be taken against "invasive alien species," as defined by the Convention on Biological Diversity, as far as they are plant pests.[4] The revival of the European ideals following the trauma caused by the two World Wars triggered, as far as plant protection is concerned, the formation of the European Plant Protection Organization (EPPO) in 1951. EPPO was given the task to ensure cooperation between national plant protection organizations (e.g., official plant protection, plant health, or plant quarantine services) and to harmonize for plant health. EPPO grew from 15 founding members to a current membership of 50 European and Mediterranean countries (including countries from North Africa, the Middle East, and former Soviet Union Republics of Central Asia). The first objective of EPPO is "to develop an international strategy against the introduction and spread of pests that damage cultivated and wild plants in natural and agricultural ecosystems" (EPPO website http://www.eppo.org). In particular, the organization has tried to identify the main risks for Europe and made recommendations to its member countries, as to which pests should be regulated as quarantine pests (EPPO A1 and A2 Lists) and which phytosanitary measures could be taken. Another European particularity was the development of the European Union (EU), an economic and political union of 27 member states. Since 1993, lists of quarantine pests and phytosanitary measures have been harmonized in the EU Council Directive 2000/29/EC (revising Council Directive 77/93/EC). Today, approximately 300 pests have been identified as quarantine pests (largely on the basis of EPPO's recommendations), and many of them are invasive plant pathogens. Accordingly, the issue of invasive plant diseases is still a main concern in Europe, requiring significant effort in both research and management.

In this chapter, we first summarize the results of recent inventories of invasive plant pathogens in Europe. Second, we present data on the economic impact of some of the worst invasive plant pathogens, considering both direct (market cost via yield and quality loss) and indirect (detection, control, eradication, and compensation) costs. Third, we discuss the methods used to assess the impact of invasive plant pathogens on ecological services. Last, we evaluate the threat represented by plant pathogens currently emerging in Europe or not detected yet.

## 9.2   Recent inventories of invasive plant pathogens in Europe

The International Union for Conservation of Nature (IUCN), mostly concerned with wild ecosystems, restricts the definition of invasive species to "species with a potential impact on biological diversity." Accordingly, IUCN lists only three plant pathogens among "100 of the World's Worst Invasive Alien Species"[5]: *Ophiostoma novo-ulmi*, causing Dutch elm disease; *Phytophthora cinnamomi*, causing dieback, crown, and root rot in some 900 species of perennial trees; and *Cryphonectria parasitica*, causing chestnut blight. In a European perspective, the Delivering Alien Invasive Species Inventories for Europe (DAISIE) consortium lists *O. novo-ulmi*, *P. cinnamomi*, and *Seiridium cardinale*, the cause of a lethal canker disease on cypress and related conifers, among the "100 of the Worst Invasive Alien Species."[6]

Conversely, most fungal pathogens of crops would not be considered to be invasive according to IUCN definition, since they have no known impacts on biodiversity. Indeed, many crop plants are themselves exotic in the areas where they are grown, and the pathogen species that attack them followed them from their area of origin.[7] However, in several cases, pathogens were introduced in Europe decades or even centuries after the introduction of their host; such a "reencounter"[8] between a fungus and a plant that had escaped the pathogen challenge for centuries and progressively lost resistance factors is a main cause of "successful" invasions, as exemplified in the case of potato late blight. Rather than a consequence of invasion by pathogenic fungi, the dramatic narrowing of genetic diversity of most crop plants in the twentieth century is an inadvertent help to new fungal attacks. New pathogens usually emerge at the infraspecific level, as virulent pathotypes of pathogens already established, and are not taken into account in inventories of new species. Recent examples are the spread of the virulent strain Yr17 of yellow rust (*Puccinia striiformis*) all over Europe;[9] the emergence in Africa of the Ug99 strain of *P. graminis* f.sp. *tritici*, the stem rust fungus, a strain that potentially threatens wheat growing worldwide;[10] and the inadvertent introduction in Europe of the A2 mating type of *P. infestans*.[11] In a few cases, pathogenic fungi on cultivated species were shown to be able to infect wild species as well (for instance, *Ramularia collo-cygni* and *Sclerophthora macrospora* reported on grasses and cereals).

The first inventories of alien invasive species for Europe, including fungi, were recently issued by the DAISIE consortium.[12] The list of alien fungi is a compilation of available national lists and contains 688 species, among which plant pathogens represent 77%.[13] The highest numbers of alien species were found in the biggest European countries, France, the United Kingdom, Germany, and Italy. Rather than geographic characteristics, such as surface, latitude, longitude, and climate, the level of import of goods was the best predictor of the number of alien species in a given country. Most introductions of species at the European level are of North American and Asian origin. Most alien plant pathogenic fungi are assumed to have been inadvertently introduced with contaminated material, such as nursery stock (*Phytophthora ramorum*), log shipments (*O. novo-ulmi*), or even military equipment (*Ceratocystis platani*).

National inventories can provide more detailed insights into the characteristics of plant pathogen invaders. The French inventory includes 227 fungal species of presumably non-European origin recorded in France since 1800.[14] Plant pathogens are the most numerous ecological category, with 65% of all species, mycorrhizal and saprotrophic fungi representing 30% and 4%, respectively. Nearly half (46%) of the plant pathogen species have been recorded primarily on crop plants, while ornamental and forest pathogen species account for 31% and 22% of the records, respectively. More than 50% of plant pathogens attack woody plants (forest, fruit, and ornamental trees and shrubs). Three groups of diseases account for nearly 50% of the reported invasions, the downy mildews (Peronosporales), the powdery mildews (Erysiphales), and the rusts (Pucciniales), in respective proportions 2:1:1. The high multiplication rate of these pathogens; the extended dispersal potential of windborne powdery mildews, rusts, and some downy mildews; and the very visible damage they cause on plants might explain their apparent overrepresentation in the database.

The inventory of introduced, nonnative plant pathogens into Great Britain[15] was limited in time but addressed all taxonomic groups of pathogen species. The inventory includes 234 pathogens (fungi, bacteria, phytoplasma, and viruses) reported for the first time in Britain from 1970 to 2004, 79% of them being fungi. Some 60% of the fungal records were made on ornamental species, which is twice the proportion reported in France. Most of the ornamentals and a significant proportion of the horticultural crops on which new pathogens were detected grew in glasshouses or under polyethylene covers; accordingly, 50% of the new British records in the given period were found in protected environments. Here, again, transportation of contaminated plant material seems to be a main way of introduction of plant pathogens; from the limited evidence available on the origin of the infected plants, the Netherlands was pointed out as a main source of introduction. Evaluating the potential impact of the introduced species, panels of British specialists labeled 19% of the pathogens "important."

## 9.3   Economic impacts of invasive plant pathogens

Detailed data on the economic impact of invasive plant pathogens in Europe are scarce. Pimentel et al.[16] estimated that damage associated with alien plant pathogens attacking British crop species could reach US$2 billion (€1.46 billion) per year; the estimation was the product of the estimated rate of alien plant pathogens in Britain (74%) and the economic loss associated with crop pathogens (US$2.7 billion per year). No such figures are available at the European scale, for which the proportion of alien to native plant pathogens has not been assessed. We list in Sections 9.3.1 through 9.3.3 recently published assessments of economic loss due to some of the worst invasive plant pathogens; while most of the data are available at a specific country's level, we have attempted in some cases to extrapolate the economic cost of the pathogens to larger areas. Understandably, the economic impact of plant pathogens has mainly been assessed when host plants have an economic value, for example, for crop plants and trees with ornamental value.

### 9.3.1   Fungi and Oomycetes

#### 9.3.1.1   Cultivated plants

The downy mildew and the powdery mildew of grapevine, introduced from the United States in the nineteenth century, still threaten grapevine and therefore wine production in many European countries. Control of the disease using fungicides is mandatory to save grapevine yield and quality; in wet years, up to 90% of the fungicide sprays are targeted to these two diseases. According to a recent report on pesticide use in France,[17] fungicide cost in 2002 represented €287 per hectare for average quality wines, which account for 46% of the grapevine acreages; the cost increased to €398 per hectare for quality wines, which require a better protection against the mildews. Considering that grapevines cover slightly less than 600,000 ha, the total annual cost of chemical control of the downy mildews is more than €180 million. The French vineyards represent 12% of the European vineyard area, and the total cost for Europe can be extrapolated at €1.5 billion per year. However, the actual cost must be lower since most other European countries use less fungicide on vineyards than France.

Potato late blight has remained the main potato disease in Europe since its introduction in the nineteenth century. In 1991, US$223 million (€163 million) of fungicides were used worldwide against potato diseases, Europe accounting for 59% of this use. Diseases, including potato late blight, accounted for a yield loss of 15%, which could have reached 35% if no fungicide had been applied.[18]

Since then, the disease has increased in aggressivity and earliness in most European countries, a shift partially explained by the invasion of the continent by fungal populations with mating type A2.[19] In Finland, sales of fungicides used against late blight increased fourfold from the 1980s to 2002.[20] The total annual costs of the disease in Norway are about NOK 60 million (€7.4 million), including fungicide application, yield and quality loss, cost of inspection, research, advisory service, and warnings.[21] Annual losses in Ireland have been estimated at £8 million (€9.1 million) per year[22]; the value of the Irish potato pesticide market is approximately £3.5 million (ca. €4 million), of which 63% (£2.2 million, ca. €2.5 million) is spent on fungicides for the control of late blight.[23]

To control late blight, professional growers in Europe applied an average of 7.5 and 6.7 fungicide sprays in 2007 and 2008, respectively. For the four countries with the most intensive potato-growing system (the United Kingdom, Belgium, France, and the Netherlands), the average number of sprays was 12.8 and 14 in 2007 and 2008, respectively.[20] In Flanders (Belgium), the application of 10–14 sprays in most seasons costs between €200 and €400 per hectare for the fungicides depending on product choice.[24] In England, the cost of protection is £130–£200 per hectare (€148–€227),[25] whereas a cost of £167 per hectare (€190) for 13 applications was considered "relatively low."[26] Using an average number of applications, derived from the 2007–2008 data of 10, 16, 14, and 16 for the United Kingdom, Belgium, the Netherlands, and France, respectively, with potato-growing areas of 140,000 ha, 68,000 ha, 157,000 ha, and 158,000 ha for the same countries, respectively, and an average cost of €18 per hectare for application, the cost of the protection against late blight in the European intensive production systems reaches €130 million per year.

### 9.3.1.2 Forest and amenity trees

In a detailed study of the impact of invasive alien species in Europe based on the DAISIE inventory, Kettunen et al.[27] listed 125 invasive species "with existing evidence of significant environmental, social, and economic impacts in Europe. (p. 11)" Only four plant pathogens, all terrestrial fungi attacking trees, are included in the list: *Ophiostoma ulmi* (as "*Ceratocystis ulmi*"), *Ophiostoma novo-ulmi*, *Phytophthora cinnamomi*, and *Seiridium cardinale*. Figures of economic costs were only given for the second Dutch elm disease epidemics caused by *O. novo-ulmi*; extrapolating the data available in Sweden (calculated costs) and Germany (estimated costs), the cost of the disease was estimated at €124 million per year in Europe. In the same report, the authors gave an "indicative estimate" of the economic impact of "unspecified plant pathogens" (fungi and others) as €1785 million per year. However, the estimation is indeed the figure given by Pimentel et al.[16] for damage by alien pathogens to crops in Great Britain, corrected by a 0.3% annual inflation rate, using 2007 as a reference point.

Canker stain of plane trees caused by the fungus *Ceratocystis fimbriata* f.sp. *platani*, was probably introduced in southern European harbors with American military equipment during the Second World War. The disease threatens city and road plantings; in the earliest foci of disease, nearly all trees were killed by the fungus. The Roads Agency of the Department of Bouches-du-Rhône (south of France, Marseilles area) conducted a detailed evaluation of the costs associated with the disease.[28] The removal of a diseased tree costs at least €1000; however, two neighboring trees have to be removed, too, for sanitation; therefore, the actual cost is €3000. Replanting in situ costs €1135 for a plane tree resistant to the disease, or €850 for a European hackberry immune to the disease. A brand new planting is a less-expensive option, with a cost of €450 for a resistant plane tree and €500 for a European hackberry.

For the whole of Bouches-du-Rhône, 170 infected plane trees were detected and removed in 2006. Prevention and detection cost €30,000, removal costs €170,000, while replanting, not completed, would have cost an average of €120,000 (€75,000–€193,000 depending on the retained technical options). Accordingly, the fungus generates an economic impact of €1700 per tree. The cost of the disease for a private owner would be much higher. Moreover, reports of the spread of the fungus to natural forests in southern Italy[29] indicate that the impact of the fungus could increase in the future.

### 9.3.2  *Bacteria and phytoplasma*

Flavescence dorée is a quarantine disease of grapevine, caused by a phytoplasma transmitted by the insect *Scaphoideus titanus*. Steffek et al.[30] evaluated the cost of uprooting of a 800-ha vineyard in Serbia following the detection of the disease; the primary loss, due to lost investment, was €3.2 million, the income loss for the wine producers due to the decrease in grape production being assumed to be even greater.

Potato brown rot, caused by the bacterium *Ralstonia solanacearum* race 3 biovar 2, is another quarantine disease that has occasionally been found in European countries (including the Netherlands and the United Kingdom), but outbreaks are always submitted to eradication campaigns. Using a bioeconomic model simulating the spatiotemporal spread of the disease over a series of years, Breukers et al.[31] evaluated the cost of the disease to the Dutch potato industry. Analyzing the cost categories, they showed that reducing pathogen monitoring would half the structural costs but dramatically (nearly × 10) increase the export losses; accordingly, the overall cost would increase from €7.7 million to €12.5 million per year.

In the United Kingdom, trial programs have been set up to remove *Solanum dulcamara*, a common native plant from riverbanks that can be infected by the bacterium and acts as a source of inoculum; the cost of *S. dulcamara* removal was £1260 per km of river. The cost of a 4-year campaign on the River Trent was estimated at £2.06–£2.2 million (€2.3–€2.5 million), including removal of *S. dulcamara* and irrigation with disinfectant every year, as well as tuber testing before planting in the first year. Incidentally, the policy of removal of *S. dulcamara* was not implemented in the River Trent mostly because of its low benefit–cost ratio.[32]

Fire blight of Pomaceae is caused by the bacterium *Erwinia amylovora*. The introduction of the disease into Europe during the twentieth century has led to severe losses in pome fruit tree orchards and nurseries, as well as in the ornamental sector. In Switzerland, where the disease was first observed in 1989, the financial burden of control measures (from quarantine to diagnostics), together with compensation payments for destroyed plants, were estimated as follows: €4.5 million in 1989/97, €26.5 million in 1998/02, and more than €4 million in 2003 (i.e., a total of €35 million over a 14-year period).[33] But even for a well-documented disease such as fire blight, there are no general estimates of economic impact given for the whole of Europe.

### 9.3.3  *Viruses*

Sharka disease, caused by *Plum pox virus* (PPV), threatens the growth of *Prunus* worldwide. Cambra et al.[34] estimated the loss in European plum fruit production due to the disease at €5.4 million over the last 30 years. On peaches, PPV-M, an aggressive strain of the virus, caused a loss of €576 million over the last 20 years in Mediterranean countries. In Spain, mandatory and/or voluntary eradication programs have cost €63 million since 1989, including removal, compensation, and production loss. An extrapolation to Europe gave a cost of survey and eradication of PPV of €39 million since 1980. Worldwide cost during the same period, excluding indirect trade loss, is €10 billion. Detected in Switzerland in 1967, sharka disease was subjected to an important program of eradication, and it was believed eradicated in 1973. Afterwards, the disease only occurred sporadically until 2004, when a new outbreak was detected and again submitted to eradication and containment measures. The total cost of the first eradication campaign (from 1967 to 1973) was evaluated at CHF 500 million (€340 million), including compensation payments and the costs of research on diagnosis and epidemiology. These eradication costs were estimated to be equivalent to an annual crop loss of 10% (assuming infection induced a 25% yield loss), and if no measures had been taken, Switzerland would have had an equivalent loss after a few more years.[35]

## 9.4 Ecological impacts of invasive plant pathogens

Traditionally, the estimation of the impacts of plant pathogens relies on the estimates of crop loss and control costs.[16] This approach will underestimate, however, the actual impact of pathogens on plants growing in wild environments, for instance, tree species. The second epidemic of Dutch elm disease, caused by *O. novo-ulmi*, caused the death in 1970–1990 of 28 million mature and 20 million young elms in the United Kingdom; comparable losses were also recorded in continental Europe, central Asia, and North America. Brasier[36] pointed out that economic formulae based mainly on visual and shade impacts, as applied at the landscape scale, could only provide a guide to estimate the actual loss.

Brasier[36] further argued that the cost of irreplaceable loss of a species, a part of the historical and cultural heritage of a country, cannot be evaluated. The economic assessment of biodiversity, especially of ecological services of ecosystems, is indeed a challenge that could prevent a comprehensive evaluation of the impact of invasive plant pathogens. In their review of the impacts of alien species on ecological services in Europe,[37] the DAISIE experts stressed that ecological and/or economic impacts were documented for only ca. 10% of the alien species recorded in the DAISIE inventory. Often evaluated separately, ecological and economic impacts are likely to be highly correlated and should be assessed together in impact studies. Unfortunately, the pathogenic fungi were not explicitly addressed in the review, which focused on the "worst" invading terrestrial plants, vertebrates, and invertebrates, as well as aquatic and marine fauna and flora. However, most conclusions of the review can be applied to plant pathogens. When assessing the impacts of alien species, such as invasive plant pathogens, on ecosystem services, most available data relate to provisioning impacts (food loss, threat to endangered native species) while data on cultural impacts (changes in recreational use, effects on ecotourism, changes in the perception of landscapes and aesthetics) are scarce.

The value of the ecosystemic services potentially impacted by invasive plant pathogens is also poorly documented. A recent report delivered to the French prime minister[38] proposed guidelines to improve the precision and accuracy of the assessment of the reference values of ecosystemic services. Accordingly, the reference value of different ecosystemic services of the French forest was evaluated at €970 per hectare per year on average (ranging from €500 to more than €2000), that is, €35,000 per hectare in total actualized value. Within this reference value, only a small part (€75 per hectare per year) was attributed to wood provision, while the most important parts related to carbon capture and storage and recreational services. This assessment can provide a basis to estimate the impacts of forest pathogens, especially of alien species, if one can estimate the part of "forest services" lost due to their action. We tentatively estimated this loss as follows. The rate of fungal diseases recorded in the systematic survey of crown status on 10,000 trees belonging to the International Cooperative Programme (ICP) forest European network[39] was 7% in 2005 and 2006 in France (see Forest Health Service)[40,41] for the last available years, which corresponds to a significant, generally strong, impact of fungi on the crown status of examined trees. We estimated that 20% of the ecological services provided by those trees were lost (wood production, carbon storage, amenity value, etc.), leading to a low estimate of 1.4% loss in "forest value," that is, €208 million per year lost due to forest pathogenic fungi in France.

However, only a few visible pathogen species are recorded in the systematic survey, which, for example, does not include root pathogens. In the database of the Forest Health Service, the species recorded in the ICP network correspond to only 50% of all records. A more realistic estimate would therefore be over €400 million lost each year. This figure itself is probably an underestimation since even the Forest Health Service database only records disease symptoms and not growth loss per se. We therefore suggest an estimation of the impact of forest pathogens in France in the range of €400–€800 million per year. Because 37% of all diseases in the Forest Health Service database are attributed to alien fungi,[42] this would give an estimate of €148–€296 million per year for alien forest pathogenic fungi.

In the aforementioned study[28] on the impact of the canker stain of plane trees, the amenity value of an average plane tree was evaluated at €4200. The infection of the tree by the fungus nullifies its amenity value. Accordingly, amenity loss due to the removal of the 170 infected trees in the Marseilles area in 2006 can be estimated at €715,000.

## 9.5    Foresight of impacts of invasive plant pathogens

### 9.5.1    Pest risk analysis and the assessment of economical impact of invasive pathogens (emerging or still absent in Europe)

Pest risk analysis (PRA) as defined by IPPC[43] is "The process of evaluating biological or other scientific and economic evidence to determine whether an organism is a pest, whether it should be regulated, and the strength of any phytosanitary measures to be taken against it. (p. 2)" At the global level, guidance on how to perform PRA is given by the International Standard on Phytosanitary Measures (ISPM) no. 11.[44] At the European level, EPPO has elaborated a scheme to carry out PRAs[45] and also conduct PRAs itself by organizing expert working groups to assess the risks presented by specific pests. At the EU level, the European Food Safety Authority (EFSA) now also performs PRAs for the EU Commission. PRA can be used as a tool to predict the economic impact of invasive pathogens (whether they are regulated at the end of the process), but this remains a difficult task, as much economic data is lacking. The economic assessment is usually based mostly on expert judgment.[46,47] While representing a low-cost and efficient use of scientific knowledge, such an assessment lacks transparency and repeatability. Reliable economic data are rarely available to PRA assessors, especially when a potentially invasive pathogen is detected in a new area. However, attempts have been made to quantify the potential economic cost of the introduction of an invasive plant pathogen into Europe. The method used in these studies is the evaluation of direct economic consequences of an introduction by partial budgeting.[47]

The fungus *Tilletia indica*, which causes karnal bunt of wheat, is a quarantine pathogen in Europe (regulated in the EU Directive 2000/29, and included in the EPPO A1 List). A PRA funded under the EC 5th Framework concluded that the pathogen has the potential of establishing in the United Kingdom and many other European countries.[48,49] The economic impact of the introduction of karnal bunt of wheat was also evaluated in the PRA.[49] Assuming that a large disease outbreak (50,000 ha) would occur in the United Kingdom, the total costs at the European level would be ca. €34 million in the year of the outbreak; direct costs (yield and downgrading) would be insignificant (5% of total) compared with reaction costs (indirect quality losses, loss of exports related to its categorization as a quarantine pest in other countries, and seed industry costs; 43%) and control costs (52%). The main cost would be caused by downgrading (31% of total) or destruction (27%) of nonaffected crops due to mandatory measures. After 10 years, the cost supported by the United Kingdom would be €454 million.

*Potato spindle tuber viroid* (PSTVd), the cause of a destructive disease on potatoes worldwide, was placed on the A2 List by EPPO in the 1970s, as it was sporadically reported from a small number of European countries. However, the epidemiological situation of this disease needs to be further investigated because recent observations have shown that PSTVd could be detected in asymptomatic solanaceous ornamentals, which might act as reservoirs for the viroids. In particular, there is now epidemiological evidence that ornamental species (i.e., *Brugmansia* spp., *Solanum jasminoides*) can act as sources of PSTVd for tomato crops.[50] Soliman et al.[47] concluded that the introduction of PSTVd in the main potato-producing areas of the EU would have a direct impact of €685 million per year (control costs, €118 million; reduced revenues due to yield loss, €685 million). Further analyzing the indirect economic consequences of a PSTVd invasion by partial equilibrium modeling, Soliman et al.[47] found that the direct negative impacts would be transferred from producers (whose welfare would increase by 0.02%) to consumers (domestic prices would increase by 0.73%).

The complexity of assessing the economic impact of an invasive pathogen at a regional level is illustrated by the example of *Pepino mosaic virus* (PepMV). First described in Peru in 1974 (on *Solanum muricatum*), this virus emerged in Europe on glasshouse tomatoes in the 1990s. The virus is highly contagious (mechanically transmitted) and has the potential to damage tomato crops. The economic impact varies among European countries, as it seems to be particularly influenced by tomato fruit marketing systems. In the United Kingdom, only high-quality fruit is profitable, and there is no market for second-class fruit; therefore, any disease symptom will lead to downgrading and unacceptable economic losses for the growers.[51] In other European countries such as the Netherlands, a similar loss in fruit quality will not lead to unacceptable losses in terms of sale profits for Dutch growers. An EU research project (PEPEIRA) specifically dedicated to PepMV is currently underway, and one of its objectives is to assess the economic impact of PepMV on tomato crops and develop an economic model to determine the overall economic impact in EU member states, the ultimate goal being to provide an EU-wide PRA for PepMV.

To overcome the lack of sufficient data required to effectively carry out PRAs, an EU-funded project "PRATIQUE—Enhancements of Pest Risk Analysis Techniques" was launched in 2008. A particular objective of this project is to enhance techniques for assessing the economic, environmental, and social impacts[52] of quarantine pathogens.

## 9.5.2 Should we expect more invasive species in the future?

The globalization of trade and tourism and increased migrations of human populations for economic and political reasons, as well as global warming, have been highlighted as drivers of a potential increase in the rate of invasions by living organisms.[53–55]

At the British scale and over four decades, the number of pathogens introduced on the two most important plant groups, that is, ornamental and crop plants, has not significantly increased over a 5-year period (11–28 species per period on ornamental plants, 4–15 species per period for crop plants). Considering the whole data set, the average rate of introduction of plant pathogens over 35 years of the survey is 6.7 species per year, including 5.3 fungal species per year.[15] For France, lower rates have been recorded, ranging from 0 to 36 per decade. However, many species labeled "alien" in Great Britain are considered to be indigenous in continental Europe. A significant result for the French inventory was the marked increase in introductions from 1800 onwards with less than 0.5 new species of fungi recorded per year until 1930, in spite of the effort of talented mycologists eager to record new species, to two new species per year in the last decades.[13] A significant exponential pattern in the rate of introductions from 1800 onwards was also observed at the whole European scale.[13]

### 9.5.3   Emerging invasive plant pathogens in Europe

Most of the aforementioned listed economic and ecological impacts on European agro-ecosystems are caused by invasive plant pathogens established in Europe for decades. We present here selected cases of invasive pathogens currently emerging in Europe.

*Acidovorax avenae* subsp. *citrulli*, a pathogen causing bacterial watermelon fruit blotch, is not yet invasive but shows potential for invasion (e.g., in the United States). That *A. avenae* subsp. *citrulli* is a seed-borne disease probably adds to the risk. Isolated findings have been made in Greece in 2005 and Israel in 2006 but have not been followed by the establishment of the disease in watermelon crops. Detected in July 2007 in Hungary, the pathogen seems to have been introduced on grafted watermelon transplants imported from Turkey, where the disease is present.[56]

*Chalara fraxinea*, a fungus causing the dieback of common ash, is an invasive pathogen currently increasing its distribution area. After its first finding in Poland in 2006,[57] the fungus has since been detected in many countries, covering most of the distribution range of *Fraxinus excelsior* (Kirisits, pers. comm.). The situation of *C. fraxinea* in Europe needs to be further investigated; in particular the relationships between the anamorph and teleomorph stages need to be clarified. The teleomorph of *C. fraxinea*, which has recently been identified (*Hymenoscyphus albidus*), is widespread, nonpathogenic, and native to Europe, while *C. fraxinea* apparently behaves like an "exotic" disease.[58]

The pine wood nematode, *Bursaphelenchus xylophilus*, is a major threat to European forests. *B. xylophilus*, an endemic species in North America, has caused serious economic damage in Japan, China, and Korea. This pest has been intercepted in packing material shipped into Europe from North America and Asia. More critical is an outbreak of the disease on maritime pine in Portugal, since 1999.[59] The European Commission has adopted emergency measures requiring the treatment of all nonmanufactured wood packing material from contaminated areas in order to prevent introduction of the nematode throughout Europe.

A stream of *Phytophthora* spp. with impacts on wild and horticultural plants has been introduced in Europe since the 1990s.[36] *P. ramorum*, the cause of sudden oak death in the United States, is spreading in commercial nurseries, woods, and gardens in Europe where it attacks several species (especially rhododendron), although forest tree infections are still limited.[60] In the United Kingdom, rhododendrons are attacked by *P. kernoviae*, which has been recently recorded on *Vaccinium* and therefore represents a major threat to the native heathland.[61] The highly aggressive species *P. alni* subsp. *alni*, spreading to riparian alders all over Europe, has been shown to be a hybrid between two less aggressive species, *P. alni* subsp. *uniformis* and *P. alni* subsp. *multiformis*.[62] *P. alni* subsp. *alni* easily transfers with nursery stocks and has the ability to jump from host to host and to hybridize; this *Phytophthora* species represents major threats to cultivated and wild plants in Europe. In a 6-year survey of Spanish nurseries, Moralejo et al.[63] detected 17 species from 37 host plants; several host–pathogen combinations were the first reports. Moreover, most of these species are of alien origin and could spread to natural environments.

Viruses transmitted by *Bemisia tabaci* have been a matter of increased concern since the emergence in the 1990s of many new species damaging vegetable crops (beans, capsicum, cucurbits, lettuce, and tomatoes). In a review, Polston and Anderson[64] stated that the number of new whitefly-transmitted viruses infecting tomatoes in Latin America increased from 3 in the 1970s to nearly 20 in the 1990s. More virus species have been described since then, and most of them still do not occur in Europe (e.g., *Tomato mottle virus*, *Chino del tomate virus*, and *Sinaloa tomato leaf curl virus*). But in Europe, outbreaks of yellow leaf curl diseases on tomato crops (caused by *Tomato yellow leaf curl virus* and *Tomato yellow leaf curl Sardinia virus*), only sporadic in the 1960s have now become a serious economic problem. In the eastern part of the Mediterranean Basin, tomato yellow leaf curl outbreaks sometimes result in total crop failures.[65,66] These disease emergences can be linked to the spread of their insect vector *B. tabaci*, which has recently increased its distribution area by moving northwards, possibly as a consequence of global warming, but certainly aided by the international transport of plant material.

## 9.5.4    Invasive plant pathogens still absent from Europe

The EPPO A1 List contains plant pathogens undetected yet in Europe, for which regulation as quarantine pests is recommended to its member countries. Information of worldwide disease distribution, biology, and risk analysis is available on the EPPO Web site.[67] We present below a short list of the pathogens from the A1 List, which would most probably have the most severe impacts if introduced in Europe.

### 9.5.4.1    Fungi and Oomycetes

EPPO considers *Ceratocystis fagacearum*, the cause of oak wilt, which occurs in the eastern and midwestern United States, to be a threat to oak trees in Europe: the fungus is pathogenic to European oaks[68] and could find a suitable insect vector. The main measure to prevent the entry of this pathogen in Europe is the prohibition of imported oak plants. Another means of introduction is the trade of oak wood infected with fungal mycelial mat or carrying bark beetles, but specific requirements are made by European countries in their phytosanitary regulations on wood and wood products to prevent this.

*Sirococcus clavigignenti-juglandacearum*, the cause of butternut canker, could present a high risk to Europe to kill large numbers of trees used for production of wood, nuts, and oil and threaten the walnut tree as an amenity species. *S. clavigignenti-juglandacearum* is even considered more aggressive than the fungi responsible for chestnut blight and Dutch elm disease. The introduction of infected host plants is the most probable means of entry of the fungus in Europe, and EPPO has recently added it to its A1 List of pests recommended for regulations.

*Thecaphora solani*, the cause of potato smut, indigenous in Central and South America, presents a significant risk to both seed and ware potato production in Europe. Mostly spread on infected tubers, the fungus is regulated as a quarantine pest in many European countries.

*Cronartium* spp. are various rust species, known in North America as "blister rusts" infecting conifer trees. While their aeciospores can travel over long distances, the trade of conifer plants from North America could be a pathway for introducing blister rusts. These pathogens among others are the reason why imports of conifer plants from North America into Europe are prohibited.

*Diaporthe vaccini*, the cause of blueberry twig blight, has been reported a few times in Europe but does not seem to have persisted yet. *D. vaccini* is probably imported from America on blueberry vines.

*Gymnosporangium* spp. are various rust species mostly of North American origin infecting fruit trees, especially apple trees. Infection of the telial host of the fungus, *Juniperus* spp. is systemic; accordingly, *Juniperus* branches could also be a source of entry of the fungus in Europe. Regulations on the different host plants of *Gymnosporangium* spp. have been put in place in Europe to avoid their introduction.

*Puccinia hemerocallidis*, causing daylily rust, is of Siberian origin and results in severe losses to gardeners and nurseries in North America. The fungus has already been detected in imported plants in the United Kingdom, showing that trade of infested plants could be a pathway of entry. Once established, its eradication would be difficult since the fungus can survive as a latent infection. Therefore, this pathogen has been recommended recently by EPPO to be regulated as a quarantine pest.

### 9.5.4.2  Bacteria

Huanglongbing or citrus greening (associated with *Candidatus* Liberibacter asiaticus, *Ca.* L. africanus, *Ca.* L. americanus), a severe disease of citrus, presents a high risk for the Mediterranean regions of Europe, provided its insect psyllid vectors (*Diaphorina citri* and *Trioza erytreae*—also regulated as quarantine pests in many European countries), were also introduced. So far huanglongbing has never been detected in Europe, but isolated findings of the vector *T. erytreae*, were reported in Madeira (Portugal) in 1994 and in the Canary Islands (Spain) in the 2000s, stressing that particular attention should be paid to this disease.

*Xanthomonas axonopodis* pv. *citri*, the cause of bacterial canker of citrus, is recognized as a significant problem in countries where it occurs. In Europe, to protect citrus production against severe diseases such as huanglongbing or citrus canker, importation of citrus plants from outside the region is prohibited.

*Xylella fastidiosa*, the cause of Pierce's disease of grapevine and related diseases on peach trees, citrus, and other woody plants, could destroy vineyards and prevent grapevine cultivations in European countries. The insect vectors present in America are not found in Europe, but because transmission of the bacterium is not vector specific, the bacterium could most probably find a suitable vector in Europe. *X. fastidiosa*, together with other damaging grapevine pests, is one of the reasons why imports of *Vitis* plants from outside Europe are prohibited.

### 9.5.4.3  Viruses

EPPO considers potato viruses of South American origin (e.g., *Potato Andean mottle virus*, *Potato black ringspot virus*, *Potato virus T*, *Potato yellow dwarf virus*, and *Potato yellowing virus*) to be serious threats to potato seed production in Europe. If introduced, they would increase the cost and difficulty in operating the seed production schemes. As a consequence, most EPPO member countries prohibit the importation of potatoes from outside Europe.

## 9.6  Conclusion

Invasive plant pathogens still represent a main threat to cultivated and wild plants in Europe. Pathogens introduced decades ago still have a huge direct economic impact, represented by yield and quality loss and the cost of fungicides required to protect crops with high cash value, such as grapevine and potato. In the future, farmers will have to use less fungicide to follow European and national regulations and rely on more sustainable methods of disease management to decrease crop loss.

For noncrop ecosystems, there is increasing consensus toward an evaluation of impacts accounting for not only the direct, economic loss but also the damage caused to the ecosystemic value of the attacked plants. Due to the increase in international trade of nursery stocks and planted trees, native plant communities, woodlands, and landscapes face an increasing risk of invasion by pathogens.[36]

The strict enforcement of quarantine rules is critical to keep potentially invasive pathogens at bay. Quarantine lists based on PRA should be regularly updated to prevent disastrous introductions—or, at least, to increase awareness of pathogens that are poorly known outside their native areas. However, the traditional European species-targeted approach has inherent limits since it can only apply to known species. Many recent emerging diseases, especially in noncrop plants, were caused by previously undescribed species of unknown origin, such as *P. ramorum* or *C. fraxinea*. A pathway approach to prevent the movement of pests and pathogens in international trade is therefore increasingly considered, in addition to the species approach. For example, a global standard on wood packaging material[69] is now being implemented by many countries around the world, and a new standard on plants for planting is under preparation.

Solving the difficulties encountered in predicting the invasive behavior of plant pathogens and taking appropriate actions against them certainly remains a challenge for all stakeholders, from the growers, plant traders, economists and plant health policymakers to plant pathologists.

## References

1. Large, E. C. 1940. *The Advance of the Fungi.* London: Jonathan Cape Ltd. Re-issued by the American Phytopathological Society, St. Paul, 2003.
2. Eriksson, J. 1913. *Les maladies cryptogamiques des plantes agricoles et leur traitement* [Cryptogamic diseases of plant crops and their control]. (French translation of the Swedish original). Paris: Librairie Agricole de la Maison Rustique.
3. Castonguay, S. 2005. Biorégionalisme, commerce agricole et propagation des insectes nuisibles et des maladies végétales: les conventions internationales phytopathologiques [Bioregionalism, agricultural trade and spread of plant pests and diseases: The international phytopathological conventions]. 1878–1929, *Ruralia*, 16/17. http://ruralia.revues.org/document1074.html (accessed January 31, 2010).
4. Schrader, G., and J. G. Unger. 2003. Plant quarantine as a measure against invasive alien species: The framework of the International Plant Protection Convention and the plant health regulations in the European Union. *Biol Invasions* 5:357.
5. Lowe, S., M. Browne, S. Boudjemas, and M. De Poorter. 2004. *100 of the World's Worst Invasive Alien Species: A Selection from the Global Invasive Species Database.* Auckland: Invasive Species Specialist Group (ISSG). http://www.issg.org/pdf/publications/worst_100/english_100_worst.pdf (accessed January 31, 2010).
6. DAISIE European Invasive Alien Species Gateway. 2008. 100 of the worst. http://www.europe-aliens.org/speciesTheWorst.do (accessed January 31, 2010).
7. Desprez-Loustau, M. L., C. Robin, M. Buée, R. Courtecuisse, J. Garbaye, F. Suffert, I. Sache, and D. M. Rizzo. 2007. The fungal dimension of biological invasions. *Trends Ecol Evol* 22:472.
8. Robinson, R. A. 1996. *Return to Resistance: Breeding Crops to Reduce Pesticide Dependence.* Davis, CA: agAccess. http://www.idrc.ca/openebooks/774-4/ (accessed January 31, 2010).
9. Bayles, R. A., K. Flath, M. S. Hovmøller, and C. de Vallavieille-Pope. 2000. Breakdown of the Yr17 resistance to yellow rust of wheat in northern Europe. *Agronomie* 20:805.
10. Singh, R. P., D. P. Hodson, Y. Jin, J. Huerta-Espino, M. G. Kinyua, R. Wanyera, P. Njau, and R. W. Ward. 2006. Current status, likely migration and strategies to mitigate the threat to wheat production from race Ug99 (TTKS) of stem rust pathogen. *CAB Rev Perspect Agric Vet Sci Nutr Nat Res* 1(054):1.
11. Drenth, A., L. J. Turkensteen, and F. Govers. 1993. The occurrence of the A2 mating type of *Phytophthora infestans* in the Netherlands; significance and consequences. *Netherlands J Plant Pathol* 99(Suppl. 3):57.

12. DAISIE. 2009. *Handbook of Alien Species in Europe*, ed. Delivering Alien Invasive Species Inventories for Europe (DAISIE). Dordrecht: Springer.

13. Desprez-Loustau, M. L. 2009. Alien fungi of Europe. In *Handbook of Alien Species in Europe*, ed. Delivering Alien Invasive Species Inventories for Europe (DAISIE), 15. Dordrecht: Springer.

14. Desprez-Loustau, M. L., R. Courtecuisse, C. Robin, C. Husson, P. A. Moreau, D. Blancard, M. A. Selosse, B. Lung-Escarmant, D. Piou, and I. Sache. 2010. Species diversity and drivers of spread of alien fungi (*sensu lato*) in Europe with a particular focus on France. *Biol Invasions* 12:157.

15. Jones, D. R., and R. H. A. Baker. 2007. Introductions of non-native plant pathogens into Great Britain, 1970–2004. *Plant Pathol* 56:891.

16. Pimentel, D., S. McNair, J. Janecka, J. Wightman, C. Simmonds, C. O'Connell, E. Wong et al. 2001. Economic and environmental threats of alien plant, animal, and microbe invasions. *Agric Ecosyst Environ* 84:1.

17. Aubertot, J. N., J. M. Barbier, A. Carpentier, J. J. Gril, L. Guichard, P. Lucas, S. Savary, I. Savini, and M. Voltz, eds. 2005. *Pesticides, agriculture et environnement: réduire l'utilisation des pesticides et limiter leurs impacts environnementaux* [Pesticides, agriculture and environment: Decreasing pesticide use and limiting their environmental impact]. Paris: INRA—CEMAGREF.

18. Oerke, E. C., H. W. Dehne, F. Schönbeck, and A. Weber. 1994. *Crop Production and Crop Protection: Estimated Losses in Major Food and Cash Crops*. Amsterdam: Elsevier.

19. Euroblight—A potato late blight network for Europe. 2009. http://www.euroblight.net (accessed January 31, 2010).

20. Hansen, J. G., and the EuroBlight group. 2009. The development and control of late blight (*Phytophthora infestans*) in Europe in 2007 and 2008. *PPO-Spec Rep* 13:11.

21. Hermansen, A., and R. Nærstad. 2008. Forecasting potato blight in Norway. *PPO-Spec Rep* 12:301.

22. Dowley, L. J., R. Leonard, B. Rice, and S. Ward. 2002. Efficacy of the NegFry decision support system in the control of potato late blight in Ireland. *PPO-Spec Rep* 8:81.

23. Leonard, R., L. Dowley, B. Rice, and S. Ward. 2001. The use of decision support systems in Ireland for the control of late blight. *PAV-Spec Rep* 7:91.

24. Heremans, B., and G. Haesaert. 2004. Late blight on potato in Flanders, Belgium: Field trials and characteristics of the *Phytophthora infestans* population. *PPO-Spec Rep* 10:247.

25. Hinds, H. 2000. Using disease forecasting to reduce fungicide input for potato blight in the UK. *PAV-Spec Rep* 6:82.

26. Hinds, H. 2001. Can blight forecasting work on large potato farms? *PAV-Spec Rep* 7:99.

27. Kettunen, M., P. Genovesi, S. Gollasch, S. Pagad, U. Starfinger, P. ten Brink, and C. Shine. 2008. *Technical Support to EU Strategy on Invasive Species (IAS)—Assessment of the Impacts of IAS in Europe and the EU (Final Module Report for the European Commission*. Brussels: Institute for European Environmental Policy (IEEP). http://ec.europa.eu/environment/nature/invasivealien/docs/Kettunen2009_IAS_Task%201.pdf (accessed January 31, 2010).

28. Mollet, J. M. 2007. Les dégâts dus au chancre coloré du platane. Coûts économiques, esthétiques et symboliques, in *Colloque national "Chancre coloré du platane."* [Damages caused by canker stain of plane trees. Economical, esthetical and symbolical costs, in *National Symposium "Canker Stain of Plane Trees."*] Toulouse: AFPP.

29. Panconesi, A., S. Moricca, I. Dellavalle, and G. Torraca. 2003. The epidemiology of canker stain of plane tree and its spread from urban plantings to spontaneous groves and natural forests. *Mitteilungen aus der Biologischen Bundesanstalt für Land- und Forstwirtschaft* 394:84.

30. Steffek, R., H. Reisenzein, and N. Zeisner. 2007. Analysis of the pest risk from Grapevine flavescence dorée phytoplasma to Austrian viticulture. *EPPO Bull* 37:191.

31. Breukers, A., W. Van der Werfe, M. Mourits, and A. O. Lansink. 2007. Improving cost-effectiveness of brown rot control: The value of bio-economic modelling. *EPPO Bull* 37:391.

32. Macleod, A. 2007. The benefits and costs of specific phytosanitary campaigns in the UK. In *New Approaches to the Economics of Plant Health*, ed. A. G. J. M. Oude Lansink, 163. Berlin: Springer.

33. Duffy, B., H. J. Schärer, M. Bünter, A. Klay, and E. Hollinger. 2005. Regulatory measures against *Erwinia amylovora* in Switzerland. *EPPO Bull* 35:239.

34. Cambra, M., N. Capote, A. Myrta, and G. Llácer. 2006. Plum pox virus and the estimated costs associated with sharka disease. *EPPO Bull* 36:202.

35. Ramel, M. E., P. Gugerli, and M. Bünter. 2006. Control and monitoring: Eradication of Plum pox virus in Switzerland. *EPPO Bull* 36:312.

36. Brasier, C. M. 2008. The biosecurity threat to the UK and global environment from international trade in plants. *Plant Pathol* 57:792.

37. Vilà, M., C. Basnou, P. Pyšek, M. Josefsson, P. Genovesi, S. Gollasch, W. Nentwig et al. and DAISIE partners. 2010. How well do we understand the impacts of alien species on ecosystem services? A pan-European cross-taxa assessment. *Front Ecol Environ* 8:135.

38. Chevassus-au-Louis, B., ed. 2009. *Approche économique de la biodiversité et des services liés aux écosystèmes. Contribution à la décision publique.* Paris: Rapports et documents, Centre d'analyse stratégique. [Economical analysis of biodiversity and ecosystemic services. Paris: Reports and Documents, Centre for Strategic Analysis]. http://www.strategie.gouv.fr/IMG/pdf/04Rapport_biodiversite_28avril2009_.pdf (accessed January 31, 2010).

39. International Co-operative Programme on Assessment and Monitoring of Air Pollution Effects on Forests. http://www.icp-forests.org/ (accessed January 31, 2010).

40. Département Santé des Forêts [Forest Health Department]. 2006. Bilan de la santé des forêts en 2005. [Forest Health in 2005]. http://agriculture.gouv.fr/sections/thematiques/foret-bois/sante-des-forets (accessed January 31, 2010).

41. Département Santé des [Forêts Forest Health Department]. 2007. Bilan de la santé des forêts en 2006. [Forest Health in 2006]. http://agriculture.gouv.fr/sections/thematiques/foret-bois/sante-des-forets (accessed January 31, 2010).

42. Vacher, C., J. J. Daudin, D. Piou, and M. L. Desprez-Loustau. 2010. Ecological integration of alien species into a tree-parasitic fungus network. *Biol Invasions* 12:3249.

43. IPPC. 2007. *International Plant Protection Convention.* Food and Agriculture Organization of the United Nations, Rome, 1951 (amended 1979, 1997). https://www.ippc.int/file_uploaded//publications/13742.New_Revised_Text_of_the_International_Plant_Protectio.pdf (accessed January 31, 2010).

44. FAO. 2004. *ISPM No. 11. Pest Risk Analysis for Quarantine Pests Including Analysis of Environmental Risks and Living Modified Organisms.* Food and Agriculture Organization of the United Nations. https://www.ippc.int (accessed January 31, 2010).

45. European and Mediterranean Plant Protection Organization. 2009. Standard on PRA PM 5/3(4). Decision-Support Scheme for Quarantine Pests. http://archives.eppo.org/EPPOStandards/PM5_PRA/PRA_scheme_2009.doc (accessed January 31, 2010).

46. Sansford, C. 2002. Quantitative versus qualitative: Pest risk analysis in the UK and Europe including the European and Mediterranean Plant Protection (EPPO) system. In *NAPPO International Symposium on Pest Risk Analysis*, Puerto Vallarta. http://www.nappo.org/PRA-Symposium/PDF-Final/Sansford.pdf (accessed January 31, 2010).

47. Soliman, T. A. A., M. C. M. Mourits, A. G. J. M. Oude Lansink, and W. Van der Werf. 2010. Economic impact assessment in pest risk analysis. *Crop Prot* 29:517.

48. Baker, R. H. A., C. E. Sansford, B. Gioli, F. Miglietta, J. R. Porter, and F. Ewert. 2005. Combining a disease model with a crop phenology model to assess and map pest risk: Karnal bunt disease (*Tilletia indica*) of wheat in Europe. In *Plant Protection and Plant Health in Europe: Introduction and Spread of Invasive Species*, DPG-BCPC Symposium Series 81, eds. D. Alford and G. Backhaus, 89. Brighton: British Crop Protection Enterprises.

49. Sansford, C., R. Baker, J. Brennan, F. Ewert, B. Gioli, A. Inman, P. Kelly et al. 2006. Report on the risk of entry, establishment and economic loss for *Tilletia indica* in the European Union. Deliverable Report, D. L., 6.1. EC Fifth Framework Project QLKS-1999-01554 Risks Associated with Tilletia indica, the Newly Listed, E. U., Quarantine Pathogen, the Cause of Karnal Bunt of Wheat. http://lmt.planteforsk.no/pfpntr/karnalpublic/files/eu_karnalbunt_pra.pdf (accessed January 31, 2010).

50. Verhoeven, J. T. J., C. C. C. Jansen, M. Botermans, and J. W. Roenhorst. 2010. Epidemiological evidence that vegetatively propagated, solanaceous plant species act as sources of *Potato spindle tuber viroid* inoculum for tomato. *Plant Pathol* 59:3.

51. Spence, N. J., J. Basham, R. A. Mumford, G. Hayman, R. Edmondson, and D. R. Jones. 2006. Effect of *Pepino mosaic virus* on the yield and quality of glasshouse-grown tomatoes in the UK. *Plant Pathol* 55:595.

52. Baker, R. H. A., A. Battisti, J. Bremmer, M. Kenis, J. Mumford, F. Petter, G. Schrader et al. 2009. PRATIQUE: A research project to enhance pest risk analysis techniques in the European Union. *EPPO Bull* 35:239.

53. Anderson, P. K., A. A. Cunnigham, N. G. Patel, F. J. Morales, P. R. Epstein, and P. Daszak. 2004. Emerging infectious diseases of plants: Pathogen pollution, climate change and agrotechnology drivers. *Trends Ecol Evol* 19:535.

54. Levine, J. M., and C. M. d'Antonio. 2003. Forecasting biological invasions with increasing international trade. *Conserv Biol* 17:322.

55. Jones, K. E., N. G. Patel, M. A. Levy, A. Storeygard, D. Balk, G. L. Gittleman, and P. Daszak. 2008. Global trends in emerging infectious diseases. *Nature* 451:990.

56. Palkovics, L., M. Petróczy, and B. Kertész. 2008. First report of bacterial fruit blotch of watermelon caused by *Acidovorax avenae* subsp. *citrulli* in Hungary. *Plant Dis* 92:834.

57. Kowalski, T. 2006. *Chalara fraxinea* sp. nov. associated with dieback of ash (*Fraxinus excelsior*) in Poland. *For Pathol* 36:264.

58. Kowalski, T., and O. Holdenrieder. 2009. The teleomorph of *Chalara fraxinea*, the causal agent of ash dieback. *For Pathol* 39:304.

59. Mota, M., H. Braasch, M. A. Bravo, A. C. Penas, W. Burgermeister, K. Metge, and E. Sousa. 1999. First report of *Bursaphelenchus xylophilus* in Portugal and in Europe. *Nematology* 1:727.

60. Tracy, D. R. 2009. *Phytophthora ramorum* and *Phytophthora kernoviae*: The woodland perspective. *EPPO Bull* 39:161.

61. Beales, P. A., P. G. Giltrap, A. Payne, and N. Ingram. 2009. A new threat to UK heathland from *Phytophthora kernoviae* on *Vaccinium myrtillus* in the wild. *Plant Pathol* 58:393.

62. Ioos, R., A. Andrieux, B. Marçais, and P. Frey. 2006. Genetic characterization of the natural hybrid species *Phytophthora alni* as inferred from nuclear and mitochondrial DNA analyses. *Fungal Genet Biol* 43:511.

63. Moralejo, E., A. M. Pérez-Sierra, L. A. Álvarez, L. Belbahri, F. Lefort, and E. Descalsa. 2009. Multiple alien *Phytophthora* taxa discovered on diseased ornamental plants in Spain. *Plant Pathol* 58:100.

64. Polston, J. E., and P. K. Anderson. 1997. The emergence of whitefly-transmitted geminiviruses in tomato in the Western Hemisphere. *Plant Dis* 81:1358.

65. Garcia-Andrés, S., J. P. Accotto, J. Navas-Castillo, and E. Moriones. 2007. Founder effect, plant host and recombination shape the emergent population of begomoviruses that cause the tomato yellow leaf curl disease in the Mediterranean basin. *Virology* 359:302.

66. Oliveira, M. R. V., T. J. Henneberry, and P. Anderson. 2001. History, current status, and collaborative research projects for Bemisia tabaci. *Crop Prot* 20:709.

67. European and Mediterranean Plant Protection Organization. 2009. EPPO A1 List of pests recommended for regulation as quarantine pests. http://www.eppo.org/QUARANTINE/listA1.htm (accessed January 31, 2010).

68. MacDonald, W. L., J. Pinon, F. H. Tainter, and M. L. Double. 2001. European oaks-susceptible to oak wilt? In *Shade Tree Wilt Diseases*, ed. C. L. Ash, 131. St. Paul, MN: APS Press.

69. FAO. 2009. *ISPM No. 15 (Revision). Regulation of Wood Packaging Material in International Trade.* Food and Agriculture Organization of the United Nations. https://www.ippc.int (accessed January 31, 2010).

*section five*

---

*India*

*chapter ten*

# Invasive plants in the Indian subcontinent

*Daizy R. Batish, R. K. Kohli, and H. P. Singh*

## Contents

## 10.1 Introduction

Biological invasion is a global phenomenon that occurs when species move from one geographic region to another, establish and proliferate there, and harm the native species. The migration or movement of species from one biogeographic region to another has occurred since ancient times; however, of late, the frequency of migration or movement of species has accelerated, and humans are largely responsible for this. In fact, human-driven movement of species is accidental as well as intentional, and an intrinsic part of our history.[1] Better transport facilities have further facilitated the movement of species across biogeographic boundaries. There is hardly any barrier to prevent the movement of species from one place to other. In fact, this movement has resulted in the homogenization of biota across the world.[2] Invasive species are not restricted to any taxonomic group; they can be plants, animals, and microbes and thus cannot be classified on taxonomic lines.[3] Among these, the movement of plants from one part of the world to another has been of interest to man as these provide a multitude of benefits such as food, medicine, timber, and furniture, and also because of their ornamental value. According to Myers and Bazely,[4] because of the interest of horticulturists and gardeners for ornamental plants, there is a great demand for international trade in plants and their propagules. *Lantana camara* is a well-known example of intentional plant introduction for ornamental value from tropical America to different parts of the world before becoming one of the worst plant invaders in different ecosystems, especially affecting forests. *Parthenium hysterophorus*, on the other hand, is an example of human-aided accidental entry into different regions of the world from its native region of tropical America. Environmental problems such as climate change, habitat disturbances, and changing landscape patterns have further escalated the invasion process.

Invasive plants pose a serious threat to native ecosystems by altering the composition of plant communities, reducing biodiversity, changing soil structure, and affecting the health of human beings, thereby causing enormous economic costs.[5–8] In fact, the impact of invasive species on biodiversity is rated second only to habitat fragmentation and even greater than pollution.[2] The problem of biological invasion is thus globally recognized and threatens the ecological and economic integrity of the invaded area.[9,10] However, the impact varies from one part of the world to another depending on the geographic features, diversity of landscapes, and human population density.[11,12] Sakai et al.[13] pointed out that invasive plants quickly respond to anthropogenic changes and have the ability to undergo genetic changes owing to selection pressure.

Various terms such as "introduced," "nonnative" or "nonindigenous," "exotic," "alien," "foreign" or "invasive alien" refer to plants that evolved in a biogeographic area different from the new locales they have been moved to. The term *invasive* refers to those species that are naturalized in the alien environment and produce large number of offspring over a large area.[14] Colautti and MacIsaac[15] have provided a framework for different terms used in invasion ecology. According to their framework, a potential invader has to pass through a series of stages, which differentiate more or less the ambiguous terms like "invasive," "introduced," "weedy," or "naturalized species." Such naturalized species are called invasive alien species (IAS) and have been referred to under article 8(h) of the CBD. The problem of biological invasion is well addressed by organizations such as the International Union for Conservation of Nature and Natural Resources (IUCN), Convention on Biological Diversity (CBD), and the Global Invasive Species Programme (GISP). According to the CBD, IAS are those species whose introduction and/or spread outside their natural past or present distribution threaten the biological diversity of their new habitat. The Indian government duly recognizes the threats presented by these species and in accordance with article 8(h) of the CBD is committed to manage them.[16]

## 10.2   *The Indian subcontinent and plant invasion*

The Indian subcontinent, or South Asia, refers to the region now occupied by India, Pakistan, Bangladesh, Sri Lanka, Nepal, Bhutan, and the Island of Maldives. It is a peninsular region with the Himalayan Mountains to the north and Indian Ocean to the south. This region, being largely tropical, has rich floral and faunal diversity. India is the major and dominant country of this region and is among the megadiverse countries of the world. In addition, India is one of the eight Vavilovian centers of origin of crop plants and has a great repository of biological resources and traditional knowledge. In India, there are three major hot spots of biodiversity: the Himalayas, the Western Ghats (up to Sri Lanka), and the Indo-Burma region, all regions rich with endemic species diversity. In terms of plant diversity, India ranks tenth in the world and fourth in Asia and has 11% of the world's plant diversity.[16] Several features, such as its variable edaphoclimatic conditions, topography, and geologic makeup, account for its rich biodiversity. There is a great diversity of ecosystems and habitats, such as grasslands, forests, wetlands, and aquatic and marine habitats.

Indian flora shows great affinity with the Indo-Malayan and Indo-Chinese region, though there is considerable endemism too. There is no well-established reporting with regard to the exact nature of Indian flora. According to Nayar,[17] nearly 18% of Indian flora (both cultivated and naturalized) are aliens, of which 55% are of New World origin. Per a recent report, there are a total of 173 IAS that belong to 117 genera under 44 families.[18] The majority of these species are from tropical America (74%), followed by tropical Africa (11%).

Many of the introduced plants, such as crops, have proved very useful and contributed a great deal to India's economy.

However, several exotic species, whether introduced deliberately or accidently, are troublesome and pose a threat to native ecosystems. Raghubanshi et al.,[19] based on the deliberations of a workshop, concluded that plants like *Parthenium hysterophorus*, *Lantana camara*, *Eupatorium adenophorum*, *Eupatorium odoratum*, *Mikania micrantha*, and *Ageratum conyzoides* are very harmful invasive species in terrestrial ecosystems, whereas *Eichhornia crassipes*, *Salvinia molesta*, and *Ipomoea* species are nuisance species in aquatic ecosystems. A study conducted in the Shiwalik region of the northwestern Himalayas revealed that weeds like *Ageratum conyzoides*, *Parthenium hysterophorus*, and *Lantana camara* have established themselves very well and cause considerable harm to the region's precious native diversity.[20] Further studies have revealed that several biological and ecological features of these weeds, besides favorable habitat and climatic conditions, are responsible for their invasiveness.[21,22] In the Kashmir Himalayas, a geographically distinct South Asian region, Khuroo et al.[23] reported 571 alien species belonging to 352 genera and 104 families; most of these alien species originated in Europe. In the Doon Valley of the northwestern Himalayas, Negi and Hajra[24] reported 308 woody and 128 herbaceous exotic species, of which nearly 38% are from the New World alone. With a few exceptions, the majority of these are not considered safe for the native flora of the region, which is a part of the Himalayan biodiversity hot spot that requires priority conservation owing to its rich endemic flora.

## 10.3    Pathways of invasion in the context of the Indian subcontinent

There are definite pathways by which a species moves from one geographic area to another. Globalization has increased international trade and travel and has facilitated much faster movement of invasive species. Hulme et al.[25] have stated three broad ways by which alien species migrate: through importation of a commodity, arrival of a transport vector (both of which are facilitated by humans), or natural spread from a neighboring region. However, human-aided movement of plants has been frequent and efficient in their movement across geographic boundaries.[6] Alien species may also enter different geographic areas via ship ballasts, contaminated agricultural seeds, adhering to people or animals, intentional introductions by people, or via translocation of contaminated machinery and equipment.[26] All these pathways have contributed a lot to the successful establishment of invasive plants in the Indian subcontinent. In fact, this part of the world has been a very active site for trade and transport (being the second most populated country in the world) for centuries because of the richness of its biological resources. In the present era, both trade and transport have become extremely common and frequent because of recent rapid economic development. This has resulted in the introduction of many exotic species of plants and animals. Westphal et al.[27] have shown a direct correlation between international trade and the global distribution of invasive species. This seems to be true in the Indian context as well.

## 10.4    Some prominent invasive plants in the Indian subcontinent

The exact status of invasive alien plants in the Indian subcontinent is not certain. However, based on the available databases and some reports, over 55 invasive alien plants that have caused havoc in this region have been identified and listed in this chapter (Table 10.1). These invasive alien plants belong to all the plant forms, including herbs, shrubs, trees, grasses, vines and climbers, and aquatic plants. Most of the plants listed in Table 10.1 originate in tropical America and belong to the daisy family, that is, Asteraceae. Some plants (e.g., *Zephyranthes citrina* and *Arundo donax*) mentioned in the table are based on the personal observations of the authors. *Z. citrina* is spreading in grassy areas of Chandigarh, a modern city in North India. Likewise, *A. donax*, which is primarily being promoted for prevention of erosion in protected areas, may assume invasive proportions due to its fast growth and huge biomass. Some prominent invasive plants that spread over large areas and created a nuisance in a number of landscapes and habitats are *Ageratum conyzoides* and *Parthenium hysterophorus* (terrestrial weeds), *Lantana camara* and *Chromolaena odorata* (shrubs), *Prosopis juliflora* and *Leucaena leucocephala* (trees), *Mikania micrantha* (vines), *Imperata cylindrica* (grass), and *Salvinia molesta*, *Eichhornia crassipes,* and *Pistia stratiotes* (aquatic plants).

***Table 10.1*** List of Prominent Invasive Alien Plants in the Indian Subcontinent

| Botanical name and family | Common English name | Nativity | Life form |
|---|---|---|---|
| *Acacia farnesiana* (L.) Willd. Fabaceae | Sweet acacia | Tropical America | Tree/shrub |
| *Acacia mearnsii* De Wild. Fabaceae | Black wattle | Australia | Tree/shrub |
| *Acacia melanoxylon* R. Br. ex Ait. f. Fabaceae | Blackwood acacia | Australia | Tree |
| *Ageratum conyzoides* L. Asteraceae | Billy goat weed | Tropical America | Herb |
| *Alternanthera philoxeroides* (Mart.) Griseb. Amaranthaceae | Alligator weed | South America | Aquatic herb |
| *Ambrosia artemisiifolia* L. Asteraceae | Small ragweed | USA, Canada, and Mexico | Herb |
| *Anthemis cotula* L. Asteraceae | Stinking mayweed | Europe | Herb |
| *Arundo donax* L. Poaceae[a] | Giant cane | Indian Subcontinent | Perennial grass |
| *Asparagus densiflorus* (Kunth) Jessop Asparagaceae | Asparagus fern | South Africa | Herb |
| *Azolla pinnata* R. Br. Azollaceae | Mosquito fern | Not specific[b] | Aquatic plant |
| *Broussonetia papyrifera* (L.) Hert. ex Vent. Moraceae | Paper mulberry | China | Tree |
| *Cannabis sativa* L. Cannabaceae | Indian hemp | Central Asia | Herb |
| *Chenopodium album* L. Amaranthaceae | Lamb's quarters | Europe | Herb |
| *Chromolaena odorata* (L.) King and Robinson Asteraceae | Siam weed | Central and South America | Shrub |
| *Cirsium arvense* (L.) Scop. Asteraceae | Creeping thistle | Europe | Herb |
| *Clidemia hirta* (L.) D. Don Melastomataceae | Koster's curse | South America | Shrub |
| *Cabomba caroliniana* Gray Cabombaceae | Green cabomba | South America (Brazil) | Aquatic plant |

*Table 10.1* List of Prominent Invasive Alien Plants in the Indian Subcontinent (*Continued*)

| Botanical name and family | Common English name | Nativity | Life form |
|---|---|---|---|
| *Coffea arabica* L. and *C. canephora* L. Rubiaceae | Arabica coffee and Robusta coffee | Africa | Shrub |
| *Cryptostegia grandiflora* (Roxb. ex R. Br.) R. Br. Apocynaceae | Rubber-vine | Madagascar | Vine-climber |
| *Eicchornia crassipes* (Mart.) Solms Pontederiaceae | Water hyacinth | South America | Aquatic plant |
| *Elaeagnus umbellate* Thunb. Elaeagnaceae | Japanese Silverberry | China, Korea, and Japan | Tree/shrub |
| *Eugenia uniflora* L. Myrtaceae | Surinam cherry | South America | Tree/shrub |
| *Eupatorium adenophorum* (Spreng.) King and H. Rob. Asteraceae | Crofton weed | Central America | Shrub |
| *Eupatorium cannabinum* L. Asteraceae | Hemp-agrimony | British Isles | Herb (woody perennial) |
| *Gymnocoronis spilanthoides* (D. Don ex Hook. & Arn.) DC. Asteraceae | Senegal tea plant | South America | Aquatic plant |
| *Hydrilla verticillata* (L. f.) Royle Hydrocharitaceae | Water thyme | Asia and North Australia | Aquatic plant |
| *Imperata cylindrica* (L.) Beauv. Poaceae | Cogon grass | Asia/Africa doubtful | Grass |
| *Ipomoea aquatica* Forsk. Convolvulaceae | Water spinach | China | Vine-climber |
| *Lantana camara* L. Verbenaceae | Wild sage | Tropical America | Shrub |
| *Leucaena leucocephala* (Lam.) De wit Fabaceae | Wild tamarind | Tropical America | Tree |
| *Leucanthemum vulgare* Lam. Asteraceae | Oxeye daisy | Europe | Herb |
| *Limnocharis flava* (L.) Buchenau Limnocharitaceae | Yellow velvet leaf | South America | Aquatic plant |
| *Ludwigia peruviana* (L.) Hara Onagraceae | Peruvian primrose-willow | South America | Aquatic plant |
| *Macfadyena unguis-cati* (L.) A. H. Gentry Bignoniaceae | Cat's claw vine | Central America | Vine-climber |
| *Merremia peltata* (L.) Merr. Convolvulaceae | Merremia | Africa | Vine-climber |
| *Miconia calvescens* D.C. Melastomataceae | Velvet tree | Tropical America | Tree |
| *Mikania micrantha* (L.) Kunth. Asteraceae | Mile-a-minute weed | Central or South America | Vine-climber |
| *Mimosa diplotricha* C. Wright ex Sauvalle Fabaceae | Giant sensitive plant | South America | Vine-climber/shrub |

(*Continued*)

**Table 10.1** List of Prominent Invasive Alien Plants in the Indian Subcontinent (*Continued*)

| Botanical name and family | Common English name | Nativity | Life form |
|---|---|---|---|
| *Mimosa pudica* L. Fabaceae | Touch-me-not | South America | Herb |
| *Parthenium hysterophorus* L. Asteraceae | Ragweed parthenium | Tropical America | Herb |
| *Paspalum vaginatum* Sw. Poaceae | Seashore paspalum | North America | Grass |
| *Pennisetum clandestinum* Hochst. ex Chiov. Poaceae | Kikuyu grass | Tropical Eastern Africa | Grass |
| *Phalaris arundinacea* L. Poaceae | Reed canary grass | Europe | Grass |
| *Physalis peruviana* L. Solanaceae | Cape gooseberry | South America | Shrub |
| *Pistia stratiotes* L. Araceae | Tropical duckweed | South America | Aquatic plant |
| *Prosopis juliflora* (Sw.) DC. Fabaceae | Mesquite | Central and South America | Tree |
| *Psidium guajava* L. Myrtaceae | Apple guava | Central and South America | Tree |
| *Ricinus communis* L. Euphorbiaceae | Castor bean | Northeastern Africa | Tree/shrub |
| *Salvinia molesta* D. S. Mitchell Salviniaceae | Water fern | South America | Aquatic plant |
| *Sapium sebiferum* (L.) Roxb. Euphorbiaceae | Chinese tallow | China | Tree |
| *Solanum mauritianum* Scop. Solanaceae | Wild tobacco tree | South America | Tree/shrub |
| *Solanum sisymbriifolium* Lam. Solanaceae | Sticky nightshade | South America | Herb |
| *Solanum viarum* Dunal Solanaceae | Tropical soda apple | South America | Shrub |
| *Spartina alterniflora* Loisel. Poaceae | Smooth cordgrass | South America | Grass |
| *Sphagneticola trilobata* (Rich.) Pruski Asteraceae | Singapore daisy | Central America | Herb |
| *Synedrella vialis* (Less.) A. Gray Asteraceae | Straggler daisy | South America | Herb |
| *Tagetes minuta* L. Asteraceae | Mexican marigold | South America | Herb |
| *Ulex europaeus* L. Fabaceae | Gorse | Europe | Tree/shrub |
| *Zephyranthes citrina* Baker Amaryllidaceae | Yellow rain lily | Central and South America | Herb |

*Sources:* Reddy, C. S., *Life Sci J* 5:8, 2008; Raghubanshi, A. S., L. C. Rai, J. P. Gaur, and J. S. Singh, *Curr Sci* 88:539, 2005; Kohli, R. K., S. Jose, H. P. Singh, and D. R. Batish, *Invasive Plants and Forest Ecosystems*, Boca Raton, FL: CRC Press, 2009; Holzmueller, E. J., and S. Jose, *J Trop Agric* 47:18, 2009; GISD, Global Invasive Species Database, 2010. http://www.issg.org/database (accessed June 3, 2010).

[a] A native of the Indian subcontinent but may be a potential invader due to its fast growth rate, huge biomass, and large-scale promotion for various purposes like erosion control along semiaquatic and aquatic areas.

[b] Africa and Madagascar, India, Southeast Asia, China and Japan, Malaysia and the Philippines, and the New Guinea mainland and Australia as per GISD, 2010.

*Parthenium hysterophorus* (Asteraceae) is one of the most invasive noxious weeds of tropical America that has spread to various tropical and subtropical parts of the world, including India, and been listed by the Global Invasive Species Database of IUCN.[28] The weed seems to have entered India via contaminated cereal grains (PL-480) imported from the United States. It colonizes wastelands, agricultural areas, grasslands, urban areas, forest, and plantation areas and makes its own huge monoculture stands and has spread to almost every state in India except to areas at very high altitudes. *Parthenium hysterophorus* (Asteraceae) is rated as one of the worst weeds of the subcontinent because of the huge land area covered by it, nearly 2,025,000 ha of land.[29,30] The major hazards of *P. hysterophorus* include the following:

- Reduces native floral diversity
- Causes fodder famine
- Is toxic to livestock
- Causes health problems like allergies, asthma, rhinitis, and so on to humans

*P. hysterophorus* is also called a biological pollutant because of its ill effects on human health and livestock. Various survival strategies of this weed include rapid growth (both vegetative and reproductive), adaptability to diverse environmental conditions, quick regeneration upon removal of plant parts, and an unpalatable nature because of toxic principles and its allelopathic nature.[21] The weed produces an enormous number of seeds that are disseminated by wind and water, but human-driven dissemination of these seeds cannot be ruled out.[30]

Another terrestrial weed that has spread to most plains and hilly regions is *Ageratum conyzoides* (Asteraceae) commonly known as billy goat weed. It was introduced in India in 1860, probably as an ornamental plant.[31] *A. conyzoides* has spread over a huge area in the tropical and subtropical regions of the world. In India, it is one of the worst invaders of agricultural land, but is also found in grasslands, rangelands, and forests, especially along margins and water channels. It shares features with *P. hysterophorus* like fast growth rate, high reproductive ability, and production of a large number of small seeds, besides being allelopathic, that provide invasive character.[21] *A. conyzoides* also reproduces vegetatively by stolons, which allows the weed to spread quickly.

*Lantana camara* L. (Verbenaceae) is one of the worst invasive species identified by the Global Invasive Species Database and is rated among the top 100 invasive species of the world.[28,32] *L. camara* was probably introduced to India in the beginning of the nineteenth century as an ornamental plant, but later assumed invasive proportions in different ecosystems. It is now widespread on the Indian subcontinent and is also rapidly spreading into high altitudes, particularly the Himalayas (>1700 m).[21] *Lantana camara* is a serious invader of forests, grasslands, agricultural land, and any vacant areas in urban areas and is also the worst invader of protected areas.[33,34] The invasion of *L. camara* causes serious implications on the structure and dynamics of native forests and understory vegetation, especially in the dry deciduous forests and the lower Shiwalik hills of the northwestern Himalayas.[20,35] Seed recruitment of native species rapidly declines with the invasion of *Lantana*. *L. camara* is also strongly allelopathic and interferes with the growth and development of a wide range of plants due to the presence of a number of volatile and nonvolatile allelochemicals.[21,36]

Another weed from the Asteraceae family is *Chromolaena odorata* (L.) King and Robinson (= *Eupatorium odoratum* L.), a globally important invasive plant holding a place in the list of top 100 invasive species by the Global Invasive Species Database.[28] It was also introduced to India as an ornamental plant[31] in 1840. In India, *C. odorata* is a most obnoxious weed, especially in the northeastern region and the Western Ghats and interferes with several useful plantation crops such as coconut, rubber, coffee, teak, and so on.[37] Its spread to northwestern parts of India is a cause for concern.[38]

Another invasive weed that is a cause of concern is *Mikania micrantha* (Asteraceae), commonly known as mile-a-minute, a creeping vine that rapidly encroaches upon the native vegetation in invaded areas. *M. micrantha* is of Neotropical origin and was introduced in India intentionally for camouflaging airfields; after the Second World War, it spread rapidly without any natural enemies.[39] It is now a very noxious weed in plantations and forests and is difficult to eradicate.[38]

Among the aquatic weeds, *Eichhornia crassipes*, *Salvinia molesta*, and *Ipomoea* sp. have done much harm to freshwater ecosystems and wetlands. *Alternanthera philoxeroides* (Mart.) Griseb. (Amaranthaceae), a native of temperate regions of South America, is another serious invasive weed of aquatic ecosystems including wetlands and is known to cause various environmental, economic, and health problems.[18]

Among tree species, *Leucaena leucocephala* (Lam.) de Wit of the family Mimosaceae is one of the worst invaders listed.[28] *L. leucocephala* was introduced to various parts of the world including India as an agroforestry tree. It is a fast-growing tree that produces enormous seeds, which germinate quickly and form monoculture stands.

*Prosopis juliflora* (Sw.) DC of the family Mimosaceae, another tree with potential to invade forest areas and grasslands, was introduced to India under afforestation programs. *P. juliflora* is a prolific seed producer and forms dense monocultures, which hamper the growth of native vegetation. *P. juliflora* also figures in the list of invasive plants as given by the Invasive Species Specialist Group (ISSG) of IUCN.[28]

## 10.5   *Alien plants in the process of establishment*

A number of plants have been purposefully introduced in the Indian subcontinent for some utility such as ornamental value, essential oils, timber, or medicine. One among such is *Tagetes minuta* L. (Asteraceae), a native of South America introduced to India and various other parts of the world for its essential oil and medicinal value. However, in several locations, *T. minuta* has become a noxious invasive weed, also known to possess toxic properties.[40,41] In India, it seems to have escaped cultivation and can be seen growing luxuriantly in the disturbed sites of the northwestern Himalayas at lower altitudes.[22]

*Anthemis cotula* L. (Asteraceae), an ornamental plant native to Europe, is quickly spreading in the Kashmir Valley, a part of the Himalayan biodiversity hot spot. Shah and Reshi[42] have pointed out that the invasive nature of this weed may be attributed to the absence of natural enemies and a favorable environment, such as friendly mycorrhizal associations, in the invaded areas.

*Leucanthemum vulgare* Lam., another ornamental plant from Europe belonging to the family Asteraceae, is fast assuming invasive proportions in the Kashmir Valley and has the potential to inhibit regeneration of native forest floor vegetation.[43] The investigators stressed that effective steps should be taken to control its further spread (as it is restricted to tourist locales so far) before it becomes difficult to control.

Among tree species, *Sapium sebiferum* Roxb. (Euphorbiaceae), a native of China, is reported to be spreading fast in the Indian Himalayan region.[44] Though listed as one of the worst invaders by the Global Invasive Species Database, as yet there is no immediate risk from this plant in the Indian region. However, risk assessment should be undertaken.

*Broussonetia papyrifera*, another woody species belonging to the family Moraceae, commonly known as paper mulberry, exhibits invasive tendencies. *B. papyrifera* is a shrubby deciduous tree of Chinese origin (native of China) and has been introduced in various other parts of the world for paper and as a shade or ornamental tree, particularly in Europe and America. *B. papyrifera* has an invasive character and is spreading at a fast rate in northern India.

## 10.6 Environmental and economic harm of invasive aliens in the Indian Subcontinent

### 10.6.1 Environmental harm

Invasive plants do enormous environmental harm. These plants grow fast and rapidly spread through the invaded area owing to their very high density in terrestrial and aquatic ecosystems. These plants are known to alter ecosystem community structure and adversely affect functions such as nutrient cycling and fire regimes in native ecosystems.[45] A number of invasive alien plants cause health problems to humans and livestock (e.g., *Parthenium hysterophorus*).[30] Though many invasive plant awareness programs are being held in India on the international concerns raised by IUCN, CBD, and other organizations, there are only a few reports in peer-reviewed journals showing scientific data on the impact of invasive plants in India.

In general, the environmental harm caused by invasive plants can be summarized as follows:

1. Direct interference with agricultural crops in the agroecosystems leading to direct monetary loss due to poor production and management costs
2. Loss of native species that may be useful directly or otherwise threatened or endemic
3. Change in the structure and composition of native plant communities (grasslands, urban areas, aquatic systems, forests, etc.)
4. Overall deterioration of ecosystems in both structure and function
5. Health problems to humans and livestock such as allergies, asthma, bronchitis, poisonings, etc.
6. Scarcity of native fodder and commonly found native medicinal plants (India relies greatly on the indigenous medicine system and traditional knowledge)
7. Deterioration of water quality and clogging of waterways affecting both humans and livestock
8. Reduction in recreational value of the ecosystem thus affecting tourism

Some specific examples of the environmental impact of invasive plants are described below.

In the Shiwalik region of the northwestern Himalayas, studies by Kohli et al. indicate that there are three major invasive weeds: *Lantana camara*, *Parthenium hysterophorus*, and *Ageratum conyzoides* in the terrestrial areas, which cause enormous harm to the native vegetation.[20,21] These weeds occupy a variety of habitats ranging from wastelands to agricultural areas and interfere with native plants and crops. These weeds reduce the density and biomass of native species and the diversity and richness of native flora at various altitudes.[20] All three of these possess a combination of biological traits and characteristics that enhance their capacity to invade native communities and aid their spread over large areas.[21] Besides the biological traits of the invasive plant, the region itself is favorable to alien invasive plants owing to greater disturbances posed by human settlement and live-stock grazing pressure.[22] As a result, there are a few other plants that are fast making their impact realized, such as *Tagetes minuta*, *Broussonetia papyrifera*, and *Chromolaena odorata*.

In central India, dry deciduous forests are prominent. These forests are under constant biotic stress and are heavily infested with *L. camara*. As a result of *L. camara* infestation, a significant alteration in the structure and floristic composition has been reported.[46] Further, in the southern part of India, there are serious environmental problems imposed by invasive plants. In this region of the Indian subcontinent, the Western Ghats extending to Sri Lanka constitutes one of the 34 hot spots of biodiversity with a high degree of endemism and species diversity.[47] Gunawardene et al.[48] have pointed out that besides other factors threatening biodiversity, exotic invasive species also pose a serious threat to the biodiversity of this important hot spot. Muniappan and Viraktamath[38] reported that the Western Ghats are very much prone to invasive alien plants such as *Lantana camara*, *C. odorata*, *Ageratina adenophora*, *Mikania micrantha*, *Mimosa invisa*, and *Prosopis juliflora*. Of late, commercial coffee plantations have exhibited invasive tendencies as they are encroaching upon the adjoining rain forests. These invasive plants have wreaked havoc with the rich native flora of the region.

A new plant introduction, Jatropha, by the Indian government is a major threat to agriculture and livestock. The Indian government plans on producing 20% of the country's diesel fuel, mostly from Jatropha by 2017.[49] This would require 30–40 million ha of land to grow the shrubs. The plant is highly toxic to humans and livestock. Another major problem is that jatropha does not produce the oil that is claimed. A worker picking jatropha beans for 8 hours can only collect US$2.86 worth of beans (D. Pimentel, pers. comm.).

## 10.6.2   Economic harm

Problems caused by invasive plants are not restricted to ecology but also have economic consequences.[50] In general, the economic costs of invasive species include both losses in agricultural production and costs of managing these invasive species, beyond their damaging impact on the environment. With a knowledge of economics, both market and nonmarket impacts of invasive plants can be determined. According to Pimentel et al.[9] the total economic cost due to alien species in India amounts to US$91 billion per year, and out of this, weeds alone account for US$38.7 billion. Among the various alien invasive weeds, the loss due to *Lantana camara* has been estimated to be US$924 million per year.[9] Pimentel et al.[7] opined that the actual economic cost due to IAS is much higher if we take into account their impacts on the environment, including habitat damage, loss of rare and endangered species, extinctions, and/or ecosystem or environmental services. Olson[51] has given a comparative account on the economics of invasive species in different countries but none from Asia. There are hardly any studies that estimate the economic costs of invasive plants in India, though there may be some scanty reports or assumptions. However, economic costs caused by invasive plants are important and require urgent attention.

## 10.7   Conclusions and the way forward

Clearly, several invasive species have successfully established themselves in the Indian subcontinent, whereas still others are in the process of establishing themselves. Some efforts are being undertaken by the Indian government to control these species as they threaten biodiversity. These efforts include regulatory checks during import of germplasm and following international quarantine regulations. At the local level, community participation in these efforts could be of great help in managing these invasive plants.[52] Some of these efforts could be:

1. Preparing a national database on invasive plants.
2. Raising awareness about the potential invasive plants from other parts of the world.
3. Assessing the environmental and economic impacts of the introduced plants. This could be a very helpful tool in managing the initial stages of introduction and establishment of invasive species.
4. Determining the pathways of invasion. If information regarding possible pathways by which invasive species enter into the alien environment are known, appropriate steps can be taken to prevent their entry and spread.
5. A collaborative approach integrating various government and nongovernment agencies and community participation for detecting harmful invasive species to developing an early warning system for their management.
6. Following international standards and quarantine measures based on the International Plant Protection Convention (IPPC) or the Convention on International Trade on Endangered Species of Flora and Fauna (CITES) to prevent the entry of potential plant invaders.

## *References*

1. di Castri, F. 1989. History of biological invasion. In *Biological Invasions: A Global Perspective*, ed. J. A. Drake, H. A. Mooney, F. di Castri, R. H. Groves, F. J. Kruger, M. Rejmanek, and M. Williamson, 1. Chichester: John Wiley & Sons.
2. Drake, J. A., H. A. Mooney, F. di-Castri, R. H. Groves, F. J. Kruger, M. Rejmanek, and M. Williamson. 1989. *Biological Invasions: A Global Perspective*. Chichester: John Wiley & Sons.
3. Crawley, M. J. 1997. Biodiversity. In *Plant Ecology*, ed. M. J. Crawley, 595. Oxford: Blackwell Scientific Publications.
4. Myers, J. H., and D. R. Bazely. 2003. *Ecology and Control of Introduced Plants*. Cambridge: Cambridge University Press.
5. Vitousek, P. M., C. M. D'Antonio, L. L. Loope, and R. Westbrooks. 1996. Biological invasions as global environmental change. *Am Sci* 84:218.
6. Mack, R. N., D. Simberloff, W. M. Lonsdale, H. Evans, M. Clout, and F. A. Bazzaz. 2000. Biotic invasions: Causes, epidemiology, global consequences and control. *Ecol Appl* 10:689.
7. Pimentel, D., R. Zuniga, and D. Morrison. 2005. Update on the environmental and economic costs associated with alien invasive species in the United States. *Ecol Econ* 52:273.
8. Herron, P. M., C. T. Martine, A. M. Latimer, and S. A. Leicht-Young. 2007. Invasive plants and their ecological strategies: Prediction and explanation of woody plant invasion in New England. *Divers Distrib* 13:633.
9. Pimentel, D., S. McNair, J. Janecka, J. Wightman, C. Simmonds, C. O'Connell, E. Wong et al. 2001. Economic and environmental threats of alien plant animal and microbe invasions. *Agric Ecosyst Environ* 84:1.
10. Simberloff, D. 2003. Confronting introduced species: A form of xenophobia? *Biol Invasions* 5:179.
11. Rejmanek, M. 2000. Invasive plants: Approaches and predictions. *Austral Ecol* 25:497.
12. Pysek, P., and K. Prach. 2003. Research into plant invasions in a crossroads region: History and focus. *Biol Invasions* 5:337.
13. Sakai, A. K., F. W. Allendorf, J. S. Holt, D. M. Lodge, J. Molofsky, K. A. With, S. Baughman et al. 2001. The population biology of invasive species. *Annu Rev Ecol Syst* 32:305.
14. Richardson, D. M., P. Pysek, M. Rejmánek, M. G. Barbour, F. D. Panetta, and C. J. West. 2000. Naturalization and invasion of alien plants: Concepts and definitions. *Divers Distrib* 6:93.

15. Colautti, R. I., and H. J. MacIsaac. 2004. A neutral terminology to define invasive species. Divers Distrib 10:135.

16. Ministry of Environment and Forests. 2009. India's fourth national report to the convention on biological diversity. Ministry of Environment and Forests, Government of India, New Delhi.

17. Nayar, M. P. 1977. Changing patterns of the Indian flora. *Bull Bot Surv India* 19:145.

18. Reddy, C. S. 2008. Catalogue of invasive alien flora of India. *Life Sci J* 5:8.

19. Raghubanshi, A. S., L. C. Rai, J. P. Gaur, and J. S. Singh. 2005. Invasive alien species and biodiversity in India. *Curr Sci* 88:539.

20. Kohli, R. K., K. S. Dogra, D. R. Batish, and H. P. Singh. 2004. Impact of invasive plants on the structure and composition of natural vegetation of northwestern Indian Himalayas. *Weed Technol* 18:1296.

21. Kohli, R. K., D. R. Batish, H. P. Singh, and K. S. Dogra. 2006. Status, invasiveness and environmental threats of three tropical American invasive weeds (*Parthenium hysterophorus* L., *Ageratum conyzoides* L., *Lantana camara* L.) in India. *Biol Invasions* 8:1501.

22. Kohli, R. K., S. Jose, H. P. Singh, and D. R. Batish. 2009. *Invasive Plants and Forest Ecosystems*. Boca Raton, FL: CRC Press.

23. Khuroo, A. A., I. Rashid, Z. Reshi, G. H. Dar, and B. A. Wafai. 2007. The alien flora of Kashmir Himalaya. *Biol Invasions* 9:269.

24. Negi, P. S., and P. K. Hajra. 2007. Alien flora of Doon Valley, Northwest Himalaya. *Curr Sci* 92:968.

25. Hulme, P. E., S. Bacher, M. Kenis, S. Klotz, I. Kühn, D. Minchin, W. Nentwig et al. 2008. Grasping at the routes of biological invasions: A framework for integrating pathways into policy. *J Appl Ecol* 45:403.

26. Holzmueller, E. J., and S. Jose. 2009. Invasive plant conundrum: What makes the aliens so successful? *J Trop Agric* 47:18.

27. Westphal, M. I., M. Browne, K. MacKinnon, and I. Ian Noble. 2008. The link between international trade and the global distribution of invasive alien species. *Biol Invasion* 10:391–8.

28. GISD. 2010. Global Invasive Species Database. http://www.issg.org/database (accessed June 3, 2010).

29. Aneja, K. R., S. R. Dhawan, and A. B. Sharma. 1991. Deadly weed *Parthenium hysterophorus* L. and its distribution. *Indian J Weed Sci* 23:14.

30. Kohli, R. K., and Rani, D. 1994. *Parthenium hysterophorus*—a review. *Res Bull (Sci) Panjab Univ* 44:105.

31. National Focal Point for APFISN, India. 2005. Stocktaking of national forest invasive species activities, India (India Country Report 101005). Ministry of Environment and Forests, New Delhi.

32. Lowe, S., M. Browne, S. Boudjelas, and M. De Poorter. 2004. 100 of the world's worst invasive alien species, a selection from the Global Invasive Species Database, 12. Auckland, New Zealand: Invasive Species Specialist Group (ISSG)—a specialist group of the Species Survival Commission (SSC) of the World Conservation Union (IUCN).

33. Sahu, P. K., and J. S. Singh. 2008. Structural attributes of lantana-invaded forest plots in Achanakmar-Amarkantak Biosphere Reserve, Central India. *Curr Sci* 94:494.

34. Love, A., S. Babu, and C. R. Babu. 2009. Management of *Lantana*, an invasive alien weed, in forest ecosystems of India. *Curr Sci* 97:1421.

35. Sharma, G. P., and A. S. Raghubanshi. 2007. Effect of *Lantana camara* L. cover on local depletion of tree population in the Vindhyan tropical dry deciduous forest in India. *Appl Ecol Environ Res* 5:109.

36. Ambika, S. R., S. Poornima, R. Palaniraj, S. C. Sati, and S. S. Narwal. 2003. Allelopathic plants. 10. *Lantana camara* L. *Allelopathy J* 12:147.

37. Singh, S. P. 1998. A review of biological suppression of *Chromolaena odorata* K & R in India. In *Proceedings of the 4th International Workshop on Biological Control and Management of Chromolaena odorata*, ed. P. Ferrar, R. Muniaapan, and K. P. Jayanth, 86. Guam: University of Guam.

38. Muniappan, R., and C. A. Viraktamath. 1993. Invasive alien weeds in the Western Ghats. *Curr Sci* 64:555.

39. Randerson, J. 2003. Fungus in your tea, sir? *New Sci* 178:10.

40. Soule, J. A. 1993. *Tagetes minuta*: A potential new herb from South America. In *New Crops*, ed. J. Janick and J. E. Simon, 649. New York: Wiley.
41. Holm, L., J. Doll, E. Holm, J. Pancho, and J. Herberger. 1997. *World Weeds. Natural Histories and Distribution*. New York: John Wiley & Sons.
42. Shah, M. A., and R. Zafar. 2007. Invasion by alien *Anthemis cotula* L. in a biodiversity hotspot: Release from native foes or relief from alien friends? *Curr Sci* 92:21.
43. Khuroo, A. A., A. H. Malik, Z. A. Reshi, and G. H. Dar. 2010. From ornamental to detrimental: Plant invasion of *Leucanthemum vulgare* Lam. (ox-eye daisy) in Kashmir valley, India. *Curr Sci* 98:600.
44. Jaryan, V., S. Chopra, S. K. Uniyal, and R. D. Singh. 2007. Spreading fast yet unnoticed: Are we in for another invasion? *Curr Sci* 93:1483.
45. Levine, J. M., M. Vilà, C. M. D'Antonio, J. S. Dukes, K. Grigulis, and K. Lavorel. 2003. Mechanisms underlying the impacts of exotic plant invasions. *Proc R Soc Lond B* 270:775.
46. Raghubanshi, A. S., and A. Tripathi. 2009. Effect of disturbance, habitat fragmentation and alien invasive plants on floral diversity in dry tropical forests of Vindhyan highland: A review. *Trop Ecol* 50:57.
47. McGinley, M., topic ed. 2008. South Western Ghats moist deciduous forests. In *Encyclopedia of Earth*, ed. C. J. Cleveland. Washington, DC: Environmental Information Coalition, National Council for Science and the Environment. http://www.eoearth.org/article/South_Western_Ghats_moist_deciduous_forests (accessed June 3, 2010).
48. Gunawardene, N. R., J. D. Majer, and J. P. Edirisinghe. 2010. Investigating residual effects of selective logging on ant species assemblages in Sinharaja Forest Reserve, Sri Lanka. *For Ecol Manage* 259:555.
49. Friends of the Earth (FoE)—Europe. 2009. Losing the plot: The threats to community land and the rural poor through the spread of the biofuel jatropha in India. http://www.foeeurope.org/agrofuels/jatropha_in_india.pdf (accessed September 1, 2010).
50. Evans, E. A. 2003. Economic dimensions of invasive species. *Choices: Mag Food Farm Resour Issues* 18:5.
51. Olson, L. J. 2006. The economics of terrestrial invasive species: A review of the literature. *Agric Resour Econ Rev* 35:178.
52. Batish, D. R., H. P. Singh, R. K. Kohli, V. Johar, and S. Yadav. 2004. Management of invasive exotic weeds requires community participation. *Weed Technol* 18:1445.

*chapter eleven*

# Invasive invertebrates in India
## Economic implications

*T. N. Ananthakrishnan*

### Contents

## 11.1  Introduction

Many human activities, such as agriculture, horticulture, forestry, and transportation, promote intentional or accidental spread of species, resulting in devastating economic impacts, often threatening native biodiversity and ecosystem function.[1] Such species increase their probability of getting established by being in habitats that provide food and climate that the species are adapted to. Invasive species, as these are designated, upon entering a new environment or habitat reproduce until their numbers predominate, subsequently causing substantial damage and resulting more often in epidemics, ecological disaster, and the extinction of native species in many cases.[1] In recent years, efforts to understand the ecology of such biological invasions have increased substantially as a result of international trade, which brings about the movement of seeds and plant materials, including exotic species. To be described as invasive, a species has to pass through the introductory stage, establishment of viable populations, and cause economic and environmental harm. The rate at which invasive insects colonize depends on the capacity of the insects to disperse by their own ability or through human activities. The international introduction of many species, such as those involved in biological control, is a practice in vogue today. Two of the most famous examples of such introductions into India from Australia are the beetle *Rodolia cardinalis* and the fly *Cryptochaetum iceryae*, as well as the cotton cushion scale *Dactylopius tomentosus* to control *Opuntia*, the prickly pear. These introductions have laid the foundation for the promotion of interaction of invasive species.[2] Successful introduction is possible only if the invasive invertebrates can survive the extremes of temperature, although other environmental factors such as relative humidity and availability of nutrients also affect their survival. Nevertheless, invasions of exotic insects and other invertebrates have a great impact on agricultural crops. In this era of international trade, travel and exchange of biological material for commerce and research purposes increasingly take place; with this increased travel and commerce, accidental introductions of alien species into new agricultural environments have become more frequent.[3] Invasions by insect species could also be facilitated by phenotypic plasticity for resistance or competitive ability. By adjusting its phenotype to the local environment, a plastic genotype might maintain fitness across a wide range of environments.[4] Climate is of key importance in limiting the distribution of invasive species, although other abiotic factors also tend to influence survival.

## 11.2    Ecodynamics of invasive species

Establishment of invasive species occurs when newly arrived species maintain a population aided by suitable climate, nutrients, and reproductive availability. Invasive species are often considered among the top two or three forces driving other species to extinction.[5] They can have grave effects on human health, devastating economic impacts, and can threaten native biodiversity and ecosystem functions.[6] Among the earliest in this area of research is Elton,[6] whose book *The Ecology of Invasions of Animals and Plants* became an eye-opener and a source of interest and inspiration in this field of research. Invasive species pass through four phases: arrival, establishment, integration, and spread. When a new invasive species maintains and increases its population through reproduction, it tends to become established by active or passive means given the proper climate and nutrients.[7,8] Invasive species function in a wide range of conditions with a high capacity for population increase and a strong capacity for dispersal.[9]

## 11.3    Economic implications

In the absence of natural biocontrol agents, invasive species increase to high and economically damaging levels in their new environment.[6] With increasing intensity of traffic along international trade routes, there has been an ever-growing invasion of alien species in environmentally stable areas. A species with genetic variability, which allows it to withstand environmental fluctuations, adapting to wider habitats, is a good invader.[10] Habitats susceptible to more invasions are as follows: (1) habitats that provide food and climate similar to those to which invasive species are adapted, (2) disturbed habitats, (3) habitats that provide less biotic resistance, and (4) islands, which have fewer species.[6] Massive seasonal invasions of huge aggregations of the litter dwelling detritivorous beetle *Luprops tristis* (Tenebrionidae; Figure 11.1) numbering to 0.5–4 million per residential building following summer showers is a good example.[11] Their staggering abundance after the introduction of rubber plantations in the moist western slopes of the Western Ghats in the 1970s is phenomenal. Man-mediated intracontinental movement of this species has resulted in its establishment as a serious nuisance pest. Invasion of *Luprops tristis* to the northeastern region of India is not far off in view of the introduction of rubber plantations there. Predicting the time and appearance of various developmental stages of *L. tristis* is possible by tracking the phenology of leaf fall in rubber plantations.[11]

Eucalyptus plantations and nurseries in several localities in Tamil Nadu, Karnataka, and Andhra Pradesh have been invaded in recent years by *Leptocybe invasa*, a tiny hymenoptera wasp from Australia, inducing galls on shoot terminals, on petioles and midribs, and on saplings and trees.[12] Successful establishment has occurred on clones of *Eucalyptus camaldulensis* (Figure 11.2); being a long-rotation crop, impact of galls on productivity is considerable.[12] Because of the importance of trees as key species in forest ecosystems, the impact of invasive species on trees is more obvious. Extensive plantations using such clones of *E. camaldulensis* facilitate spread of *Leptocybe* in states like Tamil Nadu and

*Luprops tristis*

***Figure 11.1*** The rubber beetle *Luprops tristis.* (Courtesy of Sabu Thomas.)

***Figure 11.2*** Eucalyptus gall induced by *Leptocybe invasa.* (Courtesy of J. Prasanth Jacob.)

Andhra Pradesh, with continuous enlargement of the infected area. Variations of incidence of galls and intensity of injury and damage exist among these clones.[12] Accidental transportation of infested seedlings into uninfected areas for plantation purposes increased the likelihood of spread of *Leptocybe* to new areas. The coconut eriophyid mite *Aceria guerreronis*, introduced in the 1990s, is still causing serious damage to coconut, not to mention the threat of invasion by new pests like the *Brontispa* beetle.[13]

Recently, the cotton mealy bug *Phenacoccus solenopsis* has caused devastating damage to cotton in nine Indian states[14] and also damage to agricultural and horticultural crops.[15] The most recent introduction is the papaya mealy bug *Paracoccus marginatus*, having a wide host range including some economically important crop plants. *Paracoccus marginatus*

attacks papaya as well as guava, cotton, mango, tomato, tapioca, and other fruits and vegetables. In papaya, the level of infestation is very high, affected trees tending to drop fruits.[16]
In 2008, this species was noticed in India and Indonesia (Figure 11.3). An outbreak of the
wooly aphid of sugarcane *Ceratovacuna lanigera*, a pest believed to have moved from the
northeastern state to Maharashtra, Karnataka, and Tamil Nadu, has in recent years caused
considerable damage to sugarcane.[13]

The coral tree *Erythrina* sp. is planted as an ornamental and avenue tree in India,
besides acting as a useful stand for the trailing black pepper vines and as a shade tree in
coffee plantations. The *Erythrina* gall wasp *Quadrastichus erythrinae*, damaging this plant
in Taiwan, was recorded in India in 2004 in the southern districts of Kerala, subsequently
spreading to Maharashtra (Figure 11.4). The larvae, which develop from eggs laid on ten-
der leaves and stems, induce the formation of thick-walled, globular galls in leaves and
shoots, and severe infestations cause defoliation, shrinking, and mortality of the tree. This
ultimately affects the survival and growth of pepper vines.[17]

An example of an invasive alien species in rice ecosystems is the golden apple snail,
*Pomacea canaliculata*, a native of Brazil (Figure 11.5). This snail was deliberately introduced from South Africa into China, Taiwan, and other Asian countries as a source of
higher-protein food. Besides damaging the rice crop, these snails also pose a threat to
human health by hosting the nematode *Angiostrongylus cantonensis* that causes eosino-
philic meningoencephalitis in humans.[18] The rice water weevil *Lissorhoptrus oryzophilus*,
a native of the eastern United States and Cuba, has also become a major problem.[19] Many
other minor pests of rice like the rice black bug *Scotinophora coarctata* and the panicle mite
*Steneotarsonemus spinki* are also becoming established in India.[18]

***Figure 11.3*** The papaya mealy bug, *Paracoccus marginatus*. (Courtesy of V. V. Ramamurthy.)

***Figure 11.4*** Galls on Erythrina induced by *Quadrastichus erythrinae*. (Courtesy of J. Prasanth Jacob.)

***Figure 11.5*** Golden apple snail, *Pomacea canaliculata*. (Courtesy of J. S. Bentur.)

A number of important agricultural pests spread at varying rates, as exemplified by *Sogatella furcifera* and *Nilaparvata lugens*, which migrate long distances to occupy virtually all suitable habitats with favorable conditions.[20] Both ecological and intrinsic factors influence the distribution of crop pests. Although climate, soil, host plants, vectors, natural enemies, and competition are the main ecological factors, genetic adaptability and reproductive diversity are intrinsic to success as an invasive species. Warmth-loving species such as *Frankliniella*, *Liriomyza*, and *Bemisia* are unable to complete their life cycles outside protected environments.[21] Knowledge of both phenotypic and genetic variation of species becomes useful because those with genetic variability withstand environmental variation and adapt to wider habitats, making them good invaders.[10]

In recent years, the threats posed by several invasive species like *Heteropsylla cubana*, *Liriomyza trifolii*, coffee berry borer *Hypothenemus hampei*, and the silver whitefly *Bemisia tabaci* (Figure 11.6), which have spread through transport of plant products, are well known.[22-27] The spiraling whitefly *Aleurodicus dispersus* and the silver whitefly *Bemisia tabaci* B have led to increased interest in invasive species, not to mention several thrips species; *Bemisia tabaci* B is known to be one of the worst invaders, infecting more than 900 host plants and transmitting more than 10 viruses, and has become invasive through the transport of plant products.[28] First noticed in the Kolar district of Karnataka, it subsequently spread all over India. Similarly, *Aleurodicus dispersus* (Figure 11.7) was introduced into India in 1994. The stalked eggs are laid in an irregular spiral and adult populations build up on avenue trees. Copious white, waxy, flocculent material secreted by nymphs readily spreads through wind, and the sticky honeydew serves as a substrate for dense growth of sooty mold, interfering with photosynthesis.[28]

The serpentine leaf miner *Liriomyza trifolii*, an agromyzid, entered India accidentally in 1990 and has since invaded Andhra Pradesh and Karnataka, infesting several host plants including castor. It has affected more than 100 hosts and is a serious pest of greens, cucurbits, tomatoes, castor, and ornamental plants. A number of important agricultural pests spread at varying rates, such as *Sogatella furcifera*, the green leafhopper, and *Nilaparvata lugens*, the brown planthopper, migrate long distances to occupy virtually all suitable habitats under favorable conditions. Warmth-loving species, such as the thrips species *Frankliniella sulphurea*, *Thrips flavus*, *Phibalothrips peringueyi*, and cashew thrips *Selenothrips rubrocinctus*, are good invaders.[21]

*Figure 11.6* Adults of whitefly, *Bemisia tabaci*. (Courtesy of R. W. Alexander Jesudasan.)

***Figure 11.7*** Adults of spiraling whitefly. (Courtesy of V. V. Ramamurthy.)

Large-scale ecosystem perturbation of diverse intensities and land transformations provides ideal conditions for invasive species to colonize and multiply.[6] Chronic invasive species like the oriental fruit fly *Bactrocera dorsalis* are known in such susceptible habitats. *Leucaena leucocephala* was introduced into India from Central America and the psyllid *Heteropsylla cubana* has posed a severe threat to the cultivation of *Leucaena* all over the tropics.[19] One of the worst invaders causing damage to agricultural crops is the giant snail *Achatina fulica* (Figure 11.8).[19] The ladybird beetle *Curinus coeruleus* was introduced from Mexico to Thailand; this predator has since become established in various parts of South India.

Many times ecologists refer to exotic species as "ecological malignancies" in an ecosystem. Nevertheless, one cannot deny that exotic species (invasive species) are an important structural force in natural communities. Invertebrate species invasions pose a threat to current ecological communities and global biodiversity. Biological differences in the invasion dynamics of various taxa are also an aspect deserving consideration. Because some exotic species tend to cause extinction of native species, increasing introduction will lead to further reduction in global biodiversity.[29] Stressing the need for a regular introduction of predators as biocontrol agents, Strong[30] said, "biological control is an important arrow in the quiver of plant management, perhaps the only arrow in some cases of pests of grave environmental concern" (p. 1059).

No account of invasive invertebrates can be complete without knowledge of gall insects. Gall-inducing ability is best known among thrips, Hemiptera, Diptera, and Hymenoptera. Patterns of radiation among the gall-inducing insects appear more complex, the patterns depending on the dispersal of host plants. Mani[31] states that gall-inducing insects in peninsular India have morphological affinities to elements found in Afro-Asian and Mediterranean regions. Gall-inducing insects in the northwestern Himalayas, on the contrary, include a high proportion of Asian and European elements. This is evident by the fact that gall-inducing saw flies, cynipids, and aphids occur only along the Himalayan slopes, whereas most of the gall thrips and coccids occur only in the tropical stretches of peninsular India.[32,33] Because of the increasing introduction through cut flowers, twigs, and branches, many species of thrips and other gall-inducing insects have spread into India. As the number of species being transported beyond their native range has increased with globalization, the need for increased understanding of the ecology of biological invasions cannot be overemphasized.

**Figure 11.8** *Achatina fulica*. (Courtesy of G. Thirumalai.)

## Acknowledgments

My sincere thanks are due to Dr. V. V. Ramamurthy of the Indian Agricultural Research Institute, New Delhi, Dr. Prasanth Jacob of the Institute of Forest Genetics and Tree Breeding, Coimbatore, and Dr. J. S. Bentur of the Rice Research Institute, Hyderabad, for photographs and other courtesies.

## References

1. Kolar, C. S., and D. M. Lodge. 2001. Progress in invasion biology: Predicting invaders. *Trends Ecol Evol* 16:199.
2. Debach, M. D., and D. Rosen. 1991. *Biological Control by Natural Enemies.* 2nd ed. Cambridge: Cambridge University Press.

3. Venette, R. C., and R. L. Koch. 2009. IPM for invasive species. In *Integrated Pest Management*, ed. E. B. Radcliffe, W. C. Hutchinson, and R. E. Cancelado, 424. Cambridge: Cambridge University Press.

4. Franks, S. J., P. D. Pratt, F. A. Dray, and E. L. Simms. 2008. Selection on herbivory resistance and growth rate in an invasive plant. *Am Nat* 171:678.

5. Pimentel, D. Forthcoming. Environmental and economic costs associated with alien invasive species in the United States. In *Invasive Species*, ed. D. Pimentel, Boca Raton, FL: Taylor & Francis.

6. Elton, C. S. 1958. *The Ecology of Invasion of Animals and Plants*. London: Metheun.

7. Liebhold, A. M., W. L. MacDonald, D. Bergdahl, and V. C. Mastro. 1995. Invasions by exotic forest pests: A threat to forest ecosystem. *For Sci Monogr* 30:1.

8. Williamson, M. 1996. *Biological Invasions*. London: Chapman and Hall.

9. Warner, S. P. 2002. Predicting the invasive potential of exotic insects. In *Invasive Arthropods in Agriculture: Problems and Solutions*, ed. G. A. Hallman and C. P. Schwalbe, 119. Enfield: Science Publishers.

10. Di Castri, F., A. J. Hanson, and M. Debussche. 1990. *Biological Invasions in Europe and the Mediterranean Basin, (Monographiae Biologicae)*. New York: Springer.

11. Sabu Thomas, K. 2008. Invasion of litter beetle *Luprops tristis* and its emergence as a serious nuisance pest in the moist Western slopes of the Western ghats. In *Annual Discussion Meetings in Entomology, Invasive Insects in Agriculture, Forestry and Medicine* (Abstract). Chennai Series IX:23.

12. Prasanth Jacob, J. 2008. Invasive insect pests in monoculture forest ecosystem with specific reference to Eucalyptus gall problem in India. In *Annual Discussion Meetings in Entomology, Invasive Insects in Agrilculture, Forestry and Medicine*. Chennai Series IX:11.

13. Rabindra, R. J. 2008. Challenges in the biological control of alien insects. In *Annual Discussion Meetings in Entomology Invasive insects in Agrilculture, Forestry and Medicine* (Abstract). Chennai Series IX:8.

14. Nagrare, V. S., S. Kranthi, V. K. Biradar, N. N. Zade, V. Sangode, G. Kakde, R. M. Shukla, D. Shivare, B. M. Khadi, and K. R. Kranthi. 2009. Widespread infestation of the exotic mealybug species, *Phenacoccus solenopsis* (Tinsley) (Hemiptera: Pseudococcidae), on cotton in India. *Bull Entomol Res* 99:537.

15. Central Institute for Cotton Research, Indian Council of Agricultural Research. 2009. Package of practices for managing mealybug on cotton. http://www.cicr.org.in/Package_of_practices_for_managing_mealybug_on_cotton.pdf (accessed April 15, 2010).

16. Muniappan, R., B. M. Shepard, G. W. Watson, G. R. Carner, D. Sartiami, A. Rauf, and M. D. Hammig. 2008. First report of papaya mealybug, *Paracoccus marginatus* (Hemiptera: Psudococcidae), in Indonesia and India. *J Agric Urban Entomol* 25:37.

17. Devasahayam, S., and T. K. Jacob. 2008. Status of Erythrina gall wasp (*Quadrastichus erythrinae* Kim) on invasive insect pest on Erythrina sp. in major black pepper areas in Kerala and Karnataka. In *Annual Discussion Meetings in Entomology, Invasive Insects in Agriculture, Forestry and Medicine* (Abstract). Chennai Series IX:20.

18. Chitra, S., and J. S. Bentur. 2008. Invasive species in rice—A global perspective. In *Annual Discussion Meetings in Entomology, Invasive Insects in Agriculture, Forestry and Medicine* (Abstract). Chennai Series IX:18.

19. Nair, K. S. S. 2001. *Pest Outbreaks in Tropical Forest Plantations: Is There a Greater Risk for Exotic Tree Species?* Bogor, Indonesia: Center for International Forestry Research, (CIFOR).

20. Miller, E. R., D. J. Williams, and A. B. Hamon. 1999. Notes on a new mealy bug pest in Florida and Caribbean, the papaya mealy bug *Paracoccus marginatus*. Williams and Granara. *Insecta Mundi* 13:3–4.

21. Ananthakrishnan, T. N. 2009. Invasive insects. In *Ecodynamics of Insect Communities*, ed. T. N. Ananthakrishnan, 83. Jodhpur, India: Scientific Publishers.

22. Geiger, C. A., and A. P. Gutierrez. 2000. Ecology of *Heteropsylla cubana* (Homoptera: Psyllidae): Psyllid damage, tree phenology, thermal relations and parasitism in the field. *Environ Entomol* 29:76.

23. Cikman, E., and N. Comlekcioglu. 2006. Effects of *Bacillus thuringiensis* on larval serpentine leafminers *Liriomyza trifolii* (Burgess) (Diptera: Agromyzidae) in bean. *Pak J Bio Sci* 9:2082.

24. Vega, F. E., R. A. Franqui, and P. Benavides. 2002. The presence of the coffee berry borer, *Hypothenemus hampei*, in Puerto Rico: Fact or fiction? *J Insect Sci* 2:13.

25. Gould, J., K. Hoelmer, and J. Goolsby, eds. 2008. Classical biological control of *Bemisia tabaci* in the United States: A review of interagency research and implementation. In *Progress in Biological Control*, vol. 4, 343. New York: Springer.

26. Banjo, A. D. 2010. A review of an *Aleurodicus disperses* Russel (spiraling whitefly) (Hemiptera: Aleyrodidae) in Nigeria. *J Entomol Nematol* 2:1.

27. Neuenschwander, P. 1994. Spiraling whitefly *Aleurodicus dispersus*, a recent invader and a new cassava pest. *Afr Crop Sci J* 2:419.

28. Alexander Jesudasan, R. W. 2008. Implication of two invasive adventives white flies *Bemisia tabaci* (Gennadius) and *Aleurodicus dispersus* Russel (Aleyrodidae: Homoptera). In *Annual Discussion Meetings in Entomology, Invasive Insects in Agriculture, Forestry and Medicine*. Chennai Series IX:19.

29. Lodge, G. M. 1993. Biological invasions: Lessons from ecology. *Trends Ecol Evol (TREE)* 8:133.

30. Strong, D. R. 1997. Fear no weevil? *Sci* 277:1058.

31. Mani, M. S. 2000. *Plant Galls of India*. 2nd ed. Enfield: Science Publishers.

32. Raman, A., and T. N. Ananthakrishnan. 1985. Cecidogenous *Crotonothrips* (Thysanoptera) and *Memecylon* interactions: Host relations, nutritive tissue, tissue dynamics, and cecidogenetic patterns. *Proc Indian Acad Sci* 95:103.

33. Raman, A. 2007. Insect induced plant galls of India: Unresolved questions. *Curr Sci* 92:748.

## *Further reading*

Ananthakrishnan, T. N. Introduction. In *Annual Discussion Meetings in Entomology, Invasive Insects: Implications in Agriculture, Forestry and Medicine* [Abstracts]. Chennai Series IX (2008): 5.

Atwal, A. S., and G. S. Dhaliwal. *Agricultural Pests of South Asia and their Management*. New Delhi: Kalyani Publishers, 2003.

Crawley, M. J. "The Population Biology of Invaders." *Philosophical Transactions of the Royal Society of London* 314 (1986): 711.

Erserink, M. "Predicting Invasions: Biological Invasions Sweep In." *Science* 285 (1999): 1834.

Groves, R. H., and J. J. Burdon. *Ecology of Biological Invasions*. Cambridge: Cambridge University Press, 1986.

Hallman, G. J., and C. P. Schwalbe. *Invasive Arthropods in Agriculture: Problems and Solutions*. Enfield: Science Publishers, 2003.

Hengeveld, R. *Dynamics of Biological Invasion*. London: Chapman and Hall Ltd, 1989.

Liebhold, A. M., T. T. Work, D. G. McCullough, and J. F. Cavey. "Airline Baggage as a Pathway for Insect Species Invading the United States of America." *American Entomologist* 52 (2006): 48.

Nayar, K. K., T. N. Ananthakrishnan, and B. V. David. *General and Applied Entomology*. New Delhi: Tata McGraw Hill and Publishing Company, 1976.

Newsome, A. E., and I. R. Noble. "Ecological and Physiological Characters of Invading and Species." In *Ecology of Biological Invasions*, edited by R. H. Grover and J. J. Burdon, 1. Cambridge: Cambridge University Press, 1986.

Parsons, P. A. *The Evolutionary Biology of Colonizing Species*, 1st ed. Cambridge: Cambridge University Press, 1983.

Perrings, C., K. Schmutz, J. Tonza, and M. Williamson. "How to Manage Biological Invasions under Globalization." *Trends in Ecology and Evolution* 20 (2005): 212.

Pimentel, D. "Biological Invasions of Plants and Animals in Agriculture and Forestry." In *Ecology of Biological Invasion in North America and Hawaii*, edited by H. A. Mooney and J. A. Drake, 149. New York: Springer Verlag, 1986.

Pimentel, D. "Habitat Factors in New Pest Invasions." In *Evolution of Insect Pests: Patterns of Variation*, edited by K. C. Kim and B. A. McPheron, 165. New York: John Wiley & Sons, 1993.

Ramakrishnan, P. S. ed. *Ecology of Biological Invasions in the Tropics.* New Delhi: International Scientific Publications, for the National Institute of Ecology, 1991.

Ramakrishnan, P. S., and P. M. Vitousek. "Ecosystem-Level Processes and the Consequences of Biological Invasions." In *Biological Invasions: A Global Perspective—SCOPE 37*, Published on Behalf of the Scientific Committee on Problems of the Environment (SCOPE), 281. Chichester: John Wiley & Sons, 1989.

Sutherst, R. H. "Climate Change and Invasive Species: A Conceptual Framework." In *Invasive Species in a Changing World*, edited by H. A. Mooney and R. J. Hobbs, 211. Washington, DC: Island Press, 2000.

Vermej, G. L. "An Agenda for Invasion Biology." *Biological Conservation* 78 (1996): 3.

Vitousek, P. M., C. D'Antonio, L. L. Loop, and R. Westbrooks. "Biological Invasions as Global Environment Change." *American Scientist* 84 (1996): 468.

Williamson, M. H., and A. Fitter. "The Characters of Successful Invaders." *Biological Conservation* 78 (1996): 163.

*section six*

---

# New Zealand

## chapter twelve

# Economic impacts of weeds in New Zealand

## Some examples

*Peter A. Williams and Susan M. Timmins*

### Contents

## 12.1   Introduction

New Zealand ranks among the most highly invaded areas on Earth. After 200 years of European colonization and the attendant processes of animal and plant introductions, New Zealand has the largest number of introduced mammals of any country in the world and the second highest number of birds.[1] At least 2200 exotic plant species have become naturalized, resulting in an equal proportion of exotic to native species in the wild flora.[2] The primary industries of agriculture, horticulture, and forestry are based on a total of only 140 plant species,[3] nearly all of them introduced. In contrast, there are about 500 introduced plant species that threaten these industries and the native biodiversity. There appears to be some slackening of the rate at which agricultural weeds are naturalizing,[2] but overall there has been an average of one new species every 39 days since Captain James Cook first sighted land in 1769.[2] Statistics such as these, and the huge annual costs for the control of animal pests such as rabbits (*Oryctolagus cuniculus*) and the Australian brush-tailed possum (*Trichosurus vulpecula*), which destroys forests and is the major vector of tuberculosis (TB) in cattle, have prompted two attempts at quantifying the costs of weeds and pest animals to New Zealand. The first was published in 1999,[4] and we made great use of this study in our first edition of this chapter.[5] A 2009 review by the New Zealand Ministry of Agriculture and Fisheries[6] revised these figures.

This latter study showed that the total economic cost of weeds and pest animals to New Zealand, including both direct control costs and production losses with multipliers, but excluding an estimate of cost to public conservation land, was NZ$3424 million per year. (Note: All figures in this chapter are for 2008 unless otherwise stated.) This figure represents about 1.93% of the gross domestic product (GDP). These figures do not distinguish between all phyla such as viruses and fungi, but at least the data for plants and animals are presented in the *Ministry of Agriculture and Forestry* (MAF) study.[6] These latter data have enabled us to be more precise about the costs of weeds alone, so wherever appropriate we have substituted them for our first analysis.[5]

In this chapter, we first present the economic costs of weeds, "plant pests" in the parlance of the New Zealand biosecurity legislation. Second, we describe the cost of weed management on public conservation land in New Zealand. Then, we provide examples of the economic costs of individual weed species to other sectors. We acknowledge that some of these same plant species also deliver economic benefits. These range from *Pinus* species, which are the mainstay of the forestry industry (4% gross national product [GNP]), to the pollen provided by scrub weeds to the honey industry, but we do not consider these benefits further.

## 12.2   The national total

In analyzing the impact of weeds and pest animals on the New Zealand economy, Bertram[4] distinguished between two major components: (1) *defensive expenditure*, the financial cost of resources devoted to preventing pest plants from entering the country and controlling the populations of those already here and (2) *loss of economic output*, the production that is foregone each year as a result of weed infestation. They acknowledged, but did not attempt to quantify, the *"welfare* loss" of existing pest populations on such values as damage to indigenous biodiversity. The MAF study[6] took a similar approach, as summarized below.

Defensive expenditure by the central government includes border quarantine, pest surveillance and response, pest control on conservation land for the protection of indigenous habitats, and central government expenditure on scientific research aimed at animal and plant pests. For the year 2008, this amounted to NZ$418.6 million.[6]

The legislation covering pests underwent major changes in the late 1980s, culminating with the Biosecurity Act in 1993. This legislation redefined the role of the central government, local government, and property owners in the control of all pests, including pest plants (weeds hereafter) and pest animals, largely by removing subsidies to property owners and making them primarily responsible for pests on their land. Some of the monies spent by local government are derived from central government subsidies to the regions, but the net figure for local government defensive expenditure in 2008 was estimated at about NZ$36.9 million for all pests. Most monies are spent implementing the Regional Pest Management Strategies (RPMS), wherein the councils state their objectives with respect to a defined list of pests. Most of this covers monitoring, educating, enforcing the regulations, and reviewing the strategies themselves. Very little is spent by regional councils on actually killing weeds or pest animals, which is the responsibility of the landowner and is an expenditure captured under the private sector.

The importance of agriculture to the New Zealand economy means that industry spends large sums on pest control (defensive expenditure), most of which is from the private sector. In 2008–2009, this amounted to NZ$365 million. Households were estimated to spend NZ$42 million in the same year controlling pests, primarily in urban areas. The total figure for the central government, regional councils, and the private sector on defensive expenditure on all pests in 2008/09 was therefore NZ$862.5 million. How much of this was for weeds?

In our earlier account,[5] we calculated that the proportion of the total pest budget spent on weeds by regional councils was 14.7%, whereas the MAF study[6] calculated this to be 50%, although they actually chose to use the midpoint of these two figures, that is, 32%. The MAF study did not apply this ratio to any of the national figures apart from private sector expenditure. If we do so, the annual *defensive expenditure*, the amount spent for protecting the country against weeds, comes to about NZ$276 million. This is likely to be an overestimate, but even if we use our earlier ratio of 14.7%,[5] the figure is NZ$128.6 million.

To determine the economic damage directly attributable to pests, Bertram[4] gave detailed accounts of individual species of plants and animals to produce a total figure of NZ$452 million. He suggested that this might be an underestimate, and this view is supported by the examples of individual weeds, which we discuss in Sections 12.4 through 12.8. For example, no account was taken of the direct losses caused by weeds to the forestry industry, as described in Section 12.4. As anticipated, this figure for output losses was magnified threefold to NZ$1277 million for all terrestrial weeds and pest animals, including invertebrates, in the MAF study.[6] Weeds alone contributed NZ$302 million. When added to the NZ$276 million for defensive expenditure, the total cost of weeds to New Zealand is NZ$578 million per annum. This is over five times our earlier estimate—a function of better measurement and a growing problem.

## 12.3   Public conservation land

The Department of Conservation (DOC) manages 8 million (8,593,371) ha of protected land including terrestrial sites as well as wetland, freshwater, and coastal marine and island sites amounting to 30% of the New Zealand land area.[7] There are several acts of parliament that set out objectives, approaches, and criteria for protecting these areas, and the control of invasive weeds is but one objective. DOC is not required to manage weeds other than on public conservation land, but it contributes on a wider scale through involvement in strategies drawn up by the central and local government under the Biosecurity Act.

The public conservation land under weed threat includes most of the 20 types of land environment* ranging from tussock lands to other kinds of grassland, montane to alpine communities, forest and scrub as well as a range of coastal and duneland communities, freshwater and saline wetlands, geothermal sites, and arid sites. Many components of the ecosystem are impacted by these weeds, especially threatened plant species. Weeds are the main risk to 61 of the 125 indigenous species with a high priority for management.[8] Of the 8,593,371 ha of public conservation land, over 1,496,788 ha[7] (17%) is formally reported as under weed threat and requires ongoing weed control to protect biodiversity values. However, the hectare figure reflects how far resources will stretch in controlling weeds at the sites with the highest biodiversity value and urgency of control. The real extent of the conservation land under weed threat is open to debate and interpretation but is likely to be much larger.

Total expenditure by DOC on weed control has risen from NZ$1.76 million in 1994–1995 to NZ$20.27 million in 2008–2009. This is all defensive expenditure[5] and does not account for the economic cost of biodiversity loss. In 2008–2009, DOC conducted active weed control over 482,193 ha using a site-led approach, which represents 5.6% of DOC-managed land and 32% of the area described above as under immediate weed threat. Weed control is also conducted using weed-led plans. Work under 99 weed-led plans was reported for

---

* Land Environments of New Zealand (LENZ) classification system developed by Landcare Research and the Ministry for the Environment.

2008–2009,[7] which is likely to be an underestimate; the number of active weed-led plans is likely to be upwards of 150 (T. Belton and C. Howell, unpublished data).

Further, the current level of investment is insufficient to achieve the objectives of some of the site-led and particularly the weed-led plans (T. Belton and C. Howell, unpublished data). For example, of seven DOC *Spartina* control programs, their success or otherwise appears to be more related to allocation of sufficient resources than the size of the control area (T. Belton and C. Howell, unpublished data). This suggests the disparity between the size of the problem and the available money is greater than is apparent, which is not surprising given the expense of weed control. This disparity will worsen over time as introduced plants continue to naturalize and some subsequently become environmental weeds.[2]

Taking all pest (plant and animal) expenditure, both that are classified as pest control and that are done under the guise of a restoration or species protection program, DOC spent a total of NZ$78,659 in 2008–2009. Thus, that spent on weeds represents 26% of the pest total and 12% of DOC's total expenditure on the management of natural heritage. The comparison by area tells a similar story. In 2008–2009, DOC conducted 101 possum control operations over 187,562 ha, deer control over 409,294 ha, and goat control over 1,466,340 ha. As some sites would have been treated for multiple pest animals, the hectares cannot be summed. However, the figures suggest that in excess of 20% of DOC-managed land was treated for one or more pest animals in 2008–2009. While the level of weed activity is less than that for pest animals, these percentages are toward the top of the range spent by regional governments and reflect a greater emphasis by DOC on protecting biodiversity, in contrast to the greater focus on pest animals of agricultural land. The way progress in both weed and pest control on public conservation land is measured and reported is likely to change in the near future with the introduction of the Natural Heritage Management System.

An analysis of 58 weed control projects throughout New Zealand conducted by DOC and involving a range of ecosystems and weed densities defined the costs of weed control on public conservation land.[9] Light infestations of weeds, where monitoring with or without some control can be conducted over large areas, can be treated for as little as NZ$11 per hectare. Once the infestations become well established, the costs increase tenfold, to NZ$110–NZ$225 per hectare, and also require NZ$50–$100 per hectare for follow-up and maintenance. Very dense infestations in difficult habitats necessitating helicopters, intensive labor, expensive chemicals, or a combination of all these, require a further tenfold increase in expenditure from NZ$1150 to NZ$2800 per hectare.[9] These logarithmic increases in costs can occur over as little time as 5–8 years, in the case of woody species such as broom (*Cytisus scoparius*) and lodgepole pine (*Pinus contorta*) with either a persistent seed bank or seeds readily dispersed over long distances. Many species on the current list of 328 weeds controlled on public conservation land have these features, and more species are added to the list each year (T. Belton and C. Howell, unpublished data). Thus, the present expenditure of only NZ$20.27 million is insufficient to even maintain the status quo, let alone reduce the impacts of conservation weeds. This means that the costs of treating many infestations will be increasing logarithmically and new weed infestations will continue to arise and threaten biodiversity values. For example, the untreated lodgepole pine invasions of montane basins in the South Island, which presently require hundreds of thousand dollars per annum to prevent canopy closure over huge areas, will require tens of millions of dollars per year to treat in a few decades.[10]

The above figures are just the cost of trying to restrict the distribution of weed populations (*defensive expenditure*); they do not accommodate the cost of biodiversity loss (*loss of economic output*). The total (direct, indirect, and passive) value of New Zealand's land-based biodiversity was estimated to be NZ$43 billion in 1994 (nonadjusted dollar), of which

NZ$26 billion is attributable to conservation.[11] Using the above figure, that is, that high conservation value areas directly threatened by weeds represent 17% of the land managed by DOC, then very conservatively, we can estimate that weeds cause a loss of native biodiversity of NZ$4.42 billion.

## 12.4    Gorse

The spiny European shrub gorse (*Ulex europaeus*) was introduced into New Zealand with the first colonists of the early nineteenth century. Gorse spines deter grazing, adult bushes recover from damage, and a persistent seed bank gives rise to a mat of seedlings following fire. These attributes have enabled gorse to cover large areas of farmland, particularly in the hill country. On conservation land, it has often replaced the native myrtaceous species, as the first woody species in vegetation successions back to forest.

Gorse was widely planted for hedges, as noted by Charles Darwin on his first visit here in 1835, and no doubt it escaped soon after. Gorse has been the major weed of concern to farmers for almost 100 years, for it was ranked as an "important weed" in the first national surveys of weeds in 1917.[12] By 1980, it contributed to scrub that covered 941,300 ha, some 3.5% of the total land area of the country.[13] A postal inquiry reported gorse to be a serious problem for 34% of all South Island farmers, and a minor problem on a further 40%.[14] Gorse and broom are among the five most serious weeds impacting the survival of threatened plant species in New Zealand.[8]

The costs of gorse control are available from a period in New Zealand history prior to the mid-1980s when weed control on farmland was subsidized by the state, and these data were summarized.[15] During the early 1980s, the annual subsidy for spraying gorse averaged NZ$9.5 million. Combined with on-farm costs, and the expenditure of others not eligible for the subsidy such as territorial authorities, the costs of gorse control for 1984–1985 were NZ$25.7million.

Much of the weed control on public conservation land is done using a site-led approach, that is, suites of weeds are controlled at high-value sites only. As a consequence, the costs of control of a single species are not usually reported separately. However, the cost of gorse control is likely to run into many thousands of dollars, still less than might be expected from the extent of the gorse distribution on conservation land. One reason for this is that once gorse occupies a site that formerly supported native forest, the vegetation usually regenerates back to indigenous vegetation in about three decades if it is not burnt again. But the economic impacts of gorse have been underestimated[5] because there are undesirable effects of gorse on nonforest ecosystems, which are not naturally involved in a succession, such as open land with endangered species.[8]

Some of the best data on the direct and indirect costs of weeds in New Zealand have been produced by the plantation forestry industry, which relies heavily on radiata pine (*Pinus radiata*). Gorse is a major weed of plantations, particularly between rotations when the seedlings are planted. Furthermore, the spread of gorse and other scrub weeds was, until recently, a major factor in large areas of high country changing from pastoral farming to forestry. The costs of clearing gorse are incurred during the initial site preparation phase and over the first few years immediately following planting. These costs and the additional costs incurred during subsequent thinning regimes summed to NZ$14.8 million in the 1980s (G. G. West and R. Van Rossen, unpublished). Since the 1980s, the annual area of cutover in New Zealand has approximately doubled (G. G. West and R. Van Rossen, unpublished), with the result that the total gorse control costs to forestry in New Zealand will now be greater.

The cost of oversowing cutover land with introduced legumes and grasses to suppress weed growth and reduce forest establishment costs during the first 2 years ranges from NZ$310 per hectare to NZ$400 per hectare. This was practiced over 12,850 ha of cutover (48% of the national total) in New Zealand in 1993,[16] which equates to some NZ$6 million over a 2-year period. No data are provided on the treatment of the remainder. However, while oversowing is not as widely practiced today, virtually all forestry land is now given some form of weed control (B. Richardson, pers. comm.). The cost to forestry for weed control in postharvest site preparation, and in the subsequent 2 years, would be in the order of NZ$7 million annually. This does not include the initial costs of clearing the land of weeds, maintaining access tracks, and regulatory compliance costs. These costs are substantial and resulted in the largest forestry company in New Zealand spending NZ$9 million on all aspects of scrub weed control in 1999 (P. Stevens, pers. comm.). Scrub weeds other than gorse are included here too, but gorse is a major contributor. These data are mostly *defensive expenditure*, in the sense of Bertram.[4]

Estimates of *loss of economic output* to forestry attributable to weeds are derived from the work by B. Richardson and G. West, (unpublished). They calculated the economic benefit of weed control in the early stages of tree growth as reflected in gains in timber volume. Data are available only up to the midrotation (ca.15–18 years) stands when the gains for controlling weeds are likely to be greatest when applied to tall weeds such as shrubs (*Buddleja davidii*, *Cytisus scoparius*, and *Cortaderia* species). Depending on the site, these may be less competitive with trees in the very early stages of growth, but they influence tree growth for a longer period of time than short-lived herbaceous weeds. Gains from weed control amounted to approximately 23 m$^3$ per hectare per year of wood, and the relative value of gains was greatest on low-productivity sites because of the effect on the proportion of high-value logs. Overall, weed control may be equivalent to 1–4 years extra growth. Based on the economics of log production in 1993, this equates to between NZ$377 per hectare and NZ$2022 per hectare net present value (NPV), and depending on a range of other factors, the cost–benefit of these gains from weed control ranges from marginal to extremely economic (B. Richardson and G. West, unpublished). If we take the figures and assume 50% of the remaining cutover was also treated, then the NPV gain from weed control in that year for increased wood production alone was between NZ$6.5 million and NZ$34 million, with a mean of NZ$23.3 million for New Zealand as a whole. Conversely, had the control work not been undertaken, this figure would represent the loss of wood value due to weeds. Interestingly, this figure for lost production is close to that of *defensive expenditure* (NZ$14.7 million) spent on gorse for the initial site preparations and release cutting costs calculated a decade earlier.[15]

## 12.5   Broom

Scotch broom (*Cytisus scoparius*) is the second most important woody weed after gorse and infests a wide range of land-use classes totaling approximately 1% of New Zealand.[16] Jarvis et al. conducted an economic study of broom as a prerequisite to allowing the entry of a beetle (*Gonioctena olivacea*) as a biocontrol agent. A measure of the importance of this weed is that regional council expenditure and the application costs in 1999 summed to NZ$0.56 million, predominantly in the South Island.[16] They derived an estimate of the total costs and benefits likely to accrue to a biological control program using data for central government subsidies on broom control similar to those used in the gorse study. These summed to about NZ$0.7 million per annum for the period 1981–1985, and together with all associated nonsubsidy on-farm costs summed to NZ$1.8 million.[16] Most on-farm costs are incurred during land development, which increases farm productivity. Estimates of net benefits per year that could be derived if all the grazing land in New Zealand was developed for agriculture were calculated to be the savings on the costs of control (NZ$2 million), plus the share of the increased productivity attributed to this control (NZ$7.3 million), a total of NZ$9.3 million. As an alternative means of assessing the costs of broom, changes in the market value of all land in New Zealand infested with broom before and after development were calculated. Combined with the savings in ongoing broom maintenance, this gave a figure of NZ$4.8 million, which was an estimate of the lower boundary of the benefits of broom control on farmland.[16]

The economic importance of broom to forestry is indicated by the annual expenditure in 1999 of the 17 major forestry companies managing a total of 1.6 million ha. They spent NZ$1.5 million on broom control.[16] Broom has a particularly severe impact on production volumes in 8000 ha of dry production forests of the eastern South Island. If broom causes a similar productivity loss to gorse, that is, a loss of 2 years production, then the annual production loss, net of harvesting and freight costs, is NZ$0.7 million.[16] These estimates show that even when weed species such as gorse and broom have only slightly different ecologies, and thus control regimes, their costs to the forestry industry are additive.

Broom is widespread on conservation land. Like gorse, broom is among the five most serious weeds jeopardizing the survival of threatened plant species in New Zealand changing naturally open communities into woody cover. The cost to control all the broom would be enormous, so DOC controls it only on sites of high biodiversity value. Further, the broom must be compromising the conservation values at these sites, and the achievable level of broom control must result in an increase in biodiversity at the sites. In 1999, DOC spent NZ$0.6 million on broom control. This expenditure is an underestimate of the true conservation cost of broom; the true figure would be orders of magnitude higher. An additional NZ$0.8 million was spent on controlling broom in 1999 by other crown agencies with land under their control and utilities controlling power lines and highways.[16] Summing across all sectors, a total of NZ$5.3 million is spent each year on controlling broom.

These analyses of both gorse and broom rely heavily on data from the early 1980s when the government subsidized clearing land of weeds and the cost of chemicals for ongoing control. Much land thus cleared was certainly uneconomic for pastoral farming, and it has since reverted back to weeds or been subsequently cleared again without direct subsidies and has been planted in commercial forests. Because most of the data on weed expenditure in forests was derived after the 1980s, there is in effect an overestimate of the amount spent on gorse and broom when projected to the present land-use pattern in New Zealand. While we are uncertain as to the extent of this bias, the total cost of shrub weeds would be much higher if we were able to place a dollar value on its impact on conservation land.

## 12.6   Old man's beard

Old man's beard (*Clematis vitalba*) is a vine that has proved particularly destructive of native forest in many parts of New Zealand, damaging revegetation plantings, and could potentially be a weed of pine plantations. An investigation into the economics of a biological control program involved a compilation of the existing expenditure on the weed and a study of the benefits of such research to New Zealand society,[17] the only such study we know of in New Zealand.

There is no market value for native forests in New Zealand as they are mostly not harvested. Greer et al.[17] used a nonmarket valuation technique known as contingency valuation to elicit the community's willingness to pay for the research as a proxy for the benefits they would expect to receive. A postal survey conducted of randomly selected adult New Zealanders suggested that the adult population was prepared to pay between NZ$50 million and NZ$124 million. In contrast, the total identified costs of *Clematis vitalba* control by the territorial authorities and DOC for the 5 years immediately prior to the study (1985–1990) had an NPV of NZ$4.3 million (1990; at a 10% discount rate). The cost of a biocontrol research program for old man's beard was estimated at NZ$1.1–NZ$2.2 million over 6 years. The survey figures also exceed the amount being spent by DOC and local authorities on *all* environmental weeds. These very large sums of money for a single weed contrast with the amount being spent on controlling it, and the smaller amount again that would be required for research into biological control.

Although the figures from the survey may be an overestimate of what the public would really be prepared to pay, they show that New Zealanders place a high nonmarket value on their native forests and would seek to avoid any biodiversity loss. Understandably, willingness to pay was highest among those who visited native bush and were aware of the problem prior to the questionnaire (68%) and lowest among those who had not visited bush and were unaware of the problem (5%). Interestingly, the willingness to pay was not confined to the regions currently affected by old man's beard; the public considered the weed to be a national problem.[17]

## 12.7   Thistles

Thistles are a major problem for agriculture in many areas of New Zealand. Californian thistle (*Cirsium vulgare*) ranks as one of the top six weeds in the South Island, infesting a third of the farms of the southern regions.[18] The cost of control and loss of production was estimated at NZ$13 million for the region.[18] These two components were not separated; and in the absence of data on their relative weightings, Bertram[4] ascribed a factor of 50% to the loss of production to give a figure of NZ$11 million. There are no data for the country as a whole, but Bertram considered it must lie above NZ$11 million and possibly as high as NZ$20 million.

## 12.8   Giant buttercup

Giant buttercup (*Ranunculus acris* L.) is an important weed of dairy pastures, and dairying itself is one of the largest earners of overseas exchange for New Zealand. A study in one South Island area estimated the annual percentage loss in pasture production due to a typical infestation of the weed subjected to typical control measures. Extrapolation to other infested regions and districts provided a national estimate of milk solids revenue loss due to giant buttercup in New Zealand dairy pastures of NZ$175 million in the 2001–2002 season.[19]

## 12.9   Alternative funding

The huge sums required to control weeds, especially on conservation land where there is no measurable economic benefit, require alternatives to general taxation. One hope of meeting more conservation objectives for weed control in New Zealand is through volunteer labor, where DOC subsidizes travel expenses. One such scheme has been running on the volcanic mountains of Central North Island for decades. Lodgepole pine is hand pulled over about 500 ha on a 3-year rotation. Here, the direct costs, apart from those borne by the volunteers, amount to only about NZ$1.1 per hectare for very light infestations and up to NZ$14.7 per hectare for heavy infestations (50 trees per hectare), which is less than 10% of the full commercial costs. There are now several community-based pine-clearing projects in the South Island high country that pool money from various sources, but the total pool has not been calculated.

Many conservation weeds originate from horticulture,[20] including those that have only recently become invasive, for example, kiwi fruit (*Actinidia deliciosa*). Where the sale and distribution of highly valuable crops is controlled through centralized commercial agencies, it would be a simple matter to levy the growers. A mere one quarter of 1 cent on every tray of kiwi fruit produced in New Zealand per year would yield NZ$170, which would be more than sufficient to control all known wild populations of the species (W. Stahl, pers. comm.). If such responsibilities were accepted before new species were allowed to be released in New Zealand, there may be a greater opportunity to maintain the diversity of agricultural plant species that the country requires,[3] while at the same time protecting the indigenous biodiversity.

## 12.10   Conclusions

The cost to New Zealand of defending its borders against new weeds and managing or controlling those already here amounts to about NZ$276 million per year. Those that are not successfully controlled and that directly affect the nation's productive output cost a further NZ$302 million per year. This represents about 17% of all the economic costs of all classes of pests as estimated by the MAF study,[6] which is similar to the 12% we had calculated previously from Bertram.[4] Many individual weed species have direct control costs of many millions of dollars per year. Assuming 200 weed species are controlled in New Zealand, the average cost per weed species in measurable lost production amounts to NZ$1.5 million per annum.

DOC spends NZ$20.27 million per year defending the one-third of New Zealand it manages from 328 invasive weed species. This is less than is required to protect high-value conservation sites, and the threat to biodiversity on public conservation land continues to increase. The costs of weed control can increase very steeply as populations spread, and on conservation land and much private land, weeds are undoubtedly increasing beyond the capacity of the current allocated resources. Some relief could come if part or all of the cost of weed control was borne by those who gain from the cultivation of a plant species that has subsequently become weedy on conservation land. In theory, the public is prepared to pay considerably more than at present to protect conservation land from weeds. The public also seems prepared to volunteer their labor to control weeds of conservation concern.

The dollar costs of controlling weeds on conservation land are but a fraction of the true costs of environmental weeds to New Zealand. Not accounted for are the many more weed invasions that are not controlled for lack of resources. Nor do the costs take account of the loss of biodiversity values wrought by weeds. The true economic cost of weeds in this weedy island country is enormous.

## References

1. Department of Conservation and Ministry for the Environment. 2000. *The New Zealand Biodiversity Strategy.* Wellington: Department of Conservation and Ministry for the Environment.
2. Williams, P. A., and E. K. Cameron. 2006. Creating gardens: The diversity and progression of European plant introductions. In *Biological Invasions in New Zealand*, ed. R. B. Allen and W. G. Lee, 33. Heidelberg: Springer-Verlag.
3. Halloy, S. R. P. 1999. The dynamic contribution of new crops to the agricultural economy: Is it predictable? In *Perspectives on New Crops and New Uses*, ed. J. Janick, 53. Alexandria: ASHS Press.

4. Bertram, G. 1999. The impact of introduced pests on the New Zealand economy. In *Pests and Weeds*, ed. K. Hackwell and G. Bertram, 45. Wellington: New Zealand Conservation Authority.

5. Williams, P. A., and S. Timmins. 2002. Economic impacts of weeds in New Zealand: Some examples. In *Biological Invasions: Environmental and Economic Costs of Alien Plant, Animal and Microbe Invasions*, ed. D. Pimentel, 175. Boca Raton: CRC Press.

6. Gierra, N., and R. Bell. 2009. *Economic Costs of Pests to New Zealand*. Wellington: Ministry of Agriculture and Forestry.

7. Department of Conservation. 2009. *Department of Conservation Annual Report for the Year ended 30 June 2009*. Wellington: Department of Conservation.

8. Reid, V. A. 1998. *The Impact of Weeds on Threatened Plants*. Wellington: Department of Conservation.

9. Harris, S., and S. M. Timmins. 2009. Estimating the benefit of early control of all newly naturalized plants. Science for Conservation 292, Publishing Team, Department of Conservation, Wellington.

10. Stevens, T., D. Brown, and N. Thornley. 2000. Conservation achievement: The Twizel area. Unpublished report, Department of Conservation, Wellington.

11. Patterson, M., and A. Cole. 1999. Assessing the value of New Zealand's biodiversity. Occasional paper no. 1, School of Resource and Environmental Planning, Massey University, Palmerston North.

12. Cockayne, A. H. 1917. Noxious weeds in New Zealand: Notes on a recent enquiry. *J. Agric (VIC)* 14:339.

13. Blaschke, P. M., G. G. Hunter, G. O. Eyles, and P. R. Van Berkel. 1981. Analysis of New Zealand's vegetation cover using land resource inventory data. *N.Z.J.Ecol* 4:1.

14. Bascand, L. D., and G. H. Jowett. 1981. Scrubweed cover of South Island agricultural and pastoral land. *N.Z.J.Exp.Agric* 9:307.

15. Sandrey, R. A. 1985. Biological control of gorse: An ex-ante evaluation. Research report, Agricultural Economics Research Unit, no. 172, Lincoln College (University of Canterbury). Lincoln.

16. Jarvis, P. J., S. V. Fowler, and P. Syrette. 2000. The economic benefits and costs of introducing a biological control agent, *Gonioctena olivacea*, for broom. Landcare Research Contract Report C09X0210, Lincoln.

17. Greer, G., and R. L. Sheppard. 1990. An economic evaluation of the benefits of the research into biological control of *Clematis vitalba*. Research report no. 203, Agribusiness and Economic Research Unit, Lincoln University, Lincoln.

18. Mitchell, R. B., and R. J. Abernethy. 1993. Integrated management of California thistle in pasture. In *Proceedings of the 46th New Zealand Plant Protection Conference*. http://www.nzpps.org/journal/46/nzpp_460240.pdf (accessed January 27, 2011).

19. Bourdôt, G. W., D. J. Saville, and D. Crone. 2003. Dairy production revenue losses in New Zealand due to giant buttercup (*Ranunculus acris*). *N.Z.J.Ag.Res* 46:295.

20. Timmins, S. M., and P. A. Williams. 1987. Characteristics of problem weeds in New Zealand's protected natural areas. In *Nature Conservation: The Role of Remnants of Native Vegetation*, ed. D. A. Saunder, G. W. Arnold, A. A. Burbidge and A. J. M. Hopkins, 241. Chipping Norton: Surrey Beatty, in association with CSIRO and CALM.

*chapter thirteen*

# Ecological and economic costs of alien vertebrates in New Zealand

**M. N. Clout**

## Contents

## 13.1 Introduction

New Zealand is one of the most isolated ancient landmasses on Earth, having been separated for over 65 million years. This isolation resulted in the evolution of high levels of endemism in both the flora and fauna. The terrestrial vertebrate fauna was particularly unusual. Dominated by birds and reptiles, it contained no indigenous land mammals, apart from some small bats. New Zealand was among the last habitable landmasses to be settled by humans; the first settlers were Maori people from Polynesia who arrived only ca.730 years ago.[1]

An ecological catastrophe followed the arrival of people and the alien mammals that they introduced. The Maori settlers brought dogs (*Canis familiaris*) and Polynesian rats (*Rattus exulans*) with them. At least 58 endemic bird species were lost in this initial settlement phase, including several large flightless species such as moa (Dinornithidae), which were probably hunted to extinction within less than 100 years.[2,3] The Polynesian rat seems to have eliminated several species of small birds, flightless insects, and reptiles.[4]

European settlement of New Zealand started about 200 years ago, and the initial trickle of alien species rapidly became a flood. In the past two centuries, Europeans have successfully introduced over 90 species of alien vertebrates, including 32 mammals, 36 birds, and 19 fish (Table 13.1). Among the introduced mammals are three further species of rodents, three mustelids, six marsupials, and seven deer species.

Some alien vertebrates, such as sheep (*Ovis aries*) and cattle (*Bos taurus*), have proved to be economically beneficial, forming the basis of profitable farming and export industries. New Zealand's economy still depends heavily on the export of primary produce, including

*Table 13.1* Numbers of Vertebrate Species Introduced to New Zealand and Their Present Status

| Group | Number introduced | Number established | Number of pests |
|---|---|---|---|
| Terrestrial mammals | 55 | 32 | 28 |
| Freshwater fish | 40 | 19 | 9 |
| Birds | 137 | 36 | 6 |
| Frogs and reptiles | 6 | 4 | 0 |

*Note:* "Number introduced" refers to species released to the wild in New Zealand. "Number established" refers to species with self-sustaining wild populations. "Number of pests" refers to species listed in current legislation or policy documents as pests, and/or under some form of control to reduce their abundance.[5–9]

meat, wool, and dairy produce. Several other introduced vertebrates have, however, proved to be both ecologically and economically damaging.

## 13.2   Ecological costs

There have been major ecological costs flowing from many of the vertebrate introductions to New Zealand. In addition to the early losses following Polynesian settlement, at least nine more endemic bird species have become extinct in the past 150 years, primarily because of predation by European mammals such as Norway rats (*Rattus norvegicus*), ship rats (*R. rattus*), cats (*Felis catus*), and stoats (*Mustela erminea*). Stoats, along with two other mustelids, were introduced in the 1880s in a failed attempt to control rabbits (*Oryctolagus cuniculus*), which had become established two decades before and were proving to be severe pasture pests. Predation and competition by introduced mammals continue to threaten the extinction of many endemic species of reptiles, frogs, and birds, several of which are now restricted to mammal-free islands.[10,11]

In freshwaters, alien fish introduced and established in the nineteenth century included brown trout (*Salmo trutta*) and rainbow trout (*S. gairdneri*), which now form the basis of important recreational fisheries. Introduced trout are widely recognized as being invasive predators in many parts of the world. In New Zealand, there is growing evidence of their negative effects on native fishes and invertebrates, including declines and local extinctions of several endemic galaxiid fish.[12] Despite this, the negative effects of trout on New Zealand freshwater ecosystems have received little publicity and have caused virtually no official concern, largely because of the popularity of trout fishing. Other invasive fish, namely rudd (*Scardinius erythrophthalmus*) and koi carp (*Cyprinus carpio*), were illegally introduced to New Zealand in the 1960s.[13] Like trout, they can become the dominant fish species in waters where they thrive and have negative effects on native freshwater ecosystems. Unlike trout, they are not valued as sport fish, are officially classed as noxious, and are the subject of pest management strategies.

Brushtail possums (*Trichosurus vulpecula*), deer (*Cervus elaphus* and others), and goats (*Capra hircus*) are all significant agents of floristic change in native forests, through their selective browsing and inhibition of regeneration of many native plants.[14,15] Possums are also significant nest predators of threatened birds,[16] so these Australian marsupials have multiple impacts on native ecosystems.[17] Rats, cats, and stoats are important agents of faunal change through their predation on behaviorally vulnerable and slow-breeding native animals such as large-bodied invertebrates and ground-feeding or hollow-nesting birds. Larger flightless birds have proved especially susceptible to mammal predation, with surviving species such as takahe (*Porphyrio mantelli*) and kakapo (*Strigops habroptilus*), now critically endangered as a result.[18] Even New Zealand's national bird, the brown kiwi (*Apteryx mantelli*), is declining at a rapid rate,[19] due to predation by alien mammals.

New Zealand's Department of Conservation (DOC) classes 2788 New Zealand taxa (species, subspecies, and forms) as threatened,[20] including 868 vascular plants, 943 terrestrial invertebrates, and 229 terrestrial vertebrates. The International Union for Conservation of Nature (IUCN) Red List[21] shows that 69 of 287 surviving New Zealand bird species are internationally listed as threatened—a higher proportion than in any other country. Most of these threatened bird species are endemic, and many of them now occur only on mammal-free islands.

Native vertebrates (especially birds, but also fish and reptiles) have suffered disproportionate rates of extinction and endangerment. The primary cause in most cases seems to be predation or competition from alien vertebrates. Alien vertebrates have clearly caused major ecological costs, through extinctions and declines of many endemic species and changes in the composition and structure of native ecosystems. Putting a value on this damage is difficult; for example, how does one place a monetary value on the extinction of an endemic bird such as the huia (*Heteralocha acutirostris*) a century ago? It is, however, possible to estimate the ongoing costs of managing vertebrate pests for conservation purposes.

## 13.3 Economic costs

Bertram[22] classed the measurable economic costs of pests in New Zealand into two major components: defensive expenditures (i.e., the costs of controlling pests) and production losses (i.e., foregone economic output). The annual defensive expenditures calculated by Bertram have since been updated (Table 13.2) to 2008 values.[23] Another category of costs is the "welfare loss" caused by the existence of pests, additional to their impact on commercial activities. As an example of the latter, Bertram[22] cited the continuing damage to native forests caused by possums, deer, and goats due to the fact that control, rather than eradication, is the purpose of defensive expenditures. Since conservation lands are largely held out of market production, the foregone values attributable to continuing pest presence can only be measured indirectly (e.g., by contingent valuation), but research on such values is generally inadequate.[22]

### 13.3.1 Defensive expenditures

Defensive expenditures for vertebrate pests include quarantine and border control costs, surveillance, research, pest control, and eradication attempts. For expenditure on

*Table 13.2* Defensive Expenditure (in NZ$ millions) on Pests in New Zealand in 2008

|  | Central government | Regional government | Agricultural sector | Other sectors |
|---|---|---|---|---|
| Quarantine and border control | 184.4 | | | |
| Surveillance and response | 45.3 | | | |
| Pest control | 76 | 36.9 | 365 | 42 |
| Research | 37 | | | |
| Other (including policy advice, etc.) | 73.4 | | | |

*Note:* Figures shown are estimated totals for all alien species, not only vertebrates.[23]

quarantine and border control and on pest surveillance, it is difficult to separate the expenditure on vertebrate pests. The figures in Table 13.2 therefore relate to all alien pest and weed species.

Despite concerns about the potential risks posed by animals such as snakes, incursions of new alien vertebrates to New Zealand are relatively uncommon. Most of the NZ$184.4 million spent on quarantine and border control and the NZ$45.3 million spent on pest surveillance and response in 2008[23] was therefore expended on the prevention or detection of diseases, invertebrate pests, and weeds, rather than vertebrates.

In contrast to quarantine and pest surveillance costs, a large fraction of the central government expenditure on the control of established pests is directed against alien vertebrates. Especially large amounts are expended on control of brushtail possums, which are primary vectors of bovine tuberculosis (TB). In 2008–2009, the Animal Health Board (AHB) spent over NZ$82 million to control bovine TB.[24] Over $55 million of this was spent on management of TB vectors (mainly possums), over an area of 3.7 million ha.

The New Zealand DOC administers nearly one-third of the land area of New Zealand and spends a large proportion of its budget on pest and weed control to protect biodiversity values. In 2008–2009, DOC spent NZ$29.9 million on animal pest control and NZ$20.3 million on weed control. The majority of the animal pest budget was spent on controlling alien mammals, dominated by possum control (NZ$14.9 million) and goat control (NZ$6.5 million).[25] Much of the regional government and agricultural sector expenditure on pest control (Table 13.2) is also directed against possums and other vertebrate pests.

Of the NZ$45 million spent annually by DOC on protecting species and managing island reserves, much is expended on managing alien vertebrates (especially predators), which are major threats to native biodiversity. The DOC "mainland islands" program (costing NZ$2.1 million in 2008–2009) concentrates especially on reducing alien vertebrates to minimal densities at key conservation sites on the mainland of New Zealand. Most of the DOC expenditure on the management of alien vertebrates is for ongoing control purposes, but some of the funds spent on managing island reserves (NZ$7.3 million in 2008–2009) are targeted at the complete eradication of mammalian pests. The eradication of alien mammals from islands has been a major advance in New Zealand conservation practice in recent years.[26] Successes on large islands include the eradication of cattle, sheep, and Norway rats from Campbell Island (11,216 ha), goats and rats from Raoul Island (2,938 ha), brushtail possums, Norway rats, and Polynesian rats from Kapiti Island (1,970 ha), possums and Polynesian rats from Codfish Island (1,350 ha), and rabbits and mice from Enderby Island (710 ha). On islands, where eradication is possible and the risks of reinvasion are low, this type of "one-off" expenditure to permanently remove a threat is more efficient than paying for perpetual control, with its attendant uncertainties.

## 13.3.2   Production losses

Production losses due to the damage caused by alien vertebrate pests are difficult to calculate, so estimation of these losses has not been attempted for most species in New Zealand. Exceptions are brushtail possums and rabbits, but the estimates are often vague.[22,23,27,28]

### 13.3.2.1   Possums

A recent estimate of total annual production losses due to possums was NZ$52 million.[23] Impacts of possum on pasture production are contentious, with Bertram[22] suggesting annual losses of NZ$12 million, while others argue that impacts of possums on pastures are negligible.[27] There is also some uncertainty about the economic significance of damage caused by possums in forestry plantations, although Butcher[27] points out that even a 5% loss at planting in a *Pinus radiata* plantation represents a loss at harvesting of over NZ$282 per hectare. Since there are over 1 million ha of pine plantations in New Zealand, even very small losses at planting may translate to millions of dollars of lost production per annum.

Although there is doubt about the precise level of production losses caused by possums, they probably barely match the defensive expenditures for this pest (amounting to ca. NZ$70 million for the combined expenditure by DOC and AHB alone). This apparent paradox is partly explained by the immense *potential* damage to the New Zealand economy caused by possums as vectors of bovine TB. Exports of dairy produce, beef, and venison may be threatened if the persistence of bovine TB in New Zealand causes its trading partners to place import restrictions on New Zealand produce. Exports of dairy produce alone earn New Zealand over NZ$3 billion per annum, so the potential damage is huge. In addition to reducing the potential for economic damage, much possum control (especially by DOC) is also justified by the protection of natural assets such as indigenous forests and wildlife.

### 13.3.2.2   Rabbits and hares
Losses due to rabbit damage are also contentious, with available estimates varying widely. Parkes[28] estimated the total cost of rabbit damage to pastoral production, horticulture, and forestry at NZ$6.8 million per annum, but Bertram[22] argued that ca. 2 million sheep are displaced by rabbits and that the total cost of rabbit damage to pastoral production was more likely to be ca. NZ$50 million per annum. This figure was also used in more recent estimates[23] of production losses due to rabbits. Calculation of overall costs of rabbit damage depends partly on the level of infestation, which has been affected by the illegal introduction of rabbit calicivirus disease (RCD) in 1996. This disease at least temporarily reduced rabbit densities across large tracts of sheep grazing country, especially in parts of the South Island. In several areas, the abundance of hares (*Lepus europaeus*) increased as rabbit numbers decreased, partially replacing one alien lagomorph with another. Hares also damage young forestry plantations, through browsing on seedlings, but the economic significance of this is unclear.

### 13.3.2.3   Other vertebrates
Among the other alien vertebrates that may cause some locally significant production losses in New Zealand are pigs (*Sus scrofa*), ferrets (*Mustela furo*), rodents, and some bird species. Wild pigs can cause local losses on some sheep farms by preying on young lambs. Along with ferrets (*Mustela furo*), they are also vectors of bovine TB. Rats and mice are pests of stored products and have at least a nuisance value to foodstuff industries and domestic households. In rural areas near to native beech (*Nothofagus* spp.) forests, the abundance of mice and ship rats can reach plague proportions in years following the mast seeding of beech trees,[11] causing major inconvenience and minor economic losses. Some alien bird species, especially starlings (*Sturnus vulgaris*), mynas (*Acridotheres tristis*), blackbirds (*Turdus merula*), rooks (*Corvus frugilegus*), sparrows (*Passer domesticus*), and various finches, can cause damage to commercially grown fruits and other horticultural crops. A combined estimate of such losses to birds in 2008 was NZ$43 million.[23]

Bertram[22] estimated the total production losses caused by minor animal pests (including insects) at around NZ$36 million per annum. Based on the fraction of his estimates of production losses caused by major vertebrate pests (possums and rabbits), the combined losses to minor vertebrate pests might perhaps total ca. NZ$10 million per annum.

## 13.4   Other losses and side effects

In addition to the costs of defensive expenditures and production losses, the presence of alien vertebrates in New Zealand continues to degrade the natural environment and threaten native species. As mentioned previously (Section 13.2), these ongoing ecological costs are extremely difficult to measure.

One side effect of the presence of alien mammals such as possums, rodents, and rabbits is that their control often involves the use of toxins, some of which may accumulate in food chains and affect other species, including native wildlife.[29] Without alien mammals, there would be no need to continue to distribute these chemicals, with their associated risks.

There are also social costs in the presence of some alien vertebrates, stemming from the conflict that can result from disagreements over the need to control pests or over the best way of conducting control. In New Zealand, there is no toxic control of deer, chamois, and Himalayan tahr, despite the damage that they cause to native ecosystems, because of opposition from hunters. It would be feasible to eradicate species such as Himalayan tahr (*Hemitragus jemlahicus*) and possibly chamois (*Rupicapra rupicapra*) from New Zealand. This has not happened because of intense lobbying pressure from hunting interests, so ongoing ecological costs are incurred by the natural environment. Similarly (but for different reasons), proposals to cull a herd of wild horses to protect fragile native plant communities in the Kaimanawa area of the central North Island met with opposition from a section of the public. Strong political pressure from horse lovers eventually caused the cull to be abandoned, against the recommendations of DOC officials, in favor of a more complex and costly "muster and sale" option.[30]

Public resistance to the repeated application of toxins, such as sodium monofluoroacetate (1080) and anticoagulants, already precludes the use of these chemicals in some areas. An example is the opposition by local Maori people to the use of toxins for vertebrate pest control in the Te Urewera "mainland island" conservation area in the eastern North Island. Because of concerns that poisoning of possums and rats might result in toxin accumulation in hunted deer and pigs and pose a risk to hunters and their dogs, DOC has now ceased using poisons in this conservation area. All possum and rat control in Te Urewera is now conducted by trapping (L. Wright, pers. comm.).

Another side effect of the presence of an alien vertebrate is that it can sometimes encourage other deliberate introductions in an attempt to control it. The classic example of this is the checkered history of attempts at biological control of rabbits. This commenced with the importation and release of mustelids in the 1880s, which was a disaster for vulnerable native wildlife and did not result in effective rabbit control. It continued with the failed introduction of myxomatosis in the 1950s and with the illegal importation and release of RCD by farmers in 1996. This latter introduction flouted laws (e.g., the Biosecurity Act 1993) that were established primarily to protect the interests of the agricultural sector of the economy.[11] Without the initial mistake of introducing rabbits, the subsequent biocontrol mistakes of introducing mustelids and myxomatosis, and the illegal introduction of RCD, would not have happened.

Finally, another set of side effects of the presence of alien vertebrates (especially mammals) is their role as vectors of human diseases. Possums, rodents, and hedgehogs (*Erinaceus europaeus*), among others, are known to carry diseases such as *Leptospirosis* and *Salmonella*,[8] which can infect water supplies and cause illness among people. The costs of these and other diseases that can be carried by alien vertebrates are difficult to calculate, but they may be significant in terms of defensive expenditures (sanitation, medicines) and production losses (lost working time). The presence of potential vectors of diseases that have yet to establish in New Zealand constitutes an ongoing health risk. For example, marsupials (e.g., possums) are the main vertebrate hosts of the Ross River virus, which is carried by the mosquito *Aedes vigilax* and can affect humans. With climate change, it is possible that *A. vigilax* could establish in New Zealand, where potential vectors already exist in the form of alien marsupials.[31]

## 13.5   Conclusions

The overall costs of alien (pest) species to the New Zealand economy has recently been estimated[23] at NZ$3424 million per annum or 1.93% of the national gross domestic product (GDP). Of this total cost, NZ$2454 million were "production losses" (with multiplier effects) and NZ$970 million were "defensive expenditures." In an earlier analysis, Bertram[22] estimated that about 25% of the production losses to alien species could be assigned to vertebrate pests (mostly possums and rabbits). If this is so, production losses caused by alien vertebrates could amount to over NZ$600 million per annum. The proportion of defensive expenditure directed against vertebrate pests is also difficult to calculate precisely, but based on overall costs for 2008 (Table 13.2) and making some brave assumptions, it is possible to derive crude estimates. Much of the central government expenditure by the New Zealand DOC on pest and weed control is directed against possums, goats, and carnivores, comprising well over 70% of total DOC expenditure in this area.[25] Expenditure on the control of alien species by the agricultural sector has a higher component of weed and insect control, but, given the concern about possums and bovine TB, expenditure on the control of alien vertebrates is likely to at least match the production losses caused by them.

Assuming that vertebrate pest control and research consumes 70% of the NZ$150 million of the central and regional government expenditure in this area, the annual cost to these sectors could total NZ$105 million. Assuming that 25% of the NZ$407 million spent annually on pest control by the agricultural community and other sectors[23] is expended on vertebrates, another NZ$102 million of costs may be incurred here. Assuming that only 5% of border control and surveillance expenditure is directed against vertebrates, a further NZ$9 million in annual costs may be incurred. Finally, if it is assumed that expenditure on policy advice and other biosecurity activities relating to vertebrate pests is in proportion to defensive expenditures against vertebrates in all other sectors combined (ca. 40% of all costs), a further NZ$29 million may be spent here. The combined, crude estimate (based on Bertram[22] and the recent[23] review of costs of pests) is that about NZ$245 million is expended per annum for defense against alien vertebrates. Added to the ca. NZ$600 million of estimated production losses, the total cost of alien vertebrates in New Zealand may therefore exceed NZ$845 million per annum or 0.48% of GDP. Regardless of the exact figures, there is no doubt that the presence of alien vertebrates represents a major drain to the New Zealand economy. Added to the direct economic costs are the ecological costs of species extinctions and continuing degradation of the unique biodiversity of the New Zealand archipelago.

The lesson to be learned from the New Zealand experience with alien vertebrates is simple: breaching natural biogeographic boundaries by introducing alien species is something that should never be undertaken lightly. The precautionary principle demands that all alien introductions (both deliberate and accidental) should be considered potentially harmful unless proven otherwise, and every effort should be made to prevent introductions, while there is any doubt about their impacts.

# References

1. Wilmshurst, J. et al. 2008. Dating the late prehistoric dispersal of Polynesians to New Zealand. *Proc Natl Acad Sci* 105:7676.
2. Anderson, A. 1989. *Prodigious Birds*. Cambridge: Cambridge University Press.
3. Holdaway, R. N., and C. Jacomb. 2000. Rapid extinction of the moas (Aves: Dinornithiformes): Model, test, and implications. *Science* 287:2250.
4. Atkinson, I. A. E., and H. Moller. 1990. Kiore. In *The Handbook of New Zealand Mammals*, ed. C. M. King, 175. Oxford: Oxford University Press.
5. Buddenhagen, C. E., S. M. Timmins, S. J. Owen, P. D. Champion, W. Nelson, and V. A. Reid. 1998. An overview of weed impacts and trends. In *Department of Conservation Strategic Plan for Managing Invasive Weeds*, ed. S. J. Owen, 11. Wellington: Department of Conservation.
6. Williams, P. A. 1997. *Ecology and Management of Invasive Weeds*. Conservation Sciences publication no. 7. Wellington: Department of Conservation.
7. Owen, S. J. 1997. *Ecological Weeds on Conservation Land in New Zealand: A Database*. Wellington: Department of Conservation.
8. King, C. M. 1990. *The Handbook of New Zealand Mammals*. Oxford: Oxford University Press.
9. Hackwell, K. 1999. Restoring an indigenous dawn chorus. In *Pests and Weeds: A Blueprint for Action*, ed. K. Hackwell and G. Bertram, 17. Wellington: New Zealand Conservation Authority.
10. Clout, M. N., and A. J. Saunders. 1995. Conservation and ecological restoration in New Zealand. *Pac Conserv Biol* 2:91.
11. Clout, M. N. 1999. Biodiversity conservation and the management of invasive animals in New Zealand. In *Invasive Species and Biodiversity Management*, ed. O. T. Sandlund, P. J. Schei and A. Viken, 349. London: Kluwer.
12. McDowall, R. M. 1990. *New Zealand Freshwater Fishes: A Natural History and Guide*. Auckland: Heinemann Reed.
13. McDowall, R. M. 1994. *Gamekeepers for the Nation: The Story of New Zealand's Acclimatisation Societies, 1861–1990*. Christchurch: Canterbury University Press.
14. Payton, I. 2000. Damage to native forests. In *The Brushtail Possum: Biology, Impact and Management of an Introduced Marsupial*, ed. T. L. Montague, 111. Lincoln: Manaaki Whenua Press.
15. Challies, C. N. 1990. Red deer. In *The Handbook of New Zealand Mammals*, ed. C. M. King, 436. Oxford: Oxford University Press.
16. Sadleir, R. M. F. 2000. Evidence of possums as predators of native animals. In *The Brushtail Possum: Biology, Impact and Management of an Introduced Marsupial*, ed. T. L. Montague, 126. Lincoln: Manaaki Whenua Press.
17. Clout, M. N. 2006. Keystone aliens? The multiple impacts of brushtail possums. *Ecol Stud* 186:265.
18. Clout, M. N., and J. L. Craig. 1995. The conservation of critically endangered flightless birds in New Zealand. *Ibis* 137:S181.
19. Heather, B. D., and H. A. Robertson. 1996. *The Field Guide to the Birds of New Zealand*. Auckland: Viking.
20. Hitchmough, R., L. Bull, and P. Cromarty, compilers. 2007. *New Zealand Threat Classification System lists, 2005*. Wellington: Department of Conservation.
21. IUCN. 2009. *IUCN Red List of Threatened Species*. Gland: The IUCN Species Survival Commission.
22. Bertram, G. 1999. The impact of introduced pests on the New Zealand economy. In *Pests and Weeds: A Blueprint for Action*, ed. K. Hackwell and G. Bertram, 45. Wellington: New Zealand Conservation Authority.
23. MAF. 2009. Economic costs of pests to New Zealand. MAF Biosecurity New Zealand Technical Paper No: 2009/31.
24. Animal Health Board. 2009. Animal Health Board annual report for the year ending 30 June 2009. Animal Health Board, Wellington.
25. Department of Conservation. 2009. Annual report for the year ended 30 June 2009. Department of Conservation, Wellington.
26. Veitch, C. R., and M. N. Clout, eds. 2002. *Turning the Tide: The Eradication of Invasive Species*. Gland: IUCN SSC Invasive Species Specialist Group.

27. Butcher, S. 2000. Impact of possums on primary production. In *The Brushtail Possum: Biology, Impact and Management of an Introduced Marsupial*, ed. T. L. Montague, 105. Lincoln: Manaaki Whenua Press.

28. Parkes, J. 1995. *Rabbits as Pests in New Zealand: A Summary of the Issues and Critical Information*. Lincoln: Landcare Research NZ Ltd.

29. Innes, J., and G. Barker. 1999. Ecological consequences of toxin use for mammalian pest control in New Zealand: An overview. *N Z J Ecol* 23:111.

30. Mack, R. N., D. Simberloff, W. M. Lonsdale, H. Evans, M. N. Clout, and F. Bazzazz. 2000. Biotic invasions: Causes, epidemiology, global consequences and control. *Ecol Appl* 10:689.

31. Clout, M. N., and S. J. Lowe. 2000. Invasive species and environmental changes in New Zealand. In *Invasive Species in a Changing World*, ed. H. A. Mooney and R. J. Hobbs, 369. Covelo: Island Press.

# *South Africa*

# chapter fourteen

# The economic consequences of the environmental impacts of alien plant invasions in South Africa

**D. C. Le Maitre, W. J. de Lange, D. M. Richardson,
R. M. Wise, and B. W. van Wilgen**

## Contents

## 14.1 Introduction

At least 9000 plant species from other parts of the world have been introduced to southern Africa[1,2] for a range of purposes—as crop species for timber and firewood, as garden ornamentals for stabilizing sand dunes, and as barrier and hedge plants.[3] About 1000 of these alien species have become naturalized and self-sustaining, and some 200 of these naturalized species have become invasive (*sensu* Richardson et al.[4]).[5] Invasive alien species are able to survive, reproduce, and spread unaided, sometimes at alarming rates, across the landscape. The invasion of newly colonized areas by alien organisms is a global problem of significant and growing proportions,[6–8] which can have serious implications for the environment and for human livelihoods and well-being.[9]

In southern Africa, much of the historic concern about invasive species has centered on the consequences for the conservation of the region's remarkable biological diversity. For example, it is home to at least 21,137 species of vascular plants, about 80% of them endemic.[10] Most of the region has been affected by alien plant invasions. South Africa has a long colonial history dating back 350 years, and it also has a well-developed infrastructure, with thriving agricultural and forestry sectors. These factors have contributed significantly to the introduction, establishment, and spread of invasive alien plants.[11,12] About 750 of the ±9000 introduced species are trees and shrubs, and they make up a disproportionate fraction of the invading species; woody plants comprise 110 of the 161 species regarded as seriously invasive, compared with 38 herbaceous and 13 succulent species.[5,13] Many more of the introduced species are likely to become important weeds in the future. Several surveys have been made of the spatial extent of alien plant invasions in South Africa. Unfortunately, each survey used a different method or concentrated on particular species or area. For these reasons, and because surveys were done at different times, these results cannot easily be merged to produce a national overview. Richardson et al.[14] reviewed available data on the extent of alien plant invasions in different parts of South Africa. The most comprehensive set of records is the South African Plant Invaders Atlas,[5,15] which can be summarized as frequencies by quarter degree (latitude and longitude) but cannot be easily converted to estimates of the extent of the invaded area. The same applies to the farmer surveys for the national desertification audit.[16]

Biological invasions have received much attention from South African scientists and land managers in the past, and useful reviews have been provided by Macdonald et al.,[17] Richardson et al.,[18] and Richardson et al.[14] Most of the South African research has been on the history, ecology, and management of invasive alien species, and studies aimed at properly quantifying the environmental and economic aspects and consequences of invasions began a while later.[19] The situation changed as people started recognizing the magnitude and wide-ranging consequences of invasions. A decisive breakthrough was the demonstration of the significant current and potential impacts of invasive alien trees and shrubs on water resources in South Africa.[20–22] This work demonstrated the economic benefits of management interventions and led to the establishment of the government-funded Working for Water program aimed at the control of invasive alien plants to protect water resources and ensure the security of water supply.[23–25] The South African government has spent more than US$657 million (US$1 = ZAR7.5) on this program between its inception in 1995 and 2009.

In this chapter, we review the environmental impacts of alien plant invasions in South Africa and their economic consequences. Unfortunately, there is insufficient quantitative information for us to include a proper assessment of their social and infrastructural impacts, such as those on human health (e.g., harboring disease vectors, causing allergies) or increased flood damage when riparian invaders are dislodged and block bridges, culverts, and storm drains. Although there have been relatively few studies of the economic impacts, those that have been done have made some significant advances and provide new perspectives on the problem. We also review the unique solutions that South Africa has pioneered in dealing with the problem and provide an overview of the economic benefits and sustainability of such approaches.

## 14.2   *Extent of the problem*

The gaps in our knowledge of the extent and distribution of invasive alien plants in South Africa have been partially filled, at least for woody shrub and tree species, by a rapid reconnaissance of the extent of invasions in South Africa undertaken in 1996–97.[26,27] This survey excluded commercial tree plantations, urban and metropolitan areas, and most nonwoody plant species. The data from this survey are now about 15 years old and need to be interpreted with caution given the limitations and uncertainties in the data and the time that has elapsed. An update of this national survey is nearing completion, and indications are that the total area invaded may be greater (A. Wannenburgh, Working for Water National Office, February 2010, pers. comm.).

The 1996–97 national survey found that about 10 million ha of South Africa had been invaded to some degree by the ±180 species that were mapped (Table 14.1). The Western Cape province had the most extensive invasions followed by the Limpopo and Mpumalanga provinces. The KwaZulu-Natal and the Eastern Cape provinces were not adequately mapped, and the true extent of the invasions in these provinces is likely to be closer to the percentage for Mpumalanga, which has similar climate, vegetation, history of colonization, and land-use patterns.[27,28] Most of the invasions are concentrated in the wetter regions of the country, and this is reflected in the number of species that have been recorded per quarter-degree square. The greatest number of species occurs in the Western Cape and along the eastern escarpment from KwaZulu-Natal through to the Limpopo province (Figure 14.1).

The data from the national survey are summarized by province, and not by the country's major biomes. However, it is possible to get indications from the distribution of the invasions and the representation of the biomes in the provinces. The Fynbos biome is the best studied and the most invaded with extensive dryland invasions in both the mountains and the lowlands, as well as invasions along all the major river systems.[18,19,27,29] Most of the Fynbos biome is located in the Western Cape, but the small portion of this biome in the Eastern Cape (10,300 km² or 6%) is also heavily invaded. The major invaders are trees and shrubs in the genera *Acacia*, *Hakea*, and *Pinus*. At the scale of the whole Cape Floristic Region (comprising mainly fynbos vegetation types but also parts of the Succulent Karoo,

*Table 14.1* Areas Invaded by Alien Plants in the Nine Provinces of South Africa, and the Mean Canopy Cover

| Province | Major biomes[1] (% of province) | Area[1] (km²) | Total area[2] invaded (km²) | (%) | Mean canopy[2] cover (%) |
|---|---|---|---|---|---|
| Eastern Cape | Grassland (40), Nama Karoo (25), Thicket (16) | 167,398 | 6,720 | 4.01 | 22.51 |
| Free State | Grassland (72), Nama Karoo (22), Savanna (6) | 129,936 | 1,661 | 1.28 | 14.56 |
| Gauteng | Grassland (78), Savanna (22) | 16,519 | 223 | 1.35 | 58.56 |
| KwaZulu-Natal | Savanna (54), Grassland (36), Thicket (8) | 94,596 | 9,220 | 9.75 | 27.21 |
| Mpumalanga | Grassland (64), Savanna (36) | 79,571 | 12,778 | 16.06 | 14.49 |
| Northern Cape | Nama Karoo (54), Savanna (30), Succulent Karoo (14) | 361,981 | 11,784 | 3.26 | 14.10 |
| Limpopo | Savanna (97), Grassland (3) | 122,143 | 17,028 | 13.94 | 15.45 |
| North West | Savanna (71), Grassland (29) | 116,010 | 4,052 | 3.49 | 13.88 |
| Western Cape | Fynbos (47), Nama Karoo (24), Succulent Karoo (24) | 129,314 | 37,274 | 28.82 | 16.80 |
| South Africa (including Lesotho) | | 1,217,467 | 100,739 | 8.07 | 17.23 |

*Sources:* Data from: (1) Low, A. B., and A. G. Rebelo, *Vegetation of South Africa, Lesotho and Swaziland*, Pretoria: Department of Environmental Affairs and Tourism, 1996; (2) Le Maitre, D. C., D. B. Versfeld, and R. A. Chapman, *Water SA* 26:397, 2000.

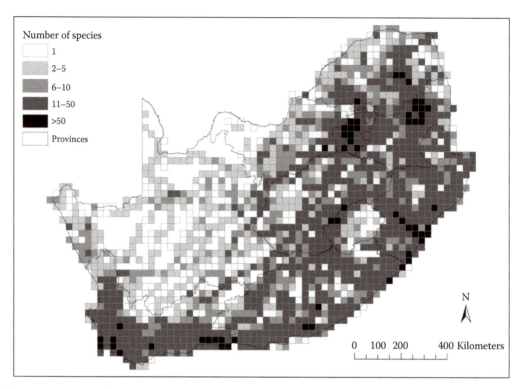

**Figure 14.1** Map of South Africa showing the number of invading alien plant species recorded per 1/4 × 1/4 degree of latitude and longitude and the provincial boundaries. Blanks indicate that there are no records from those areas yet. For more information, see Henderson.[5] (Data from the South African Plant Invaders Atlas Database, copy provided by L. Henderson, personal communication March 2010.)

Nama Karoo, thicket, and forest biomes), the data from two independent assessments of the extent of dense stands are reasonably similar (Table 14.2), but Versfeld et al.[26] clearly underestimated the extent of the lightly infested and overestimated the extent of the medium-infested invasions on the Agulhas Plain. Nevertheless, these studies all highlight the extensive invasions in the Fynbos biome.

The forest biome has been heavily invaded, but the extent cannot be quantified at present.[14] The grassland and savanna biomes have also been extensively invaded, mainly by acacias, pines, poplars, willows, and other tree species, and a variety of woody scramblers (e.g., brambles). The worst affected areas are the grasslands of the Drakensberg escarpment and the moist savanna biome along the lower escarpment and in the KwaZulu-Natal midlands and coastal belt. Most of the invasions in these biomes are found along the river banks and in the river beds; there are few, if any, river systems that have not been extensively invaded. Invading trees such as syringa (*Melia azedarach*) and jacaranda (*Jacaranda mimosifolia*) have spread into semiarid savanna by invading along perennial rivers where the freely available water allows them to survive the seasonal drought. Peruvian pepper (*Schinus molle*) has invaded large areas of semiarid savanna, especially the area around Kimberley where the species has had a major impact on natural processes.[30] The Nama Karoo (semidesert shrubland, summer rainfall) is probably the fourth most-invaded biome; woody invaders, notably mesquite trees (*Prosopis* species), have invaded at least 18,000 km² of the low-lying alluvial plains and the seasonal and ephemeral watercourses. Several cacti

*Table 14.2* A Comparison of the Extent of Invasions in the Cape Floristic Region (CFR)[a] and Agulhas Plain[b]

| Area | Source | Total study area (km²) | Density class (canopy cover) | | |
|------|--------|------------------------|-------------|----------------------|----------------|
| | | | Light (<25%) | Medium (25%–75%) | Dense (>75%) |
| Cape Floristic Region | 25 | 94,393 | 29,575 | 6,968 | 1,226 |
| | 28 | 87,892 | – | – | 1,394 |
| Agulhas Plain | 28 | 2,377 | 22 | 1,794 | 107 |
| | 98 | 2,161 | 1,081 | 167 | 302 |

*Source:*  Data from Versfeld, D. B., D. C. Le Maitre, and R. A. Chapman, Alien Invading Plants and Water Resources in South Africa: A Preliminary Assessment. Report TT99/98, Water Research Commission, Pretoria, 1998.

*Note:*  Summary statistics for the light and medium invasions for the CFR were not given by Cowling et al.[29]

[a]  Cowling, R. M. et al. 1999. *Framework for a Conservation Plan for the Cape Floristic Region.* Institute for Plant Conservation, University of Cape Town.

[b]  Cole, N. S. et al. 2000. *Framework for a Conservation Plan for the Agulhas Plain, Cape Floristic Region, South Africa.* Institute for Plant Conservation, University of Cape Town.

(*Opuntia* species) and saltbushes (*Atriplex* spp.) have invaded large areas of the Nama Karoo and Succulent Karoo (winter rainfall) biomes[31] and the thicket biome in the Eastern Cape.[14]

Several aquatic weeds have spread over large areas in South Africa, notably water hyacinth (*Eichhornia crassipes*), water lettuce (*Pistia stratiotes*), kariba weed (*Salvinia molesta*), parrot's feather (*Myriophyllum aquaticum*), and red water fern (*Azolla filiculoides*).[32] These species, in the absence of natural enemies and the presence of eutrophic waters, form large dense mats that degrade aquatic ecosystems and impact on all aspects of water utilization. Four of the five species are effectively under biological control, and the only real problem species is water hyacinth (where biological control has not been developed to its full potential). It is widespread throughout South Africa and has caused severe impacts on rivers in the Western and Eastern Cape, Kwazulu-Natal, Mpumalanga, and on the Vaal River in the Gauteng and Free State provinces.

## 14.3   *Quantifying the environmental impacts*

The development of an understanding of environmental impacts of invasive alien plants, and their consequences, is essential for the quantification of economic impacts. Unfortunately, there is no standard method for objectively quantifying the many and varied environmental impacts of invasive alien plants worldwide,[33] although an approach has been proposed.[34] As in other parts of the world, impacts of plant invasions in South Africa have been measured in a variety of ways, which makes comparisons between biomes or with other regions or countries difficult.[35] Many descriptions of impacts are anecdotal or correlative (comparing invaded sites with noninvaded sites or comparing one site at different times) or are based on the performance of the invader in other parts of the world. Very few detailed studies and manipulative experiments have been done to determine the magnitude of impacts, mechanisms of impacts, and implications of impacts of invasions of alien plants in South Africa. Nevertheless, we can draw some conclusions on the types and magnitude of impacts of the most important plant invaders (Table 14.3). The types of impacts include effects on individual plant populations or species (including genetics), effects on population dynamics of native species, effects on community dynamics (species richness/diversity, trophic structure), and effects on ecosystem processes and functioning.

Table 14.3 Examples of Environmental Effects of Invasive Alien Plants in South Africa

| Invader (life form) | Biomes affected | Disruption | Evidence | References |
|---|---|---|---|---|
| Acacia cyclops (T) | Fy, Fo, Sa | Δ in coastal sediment dynamics | O | 104 |
| | | Δ in seed dispersal dynamics | OD | 105,106 |
| | | Provides nesting habitat for rare African penguins | O | R. J. M. Crawford, unpublished data |
| | | Outcompetes native plants | OD | 107 |
| | | ↑ in biomass | OD | 108,109 |
| | | ↑ in litterfall | OD | |
| | | Δ in fire regime | OD | 110 |
| Acacia longifolia (T) | Fy, Sa | ↓ in diversity of ground-living invertebrates | OD | 111 |
| | | ↑ in streamflow after clearing | OD | 49 |
| Acacia mearnsii (T) | Gr, Fy, Fo, Sa | ↓ diversity of ground-dwelling invertebrates | OD | 111 |
| | | ↓, displace endemic dragonfly species | OD | 112 |
| | | ↑ in streamflow following clearing | OD | 39,113 |
| | | ↑ in evaporation | OD | 114,115 |
| | | Sediment accumulation; destabilization of streambanks; ↑ erosion | O | 14,116 |
| Acacia saligna (T) | Fy, Fo, Sa | ↑ in biomass | OD | 108,109 |
| | | ↑ in litterfall | OD | |
| | | Δ in nutrient chemistry in lowland fynbos | OD | 38,39,40,53 |
| | | Δ in seed dispersal dynamics | OD | 106 |
| | | ↑ in biomass; Δ in size and distribution of fuel; ↓ in moisture content = Δ in fire regime | OD | 117,110 |
| | | Attrition of seed banks of native plants with time in dense stands | OD | 53,54 |
| | | Groundwater pollution with nitrates | OD | 118 |
| Caesalpinia decapetala (CR) | Sa, Gr, Fo | Dense mats cause trees to collapse | O | 119 |
| Chromolaena odorata (SW) | Fo, Sa | ↑ in flammability in forest and riverine woodland; forms "ladders" that carry fires into crowns of fire-sensitive trees | O | 120,121,122 |
| | | ↓ in biodiversity of ecotones | | |
| | | ↓, shades crocodile nesting sites; altering the gender ratio of the embryos in the eggs | OD | 123 |

| Species | | Consequence | | Ref |
|---|---|---|---|---|
| *Eichhornia crassipes* (FFM) | Aq | ↓ in spider abundance, Δ in assemblage patterns, ↓ in diversity and estimated species richness with ↑ in density and duration of invasion | OD | 124 |
| | | ↓ in carrying capacity for livestock and game | O | 122 |
| | | ↓, degrades aquatic ecosystems | O | 125 |
| *Eucalyptus* spp. (T) | Fy, Gr, Fy, Sa, Fo | Δ in river flows ↑ in water repellency and soil erosion | OD | 126 |
| *Hakea sericea* (SW) | Fy, Fo | ↑ in biomass; Δ in size and distribution of fuel; ↓ in moisture content; net result is Δ fire regime | OD | 117 |
| | | ↑ in biomass; results in very intense fires when felled plants are burnt (water-repellent soils—erosion) | OD | 127,128 |
| | | Dense stands limit options for fire management | O | 129 |
| | | No change in indigenous mycorrhizal flora | OD | 130 |
| | | Δ in vegetation structure—↓ in abundance and diversity of native birds; Δ in arthropod community structure (some taxa ↑, some ↓); ↓ in leaf retention and % seed set in native Proteaceae | OD | 131 |
| *Lantana camara* (SW) | Fo, Sa, Fy, Gr | ↓ in diversity of ground-dwelling invertebrates | OD | 112 |
| | | ↓, suppresses regeneration via allelopathy | OO | 132 |
| | | poisons livestock (R1.7 million/y) | OD | 132 |
| *Melia azedarach* (T) | Fo, Sa | Outcompetes native plants | O | 120 |
| | | Δ in feeding dynamics of frugivorous birds | O | 120 |
| *Myriophyllum aquaticum* (FAM) | Aq | ↓, degrades aquatic ecosystems | O | 133 |
| | | Δ in river flows | | |
| | | ↑ in mosquitoes and diseases | | |
| *Nasella trichotoma* (AG) | Gr | ↓ in pasture productivity | OD | 134 |
| *Opuntia aurantiaca* (SS) | Ka, Sa | ↓ in pasture productivity | OD | 134 |
| *Opuntia ficus-indica* (SS) | Sa, Ka, Gr, Fo, Fy | ↓ in pasture productivity | OD | 134 |
| *Pereskia aculeata* (CR) | Fo | Dense mats cause trees to collapse | O | 119 |
| *Pinus* species | Gr | ↑ in streamflow after clearing | | 113 |

*(Continued)*

*Table 14.3* Examples of Environmental Effects of Invasive Alien Plants in South Africa (*Continued*)

| Invader (life form) | Biomes affected | Disruption | Evidence | References |
|---|---|---|---|---|
| *Pinus pinaster* (T) | Fy, Fo, Sa | Outcompetes native plants | OD | 107,135 |
| | | Dense stands limit options for fire management | O | 129 |
| *Pinus radiata* (T) | Fy | ↓ in streamflow | OD | 20,27,48 |
| | | No change in indigenous mycorrhizal flora | OD | 130 |
| *Prosopis* spp. (T/SW) | Sa, Ka | ↑ in biomass; Δ in vegetation structure; ↓ in access, ↓ in pasture productivity | O | 14,136 |
| | | Outcompetes native plants | O | 137 |
| | | ↓, clearing reduces declines in groundwater level | OD | F. Fourie, Water Affairs, Kimberly, unpublished data |
| | | ↓ in diversity of dung beetle assemblages | OD | 138 |
| | | ↓ in diversity of bird species, particularly raptors and frugivores | O | 139 |
| *Psidium guajava* (T) | Fo, Sa, Gr | Outcompetes native plants | O | 120 |
| *Rubus* spp. (SW) | Sa, Gr, Fo | Hybridizes with native *Rubus* sp. | O | 140 |
| *Salix babylonica* (T) | Ka, Gr | Destabilizes river banks and excludes native plants | O | 141 |
| *Salvinia molesta* (FFM) | Aq | ↓, degrades aquatic ecosystems; Δ in river flows; ↑ in mosquitoes and diseases | O | 142 |
| *Schinus molle* (T) | Sa | Δ in prevailing seed dispersal processes; ↓ in health of native trees due to superior competition for light | OD | 143 |
| | | | OD | 30 |
| *Sesbania punicea* (T) | Sa, Fo, Fy, Gr | ↓ in access, ↑ in bank erosion, ↓ in streamflow | O | 144 |
| | | poisoning of stock | O | 134 |
| *Solanum mauritianum* (T) | Sa, Fo, Gr | ↓ in diversity of ground-dwelling invertebrates | OD | 111 |
| | | Δ in feeding ecology of Rameron Pigeon (and other native birds) | OD | 145 |
| | | Outcompetes native plants | O | 120 |

Life forms: T = tree; SW = shrub (woody); SS = shrub (succulent); V = vine; AG = annual grass; CR = creeper; PG = perennial grass; AF = annual forb; PF = perennial forb; FFM = free-floating macrophyte; FAM = floating macrophyte (attached).

Biomes affected: Fo = forest; Fy = fynbos; Gr = grassland; Sa = savanna; Ka = karoo.[14]

Disruption: Δ = change; ↑ = increase; ↓ = decrease.

Evidence: OD = observation with data; O = observation without data; OO = observation in another part of the world.

When assessing the impacts of invasive plants, it is useful to consider two categories of invaders, as proposed by Chapin et al.[36] (see also Dukes and Mooney[37]): discrete-trait invaders (DTIs) and continuous-trait invaders (CTIs). DTIs add a new function such as nitrogen fixation to the invaded ecosystem, whereas CTIs differ from natives only in traits, such as litter quality or growth rates, that are distributed continuously among species. Results from several parts of the world show that DTIs generally have greater ecosystem-level impacts than CTIs, although the latter can also bring about such impacts, especially if they make up a large proportion of an ecosystem's biomass.

The most dramatic impacts of alien plants in South African systems have clearly been from DTIs. For example, the Australian *Acacia* species (notably *A. cyclops* and *A. saligna*) have radically altered nutrient-cycling regimes in the nutrient-poor systems of lowland fynbos due to their ability to fix atmospheric nitrogen (no widespread or abundant native species perform this role). Such impacts translate to marked reductions in native plant species richness.[38–41] These species and several species of *Pinus* that have invaded large areas have added a major new life form (trees) to the tree-poor fynbos. Such invasions have produced many ecosystem-level changes by altering features such as biomass distribution, plant density and vegetation height, leaf-area index, litterfall and decomposition rates, and fire behavior. These invasions are the largest threat to endangered plant species in the Cape Floristic Region.[42] Widespread tree and shrub invasions in vegetation types that previously had a low tree cover have radically altered habitats for animals. For example, increased tree cover (e.g., Macdonald[43]) has resulted in major changes in the distributions of many bird species, including species that invaded the southwestern parts of South Africa from adjacent biomes. Other DTIs that have caused major impacts in South African ecosystems include *Chromolaena odorata*, *Pereskia aculeata*, *Schinus molle*, *Prosopis* species and the suite of alien trees that have invaded watercourses in formerly tree-poor systems.

However, DTIs are not limited to the land, as is clearly the case with water hyacinth and red water fern (*Azolla filiculoides*).[44] Aquatic invaders such as these grow rapidly and form expansive colonies of interwoven submerged and/or floating plants. The impacts of these dense colonies are diverse and can include increasing water losses due to evapotranspiration and depleting oxygen in aquatic communities, which ultimately affects fisheries and biodiversity,[45] reducing the quality and quantity of potable water, obstructing navigation, and blocking water drainage patterns, which can cause flooding or even result in bridges collapsing. Aquatic weeds also hinder irrigation by impeding water flow, clogging irrigation pumps, and interfering with weirs.[46] Infestations also block access to recreational areas and decrease waterfront property values, often impacting the economies of communities that depend upon fishing and/or water sports for revenue.[44] Aquatic weeds thus impact on all aspects of water resource utilization, including fisheries, transport, hydropower generation, tourism, and the quantity and quality of potable water. Aquatic weeds, however, also have some positive uses. Water hyacinth, for example, can be processed in many ways, including feed for cattle, sheep and pigs; mulch and compost for crop production; fiber for paper making and weaving baskets and mats; biological filtration; and the production of biogas (methane).[47] Large-scale processing of water hyacinth, however, is seldom commercially viable as the plant contains 96% water and harvesting is thus very expensive. It is possible, however, to make a living by processing water hyacinth on a small scale. The promotion of water hyacinth use for these benefits should probably be avoided, however, as no utilization program will ever check the growth of water hyacinth, and promoting utilization will lead to the further spread of the weed as people become aware of its economic potential.

The most detailed work on assessing the impact of plant invasions in South Africa has focused on tree invasions in the Fynbos biome. Several studies have documented the reduction in streamflow caused by plantations of invasive tree species (mainly *Pinus* species),[20,48,49,50] and aspects of the modified nutrient regime[38,39,40,51–52] and impacts on the seed banks of native fynbos species[53,54] associated with *Acacia saligna* invasions. A country-wide study made a first attempt at assessing the effect of all woody invaders on surface water resources. The results of this study suggest that woody alien plants may be using as much as 6.71% of the total mean annual surface runoff or 9.95% of the utilizable surface runoff in South Africa.[26,27] Subsequent studies using more refined models have tended to provide similar estimates of the magnitude of reductions in river flows and the sustainable yield available from water supply schemes (e.g., dams).[55–57] Based on estimates made by Cullis et al.[56] of the current losses in high yielding catchments of South Africa (mean annual rainfall > 800 mm), the current flow reduction was about 5% of the mean annual runoff and would increase to 20% in the future if the riparian zone was allowed to become fully invaded. The gains made from clearing operations to date have been estimated to be about $34 \times 10^6$ m$^3$ per year just from the 7% of the riparian invasions that have been cleared.[58]

This chapter shows that the understanding of the environmental impacts of invasive alien plants is scattered and patchy. No overall framework for the collection and synthesis of this information has been attempted, and as a result, there are few overall syntheses of impacts at the scale of an ecosystem, province or whole country. Nonetheless, the studies that have been done could provide a useful starting point for the development of such a framework.

## 14.4   The economics of invasions

Invasions by alien plant species are one of many unintended consequences of economic activity that impose real costs on society. The globalization of the world economy and trade liberalization have facilitated invasions because many alien species are used in economic activities, especially agriculture and forestry.[12] Economic activities generally result in the large-scale conversion of natural vegetation, the promotion of trade in goods, and modern transportation systems to move them rapidly across the globe. Human behavior is one of the primary determinants and facilitators of the establishment and the spread of invasive alien plants.[12,59,60] The problem of invasions, therefore, has as much to do with economics as with ecology, and proposed solutions to these problems must, therefore, be grounded in both of these sciences.

Economic assessments of plant invasions involve the economic valuation of both market and nonmarket impacts, and the evaluation of trade-offs and risks relating to the impacts associated with various planning and management choices. Economic assessments, when implemented appropriately, improve the transparency of decision-making processes by providing welfare-based justifications for the measures being implemented. The value of an economic assessment lies, therefore, not in the precision of the numbers that are generated but in the extent to which the results enable decision makers to make the right decisions.

A key element in invasive alien plant control is the regulation of human behavior through the use of suitable behavioral instruments (e.g., policies and strategies) within a comprehensive and robust institutional framework. This requires decision and policy makers to have an understanding of how existing institutional, regulatory, and market conditions create the incentives that drive behavior, how infestations spread over time, and the costs of the externalities associated with invasive alien plants. However, there are numerous complications and challenges that impede the development of this understanding. First, it is extremely difficult to quantify the consequences of invasions in economic terms because of the lack of information on the behavior of the invasive plants, the impacts they have on their environment, and how these impacts multiply through the socioeconomic system. A second complicating factor is the human tendency to ignore and avoid a problem until it can no longer be overlooked. This behavior can be explained by the fact that humans are generally poor at identifying and accounting for low-probability, high-impact events.[61] Technically, this means that the relevant production curves are nonconvex and that an expected-utility approach to decision making will fail to predict the decisions people will actually make in such circumstances. This, in turn, implies that the positive cost–benefit ratios associated with cost-avoidance behavior will count for less in an expected utility approach and thus provide little incentive to change behavior. A third complication is the existence of invasive alien plants that have both beneficial and deleterious impacts (i.e., "conflict-of-interest" species*) as this can make the categorization of the impacts as deleterious or beneficial difficult, and raises many questions about compensation (i.e., whether to permit the species to continue invading while compensating those experiencing losses or to eradicate it and compensate those who were benefiting). A fourth complication that affects the decision-making environment is that estimates of the net economic impact of invasive species vary with the scale, scope, timing, and approach of the analysis. In such cases, it is essential that all assumptions and limitations are made explicit. Finally, the control of invasive alien plants is a public good in the sense that if it is left to the market, control measures will be underapplied. These problems justify a precautionary approach to invasive alien plant management and provide motivation for more research on the risks associated with no control and different types of control.

Cost considerations are an effective motivator for management interventions, so it is unfortunate that monetary valuation techniques still are not able to provide fully inclusive estimates of these costs. However, useful insights can be gained by examining the drivers of these costs,[66] and this can help in quantifying the magnitude of the impacts and their economic costs and in identifying the affected parties. Examples of drivers that affect costs by altering the natural provision of ecosystem goods and services include the following (after Culliney;[66] Table 14.3):

- Reduced biodiversity through competition or hybridization
- Simplified food webs
- Altered fire regimes and hydrological cycles
- Modified soil chemistry and biology through, for example, changes in pH and nutrient cycling, salt accumulation, nitrogen fixation, or altering the composition of the soil fauna and flora
- Altered geomorphologic processes through the promotion of siltation or erosion of stream banks and sand dunes
- Altered activity of pollinators
- Competition with valued plants (food and fiber crops, ornamentals, timber) for water, nutrients, and sunlight, and, additionally, reducing crop yields through allelopathy
- Reduced quality of agricultural products (i.e., contaminants and toxicity)

---

* Examples of "conflict of interest" species are found in the commercial forestry industry (*Pinus* species[62]), commercial and communal firewood industries (many *Acacia* species[63]), food production (*Opuntia* species[64]), fodder production (*Prosopis* species), or for nectar for bees (*Eucalyptus* species[65]), and where they have aesthetic or nonuse industries such as utilitarian value (ornamentals, shade trees, or windbreaks).

## 14.4.1   Approaches to economic evaluation

Invasive alien plants will have positive impacts on some members of society and negative impacts on others. The magnitudes of these impacts will also not be equally distributed. It is therefore important to identify those who are being affected by the species and whether these effects are positive or negative. This raises an important issue in valuation theory. Valuations are always done from the perspective of a beneficiary, which implies that the beneficiary needs to be defined before a valuation can be done. This is one of the reasons why valuations on invasive alien plant-derived impacts lag behind the research on the impacts. Very few studies have managed to identify a full set of beneficiaries before beginning the valuation process.

In the mid-1990s, South African scientists began to quantify the impacts of invasions by woody plants on water resources in South Africa.[20,27] These studies were pivotal in persuading the government to take the invasive alien plant issue more seriously (South Africa is a water-scarce country) while at the same time providing the impetus for further studies on the economics of plant invasions. A number of studies of the physical and economic impacts of invasive alien plants in South Africa have been carried out over the last decade (Table 14.4). They vary in terms of their economic rigor and detail, with some merely reporting the direct costs involved in clearing and others providing detailed economic analyses, which makes these studies difficult or impossible to aggregate. Most report positive returns on investment, especially those in biological control research, a topic we discuss in Section 14.5.2. A couple of case studies have attempted to provide a more comprehensive assessment: one for an ecosystem and the other for a particular invading species. In 1997, the value of a hypothetical 4 km² of mountain fynbos for water production, wildflower harvesting, recreational hiking, ecotourism, endemic species, and genetic storage was estimated at US$3 million with no alien plant management and US$50 million with effective alien plant management.[67] A study of *Acacia mearnsii* (black wattle) invasions in South Africa[68] estimated that the net present cost of the loss of water to the economy amounted to US$1.4 billion, far more than the value generated by the wood chips, tanbark, and other products of the commercial wattle growing industry.

These studies build a strong case for control because all of them show that control makes economic sense. However, it is still difficult to make the case when only the marginal differences are considered, that is, the costs of the additional impacts caused by the invasions, or invaded state, when compared with the uninvaded state. For example, what is the marginal difference in the cost of fire damage due to the impacts of invasive alien plants on fire behavior? Although we know there are differences, particularly the increases in soil loss and stormwater flows (Table 14.3), the costs of these marginal differences have yet to be estimated.

Most of the benefits of control are clearly derived from curtailing the invasives' spread and reducing their densities and thus avoiding or mitigating future negative impacts. The challenge then is to justify the public expenditure that will mitigate or avoid these costs. In this context, the "conflict-of-interest" species[68] are particularly difficult to deal with because they have both positive and negative impacts, that is, benefits and costs. For example, invasions by *Prosopis* species decrease groundwater yield and access by livestock and farmers but also provide fodder, shade, and shelter for livestock[69,70] and potential for use as a biofuel feedstock. A study in a rural community in the Eastern Cape found that alien plant species can make a substantial contribution to livelihoods by supplying food, fuelwood, and a source of income.[71] Quantifying, balancing, and allocating the true costs and benefits of such species should be explicitly addressed within comprehensive, multi-criteria frameworks, which also take livelihood strategies into account.

*Table 14.4* Studies on Some of the Economic Impacts of Alien Invasive Plants in South African Ecosystems

| Title | Findings | References |
|---|---|---|
| Invasive plants and water resources in the Western Cape province, South Africa: Modeling the consequences of a lack of management | Invading alien plants use an equivalent of up to 30% of the City of Cape Town's 1995 water demand at an estimated tariff of US$0.002/m³. | 20 |
| Valuation of ecosystem services: a case study from the Fynbos biome, South Africa | Invasive alien plant clearing can supply water at 14% of the cost of the development of new bulk water supply schemes (US$0.002 vs. US$0.012/m³). | 21 |
| The sustainable development of water resources: History, financial costs, and benefits of alien control programs | | 22 |
| An ecological economic simulation model of mountain fynbos ecosystems: Dynamics, valuation and management | Managing alien plants in mountain fynbos areas increases the value of land (particularly for lifestyle buyers). A hypothetical 4 km² area appreciated in value from US$3 million to US$50 million after it was cleared from alien plants. This could be achieved by spending a fraction of total value on clearing programs. | 67 |
| A benefit–cost analysis of removing alien trees in the Tsitsikamma mountain catchment | Clearing yields benefit:cost ratios of between 6:1 and 12:1 when focused on invaded watersheds and between 360:1 and 382:1 for clearing unprofitable plantations. | 146 |
| Economic consequences of alien infestation of the Cape Floral Kingdom's fynbos vegetation | Associated production losses due to invasion ranged from US$9.7 to US$2.3 per hectare and from US$8.3 to US$1 per hectare for recreational use. The value of water-associated losses was estimated to be US$163 per hectare. | 63 |
| Conflicts of interest in environmental management: Estimating the costs and benefits of black wattle (*Acacia mearnsii*) in South Africa | Continued cultivation of *Acacia mearnsii* without control programs yields a benefit:cost ratio of 0.4:1. Continued cultivation with clearing or with a combination of clearing and biological control of seeds yields benefit:cost ratios between 2.4:1 and 7.5:1. | 68 |

*(Continued)*

*Table 14.4* Studies on Some of the Economic Impacts of Alien Invasive Plants in South African Ecosystems (*Continued*)

| Title | Findings | References |
| --- | --- | --- |
| Economic evaluation of the successful biological control of *Azolla filiculoides* in South Africa | The study reports direct cost estimates of the biological control program against *Azolla filiculoides*. The agent has brought the problem under control at an estimated cost of US$51,000, yielding a benefit:cost ratio of approximately 1130:1. | 67 |
| A cost–benefit analysis of the Working for Water program on selected sites in South Africa | This chapter addresses the feasibility of the Working for Water program in the Eastern Cape province and regions of the southern Cape. It is shown that catchment management on all the sites carried out by this program is inefficient. Changes in key assumptions, for example, a lower discount rate, would result in a positive economic benefit. | 89 |
| The clearing of invasive alien plants in South Africa: A preliminary assessment of costs and progress | Estimates of the costs of clearing important species of invasive alien plants and of progress made with clearing, based on data from a recently developed GIS-based project information system are provided. | 73 |
| Targeting emerging invasive alien plants for biological control in South Africa: The benefits of halting the spread of alien plants at an early stage of their invasion | This chapter reviews cases where emerging weeds were targeted for biological control in South Africa, the successes that were achieved, and the prospects for enhancing this approach in the future. | 91 |
| Costs and benefits of biological control of invasive alien plants: Case studies from South Africa | This chapter describes an attempt to estimate the costs and benefits of the biocontrol of six weed species in South Africa. Benefit:cost ratios for the historical analysis (from the release of the biocontrol agent to the year 2000) ranged from 8:1 for *Lantana* (*Lantana camara*) to 709:1 for jointed cactus. When future estimates of benefits were considered, benefit:cost ratios were greater and ranged from 34:1 for *Lantana* to 4333:1 for golden wattle. | 93 |

| Title | Description | Ref. |
|---|---|---|
| The role of resource economics in the control of invasive alien plants in South Africa | This chapter reviewed the contribution of resource economics tools and methods to support the business case for IAP control. It emphasized the public good nature of IAP infestations limited opportunities and incentives to clear IAP from private lands. It emphasized the need to create suitable incentives for private decision makers to buy into IAP control and to avoid spreading. | 147 |
| The economic impact and appropriate management of selected invasive alien species on the African continent | This report attempted a total economic value assessment of selected IAP species and confirmed current limitations to such an approach. | 44 |
| Economic evaluation of biological invasions: A survey | The study assessed the suitability of economic evaluation studies in terms of their ability to support the decision process of IAP control. The various international examples included a number of South African studies. The survey confirmed some of the limitations associated with economic evaluations (mostly due to simplifying assumptions necessitated by a lack of supporting data). | 148 |
| The Working for Water program: Evolution of a payments for ecosystem services mechanism that addresses both poverty and ecosystem service delivery in South Africa | The study reported on the Working for Water program that aims to simultaneously engage in the public good challenge (as mentioned in the study by Turpie[147]) and poverty alleviation in South Africa. It excluded the latest budget figures of the program. | 24 |
| Restoration of water resources (natural capital) through the clearing of invasive alien plants from riparian areas in South Africa—costs and water benefits | This chapter reports the clearing costs and impacts of the Working for Water program since its inception in 1995 and discussed the extent, costs, and impacts of clearing IAP from riparian areas. | 58 |
| An economic assessment of the contribution of biological control to the management of invasive alien plants and to the protection of ecosystem services in South Africa: Biological invasions | The study did a cost–benefit analysis on the biological control research program in South Africa. It included a detailed costing on the research program and a valuation of the monetary impact IAPs on different South African biomes. It analyzed the sensitivity of key assumptions and confirmed that biological control remains a cost-effective option. | 94 |

*Note:* Monetary values are in U.S. dollars, where values were published in rand, we have converted to U.S. dollars at a rate of R7 = US$1.

The following equation presents the economic components of such a framework:

$$PVC = DBC - DCC + PEoBC - NEoBC$$

where PVC is the present value of integrated control; DBC is the present value of direct benefits of control (i.e., the decrease in the negative impacts of invasion); DCC is the present value of direct costs of invasive alien plant control and is calculated as the sum of the research costs (RC), implementation costs (IC), and monitoring costs (MC); PEoBC represents the present value of positive externalities of control; and NEoBC represents the present value of negative externalities of control.

The research, implementation, and monitoring costs are each calculated, respectively, as

$$RC = Y_n*(CtC*S*T)$$

$$IC = LC + EC$$

and

$$MC = Y_m*(CtC*S_m*T_m)$$

where $Y_n$ is the number of years of a research project, CtC is the annual cost to a company of a "typical" scientist, $S$ is the number of scientists involved in the project, $T$ is the proportional time allocation per research project, LC is the present value of labor cost, EC is the present value of equipment cost, OC is the overhead component of the costs, $Y_m$ is the number of years of monitoring, $S_m$ is the number of scientists involved in monitoring, and $T_m$ is the proportional time allocation to monitoring.

The direct benefits of invasive alien plant control are defined in terms of decreases in the negative impacts of the target invasive alien plant. Maintaining a research capacity on invasive alien plant control is also seen as a direct benefit. However, the value of maintaining this capacity is encapsulated by the cost of avoiding the negative impacts of alien invasive plants.

Confidence in the robustness and accuracy of the above-mentioned approach depends on the inclusivity of the measurements, particularly of the externalities. Although the studies have endeavored to be as inclusive as possible, they cannot claim full inclusiveness at this stage, so the estimates should be considered to be conservative. Economic assessment is a highly data-intensive process and depends on an accurate baseline data set before any form of valuation can be done. Experience has shown that the assumptions or estimates of the rate of spread and control efficiency have a significant impact on the results, and care is needed to ensure that these are realistic and sensitivity analyses should be used to understand their effects on cost and benefit estimates.

### 14.4.2   A first assessment of the national-scale economic impacts of alien plant invasions

Although invasive alien plants have many types of impacts on the environment, communities, and infrastructure, three categories of environmental impacts are significant in the South African context and have been quantified to an extent that facilitates more thorough economic valuation[58]: (1) surface water runoff (quantified in terms of mean annual runoff), (2) livestock carrying capacity (grazing is the most extensive land-use practice for natural vegetation in South Africa and is quantified in terms of large stock units [LSU] per hectare), and (3) biodiversity (quantified using a biodiversity index). A combination of substitution cost estimates and revealed and stated preferences were used to estimate representative prices for these types of impacts in different biomes (Table 14.4).

The impacts of invasive alien plants on mean annual runoff were initially modeled for the fynbos catchments that supply Cape Town's water by Le Maitre et al.[19] The approach developed for this study has been used in numerous ecological and economic analyses of the impacts of invasive alien plants in South Africa.[24,27,32,56,57,69,72–74] Unit-price estimates for water are normally based on modified representative value estimates for water usage.[75] A utilitarian perspective is often adopted for assessments of economic impacts on water resources since this assumes that the value of water is derived from its use.[76] As a rule, two broad water-use categories are identified: serviced (treated and generally supplied in bulk to the consumers) and unserviced (raw) water (generally supplied direct from a river, storage reservoir, or canal system for irrigation or other bulk use). The primary distinction is that the cost of the serviced water includes the treatment and delivery costs in addition to the costs of capturing and storing the water. About 64% of the total surface water supplied to consumers in South Africa is sold as unserviced water with a per-unit marginal value of US$0.019/$m^3$ while the remaining 36% is supplied as serviced water which is sold at US$0.72/$m^3$. The current weighted average value of water may be taken as US$0.25/$m^3$ for South Africa, and this value was used to estimate the value of water losses at current and potential infestation levels.

A utilitarian economic perspective on biodiversity is built on the concept that individual species in an ecosystem provide inputs in the production of useful goods and services to humans (via a set of ecological functions and processes). Thus, the value of a species is derived from the values of the goods and services it supports. From this perspective, the economic benefits of maintaining biodiversity are significant because biodiversity maintains the resilience and productivity of the underlying ecosystem and thus the ecosystem services provided by a given area of land. This is true for both a direct use (e.g., grazing, harvesting of natural products) and an indirect use (e.g., recreation, carbon sequestration, water supply and quality regulation) perspective. Some studies have included aspects of the value of biodiversity in their analyses (Table 14.4). Van Wilgen et al.,[57] for example, used the biodiversity intactness index,[77,78] which measures the change in species abundance across all well-known elements of biodiversity relative to their inferred precolonial state, to quantify the impact of invasive alien plants on biodiversity at the biome scale. These impacts can be combined with unit-value estimates of their monetary values to calculate the economic impact of invasive alien plants on biodiversity. A substantial amount of grey literature exists on the unit pricing of biodiversity and ecosystem services, based on nonmarket valuation techniques such as contingency valuation.[24,79–82] Area-specific estimates of these economic values will differ, and it is risky to extrapolate area-specific estimates to a biome or provincial scale and vice versa. Careful consideration needs to be given to differences in the methods used and the spatial and temporal contexts.

Alien infestations degrade rangelands by replacing palatable with unpalatable species and restricting access by livestock and farmers. Van Wilgen et al.[57] quantified these impacts in terms of expected changes in livestock numbers across different biomes in South Africa and argued that the carrying capacity in South African biomes decreased by at least 123,000 LSU due to invasive alien plant infestations of rangelands. These figures can be combined with unit-price estimates to calculate the value of this impact. Data available from Statistics South Africa[83] suggest a unit-price estimate of US$329.47 per LSU.

Annual water-related losses were found to amount to US$773 million, whereas those associated with grazing amount to US$45 million.[58] The annual losses due to decreases in biodiversity and ecosystem service delivery were conservatively estimated at US$57 million per year. This amounts to an estimated total annual cost of invasive alien plant-related losses of US$867 million. This figure is equal to 2.5% of the 2009 South African gross domestic product[84] and compares favorably with the estimates of Pimentel et al.[2] of just over US$1 billion per year for all weeds. These figures exclude any positive externalities of a number of the plant species involved, so the final amounts may well be slightly lower. These positive externalities, however, cannot be estimated at present because the lists of conflicting interests and hence the drivers determining their values are not complete yet. Once the list has been completed, then these can be valued and balanced by applying the Hicks–Kaldor compensation principles for achieving Pareto improvements in economic efficiency.[85–87]

The examples described in this section show that the economic consequences of invasions are substantial. However, they only include the species and species' impacts that have been studied. A thorough analysis involving more species and a more comprehensive set of the impacts would undoubtedly reveal much higher costs because many species have only negative impacts.

## 14.5   *Dealing with the alien plant problem in South Africa*

The Working for Water program is unique in that it aims to simultaneously achieve two complementary outcomes: control of invasive alien plants and social development. Achieving both these aims has required the development of innovative approaches which are described in this section.

### 14.5.1   *Labor-intensive solutions to invasive alien plant problems*

Although invasions by alien plants have significant negative environmental and economic impacts, the South African government has used the opportunities offered by the need for labor-intensive clearing programs to generate a range of benefits. By adding these benefits to the obvious environmental and economic advantages, it has been possible to justify spending approximately US$657 million on the Working for Water program between 1995 and 2008–2009. Between 1995 and 2001–2002, the investment in the Working for Water program increased rapidly but the rate of increase slowed after that (Table 14.5). Between 1995 and 2009, the program employed an average of 22,800 people (largely from rural areas) in 249 projects around the country. The areas treated by the program increased rapidly from about 30,500 ha in 1995 to 200,800 ha of initial clearing and 36,000 ha of follow-up in 1997–1998. The initial clearing then declined as increasing investment went into follow-up to give an average of 154,700 ha of initial clearing per year and 420,800 ha per year of follow-up work.

The funds have been directed through the government's Working for Water public works program, which engages unemployed people in labor-intensive clearing, follow-up, and rehabilitation projects aimed at bringing alien plant invasions under control. The program also runs a parallel social development program that seeks to maximize the opportunities for development of disadvantaged people employed by the program.[24,25,58] The Working for Water program's social development activities are an integral part of the overall strategy to clear alien plants. The social development program has several

*Table 14.5* Annual Economic Benefits Associated with the Employment of People in Alien Plant Clearing Programs in the First Full Financial Year of the Program (1996–1997), After 5 Years (2001–2002), and in 2008–2009

| Measure of benefit | 1996–97 | 2001–02 | 2008–09 |
|---|---|---|---|
| Annual expenditure | 10.7 | 49.6 | 65.2 |
| Estimated expenditure on wages | 3.8 | 17.9 | 23.5 |
| Direct employment (number of people employed) | 8,386 | 14,558 | 25,339 |
| US$/employee/year | 458 | 1,226 | 926 |
| Approximate number of dependents (assuming five per family) | 41,900 | 72,790 | 126,695 |
| US$/person/year | 76 | 204 | 154 |

*Source:* Data for 1996–1997 from the annual report and extrapolated from Marais.[88] More recent data supplied by A. Wannenburgh, Working for Water National Office, January 2010.

*Note:* All expenditures and wages are in millions of U.S. dollars unless otherwise indicated (US$1 = ZAR7.5).

main components, including (1) a childcare program, whereby children are looked after in crèches, allowing women time to earn much needed income to support their families; (2) an HIV/AIDS program, involving HIV/AIDS awareness campaigns and condom distribution among workers; and (3) partnerships, which include an ex-offender reintegration program, and several other components. Working for Water employs the poorest members of communities settled closest to the alien infested areas.[25] The trend for Working for Water has been to target women, especially single mothers, and encourage them to join teams of about 20 members with supervisors to oversee their productivity.

The main social benefits from clearing activities extend beyond employment to include the improvement of poor people's livelihoods by providing income. This should result in better nutrition for children, better clothing, and an ability to pay for education, which is essential if those children are to realize their potential when they become adults; however, often the benefits are short term as the duration of employment is limited to 2 years per individual. Additionally, the social benefits come with a contribution toward economic empowerment through small-business training and skill development for contractors, leading to better opportunities to earn a living outside the program in some cases. In one of the few detailed studies of the social impacts of the program, Marais[88] quantified benefits in the Western Cape province during 1996–1997. We have updated these estimated benefits using Working for Water's figures from the 2008–2009 financial year (Unpublished data supplied by A. Wannenburgh, Working for Water National Office, January 2010), to estimate the value of the program on a per capita basis for the employees and their direct dependants (Table 14.5). About 36% of the total program budget is spent on wages so, for example, about US$11 million was spent on wages in the 1999–2000 financial year. Average annual wages between 2000 and 2009 were US$787 per employee. If we assume five dependants per employee, this figure translates to US$131 per person per year (Table 14.5). This injection of funding into disadvantaged communities also had secondary effects in terms of suppliers and service providers. Protective clothing, tools, and mechanical equipment constituted most of the supplies bought by the project and a number of secondary jobs were linked to the project through procurement.

The program, therefore, uses financial capital to protect and conserve natural capital (and the services it provides) and to develop human capital through training and job experience and social capital from the community level (e.g., enhancing social support systems) to the national level (e.g., policies, legislation, interorganizational cooperation). The program has also pioneered the development of systems for payments for ecosystem services where, for example, water users fund the clearing programs in return for the water that is made available for other uses or for restoring degraded river systems and estuaries.[24] The program has been successful in raising the general public awareness about invasive alien plants and in reaching agreements with the nursery industry to phase out the promotion and production of species known to be invasive and to endorse the sale of alternative indigenous species. Although there has been progress in these areas, the evidence base for their effectiveness remains weak. Effective monitoring and evaluation systems and well-targeted research programs are needed to provide clear evidence that the benefits of the program outweigh the costs and justify sustaining the large investment in this program in the face of other calls on government funds.

## 14.5.2   Is biological control a cost-efficient option?

Labor-intensive public works control programs aimed at simultaneously alleviating poverty while clearing and rehabilitating invaded areas are unlikely to be sustainable in the long term because of their substantial cost and relative ineffectiveness[89] at controlling the spread and densification of invasive alien plants. Invasive alien plant management programs should, therefore, aim to develop control components that will ensure that cleared areas do not simply become reinvaded. One of the most cost-effective ways of doing this would be to use biological control, where species-specific organisms are introduced to bring or keep invasive species under control, or to reduce their invasive potential.

South Africa has been successful at finding effective biological control solutions to many invasive weed problems.[14,32,90,91] Historically, 103 biological control agents have been released in South Africa to control 46 weed species; 22 of these weed species are now under complete or substantial biological control.[32,90] Some initial assessments showed that there were very substantial returns on investment in biological control research.[92,93] A more rigorous assessment by De Lange and van Wilgen,[94] which examined groups of species with similar invasion dynamics and impacts rather than single species, found that the returns on investment from biological control research are of the same order and substantial. The returns ranged from 50:1 for subtropical, semiwoody species such as *Lantana camara*, *Caesalpinia decapetala*, and *Chromolaena odorata* (all species where current agents are not very effective), to 3726:1 for the Australian *Acacia* species (e.g., *A. mearnsii*, *A. saligna*, *A longifolia*, *A. cyclops*) where the agents have significantly reduced their fecundity and, thus, their invasiveness. The results are particularly sensitive to estimates of the effect of the biological control on the rate of spread, but the returns on investment remained positive even when the modeled rates of spread were varied by up to 75% above and below the expert's estimates.

The use of alien organisms to kill or limit invasions by other alien species is controversial. A number of studies have shown that it can significantly reduce the vigor and fecundity of invading alien plant species. Other studies suggest that the risks are simply too great for this to be considered a major component of an integrated control, and there is little consensus (see, e.g., Moran et al.[95]; Simberloff and Stiling[96]; Louda and Stiling[97]). The ongoing controversy has promoted the perception that the risks of unwanted outcomes are unacceptably high, which could result in the exclusion of biological control as an option.[98,99]

Arguments for biological control include its cost-effectiveness (Table 14.4), that is, its development costs and environmental impacts are low compared with herbicides, that it complements conventional mechanical and chemical control methods, that agents can reach remote areas that are expensive to reach with control teams, and that it is self-sustaining in most cases.[95] Opponents argue that we can never fully understand or predict the outcomes of introducing living organisms well enough to be sure that the benefits will outweigh the unintended consequences such as impacts on nontarget species or the disruption of indigenous species interactions (e.g., changing food webs). As far as the risks of introducing species go, it is very important to distinguish between agents that have been carefully selected as specialist feeders on particular plant species and their parts (e.g., flowers, fruits, seeds) and, for example, generalist herbivores.[95] The specialist agents are relatively safe, with few indications of nontarget or unexpected impacts, while the generalists often are not. Conflation of these two suites of agents can create the impression that the risks associated with all forms of agents are too high. Assessments of the risks of unpredictable and unwanted effects and outcomes must also consider the risks of excluding biological control, for example, greater impacts on ecosystems and ecosystem services and the additional investments that will be required for mechanical and chemical control to achieve the same level of control. Unlike biological control, these methods rarely limit invasions to maintenance levels so, when budgets are cut, the invading species simply recover or reinvade.

We believe that the scale and ongoing nature of invasions, the massive resources that are required to sustain national control programs, and the rarity with which eradication or full control can be achieved, all point to the fact that mechanical and chemical methods alone will not succeed in bringing invasive species under control. On their own, they are too expensive and unsustainable and the opportunity costs of not using biological control are too high.

The Working for Water program currently spends US$80 million annually on invasive alien plant control, of which only 1.6% goes to biological control (A. Wannenburgh, Working for Water National Office, January 2010, pers. comm.). This is not really sufficient to maintain the levels of expertise and experience required to undertake the research or the facilities to support it. South Africa has a chronic shortage of skilled scientists, so it would be very difficult to replace the current researchers and compensate for the loss of their knowledge and experience should this research capacity be allowed to decline. We argue that biological control should receive a greater proportion of the budget than it currently does, but how large a proportion is matter for debate. At the same time, biological control must be carefully undertaken with due cognizance of the complexities of the risk assessment process, the potential risks involved, and the potentially significant benefits that can be realized.

## 14.6  Conclusions and challenges

The studies of the environmental impacts of invasive alien plant species in South Africa have documented a very wide range of impacts including reductions in species diversity, genetic hybridization, alterations in water flows and quality, changes in fire behavior, loss of grazing, and increasing populations of disease vectors. They clearly show that the impacts of alien plant species are pervasive and have significant implications for human livelihoods and well-being. Almost all of the studies that included economic aspects of invasive alien plant invasions have indicated that these impacts can be substantial and that intervention in the form of control programs is justified. However, the accuracy and confidence in economic valuations on the impacts of invasive alien plants on the natural environment is determined by the baseline information these estimates are based on.

Currently, a significant amount of variance is present in terms of the amount of detail available on a species level, which obviously necessitates assumptions regarding key variables like spread rates and rates of increase in density. Also, a truly objective assessment of studies, which focus on economic aspects of invasive alien plants, is extremely difficult because of the variety of approaches, methods, and assumptions. An accurate aggregate of all economic impacts simply cannot be provided at this stage.

A broad overview of general features of the South African studies (Table 14.4) would include the following:

- Economic studies tend to focus on localized impacts, and there is a need for studies to evaluate the economic impacts of prevention measures on a national scale.
- The value chain of invasive alien plant impacts is poorly understood and needs to be assessed. This will enable species-specific value chain analysis, which will allow for the quantification of socioeconomic impacts with confidence.
- Ecological uncertainties about the nature of the impacts of plant invasions are not explicitly accounted for in the economic sensitivity analysis.
- Poor integration of ecological and economic assessments leads to unconvincing arguments about the social benefits of control. Cost–benefit analysis does not capture the multidimensional nature of plant invasions, and a multicriteria decision-making framework could add significant value.

Furthermore, human value systems are part of the inherent characteristics of the valuation process, but they change over time, which adds a temporal dimension to valuations. It is therefore extremely difficult to put forward fully inclusive valuations of expected economic impacts, which can be confidently extrapolated to different spatial and temporal scales.

Any debate about the control of invasive plants in South Africa is bound to result in conflicts between people with different views and interests. Many invasive species also bring benefits, and with benefits come vested interests. In such cases, the argument for or against control often becomes polarized and unbalanced (see, e.g., Johns[100] and Cellier[101]). The education of the broader public about the complexities of the problem, thereby enabling them to judge the merits of control programs, is an enormous and critical challenge that must be met if broad support is to be obtained.

One of the unique aspects of invasive plant control programs that has emerged in South Africa to date is the ability to leverage further benefits for the expensive control programs. Most of the funds for the Working for Water program have been sourced from the government's poverty relief budget (and not only from budgets aimed at protecting water resources, agricultural land, and biodiversity). This leverage has made it possible to allocate substantial funding to a program that would otherwise have struggled to obtain significant support. The links are fragile, however, and convincing economic assessments of the consequences of diverting funds from the clearing program into other necessary interventions (such as education and health care) will go a long way to maintaining the support and the benefits that it brings.

Biological control of invasive species is one solution that appears to offer considerable benefits. Biological control cannot solve the invasive species problem in all cases, but there have been a number of remarkable successes. In each of these cases, the benefits have far outweighed the costs. Biological control is nonetheless often the subject of some debate.[93] Nevertheless, we believe that biological control offers one of the best and most cost-effective interventions for addressing the problem.

# *References*

1. Glen, H. F. 2002. *Cultivated Plants of Southern Africa. Names, Common Names, Literature.* Johannesburg: Jacana Education (Pty) Ltd.

2. Pimentel, D., S. McNair, J. Janecka, J. Wightman, C. Simmonds, C. O'Connell, E. Wong et al. 2001. Economic and environmental threats of alien plant, animal, and microbe invasions. *Agric Ecosyst Environ* 84:1.

3. Wells, M. J., A. A. Balsinhas, H. Joffe, V. M. Engelbrecht, G. Harding, and C. H. Stirton. 1986. A catalogue of problem plants in Southern Africa. *Memoirs of the Botanical Survey of South Africa No. 53.* Pretoria: South African National Biodiversity Institute.

4. Richardson, D. M. et al. 2000. Naturalization and invasion of alien plants: Concepts and definitions. *Divers Distrib* 6:93.

5. Henderson, L. 2007. Invasive, naturalized and casual alien plants in Southern Africa: A summary based on the Southern African Plant Invaders Atlas (SAPIA). *Bothalia* 37:215.

6. Kaiser, J. 1999. Stemming the tide of invasive species. *Science* 285:1836.

7. Hulme, P., S. Bacher, M. Kenis, S. Klotsz, I. Kühn, D. Minchin, W. Nentwig et al. 2008. Grasping at the routes of biological invasions: A framework for integrating pathways into policy. *J appl Ecol* 45:403.

8. McGeoch, M. A., S. H. M. Butchart, D. Spear, E. Marais, E. J. Kleynhans, A. Symes, J. Chanson, and M. Hoffmann. 2010. Global indicators of biological invasion: Species numbers, biodiversity impact and policy responses. *Divers Distrib* 16:95.

9. Pyšek, P., and D. M. Richardson. in press. Invasive species, environmental change and management, and ecosystem health. *Annu Rev Environ Resour* 35.

10. Cowling, R. M., and C. Hilton-Taylor. 1994. Patterns of plant diversity and endemism in Southern Africa: An overview. In *Botanical Diversity in Southern Africa*, ed. B. J. Huntley, 31. Pretoria: National Botanical Institute.

11. Richardson, D. M. et al. 2003. Vectors and pathways of biological invasions in South Africa - Past, future and present. In *Invasive Species: Vectors and Management Practices*, ed. G. Ruiz and J. Carlton, 292. Washington, DC: Island Press.

12. Le Maitre, D. C., D. M. Richardson, and R. A. Chapman. 2004. Biological invasions in South Africa: Driving forces and the human dimension. *S Afr J Sci* 100:103.

13. Van Wilgen, B. W., and van Wyk, E. 1999. Invading alien plants in South Africa: Impacts and solutions. In *People and Rangelands: Building the Future, Proceedings of the VI International Rangeland Congress*, ed. D. Eldridge and D. Freudenberger, 566. VI Rangeland Congress Inc., Aitkenvale, Queensland, Australia.

14. Richardson, D. M. et al. 1997. Alien plant invasions. In *Vegetation of Southern Africa*, ed. R. M. Cowling, D. M. Richardson, and S. M. Pierce, 534. Cambridge: Cambridge University Press.

15. Henderson, L. 1998. South African plant invaders atlas (SAPIA). *Appl Plant Sci* 12:31.

16. Hoffman, M. T. et al. 1999. *Land Degradation in South Africa.* Kirstenbosch: National Botanical Institute.

17. Macdonald, I. A. W., F. J. Kruger, and A. A. Ferrar. 1986. *The Ecology and Management of Biological Invasions in Southern Africa.* Cape Town: Oxford University Press.

18. Richardson, D. M. et al. 1992. Plant and animal invasions. In *The Ecology of Fynbos*, ed. R. M. Cowling, 271. Cape Town: Oxford University Press.

19. Macdonald, I. A. W. 2004. Recent research on alien plant invasions and their management in South Africa: A review of the inaugural research symposium of the Working for Water programme. *S Afr J Sci* 100:21.

20. Le Maitre, D. C. et al. 1996. Invasive plants and water resources in the western Cape Province, South Africa: Modelling the consequences of a lack of management. *J appl Ecol* 33:161.

21. Van Wilgen, B. W., R. M. Cowling, and C. J. Burgers. 1996. Valuation of ecosystem services—A case study from South African fynbos ecosystems. *Bioscience* 46:184.

22. Van Wilgen, B. W. et al. 1997. The sustainable development of water resources: History, financial costs, and benefits of alien plant control programmes. *S Afr J Sci* 93:404.

23. Van Wilgen, B. W., D. C. Le Maitre, and R. M. Cowling. 1998. Ecosystem services, efficiency, sustainability and equity: South Africa's Working for Water programme. *Trends Ecol Evol* 13:378.

24. Turpie, J. K., C. Marais, and J. N. Blignaut. 2008. The working for water programme: Evolution of a payments for ecosystem services mechanism that addresses both poverty and ecosystem service delivery in South Africa. *Ecol Econ* 65:788.

25. Buch, A., and A. B. Dixon. 2009. South Africa's Working for Water programme: Searching for win–win outcomes for people and the environment. *Sustainable Dev* 17:129. http://www3.interscience. wiley.com/journal/121463591/abstract?CRETRY=1&SRETRY=0 (accessed June 11, 2010).

26. Versfeld, D. B., D. C. Le Maitre, and R. A. Chapman. 1998. Alien Invading Plants and Water Resources in South Africa: A Preliminary Assessment. Report TT99/98, Water Research Commission, Pretoria.

27. Le Maitre, D. C., D. B. Versfeld, and R. A. Chapman. 2000. The impact of invading alien plants on surface water resources in South Africa: A preliminary assessment. *Water SA* 26:397.

28. Richardson, D. M., M. Rouget, S. J. Ralston, R. M. Cowling, B. J. van Rensburg, and W. Thuiller. 2005. Species richness of alien plants in South Africa: Environmental correlates and the relationship with indigenous plant species richness. *Ecoscience* 12:391.

29. Cowling, R. M. et al. 1999. *Framework for a Conservation Plan for the Cape Floristic Region*. Cape Town: Institute for Plant Conservation, University of Cape Town.

30. Iponga, D. M., S. J. Milton, and D. M. Richardson. 2008. Superiority in competition for light: A crucial attribute defining the impact of the invasive alien tree *Schinus molle* (Anacardiaceae) in South African savanna. *J Arid Environ* 72:612.

31. Milton, S. J., H. G. Zimmermann, and J. H. Hoffmann. 1999. Alien plant invaders of the karoo: Attributes, impacts and control. In *The Karoo—Ecological Patterns and Process*, ed. W. R. J. Dean and S. J. Milton, 274. Cambridge: Cambridge University Press.

32. Zimmermann, H. G., V. C. Moran, and J. H. Hoffman. 2004. Biological control in the management of invasive alien plants in South Africa, and the role of the Working for Water programme. *S Afr J Sci* 100:34.

33. Levine, J. M., M. Vilà, C. M. D'Antonio, J. S. Dukes, K. Grigulis, and S. Lavorel. 2003. Mechanisms underlying the impacts of exotic plant invasions. *Proc R Soc Lond B* 270:775.

34. Parker, I. M. et al. 1999. Impact: Toward a framework for understanding the ecological effect of invaders. *Biol Invasions* 1:3.

35. Richardson, D. M., and B. W. van Wilgen. 2004. Invasive alien plants in South Africa: How well do we understand the ecological impacts? *S Afr J Sci* 100:45.

36. Chapin III, F. S. et al. 1996. The functional role of species in terrestrial ecosystems. In *Global Change and Terrestrial Ecosystems*, ed. B. Walker and W. Steffen, 403. Cambridge, UK: Cambridge University Press.

37. Dukes, J. S., and H. A. Mooney. 2004. Biological invaders disrupt ecosystem processes in western North America. In *Disruptions and Variability in Ecosystem Processes and Patterns*, ed. G. A. Bradshaw, H. A. Mooney, and P. Alaback. New York: Springer-Verlag.

38. Musil, C. F., and G. F. Midgley. 1990. The relative impact of invasive Australian acacias, fire and season on the soil chemical status of a sand plain lowland fynbos community. *S Afr J Bot* 56:419.

39. Witkowski, E. T. F. 1991. Effects of invasive alien Acacias on nutrient cycling in the coastal lowlands of the Cape Fynbos. *J appl Ecol* 28:1.

40. Yelenik, S. G., W. D. Stock, and D. M. Richardson. 2004. Ecosystem-level impacts of invasive alien nitrogen-fixing plants: Ecosystem and community-level impacts of invasive alien *Acacia saligna* in the fynbos vegetation of South Africa. *Restor Ecol* 12:44.

41. Gaertner, M., A. Den Breeÿen, C. Hui, and D. M. Richardson. 2009. Impacts of alien plant invasions on species richness in Mediterranean-type ecosystems: A meta-analysis. *Prog Phys Geog* 33:319.

42. Hall, A. V. 1987. Threatened plants in the Fynbos and Karoo biomes. *Biol Conserv* 40:29.

43. Macdonald, I. A. W. 1986. Range expansion in the Pied Barbet and the spread of alien tree species in Southern Africa. *Ostrich* 57:75.

44. Wise, R. M., B. W. Van Wilgen, M. P. Hill, F. Schulthess, T. Tweddle, A. Chabi-Olay, and H. G. Zimmermann. 2007. The economic impact and appropriate management of selected invasive alien species on the African continent. Report number CSIR/NRE/RBSD/ER/2007/0044/C, CSIR, Pretoria. http://www.gisp.org/publications/reports/CSIRAISmanagement.pdf (accessed June 11, 2010).

45. Midgley, J. M., M. P. Hill, and M. H. Villet. 2006. The effect of water hyacinth, *Eichhornia crassipes* (Martius) Solms-Laubach (Pontederiaceae), on benthic biodiversity in two impoundments on the New Year's River, South Africa. *Afr J Aquat Sci* 31:25.

46. Penfold, W. M., and T. T. Earle. 1948. The biology of the water hyacinth. *Ecol Monogr* 18:447.

47. Lindsey, K., and H.-M. Hirt. 1999. *Use Water Hyacinth: A Practical Handbook for the Uses of Water Hyacinth across the World*. Winnenden: Anamed Water Hyacinth Group.

48. Dye, P. 1996. Climate, forest and streamflow relationships in South African afforested catchments. *Commonw For Rev* 75:31.

49. Prinsloo, F. W., and D. F. Scott. 1999. Streamflow responses to the clearing of alien trees from riparian zones at three sites in the Western Cape. *S Afr For J* 185:1.

50. Görgens, A. H. M., and B. W. van Wilgen. 2004. Invasive alien plants and water resources in South Africa: Current understanding, predictive ability and research challenges. *S Afr J Sci* 100:27.

51. Stock, W. D., and N. Allsopp. 1992. Functional perspectives of ecosystems. In *The Ecology of Fynbos; Nutrients, Fire and Diversity*, ed. R. M. Cowling, 241. Cape Town: Oxford University Press.

52. Musil, C. F. 1993. Effect of invasive Australian acacias on the regeneration, growth and nutrient chemistry of South African lowland fynbos. *J appl Ecol* 30:361.

53. Holmes, P. M., and R. M. Cowling. 1997. The effects of invasion by *Acacia saligna* on the guild structure and regeneration capabilities of South African fynbos shrublands. *J appl Ecol* 34:317.

54. Holmes, P. M., and R. M. Cowling. 1997. Diversity, composition and guild structure relationships between soil-stored seed banks and mature vegetation in alien plant-invaded South African fynbos shrublands. *Plant Ecolog* 133:107.

55. Le Maitre, D. C., and A. H. M. Görgens. 2003. Impact of invasive alien vegetation on dam yields. Report KV141/03, Water Research Commission, Pretoria. http://www.wrc.org.za/Pages/DisplayItem.aspx?ItemID=7985&FromURL=%2FPages%2FAllKH.aspx%3F (accessed November 6, 2010).

56. Cullis, J. D., A. H. M. Görgens, and C. Marais. 2007. A strategic study of the impact of invasive alien plants in the high rainfall catchments and riparian zones of South Africa on total surface water yield. *Water SA* 33:35.

57. Van Wilgen, B. W., B. Reyers, D. C. Le Maitre, D. M. Richardson, and L. Schonegevel. 2008. A biome-scale assessment of the impact of invasive alien plants on ecosystem services in South Africa. *J Environ Manage* 89:336.

58. Marais, C., and A. M. Wannenburgh. 2008. Restoration of water resources (natural capital) through the clearing of invasive alien plants from riparian areas in South Africa—Costs and water benefits. *S Afr J Bot* 74:526.

59. McNeely, J., ed. 2001. *The Great Reshuffling: Human Dimensions of Invasive Alien Species*. Cambridge: IUCN. http://data.iucn.org/dbtw-wpd/edocs/2001-002.pdf (accessed June 11, 2010).

60. Perrings, C., M. Williamson, E. Barbier, D. Delfino, S. Dalmazzone, J. Shogren, P. Simmons, and A. Watkinson. 2002. Biological invasion risks and the public good: An economic perspective. *Conserv Ecol* 6:1.

61. Taleb, N. 2007. *The Black Swan: The Impact of the Highly Improbable*. New York: Random House.

62. Richardson, D. M. 1998. Forestry trees as invasive aliens. *Conserv Biol* 12:18.

63. Turpie, J., and B. Heydenrych. 2000. Economic consequences of alien infestation of the Cape Floral Kingdom's fynbos vegetation. In *The Economics of Biological Invasions*, ed. C. Perrings, M. Williamson, and S. Dalmazzone, 152. Cheltenham: Edward Elgar.

64. Brutsch, M. O., and Zimmermann, H. G. 1993. The prickly pear (*Opuntia ficus-indica* [Cactaceae]) in South Africa: Utilization of the naturalized weed, and of the cultivated plants. *Econ Bot* 47:154.

65. Johannesmeier, M. F. 1985. *Beeplants of the South-Western Cape*, 59. Pretoria: Department of Agriculture.

66. Culliney, T. W. 2005. Benefits of classical biological control for managing invasive plants. *Crit Rev Plant Sci* 24:131.

67. Higgins, S. I. et al. 1997. An ecological economic simulation model of mountain fynbos ecosystems: Dynamics, valuation and management. *Ecol Econ* 22:155.

68. De Wit, M., D. Crookes, and B. W. van Wilgen. 2001. Conflicts of interest in environmental management: Estimating the costs of a tree invasion. *Biol Invasions* 3:167.

69. Le Maitre, D. C. et al. 2002. Invasive alien trees and water resources in South Africa: Case studies of the costs and benefits of management. *For Ecol Manage* 160:143.

70. Zimmerman, H., and N. M. Pasiecznik. 2005. *Realistic Approaches to the Management of Prosopis Species in South Africa*. Coventry, UK: HDRA.

71. Shackelton, C. M., D. McGarry, S. Fourie, J. Gambiza, S. E. Shackleton, and C. Fabricius. 2007. Assessing the effects of invasive alien species on rural livelihoods: Case examples and a framework from South Africa. *Hum Ecol* 35:113.

72. Blignaut, J. N., C. Marais, and J. K. Turpie. 2007. Determining a charge for the clearing of invasive alien plant species (IAPs) to augment water supply in South Africa. *Water SA* 33:27.

73. Marais, C., B. W. Van Wilgen, and D. Stevens. 2004. The clearing of invasive alien plants in South Africa: A preliminary assessment of costs and progress. *S Afr J Sci* 100:97.

74. Rouget, M., D. M. Richardson, J. L. Nel, D. C. Le Maitre, B. Egoh, and T. Mgidi. 2004. Mapping the potential ranges of major plant invaders in South Africa, Lesotho and Swaziland using climatic suitability. *Divers Distrib* 10:475.

75. De Lange, W. J. 2007. Estimating the Impacts of Increased Water Tariffs on the Western Cape Economy. Report no. JECOS59, CSIR, Stellenbosch.

76. Lange, G. M., and R. M. Hassan. 2006. *The Economics of Water Management in Southern Africa: An Environmental Accounting Approach*. Cheltenham: Edward Elgar Press.

77. Biggs, R., B. Reyers, and R. J. Scholes. 2006. A biodiversity intactness score for South Africa. *S Afr J Sci* 102:277.

78. Scholes, R. J., and R. Biggs. 2005. A biodiversity intactness index. *Nature* 434:45.

79. Turpie, J., and B. W. Van Wilgen. 2004. The economic value of controlling invasive alien plant species in the Fynbos biome of South Africa. Unpublished report. Pretoria: Working for Water Programme, Department of Water Affairs and Forestry.

80. Du Plessis, L. L., and B. Reyers. 2006. The economic value of controlling invasive alien plant species in the grasslands biome of South Africa. Report to the Working for Water Programme, Department of Water Affairs and Forestry, Pretoria.

81. De Wit, M. P., and R. J. Scholes. 2006. The economic value of controlling invasive alien plant species in the savanna biome of South Africa. Report to the Working for Water Programme, Department of Water Affairs and Forestry, Pretoria.

82. Blignaut, J., S. Milton, and C. Cupido. 2006. The economic value of controlling invasive alien plant species in the succulent and Nama Karoo biomes of South Africa. Unpublished report, Working for Water Programme, Department of Water Affairs and Forestry, Pretoria.

83. Statistics South Africa. 2004. Census of commercial agriculture 2002 (summary). Statistical release P1101, Statistics South Africa, Pretoria.

84. South African Reserve Bank. 2010. Annual Report 2008/09, South African Reserve Bank, Pretoria.

85. Stewart, T. J., A. R. Joubert, L. Scott, and T. Low. 1997. Multiple criteria decision analysis: Procedures for consensus seeking in natural resources management. Report number 512/1/97, Water Research Commission, Pretoria.

86. Hicks, J. 1941. The rehabilitation of consumer's surplus. *Rev Econ Stud* 8:108.

87. Kaldor, N. 1939. Welfare propositions in economics and interpersonal comparisons of utility. *Econ J* 69:549.

88. Marais, C. 1998. "An Economic Evaluation of Invasive Alien Plant Control Programmes in the Mountain Catchment Areas of the Western Cape Province, South Africa." PhD dissertation. SunPress, University of Stellenbosch.

89. Hosking, S. G., and M. Du Preez. 2004. A cost-benefit analysis of the Working for Water programme on selected sites in South Africa. *Water SA* 30:143.

90. Olckers, T., and M. P. Hill, eds. 1999. Biological control of weeds in South Africa (1990–1998). *Afr Entomol Memoir* 1.

91. Olckers, T. 2004. Targeting emerging weeds for biological control in South Africa: The benefits of halting the spread of alien plants at an early stage of their invasion. *S Afr J Sci* 100:64.

92. McConnachie, A. J., M. P. De Wit, M. P. Hill, and M. J. Byrne. 2003. Economic evaluation of the successful biological control of *Azolla filiculoides* in South Africa. *Biol Control* 28:25.

93. Van Wilgen, B. W., M. P. De Wit, H. J. Anderson, D. C. Le Maitre, I. M. Kotze, S. Ndala, B. Brown, and M. B. Rapholo. 2004. Costs and benefits of biological control of invasive alien plants: Case studies from South Africa. *S Afr J Sci* 100:113.

94. De Lange, W. J., and B. W. van Wilgen. 2010. An economic assessment of the contribution of biological control to the management of invasive alien plants and to the protection of ecosystem services in South Africa. *Biol Invasions* 12:4113. DOI: 10.1007/s10530-010-9811-y.

95. Moran, V. C., J. H. Hoffmann, and H. G. Zimmermann. 2005. Biological control of invasive alien plants in South Africa: Necessity, circumspection, and success. *Front Ecol Environ* 3:77.

96. Simberloff, S., and P. Stiling. 1996. Risks of species introduced for biological control. *Biol Conserv* 78:185.

97. Louda, S. M., and P. Stiling. 2004. The double-edged sword of biological control in conservation and restoration. *Conserv Biol* 18:50.

98. McFadyen, R. E. 1998. Biological control of weeds. *Annu Rev Entomol* 43:369.

99. Sheppard, A. W., R. D. van Klinken, and T. A. Heard. 2005. Scientific advances in the analysis of direct risks of weed biological control agents to nontarget plants. *Biol Control* 35:215.

100. Johns, M. 1993. Are all trees green? The spotlight on forestry. *Afr Environ Wildl* 1:77.

101. Cellier, S. 1994. Are all trees green? The forest industry replies. *Afr Environ Wildl* 2:79.

102. Low, A. B., and A. G. Rebelo. 1996. *Vegetation of South Africa, Lesotho and Swaziland*. Pretoria: Department of Environmental Affairs and Tourism.

103. Cole, N. S. et al. 2000. *Framework for a Conservation Plan for the Agulhas Plain, Cape Floristic Region, South Africa*. Cape Town: Institute for Plant Conservation, University of Cape Town.

104. Avis, A. M. 1989. A review of coastal dune stabilization in the Cape province of South Africa. *Landsc Urban Plan* 18:55.

105. Fraser, M. W. 1990. Foods of redwinged starlings and the potential for avian dispersal of *Acacia cyclops* at the Cape of Good Hope Nature Reserve. *S Afr J Ecol* 1:73.

106. Knight, R. S., and I. A. W. Macdonald. 1991. Acacias and korhaans: An artificially assembled seed dispersal system. *S Afr J Bot* 57:220.

107. Higgins, S. I. et al. 1999. Predicting the landscape-scale distribution of alien plants and their threat to plant diversity. *Conserv Biol* 13:303.

108. Milton, S. J., and W. R. Siefried. 1981. Above-ground biomass of Australian acacias in the Southern Cape, South Africa. *J S Afr Bot* 47:701.

109. Milton, S. J. 1981. Litterfall of the exotic acacias in the South Western Cape. *J S Afr Bot* 47:147.

110. Van Wilgen, B. W., and D. F. Scott. 2001. Managing fires on the Cape peninsula: Dealing with the inevitable. *J Mediterr Ecol* 2:197.

111. Samways, M. J., P. M. Caldwell, and R. Osborn. 1996. Ground-living invertebrate assemblages in native, planted and invasive vegetation in South Africa. *Agric Ecosyst Environ* 59:19.

112. Samways, M. J., and N. J. Sharratt. 2009. Recovery of endemic dragonflies after removal of invasive alien trees. *Conserv Biol* 24:267.

113. Dye, P. J., and A. G. Poulter. 1995. A field demonstration of the effect on streamflow of clearing invasive pine and wattle trees from a riparian zone. *S Afr For J* 173:27.

114. Dye, P. J., G. Moses, P. Vilakazi, R. Ndlela, and M. Royappen. 2001. Comparative water use of wattle thickets and indigenous plant communities at riparian sites in the Western Cape and KwaZulu-Natal. *Water SA* 27:529.

115. Dye, P., and C. Jarmain. 2004. Water use by black wattle (*Acacia mearnsii*): Implications for the link between removal of invading trees and catchment streamflow response. *S Afr J Sci* 100:40.

116. Rowntree, K. 1991. An assessment of the potential impact of alien invasive vegetation on the geomorphology of river channels in South Africa. *S Afr J Aquat Sci* 17:28.

117. Van Wilgen, B. W., and D. M. Richardson. 1985. The effects of alien shrub invasions on vegetation structure and fire behaviour in South African fynbos shrublands: A simulation study. *J appl Ecol* 22:955.

118. Jovanovich, N. Z., S. Israel, G. Tredoux, L. Soltau, D. C. Le Maitre, and F. Rusinga. 2009. Nitrogen dynamics in land cleared of alien vegetation (*Acacia saligna*) and impacts on groundwater at Riverlands Nature Reserve (Western Cape, South Africa). *Water SA* 35:37.

119. Geldenhuys, C. J., P. J. le Roux, and K. H. Cooper. 1986. Alien invasions in indigenous evergreen forest. In *The Ecology and Management of Biological Invasions in Southern Africa*, ed. I. A. W. Macdonald, F. J. Kruger, and A. A. Ferrar, 119. Cape Town: Oxford University Press.

120. Macdonald, I. A. W. 1983. Alien trees, shrubs and creepers invading indigenous vegetation in the Hluhluwe-Umfolozi Game Reserve Complex in Natal. *Bothalia* 14:949.

121. Zachariades, C., and J. M. Goodall. 2002. Distribution, impact and management of *Chromolaena odorata* infestation in Southern Africa. In *Proceedings of the Fifth International Workshop on Biological Control and Management of Chromolaena Odorata, Durban, South Africa, 23–25 October 2000*, ed. C. Zachariades, R. Muniappan, and L. W. Strathie, 34. Pretoria, South Africa: Plant Protection Research Institute, Agricultural Research Council.

122. Zachariades, C., M. Day, R. Muniappan, and G. V. P. Reddy. 2009. *Chromolaena odorata* (L.) King and Robinson (Asteraceae). In *Biological Control of Tropical Weeds Using Arthropods*, ed. R. Muniappan, G. V. P. Reddy, and A. Raman, 130. Cambridge: Cambridge University Press.

123. Leslie, A. J., and J. R. Spotila. 2001. Alien plant threatens Nile crocodile (*Crocodylus niloticus*) breeding in Lake St. Lucia, South Africa. *Biol Conserv* 98:347.

124. Mgobozi, P. M., M. J. Somers, and A. S. Dippenaar-Schoeman. 2008. Spider responses to alien plant invasion: The effect of short- and long-term *Chromolaena odorata* invasion and management. *J appl Ecol* 45:1189.

125. Hill, M. P., and C. J. Cilliers. 1999. A review of the arthropod natural enemies, and factors that influence their efficacy, in the biological control of water hyacinth, *Eichhornia crassipes* (Mart.) Solms-Laubach (Ponterderiaceae), in South Africa. *Afr Entomol Memoir* 1:103.

126. Scott, D. F., F. W. Prinsloo, and D. C. Le Maitre. 2000. The Role of Invasive Alien Vegetation in the Cape Peninsula Fires of January 2000. Report ENV-S-C 2000-039, Division of Water, Environment and Forestry Technology, CSIR, Stellenbosch.

127. Richardson, D. M., and B. W. Van Wilgen. 1986. The effects of fire in felled *Hakea sericea* and natural fynbos and implications for weed control in mountain catchments. *S Afr For J* 139:4.

128. Breytenbach, G. J. 1989. Alien control: Can we afford to slash and burn hakea in fynbos ecosystems? *S Afr For J* 151:6.

129. Seydack, A. H. W. 1992. Fire management options in fynbos mountain catchment areas. *S Afr For J* 161:53.

130. Allsopp, N., and P. M. Holmes. 2001. The impact of alien plant invasion on mycorrhizas in mountain fynbos vegetation. *S Afr J Bot* 67:150.

131. Breytenbach, G. J. 1986. Impacts of alien organisms on terrestrial communities with emphasis on communities of the south-western Cape. In *The Ecology and Management of Biological Invasions in Southern Africa*, ed. I. A. W. Macdonald, F. J. Kruger, and A. A. Ferrar, 229. Cape Town: Oxford University Press.

132. Baars, J.-R., and S. Neser. 1999. Past and present initiatives on the biological control of *Lantana camara* (Verbenaceae) in South Africa. *Afr Entomol Memoir* 1:21.

133. Cilliers, C. J. 1999. Biological control of parrot's feather, *Myriophyllum aquaticum* (Vell.) Verd. (Haloragaceae), in South Africa. *Afr Entomol Memoir* 1:113.

134. Stirton, C. H., ed. 1978. *Plant Invaders, Beautiful but Dangerous*. Cape Town: Cape Provincial Administration.

135. Richardson, D. M., I. A. W. Macdonald, and G. G. Forsyth. 1989. Reductions in plant species richness under stands of alien trees and shrubs in the fynbos biome. *S Afr For J* 149:1.

136. Le Maitre, D. C. 1999. Prosopis and Groundwater: A Literature Review and Bibliography. Report number ENV-S-C 99077, compiled by the CSIR for the Working for Water Programme of the Department of Water Affairs and Forestry, Cape Town.

137. Brown, C. J., and A. A. Gubb. 1986. Invasive alien organisms in the Namib Desert, Upper Karoo and the arid and semi-arid savannas of western Southern Africa. In *The Ecology and Management of Biological Invasions in Southern Africa*, ed. I. A. W. Macdonald, F. J. Kruger, and A. A. Ferrar, 93. Cape Town: Oxford University Press.

138. Steenkamp, H. E., and S. L. Chown. 1996. Influence of dense stands of an exotic tree, *Prosopis glandulosa*, Benson, on a savanna dung beetle (Coleoptera: Scarabaeinae) assemblage in Southern Africa. *Biol Conserv* 78:305.

139. Dean, W. R. J., M. D. Anderson, S. J. Milton, and T. A. Anderson. 2002. Avian assemblages in native *Acacia* and alien *Prosopis* drainage line woodland in the Kalahari, South Africa. *J Arid Environ* 51:1.
140. Stirton, C. H. 1981. Notes on the taxonomy of *Rubus* in southern Africa. *Bothalia* 13:331.
141. Henderson, L. 1991. Alien invasive *Salix* spp. (willows) in the grassland biome of South Africa. *S Afr For J* 157:91.
142. Bromilow, C. 1995. *Problem Plants of South Africa*. Pretoria: Briza Publications.
143. Milton, S. J., J. R. U. Wilson, D. M. Richardson, C. L. Seymour, W. R. J. Dean, D. M. Iponga, and Ş. Procheş. 2007. Invasive alien plants infiltrate bird-mediated shrub nucleation processes in arid savanna. *J Ecol* 95:648.
144. Hoffmann, J. H., and C. M. Moran. 1988. The invasive weed *Sesbania punicea* in South Africa and prospects for its biological control. *S Afr J Sci* 84:740.
145. Oatley, T. B. 1984. Exploitation of a new niche by the Rameron pigeon *Columba arquatrix* in Natal. In *Proceedings of the Fifth Pan-African Ornithological Congress, Southern African Ornithological Society*, ed. J. Ledger, 323. Johannesburg: Southern African Ornithological Society.
146. Hosking, S. G., and M. Du Preez. 1999. A cost-benefit analysis of removing alien trees in the Tsitsikamma mountain catchment. *S Afr J Sci* 95:442.
147. Turpie, J. 2004. The role of resource economics in the control of invasive alien plants in South Africa. *S Afr J Sci* 100:87.
148. Born, W., F. Rauschmayer, and I. Brauer. 2005. Economic evaluation of biological invasions: A survey. *Ecol Econ* 55:321.

*chapter fifteen*

# Invasive vertebrates of South Africa

*Berndt J. van Rensburg, Olaf L. F. Weyl, Sarah J. Davies,
Nicola J. van Wilgen, Dian Spear, Christian T. Chimimba,
and Faansie Peacock*

## Contents

## 15.1   Introduction

Berndt J. van Rensburg

From an economic point of view, agriculture and tourism, along with mining production, are South Africa's largest economic sectors. However, South Africa is becoming increasingly reliant on ecotourism as an economic growth sector, which is, in turn, reliant on maintaining the productive and aesthetic value of South Africa's protected areas and interstitial landscapes. This, together with other important human benefits gained from maintaining biodiversity (e.g., ecological services, harvesting of wild species, aesthetically and culturally), suggests that key role players responsible for policy making in the country are faced with particular challenges where conservation and economic development needs are in conflict (see e.g., reference[1]). Yet, at the same time, vertebrate introductions that result in invasions have the potential to alter these valuable environments. For example, considering the anthropogenic movement of ungulate species (both alien and extralimital) in South Africa, known for their recreation and food source value, Spear and Chown[2] have shown that their movement has led to a 1.34% increase in the similarity of ungulate assemblages at a quarter-degree grid-cell resolution. At least 21 invasive fish species are present in South Africa. While the impact of these species on aquatic ecosystems has been profound, resulting in the extirpation or localized extinction of a number of indigenous fishes, amphibians, and invertebrates, alien invasive fishes drive a large recreational fishery that has significant economic value. As a result of these positive economic contributions of certain invasive vertebrates, the implementation of legislation governing the importation and movement of both alien and extralimital species in South Africa will require careful consideration of the trade-offs between economic gain and biodiversity loss.

Without understanding the key factors related to the likelihood of invasion and the consequences thereof, best scientific practices will not be incorporated sufficiently into the policy environment. Such lack of knowledge concerning basic principles of invasion biology, for example, the ecology behind successful dispersal, is a concern faced not only by South Africa (see reference[3]) but also globally.[4,5] Despite this lack of information, this chapter provides an evidence-based evaluation of each taxon group in as far as environmental, agricultural, and economic impacts in South Africa are concerned.

The pattern across taxa of introduced vertebrates seems to be rather inconsistent and complex for South Africa. For example, although the region is an alien fish invasion hot spot,[6] there are no introduced marine vertebrates in South Africa (for a list of introduced marine species, see references[7,8]). The number of introduced bird, reptile, and amphibian species seems to range from intermediate to low. In the latter two groups, this relatively low number might simply be as a result of undetected introductions due to failure of establishment or presence of cryptic species or populations that cannot be conventionally detected. The number of introduced mammal species is high while the number of alien invasive mammals is relatively low.[9] To further support the idea of complexity among introduced vertebrate patterns, Blackburn et al.[5] used New Zealand's South Island and continental South Africa to illustrate that, in this specific case and after controlling for colonization pressure, establishment success of alien birds is not higher on islands than on mainlands (c.f, reference,[10] p. 147). Finally, similarly for ungulates in South Africa,[9] the "rich get richer" hypothesis of Stohlgren et al.[11] does not apply to birds in South Africa.

Given the importance of scientific evidence-based evaluation, we discuss the current state of knowledge on invasion biology with regard to the impacts of alien vertebrates on the environment, agriculture, and economy of South Africa by providing general background on different vertebrate taxonomic groups that include freshwater fishes, amphibians, reptiles, birds, and mammals. Given the absence of introduced marine vertebrates in South Africa, this group of vertebrates is not reviewed further. Where published evidence is weak or nonexistent, case studies from other regions are used to predict potential impacts on South Africa. We conclude this chapter by (1) briefly considering further research directions that can be used to inform policy, and (2) discussing the development and implementation of legislation governing the importation and movement of all alien and extralimital species in South Africa in general.

For the purposes of this chapter, "introductions" are defined as introduced taxa that have been transported across a major geographic barrier[12] and released into the wild[13] while "established" is used to describe a self-sustaining population that has surmounted various abiotic and biotic barriers in order to survive and reproduce (i.e., a population that has become naturalized).[12] We treat the pet trade as an intentional pathway of introduction, although this may not always be the case, and releases from captivity may be accidental (e.g., escapes from zoos). "Invasive species" are defined as cases where species produce a reproductively viable offspring in areas distant from sites of introduction. A distinction is made between the naturalization and consequent invasion of indigenous (i.e., extralimital) species outside their natural range but within a given geopolitical boundary (in this case South Africa), as opposed to wholly alien (or extraregional) species introductions from beyond the borders of the country.[2,12] We do not assume that invasive species have clear and established impacts; they are simply those that have established viable populations and spread in the novel range.

## 15.2   Freshwater fishes

Olaf L. F. Weyl

### 15.2.1   Introduction

South Africa is an alien fish invasion hot spot.[6] Introductions of fishes into South Africa started in the early eighteenth century with goldfish *Carassius auratus* imported (most likely from Asia on Dutch ships) for ornamental purposes.[14] Subsequently, at least 24 alien and 19 indigenous fish species were introduced into South African waters. At least 21 of these species are invasive in South Africa (Table 15.1).

The introduction and the subsequent spread of these species is a result of anthropogenically facilitated movement. After their initial introduction, alien fishes were primarily propagated in the state-funded Jonkershoek hatchery[15] in the Cape from where they were widely distributed to other provincial hatcheries and spread

*Table 15.1* Summary of Invasive Freshwater Fish Species in South Africa

| Species name (common name) | Introduction date (source) | Pathway | Status | Native range | Reference |
|---|---|---|---|---|---|
| *Lepomis macrochirus* (Bluegill sunfish) | 1938 (USA) | Intentional for angling and as forage | Populations recorded from 80% of assessed river basins | North America | 15 |
| *Micropterus dolomieu* (Smallmouth bass) | 1937 (USA) | Intentional for angling | Populations recorded from 70% of assessed river basins | North America | 15 |
| *Micropterus punctulatus* (Spotted bass) | 1939 (USA) | Intentional for angling | Recorded from 80% of assessed river basins | North America | 15 |
| *Micropterus salmoides* (Largemouth bass) | 1928 (the Netherlands) | Intentional for angling | Present in all of 16 assessed river basins | North America | 15 |
| *Oreochromis mossambicus* (Mozambique tilapia) | 1936 (Extralimital) | Intentional for angling, as forage, fish culture, control of mosquitoes | Recorded from all assessed river basins outside its native range in South Africa | RSA, Limpopo to Bushmans River in the Eastern Cape | 15 |
| *Oreochromis niloticus* (Nile tilapia) | 1959 (Israel) | Intentional for angling and fish culture | Invading. Populations occur in the Limpopo and Incomati River basins and recent locality records from coastal rivers in KZN | Africa | 15,20 |
| *Tilapia sparrmanii* (Banded tilapia) | 1941 (Extralimital) | Intentional as forage fish | Invading with extralimital populations reported from all major river catchments outside its native range | RSA, Orange, Limpopo, and KZN | 15 |
| *Tilapia rendalli* (Redbreast tilapia) | 1955 (Extralimital) | Intentional for control of macrophytes | Invading with extralimital populations common in warm KZN river systems | RSA, Limpopo, and KZN | 15 |

| Species | Date (origin) introduced | Reason/pathway of introduction | Status/distribution | Native range | Reference |
|---|---|---|---|---|---|
| *Clarias gariepinus* (Sharptooth catfish) | 1975 (Extralimital) | Accidental through escapes from aquaculture and via interbasin water transfers; Intentional releases by anglers | Extralimital populations are reported from 90% assessed river basins outside its natural range | RSA, Orange, Limpopo, and KZN Rivers | 15 |
| *Pterygoplichthys disjunctivus* (Vermiculated sailfin) | 2006 | Release by aquarists | Established in the Nseleni River in KZN | South America | SAIAB collection data |
| *Carassius auratus* (Goldfish) | 1726 (India) | Control mosquitoes, releases by aquarists | Established only in urban ponds | Eastern Asia | 15 |
| *Ctenopharyngodon idella* (Grass carp) | 1967 (Malaysia) | Intentional for aquatic macrophyte control | Established and common in the Vaal River system; Present in the Limpopo and Umgeni River systems | Asia | 15 |
| *Cyprinus carpio* (Common carp) | 1859 (England) | Intentional for angling | Established in parts of all major river catchments in South Africa | Europe | 15 |
| *Hypophthalmichthys molitrix* (Silver carp) | 1975 (Malaysia) | Escapement from aquaculture | Established in the Limpopo and Incomati River systems | Asia | 15 |
| *Labeo capensis* (Orange River mudfish) | 1975 (Extralimital) | Accidental through interbasin water transfer | Extralimital populations in the Sundays, Great Fish, and Tugela Rivers. | RSA, Orange River | 15 |

*(Continued)*

**Table 15.1** Summary of Invasive Freshwater Fish Species in South Africa (*Continued*)

| Species name (common name) | Introduction date (source) | Pathway | Status | Native range | Reference |
|---|---|---|---|---|---|
| *Labeobarbus aeneus* (Smallmouth yellowfish) | 1953–1977 (Extralimital) | Intentional stocking for angling and accidental via interbasin water transfer | Relatively widespread with extralimital populations in the Kei, Great Fish, Sundays, Gouritz, and Tugela River systems | RSA, Orange River | 15,39 |
| *Tinca tinca* (Tench) | 1896 (Scotland) | Intentional for angling and forage | Well established in the Breede River system | Europe | 15 |
| *Gambusia affinis* (Mosquitofish) | 1936 | Intentional for mosquito control | Widespread with populations reported from 50% of major river systems | North America | 15 |
| *Poecilia reticulata* (Guppy) | 1912 (Barbados) | Intentional for mosquito control and releases by aquarists | Localized with populations limited to urban streams and ponds in warmer regions | South America | 15 |
| *Xiphophorus helleri* (Swordtail) | 1974 | Released by aquarists | Localized with populations limited to urban streams and ponds in warmer regions | Central America | 15 |
| *Oncorhynchus mykiss* (Rainbow trout) | 1897 (England) | Intentional for angling | Widespread occurrence in cooler, high-altitude areas Recorded in 75% of major river catchments | North America | 15 |
| *Salmo trutta* (Brown trout) | 1892 (Scotland) | Intentional for angling | Fairly common in cooler, high-altitude areas Recorded in 30% of major river catchments | Europe | 15 |

*Note:* RSA = Republic of South Africa. KZN = KwaZulu-Natal Province. Because the source of introduction often differs from their native range, the source of fish, if known, is provided in parentheses below the introduction date. The status of these fish was assessed as a function of occurrence within 16 major river basins in South Africa.

through formal stocking programs often led by government initiatives and implemented with the help of acclimatization and angling societies. In addition to angling and fish culture, the purpose for introductions includes biocontrol, the pet trade, and unintentional extralimital introductions through interbasin water transfers (IBTs; Figure 15.1).

### 15.2.1.1   Angling

Alien fish introductions during the late nineteenth and early twentieth centuries primarily focused on providing opportunities for recreational angling. Rainbow trout *Oncorhynchus mykiss* (see Box 15.1 for more information on this species) and brown trout *Salmo trutta* were imported in the late nineteenth century but because trout are intolerant

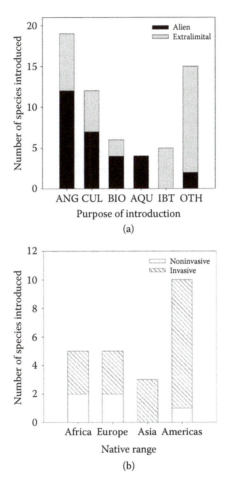

*Figure 15.1* (a) Number of fish species introduced outside their natural range into South African inland waters for the purpose of angling (ANG); fish culture (CUL); biocontrol of undesirable aquatic plants and invertebrates (BIO); introductions by aquarists (AQU); accidental introduction through interbasin water transfers (IBT); and for other reasons, including the extension of the range of rare species (OTH). Alien and extralimital indigenous introductions are shown separately. (b) Number of alien fish species introductions by native range introduced into South Africa. The proportion of subsequent invasive species is shown.

## BOX 15.1   RAINBOW TROUT (*ONCORHYNCHUS MYKISS*)

Native to the western seaboard of the United States, rainbow trout ova were imported to Jonkershoek hatchery in the Western Cape, South Africa in 1897 primarily for angling. The offspring from these fish were distributed to various areas in southern Africa, and by 1930 trout hatcheries had been established in cooler areas of the sub-region. Because rainbow trout cannot tolerate temperatures above 20°C and require flowing water over gravel beds to spawn, invasions are limited to high-altitude streams. Many other populations are, however, sustained by continual stocking from hatcheries.

Rainbow trout are the mainstay of the freshwater aquaculture industry, which produces some 1000 tons worth an estimated ZAR 63 million (ca. US$7.8 million) annually. Far more important is the specialist recreational fishery based on flyfishing. Nationally, there are about 4500 flyfishers affiliated to formal associations such as the Federation of Southern African Flyfishers, and it is estimated that at least 10 times as many anglers participate informally.[36] In the high-lying areas of the Drakensberg in the rural Eastern Cape, flyfishing is offered on more than 350 km of rivers and seven reservoirs, and makes an estimated contribution of ZAR 3.5 million (ca. US$0.5 million) to the rural economy and employs more than 50 people.[37] Such examples are not restricted to the Eastern Cape province, and the economic impact of trout fishing is likely to exceed ZAR 400 million (ca. US$5 million).[32]

In South Africa, trout inhabit upper reaches of rivers that are characterized by low species diversity but high degrees of endemism.[38] Invasion by alien fishes, including trout, has been identified as the primary threat to fishes inhabiting these areas. Trout compete with, and directly prey on indigenous fishes, amphibians, and invertebrates. This has resulted in the fragmentation of populations, severe reductions in population size, and even extirpation of indigenous biota.[15,23,24]

Economic activities focusing on this fish are, however, well established in a large proportion of river systems, and areas demarcated for the use of trout will include sections of many high-lying river catchments in South Africa. Management and control of trout will involve excluding this fish from sensitive areas and can be expected to meet with strong opposition from the public. Mitigating the future spread and impact of trout on indigenous biota will therefore require education of the public and support from angling bodies.

of temperatures exceeding 20°C successful introductions were limited to cool, clear, oxygen-rich waters in high-altitude or temperate areas. In the slower flowing, warmer lowland rivers, both common carp (*Cyprinus carpio*) and later, tench (*Tinca tinca*) were stocked.[15] These coarse fish were, however, not considered suitable for sport angling and warm water predatory game fish that could be targeted by anglers using artificial lures were needed. Perch (*Perca fluviatilis*) was imported from Britain in 1912. This species never established, leading to the introduction of four North American centrarchids between 1928 and 1939. Largemouth bass (*Micropterus salmoides*) was highly successful but favored slow flowing or static water. To fill the gap between the upland trout waters and the low-lying largemouth bass zone, smallmouth bass (*Micropterus dolomieu*) and the more turbidity and flood-tolerant spotted bass (*Micropterus punctulatus*) were introduced.[15] To provide food for these predators, a fourth centrarchid, the bluegill sunfish (*Lepomis macrochirus*) was imported in 1938. In addition, several indigenous fishes were translocated for use as fodder fish and for sport angling. As a result, some species, such as the banded tilapia (*Tilapia sparrmanii*) and Mozambique tilapia (*Oreochromis mossambicus*), now occur throughout South Africa.

### 15.2.1.2   Biological control

Mosquitofish (*Gambusia affinis*), guppies (*Poecilia reticulata*), and goldfish were distributed from state hatcheries from the 1930s and 1940s for mosquito control.[15] While feral goldfish and guppy populations are largely limited to urban streams and impoundments, mosquitofish now occur in about 50% of South Africa's river systems. Asian grass carp (*Ctenopharyngodon idella*), which were believed to be unable to breed because of its specific spawning requirements, were introduced into farm dams for aquatic macrophyte control throughout South Africa. Subsequently, this fish is now established in three river systems, and in the Vaal River it is present in such high numbers that anglers consider it a pest.

### 15.2.1.3   Aquaculture

Escape from aquaculture operations has resulted in the introduction of rainbow trout, common carp, and Nile tilapia (*Oreochromis niloticus*) and silver carp (*Hypophthalmichthys molitrix*) becoming invasive. The silver carp was imported from Israel in 1975 for aquaculture. As with grass carp, it was thought that the specific spawning requirements of the species precluded its establishment in the wild, but the species has established a viable population in the Olifants–Limpopo River system and has been reported in large numbers from the main river and various tributaries downstream in Mozambique.

### 15.2.1.4   Ornamental fish trade

The release of unwanted fish by aquarists is most likely responsible for the feral guppy, swordtail (*Xiphophorus helleri*), and goldfish (*Carassius auratus*) populations present in many urban impoundments and streams. Most recently, a number of armoured catfish (*Pterygoplichthys disjunctivus*) have also been collected from coastal drainages in South Africa. Initial assessments indicate that this fish has established but impacts are unknown.

### 15.2.1.5   Interbasin water transfers

The best example of invasion resulting from an IBT scheme is the establishment of Orange River fishes in the Sundays River and Great Fish River following the Orange-Fish River IBT scheme that was completed in 1975.[16] This IBT connects the Gariep Dam on the Orange River system with the headwaters of the Great Fish River through an 82-km tunnel, which facilitated the transfer of smallmouth yellowfish (*Labeobarbus aeneus*), Orange River mudfish (*Labeo capensis*), and sharptooth catfish (*Clarias gariepinus*). A similar transfer scheme is responsible for the transfer of at least *L. aeneus* and *L. capensis* from the Orange River to the Tugela River system.

## 15.2.2   Extent of the problem

As a result of numerous and continued introductions, alien and extralimital fishes are common components of fish assemblages in all major river systems. Formal collection records[17] show that alien and extralimital fishes comprise between 11% and 71% of fish species in 16 major river basins in South Africa (Figure 15.2). It is therefore hardly surprising that South Africa is considered a fish invasion hot spot.[14] While formal stocking is now

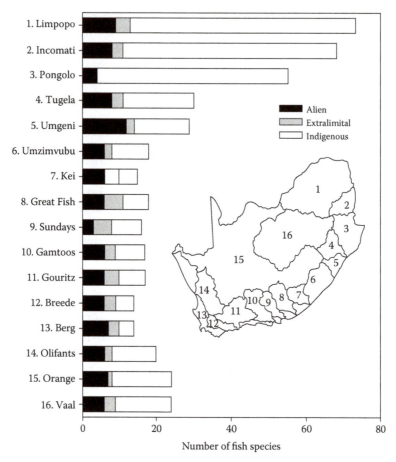

*Figure 15.2* An assessment of the fish species with formal collection records from 16 major river basins in South Africa separated into alien, extralimital, and indigenous species.

tightly controlled by national legislation,[18] continued spread by IBTs and illegal stocking by anglers continues to provide invasion opportunities. Sharptooth catfish, for example, were introduced to the Great Fish River through an IBT in 1977 but have since been translocated, presumably by anglers, to most major river systems in the Eastern and Western Cape provinces. The impact of the introduction of a large (>20 kg) predatory fish into these rivers requires urgent assessment.[19] Similarly, the Nile tilapia, first recorded in the South African reaches of the Limpopo River in 1996[20] is increasing its range, and informal records of the presence of this species in warmer areas of South Africa are now common.

## 15.2.3   *Environmental impacts*

Impacts of introduced fish on indigenous ichthyofauna include the transfer of associated parasites, direct predation, ecosystem effects, and hybridization. The tapeworm (*Bothriocephalus acheilognathi*) was introduced with grass carp from Germany[15] and has infected two indigenous yellowfish species.[21] Predation by centrarchids and salmonids on indigenous fishes has resulted in increasingly fragmented populations and extirpation of endangered minnows[22–24] and changes in fish and invertebrate community structure, which resulted in major trophic shifts in the streams of the Western Cape province.[25] The common carp is an aggressive invader that is widely considered a pest species and has been linked to habitat alterations brought about by increased turbidity, which results from its bottom-grubbing feeding behavior.[15] In the Vaal River, for example, the increased turbidity from common carp feeding was linked to the decline in a large indigenous predatory cyprinid, the largemouth yellowfish (*Labeobarbus kimberleyensis*).[26] *Oreochromis niloticus* hybridizes with the indigenous *Oreochromis mossambicus*,[27,28] which, although itself considered one of the 100 worst global invasive species,[29] now has its genetic integrity threatened by hybridization.[30] Similarly, the contact of closely related *Labeo* species has resulted in the presence of hybrids[31] in the invaded systems, which threaten the genetic integrity of Eastern Cape *Labeo umbratus* stocks.

## 15.2.4   Economic impacts

A survey undertaken in 2007 estimated that there were more than 1.5 million freshwater anglers in the country.[32] It is unlikely that this large recreational fishery would have developed in the absence of alien fishes. While indigenous species are targeted by some, the vast majority target alien carp, trout, and bass species. The expenditure by these anglers on their fishing activities is a significant contribution to the national economy. While there are few published reports on the economic impact of recreational angling, the most recent assessment, based on a questionnaire survey of anglers affiliated to angling associations, estimated that their average expenditure on related equipment and activities was ZAR 7500 (ca. US$940) per angler per year.[32] The total economic impact of these anglers, who represent about 10% of participants, was therefore in the region of ZAR 900 million (ca. US$11.2 million) per annum. Forty percent of the economic value was attributed to bank angling, which mainly targets common carp, 40% to flyfishing, where trout are the major target species, and 13% to bass angling. More recent assessments have also shown the importance of alien fishes to food security in rural communities. In South Africa's largest inland water body, Lake Gariep, common carp not only provide the basis for a recreational fishery but are also the main target species for more than 450 subsistence anglers.[33,34]

The freshwater aquaculture industry, while relatively small, is also largely reliant on alien invasive species. Apart from a wide variety of ornamental fishes for the pet trade (producing more than 1 million fish annually), invasive tilapia (10 tons) and trout (943 tons) are the main cultured fish species.[35] The current gate sale value of these fishes is estimated at about ZAR 40 million (ca. US$5 million). The sector is, however, growing, and this will likely lead to increased invasion opportunities for alien species after escaping from fish farms.

## 15.2.5   Dealing with the problem

Once established in a river system, alien fishes are almost impossible to eradicate. While localized trout and bass eradication programs are planned in small sections of Western Cape province streams using ichthyocides, successful eradication is only likely in small isolated systems. In large systems alien fishes are now so well established that their removal is neither practical, feasible or, in some cases, economically desirable. Future management of alien fishes will need to focus on their exclusion from areas of biodiversity concern and limiting further spread. This is a major objective of new legislation governing the importation and movement of alien fishes.[18]

Current alien species regulations dictate that the use of most alien invasive and many extralimital fishes (except grass carp, silver carp, and bluegill sunfish, which are considered undesirable) will be regulated by area. This legitimization of the use of alien fish for angling and/or aquaculture is widely viewed as an acknowledgement of the value of these fishes in providing sport, recreation, income, and economic benefits to the country. It also takes into account the negative impacts of these introductions on aquatic ecosystems. The demarcation of areas for use of various alien species, therefore, allows for compromises to be made using trade-offs between recreational (and economic) interests and conservation priorities.

While such legislation, demarcation, and associated permit requirements are important tools to limit the movement of alien species into the country, enforcement will be complicated by the nature of informal translocations of alien fishes within the country. In South Africa, many alien fishes are stocked illegally and are transported by anglers between watersheds in the livewells of boats, car boots, and buckets. Education campaigns to sensitize the public on the impact of alien fishes on biodiversity are therefore necessary. The involvement of organized angling bodies in the development of the regulations is likely to increase public awareness, acceptance, and compliance with the new legislation.

## 15.3   Reptiles and amphibians

Sarah J. Davies and Nicola J. van Wilgen

### 15.3.1   Introduction

There are few well-known alien reptile and amphibian invaders. Indeed, these vertebrate groups probably have the fewest current examples of species posing major threats to biodiversity, economy, or agriculture. However, a handful of species have had significant impacts in several parts of the world, and this is likely to increase. The brown tree snake (*Boiga irregularis*) on the island of Guam[40] and the cane toad (*Rhinella marina*) in Australia[41] are two of the best-known species. These species have caused considerable damage to the environment, and the invasion of the brown tree snake has had serious economic consequences.[42] One of the biggest threats associated with invaders in this group is that the drivers of their introduction are becoming more prominent, among them the pet trade (considered here deliberate introductions, though introductions outside captivity are often accidental) and the cargo and nursery trades. The latter is one of the largest sources of accidental introductions of amphibian species.[43]

South Africa has no indigenous amphibians in the orders Caudata or Gymnophiona, but a rich and diverse Anuran fauna exists with endemic and range-restricted species concentrated in two centers in the northeast and southwest of the country.[44] The southwest center has a particularly high concentration of endemic species, associated with the topographic heterogeneity and climatic and hydrologic stability of the Cape Fold Mountain system,[45] while species richness is highest in the northeast, partly due to the intrusion of tropical species into this part of the country.[44]

The southern African region is also home to a particularly diverse reptile fauna. Branch[46] and the Southern African Reptile Conservation Assessment (SARCA) recorded approximately 500 species in the region (roughly the same as the United States, Canada, and Europe combined).[47] The region has a particularly high diversity of lizards and chelonians. For example, approximately 11% of all gecko species (ca. 1054 worldwide) are native to the region, 51 of 54 Cordylid species, including many endemics, occur here, and 27% of all land tortoises (Testudinidae) occur in the region (figures calculated using SARCA data and the TIGR Reptile Database[47]). The introduction of alien species that may threaten this unique diversity is therefore cause for concern.

## 15.3.2  Extent of the problem

Though many species have been imported into South Africa, very few extralimital or alien introductions have been recorded in the country, indicating either that (1) few introductions (defined as taxa that have been released into the wild[13]) have occurred, or (2) introductions have occurred but remain undetected due to rapid failure to establish or due to cryptic populations. However, the existence of a comprehensive distribution database for South African frogs[48] makes the latter conclusion unlikely. Similarly, the ongoing SARCA project will confirm the status and extent of introduced reptiles in South Africa. Currently, the main threat is the increase in the trade of reptiles as pets, as well as a general increase in the trade that introduces both reptiles and amphibians accidentally. Table 15.2 lists the species that are known to be established or have breeding populations in the country.

### 15.3.2.1  Amphibians

The indigenous guttural toad (*Amietophrynus gutturalis*) was translocated to the Western Cape province from its historical range in northern and eastern South Africa ca. 1997,[49] a range extension of at least 800 km (Table 15.2). The species is now established over approximately 250 ha of the Cape Peninsula (City of Cape Town Biodiversity Management, unpublished data).

Range extension involving long-distance jump dispersal and multiple introductions has also occurred in the indigenous painted reed frog (*Hyperolius marmoratus*), which is now invasive on the central plateau[50] and in the Western Cape province, 200–500 km west of its historical range.[50,51] Within the Western Cape, the species is extensively distributed and spreading in agricultural areas, and recently in protected areas, although no negative impacts of its presence have yet been identified. Repeated introductions of painted reed frogs probably occur through the nursery, fruit and vegetable trades, since several specimens have been found by greengrocers in the Cape Town area, and it is likely that many have been released to the wild via this pathway (S. J. Davies, unpublished data). Interactions with a congeneric, endemic species (e.g., *Hyperolius horstockii*, arum lily frog) have not been investigated to date.

The African clawed toad (*Xenopus laevis*) has been extensively translocated within its natural range in South Africa, with unregulated introductions having been made to water bodies that did not historically hold this species.[59] Once introduced to a region, *X. laevis* can move independently over land, sometimes in mass migrations. Irrigation canals and ditches may serve as convenient corridors for such movement, with attendant impacts on water and agricultural infrastructure.[60]

Translocation of *X. laevis* has increased the range of the species and its interactions with the endemic Cape clawed toad (*X. gilli*). Levels of genetic introgression between *X. laevis* and *X. gilli* appear to be low, but there is potential for predation and resource competition to negatively affect *X. gilli* populations.[61] There is significant divergence among Southern African *X. laevis*, indicating that this may be a polytypic species complex.[54] In light of this work, *X. laevis* translocations between northern and southern populations should be avoided as they may result in the loss of differentiation between these taxa.

*Table 15.2* Successfully Established Reptile and Amphibian Species in South Africa

| Species name (common name) | Introduction date | Pathway | Status | Native range | Reference |
|---|---|---|---|---|---|
| **Amphibia** | | | | | |
| *Ambystoma mexicanum* (Axolotl) | Unknown | Deliberate, laboratory animals released to the wild | Casual, single site known: city of Bloemfontein, Free State province | Mexico (alien introduction) | M. J. Cunningham, pers. comm. |
| *Amietophrynus gutturalis* (Guttural toad) | ca. 1997 | Accidental, possibly via nursery plants | Established and possibly invading: Cape Peninsula, WC | Southern and central Africa: as far south as NC and EC (extralimital translocation) | 49 |
| *Duttaphrynus melanostictus* (Southeast Asian common toad) | 2004 | Accidental, in shipping cargo | Single, isolated record: Port of Cape Town, WC | Southeastern Asia (alien introduction) | A. Channing, pers. comm. |
| *Hyperolius marmoratus* (Painted reed frog) | ca. 1997 | Accidental and ongoing, possibly via nursery plant or fruit trade | Invasive: WC; central highlands of Gauteng and Mpumalanga provinces | Eastern sub-Saharan Africa: natural populations as far west as Tsitsikamma, WC, South Africa (extralimital translocation) | 50,51 |
| *Phrynomantis bifasciatus* (Banded rubber frog) | Unknown | Deliberate, pet trade | Unknown: Free State province | Central sub-Saharan Africa (extralimital translocation) | 52,53 |
| *Pyxicephalus adspersus* (Giant bullfrog) | Unknown | Purpose and intent unknown | Unsuccessful: recorded locality "Cape Fold Mountains" is unspecific | Southern and parts of central Africa: only entering WC in the northeast (extralimital translocation) | 53 |
| *Xenopus laevis* (African clawed toad) | Unknown | Deliberate, laboratory animals released to the wild; used as bait by anglers | Successful: WC | Endemic subspecies possibly restricted to WC and NC (extralimital translocation) | 53 (but see 54) |

**Reptilia**

| Species | Date | Means of introduction | Status | Distribution | Reference |
|---|---|---|---|---|---|
| *Acontias gracilicauda* (Slender-tailed legless skink) | <1995 | Accidental, "hitchhiker" in containers and other vehicles | Unknown: Gauteng | South Africa: Free State and EC (extralimital translocation) | 55 |
| *Afrogecko porphyreus* (Marbled leaf-toed gecko) | <1992 | Accidental in road or shipping containers; to offshore islands in lighthouse supplies | Established: Port Elizabeth and Grahamstown, EC; offshore islands | South Africa: south and southwestern Cape (extralimital translocation) | W. R. Branch, pers. comm. |
| *Afroedura pondolia* (Pondo rock gecko) | <2005 | Purpose and intent unknown | Unknown: Port Elizabeth, EC | South Africa and Swaziland (extralimital translocation) | W. R. Branch, pers. comm. |
| *Bradypodion ventrale* (Southern dwarf chameleon) | <1995 | Accidental, pet trade (escapee) | Unknown: population established in Melville Koppies, Johannesburg, Gauteng in 1995 | South Africa: EC (extralimital translocation) | 55 |
| *Chondrodactylus bibronii* (Bibron's gecko; previously: *Pachydactylus bibronii*) | <1950, 1960s | Deliberate, pet trade | Unsuccessful: Kommetjie, WC; population sustained in 1985: Fish Hoek, WC | Southern Africa: Namibia, South Africa, Swaziland (extralimital translocation) | 56 |
| *Hemidactylus mabouia* (Tropical house gecko) | 1962, 1976 | Accidental, shipping cargo; deliberate (WC) | Established (localized): East London, Port Elizabeth, EC; Simons Town, Gordon's Bay, WC; populations may be invasive but it is not known whether these will expand beyond urban areas | Sub-Saharan Africa (alien introduction) | 56 |

*(Continued)*

*Table 15.2* Successfully Established Reptile and Amphibian Species in South Africa (*Continued*)

| Species name (common name) | Introduction date | Pathway | Status | Native range | Reference |
|---|---|---|---|---|---|
| **Reptilia** | | | | | |
| *Lygodactylus capensis* (Cape day gecko) | <1980 | Accidental, stowaway in a crate from Kruger National Park | Established: between Addo Elephant National Park and Port Elizabeth, EC; Established and possibly invading: Cape Town and surrounding areas | Southern sub-Saharan Africa including northeast South Africa (extralimital translocation) | 56,57 |
| *Ramphotyphlops braminus* (Flowerpot snake) | <1800 | Accidental, along with potted plants from Indonesia | Established: Cape Town, WC; Durban, KwaZulu-Natal (KZN) | Cosmopolitan, probably originating in Asia (alien introduction) | 56 |
| *Stigmochelys pardalis* (Leopard tortoise; previously *Geochelone pardalis*) | <1930 | Deliberate | Established: Table Mountain National Park, Cape Town; De Hoop Nature Reserve, WC | Africa, including parts of central and northeastern South Africa (extralimital translocation) | 56 |
| *Trachemys scripta* (Red-eared slider; previously: *Chrysemys scripta*) | <1984 | Deliberate, pet trade | Unknown, though individuals are captured periodically; Reported in Zoo Lake, Johannesburg in 1984; individuals also captured in Pretoria and Moreleta Spruit, Gauteng and Durban, KZN | North and central America (alien introduction) | 15,53,58 |

*Note:* Not all the listed species are associated with known impacts in South Africa. EC = Eastern Cape province; KZN = KwaZulu-Natal province; NC = Northern Cape province; WC = Western Cape province.

### 15.3.2.2 Reptiles

Nearly 300 species of reptiles have been documented (imported or in the pet or zoo trade) in South Africa over the last few decades,[62,63] although not all of these species are currently likely to be present in the country. If the magnitude of the trade continues to increase as it has done over the past 30 years,[63] there could be cause for concern. Though a number of individual specimens of these species have been sighted outside captivity in both urban and natural areas,[62] only three alien species are reported to have established. The flower-pot snake (*Ramphotyphlops braminus*) has had the longest residence time of all these species, having been introduced accidentally before 1800 in association with plants brought from the East Indies by early settlers.[56] Similar introductions have occurred elsewhere in the world, although some introductions have arrived through natural dispersal on floating vegetation (e.g., reference[64]). Despite this long history of establishment, no impacts have been recorded for this species, which is often found in Cape Town gardens, and a lack of understanding of soil ecology prior to its introduction will complicate attempts to define its impacts. There are increasing reports of other species, such as geckos, for example *Lygodactylus capensis* in Cape Town and *Hemidactylus mabouia* along the Western Cape coast (E. H. W. Baard, pers. comm.) and at multiple locations on the east coast (W. R. Branch, pers. comm.) and as far inland as the Cederberg (E. H. W. Baard, pers. comm.) and Pretoria (A. A. Turner, pers. comm.). Though the full extent of *H. mabouia* populations has not been surveyed, their impact on local gecko species could be significant as they are aggressively territorial.

## 15.3.3 Impacts or potential impacts

Several investigations of South Africa's amphibian and reptile fauna have failed to identify economic, environmental, or agricultural impacts of extralimital translocations or alien introductions (e.g., references[15,56,65–68]) despite a recent systematic survey.[48] Globally, invasions are known to be closely related to trade and transport.[69,70] In particular, freshwater fish invasions have been shown to be closely related to the intensity of human activity as opposed to the characteristics of the receiving environment,[6] and a similar situation is likely to hold for amphibians and reptiles. Since the intensity of uncontrolled vectors and pathways can be expected to amplify, and the likelihood of both intentional and accidental alien introductions to increase, it is important to attempt to predict the nature and magnitude of the potential problem in South Africa, within the framework of impacts identified elsewhere.[43,71]

### 15.3.3.1 Hybridization and admixture

Due to the size and diversity of South Africa's reptile and amphibian fauna, genetic introgression through hybridization of introduced alien species with indigenous ones is of significant concern. Within the pet trade, exotic color morphs are popular (e.g., reference[72]), and the effort to produce novel forms sometimes involves hybridization of distinct genotypes. In most cases, this does not interfere with indigenous species, and breeding experiments generally do not use interspecies pairings. However, the indigenous Natal python (*Python sebae natalensis*) and the alien Burmese python (*Python molurus*) have successfully been paired in zoos.[73] Some of the hybrid eggs resulting from these crosses were infertile, and offspring seemed to be prone to disease. Whether hybridization would occur in the wild is uncertain, but if it did occur this could result in loss of unique local genetic adaptations and diversity. Indeed, hybridization in general, both inter- and intraspecific may result in the loss of evolutionary potential that is inherent in any unique species or population.

In other instances, the genetic combination of different gene pools (admixture) may result in enhanced performance of offspring. The importance of environment and colonization history in generating genetic variation and hybrid vigor following repeated introduction can be seen in the highly invasive morph of *Rhinella marina* in Australia.[74]

### 15.3.3.2   Trophic impacts and biotic homogenization

Invasive species can disrupt predator-prey relationships, with impacts ranging from devastation of local prey species (or predatory species, where the introduced species has novel defenses) to disruption of entire food webs. The brown tree snake provides an extreme example. The snake was introduced to the island of Guam, which previously lacked predators and as a result 6 of approximately 11 lizards, 16 of 22 birds, and 2 of 3 mammal species have been extirpated from the island.[40,75] The cane toad has caused similar problems in Australia due to the toxins present in its skin, against which native species had no natural defense.[76] These problems are less likely to occur in South Africa as local predators are adapted to a wide variety of prey species and are less likely to be naive.

Many amphibian species are closely associated with aquatic habitats[77] and are important predators of invertebrates, many of which are, in turn, agriculturally important. Amphibian invasions that result in biotic homogenization can be expected to affect food web structure, with related cascade effects on terrestrial and aquatic invertebrates and aquatic plants. Aquatic invasions in South Africa have already been shown to be synergistic, with invasive fish, amphibians, and plants interacting.[25,78] Simplification of food web structure can be expected to affect invertebrate community structure and populations, with attendant impacts on the agricultural sector. Direct trophic effects on indigenous amphibian species by the major anuran invader the American bullfrog (*Lithobates catesbeianus*) have been recorded in the United States, Europe, and South America (see Box 15.2).

### 15.3.3.3   Competition

Of greater concern in South Africa, however, is competition between indigenous and alien species. Because of the country's high diversity, especially of reptiles, new species are likely to compete with one or more indigenous species already occupying a specific ecological niche. See Box 15.3 for comments on the red-eared slider (*Trachemys scripta*). Vulnerable frog species could also be at risk of being outcompeted.

### 15.3.3.4   Novel pathogens

South Africa has been the source of an important amphibian invasion in the rest of the world. The widespread, economically important *Xenopus laevis* is indigenous to South Africa, and it is likely that many American and European invasive populations originated here.[54,88–90] *Xenopus laevis* is an efficient carrier of *Batrachochytrium dendrobatidis*, the causative agent of amphibian chytridiomycosis, which has been implicated in amphibian mass mortalities and population declines in several regions.[91–93] The chytrid organism appears to be endemic in South Africa, and indigenous frog species, including *X. laevis*, tend to be either resistant or asymptomatic,[91] a situation which may also hold in Australian populations postdecline.[94] Several other invasive frog species have also been shown to be efficient carriers of the organism (*Eleutherodactylus coqui*,[95] *Lithobates catesbeianus*[83]), and the introduction of any of these species carrying novel strains of *B. dendrobatidis* to South Africa could result in increased susceptibility of South African frogs to the disease.

There are a number of other diseases about which very little is known, but that could be spread through the pet trade and casual or established species. Inclusion body disease is of specific concern for boid snakes[96] and could have potentially catastrophic effects on

## BOX 15.2 AMERICAN BULLFROG (*LITHOBATES CATESBEIANUS*) POTENTIAL INTRODUCTION TO SOUTH AFRICA

The American bullfrog is invasive in the western United States, Europe, South America, and Asia. This species is not present in South Africa, but the possibility of its introduction is of concern since it is invasive in more than 25 countries.[79] It preys on both tadpoles[80] and adult frogs,[81] and its invasion into river systems has resulted in the decline of indigenous amphibian populations and alteration of species distributions[78] and changes in community structure.[82]

Introduction of such an abundant, large-bodied species which competes with local species for calling and breeding sites and is capable of consuming many smaller anurans could result in range reductions of indigenous species. The American bullfrog is also an effective carrier of the chytrid organism,[83] and introduction of novel forms of the fungus could result in increased virulence of the organism in South Africa.

## BOX 15.3 RED-EARED SLIDER (*TRACHEMYS SCRIPTA ELEGANS*), INTRODUCED, POTENTIALLY INVASIVE

The red-eared slider is a popular pet species as it is particularly attractive when young, and the turtle has been introduced as a pet to over 50 countries. However, sliders lose their bright colors and patterns as they age,[58] which reduces their desirability as pets, and owners therefore often release them into the wild. Through this activity, the terrapin has established in roughly 37 countries as well as in 20 states of the United States to which it is not indigenous.[43,84] Its impacts in these regions range from negligible to severe, including competition and the spread of disease to indigenous species as well as humans.[85–87] In France, sliders outcompete European pond turtles for preferred basking sites, while native species also have higher mortality and lower growth rates in the presence of the invader.[86,87]

The slider was introduced to South Africa via illegal pet trade imports.[58] Though possession of this species is not legal in the country, many enthusiasts are still thought to possess this species, and conservation agencies are contacted sporadically to collect feral individuals (A. A. Turner, pers. comm.). There are concerns that these feral individuals could establish breeding populations that would undoubtedly impact on the five native terrapin species in the southern African subregion through competition for food, basking sites, and through spread of novel pathogens and pathogen strains.

the indigenous Natal python as well as boid species in the pet trade. Ticks associated with reptiles and amphibians are another concern with potentially far-reaching effects. For example, some ticks are associated with diseases that affect not only reptiles but also indigenous ruminant species and livestock, for example, heartwater.[97,98]

Ranaviruses of the family Iridoviridae affect fish, amphibians, and reptiles and thus form a link between aquatic and terrestrial organisms and environments. Ranaviruses have been implicated in amphibian declines worldwide.[99] *Xenopus laevis* may act as a reservoir of disease, as infection in this species is often asymptomatic.[100] The high frequency of introduction of fish and reptiles into and within South Africa, through the aquaculture and pet trades, respectively, poses a risk of disease outbreaks among these taxa.

Alien species may also pose a health risk to humans. There are risks associated with diseases such as *Salmonella*, where spread has been documented from pet slider turtles to humans.[85] Other diseases include those spread by ticks, parasites, and helminthes (e.g., tape worms, which have been associated with cane toads[101]). Venomous snake species for which there is no antivenom in South Africa are also a health risk, as are large constricting species. A total of eight people were killed by pet pythons in the United States between 1993 and 2008,[102] with a subsequent fatality in 2009.[103] Three different species, the African rock python (*Python sebae*; one fatality), the Burmese python (*P. molurus bivittatus*; five fatalities), and the reticulated python (*Broghammerus reticulatus*; three fatalities), have caused these deaths, and these species should not be sold to inexperienced reptile keepers without the correct housing facilities and feeding protocols.

### 15.3.3.5 Opportunity cost

One of the biggest challenges of dealing with reptile and amphibian invasions is that we do not know enough about the species ecology to detect the impacts of invasion. Many of the systems in which these species interact are poorly studied and understood, which means that it may take some time for impacts to be recognized, if indeed they ever are. Furthermore, in most instances, invasions are irreversible, and therefore once impacts are recognized, minimizing these will incur very high management costs. Kraus[43] draws attention to the cost of scientific loss—if one does not understand how species interacted before new species were introduced, the opportunity to do so is lost, perhaps forever.

Finally, as Kraus[43] also points out, the loss of beauty is something important to consider. However, it is very difficult to put a price on the aesthetic value of a species or an ecosystem. Such a discussion quickly enters the purely philosophical realm and is complicated once again by the role that invasive and indigenous species play in the advancement of scientific knowledge, evolutionary potential, and human adaptive capability.

### 15.3.4 Dealing with the problem

South Africa's regulatory framework is comprehensive and has involved extensive negotiation with interested and affected parties such as the pet, aquaculture, and nursery industries. In addition to the direct impacts of invasive species introductions, it is also important to understand the potential economic impacts that enhanced regulation and control of introductions may have. For example, curtailment of the movement of nursery plants around South Africa, as proposed by the National Environmental Management: Biodiversity Act (NEM:BA; Act No. 10, 2004) will affect the volume and nature of plants traded, with accompanying reduction in the numbers of amphibians, invertebrates, parasites, and pathogens passing through this pathway. The desirable result of reduced propagule pressure is traded off against the loss of trade volume, and an increase in costs due to quarantine procedures. One should also bear in mind that the costs of controlling invasive animals or associated parasites and pathogens can be expected to exceed the costs of prevention.

Reptiles are rapidly increasing in popularity as pets in South Africa. Trends in the importation of reptiles listed by the Convention on International Trade in Endangered

Species (CITES) show that there has been a linear increase in the number of countries exporting species to South Africa and in the number of species introduced over the last 30 years,[63] while the number of individuals of these species being imported annually has increased exponentially. However, the trade in South Africa remains much smaller than in other parts of the world,[104] and despite these increases, very few species have established in South Africa as yet, perhaps due to relatively short residence times. This does not mean that invasions will not take place in the future, however. It is likely that some of the species currently kept as pets in South Africa are well adapted to local conditions and could establish feral populations if released from captivity in large numbers. Many incidents where pet species have established in Florida have been associated with pet store owners starting feral breeding populations on their grounds or where large numbers of individuals of the same species have been released following hurricanes.[105]

As the problem of invasions in this group has not fully manifested in South Africa yet, focus needs to be placed on the drivers and pathways responsible for invasions and the development of a strategy to prevent invasions from occurring. For reptiles, the greatest threat comes from the pet trade, where research has shown that there are distinct characteristics that increase species' popularity in the pet trade[63] and that pet stores tend to stock charismatic species that are easy to handle and breed. Species that are popular generate higher propagule pressure and are therefore more likely to establish than species that are traded in low numbers.[63,106] Therefore, we should be particularly wary of species that possess both qualities likely to make them popular pets and invasive threats.

Kraus[107] notes a strong reporting bias toward introduction events that resulted in establishment, particularly for the pet and cargo or nursery trade pathways, which are not generally under scientific scrutiny. The contemporary pattern in South Africa is likely to be similar, although the implementation of the NEM:BA regulations is expected to bring more introductions to the attention of regulators. Overall, introductions of alien taxa to South Africa will likely occur through either the pet trade or cargo/nursery trade pathways or for culinary purposes as food security decreases. Other economically and agriculturally important pathways, such as biological control, are unlikely, given global awareness of the potential impacts of vertebrate biological control agents such as the cane toad (*Rhinella marina*) and existing South African policy and practice (see Section 15.6).

A review of past introductions shows that a high percentage of extralimital translocations within South Africa (four out of five for amphibians, and at least two out of seven for reptiles) have been intentional, with a further two intentional introductions of alien species. However, given that introductions originating in the pet and cargo/nursery trade pathways are "grossly underreported,"[107] and the increasing anecdotal reports of lone escapee pets, this is unlikely to be an accurate reflection of the true extent of accidental introductions.

## 15.4  Birds

Berndt J. van Rensburg and Faansie Peacock

### 15.4.1  Introduction

Given the country's location on important shipping routes and its history of European settlement, South Africa has suffered a long history of alien bird introductions. The first rock doves (*Columba livia*) arrived in Cape Town in 1652 on board the ships of the first European colonists to reach the country.[56] To date, at least 77 alien species (see Table 15.3) have been recorded in South Africa, including offshore islands, for example, Robben Island (Faansie Peacock, personal observation).[108] Of these, 70 species (91%) have enjoyed a short-lived or spatially restricted presence and while many have attempted to breed, they have achieved only limited success (see Richardson et al.[109] for potential factors explaining their low success rate), although some are probably permanently established in small numbers. At least 12 of these 70 species (Table 15.3) have the potential of becoming invasive given their current numbers and spread, and they should be monitored.

The seven remaining species—mallard (*Anas platyrhynchos*), rose-ringed parakeet (*Psittacula krameri*), rock dove (*Columba livia*), Indian house crow (*Corvus splendens*), common starling (*Sturnus vulgaris*), common myna (*Acridotheres tristis*), and house sparrow (*Passer domesticus*)—have all become irreversibly established, or nearly so. As suggested by Brooke et al.,[56] these species are mostly commensal with human habitation, and none have been shown to impact significantly on natural systems in South Africa. However, with these species being competitively dominant in human-made habitats, land transformation coupled with an increasing human population (e.g., an increase in small-scale and subsistence farming) is likely to increase the available habitat and foraging areas for commensal alien birds.[109] It is expected that with such landscape-level changes, together with known spatial congruence between areas of high human activity and high indigenous species richness at macroecological scales (see, e.g., references[1,110] for a study on birds), in the future, alien bird species are expected to constitute a further economic, agricultural, and environmental burden.

Although rather complex, a secondary question relates to extralimital invasion by indigenous bird species. Conclusions drawn from the available data could lead to skewed interpretations, particularly if inconsistencies in spatial data or comparisons across long periods and unbalanced sampling efforts by different observers are not accounted for.[111] Even if allowance is made for such factors, there are a multitude of species that have unquestionably undergone extralimital range extensions in South Africa. This pertains particularly to families that are able to successfully exploit anthropogenic activities to their advantage. Such activities typically include (1) urbanization (see e.g., references[112,113] for a general framework related to urban exploiters, adapters and avoiders) that often favors cliff nesters such as swallows, martins, swifts, and some raptors and starlings; (2) afforestation favoring forest or woodland species such as raptors and various doves; (3) artificial water impoundments often exploited by, for example, large waterbirds,[114] waders, kingfishers, and wagtails; and (4) some birds, for example seabirds, have also undergone interesting and well-documented range shifts as a result of climate change.[115] Although most of these extralimital species are likely to have only limited economic, agricultural, and environmental impacts, there are a few species of potential concern that require monitoring often coupled with intense management. These include the now ubiquitous helmeted guineafowl (*Numida meleagris*), hadeda ibis (*Bostrychia hagedash*), Egyptian goose (*Alopochen aegyptiaca*), and red-billed quelea (*Quelea quelea*). Abundance, sociability, and large biomass (or large communal biomass in the case of the flocking quelea) identify these species as potential pests outside, as well as within, their natural range.

Compared to the extralimital species that are mostly self-driven facilitated invaders, the alien species have all been deliberately or accidentally introduced by man, with the exception of the voluntarily ship-assisted Indian house crow.[56,116] Of the 77 alien species recorded to date, at least 53 (69%), including the common myna, rose-ringed parakeet and mallard, were imported for the pet trade and subsequently escaped or were deliberately released. Apart from the three major alien species mentioned above, all escaped pet

*Table 15.3* Alien Bird Species Recorded in South Africa

| Species name (common name)[a] | Introduction date[b] | Pathway | Status | Native range[c] | References |
|---|---|---|---|---|---|
| *Gallus gallus* (Red junglefowl) | ca. 1950s ++ | Deliberate introductions, escapes | Various scattered semiferal populations of domestic stock | Southeast Asia | 117 |
| *Lophura nycthemera* (Silver pheasant) | | Introduced Ceres, WC | Probably extinct | Southeast Asia | 117 |
| *Phasianus colchicus* (Common pheasant) | ca. 1900s–1950s | Multiple introductions for hunting purposes | Probably extinct | Northern Asia | 108 |
| *Pavo cristatus*[a] (Common peacock) | 1968 | Deliberate introductions, escapes | Stable population on Robben Island; scattered feral populations throughout South Africa | Southcentral Asia | 108 |
| *Alectoris chukar* (Chukar partridge) | 1964 | Confiscated at customs port and released | Small but stable population on Robben Island | Central Eurasia | 108 |
| *Coturnix chinensis*[a] (Asian blue quail) | ++ | Accidental escape or deliberate release (pet trade/ornamental) | | Southeast Asia, Oceania, Australia | |
| *Colinus virginianus* (Northern bobwhite) | | Introduced for hunting | Probably extinct | North America, Caribbean | 117 |
| *Dendrocygna autumnalis* (Black-bellied whistling duck) | 1997 | Escape | Not established | Central America, South America | 117 |
| *Cygnus olor* (Mute swan) | 1918, 1941 + | Multiple introductions | No known self-sustaining populations | Eurasia | 108 |

*(Continued)*

*Table 15.3* Alien Bird Species Recorded in South Africa (*Continued*)

| Species name (common name)[a] | Introduction date[b] | Pathway | Status | Native range[c] | References |
|---|---|---|---|---|---|
| *Cygnus atratus* (Black swan) | 1926 + | Introduced, occasional escape | Not established | Australia | 108 |
| *Tadorna tadorna* (European shelduck) | 1974, 1985, 1989, 1995 | Occasional escape | Not established | Eurasia | 117 |
| *Aix galericulata*[a] (Mandarin duck) | ca. 1980 ++ | Frequent escape | Small, localized breeding populations | Eastern Asia | 117 |
| *Aix sponsa*[a] (Wood duck) | ca. 1997 ++ | Frequent escape | Small, localized breeding populations | North America | 117 |
| *Callonetta leucophrys* (Ringed teal) | 1985 + | Occasional escape | Not established | South America | 117 |
| *Anas platyrhynchos* (Mallard) | ++ | Frequent escape; commercial production | Established in two core areas (WC and GP), small, localized populations scattered throughout South Africa | North America, Eurasia | 108 |
| *Anas rubripes* (American black duck) | 1975 | Escape | Not established | North America | 117 |
| *Anas discors* (Blue-winged teal) | ca. 2000s + | Escape | Not established | Americas | |
| *Anas clypeata* (Northern shoveler) | + | Escape; some true vagrants | Not established | America, Eurasia, Africa, southeast Asia | 117 |
| *Anas acuta* (Northern pintail) | + | Escape; some true vagrants | Not established | America, Eurasia, Africa, Southeast Asia | 117 |

| Species | Date | Mode | Status | Native range | Ref |
|---|---|---|---|---|---|
| *Anas querquedula* (Garganey) | + | Escape; some true vagrants | Not established | Eurasia, Africa, Australasia | 117 |
| *Netta rufina*[a] (Red-crested pochard) | 1986 ++ | Frequent escape | Established and spreading in several areas | Eurasia, Africa | 117 |
| *Aythya ferina* (Common pochard) | | Escape | Not established | Eurasia | 117 |
| *Aythya nyroca* (Ferruginous duck) | 1994 | Escape | Not established | Eurasia | 117 |
| *Aythya fuligula*[a] (Tufted duck) | ca. 1970s ++ | Frequent escape | | Eurasia | 117 |
| *Coracias cyanogaster* (Blue-bellied roller) | 2003 | Escape | Not established | West Africa | 117 |
| *Merops malimbicus* (Rosy bee-eater) | 2003 | Escape (or true vagrant) | Not established | West Africa | 117 |
| *Cacutua sulphurea* (Yellow-crested cockatoo) | 1976–83 | Escape | Failed to establish viable population | Southeast Asia | 117 |
| *Nymphicus hollandicus* (Cockatiel) | ++ | Frequent escape | Failed to establish viable population | Australia | 117 |
| *Melopsittacus undulatus*[a] (Budgerigar) | ++ | Frequent escape | Probably locally established | Australia | 117 |
| *Agapornis roseicollis*[a] (Rosy-faced lovebird) | ++ | Frequent escape; native range extends marginally into western South Africa | Probably locally established | Southwest Africa | 117 |
| *Agapornis cana* (Gray-headed lovebird) | 1890s | Deliberately introduced | Extinct | Madagascar | 108 |

*(Continued)*

*Table 15.3* Alien Bird Species Recorded in South Africa (*Continued*)

| Species name (common name)[a] | Introduction date[b] | Pathway | Status | Native range[c] | References |
|---|---|---|---|---|---|
| *Psittacula cyanocephala* (Plum-headed parakeet) | 1979 | Escape | Not established | Southcentral Asia | 117 |
| *Psittacula krameri* (Rose-ringed parakeet) | 1850s ++ | Frequent escape | Well established in two urban cores (Durban and GP), spreading | Southcentral Asia, Africa | 108 |
| *Poicephalus rufiventris* (African orange-bellied parrot) | + | Escape | Not established | Northeast Africa | |
| *Poicephalus rueppellii* (Rüppell's parrot) | + | Escape | Not established | Southwest Africa | |
| *Aratinga jandaya* (Jandaya conure) | + | Escape | Not established | South America | 117 |
| *Aratinga weddellii* (Dusky-headed conure) | + | Escape | Not established | South America | 117 |
| *Aratinga pertinax* (Brown-throated conure) | Early 1980s + | Escape | Not established | South America, central America | 117 |
| *Nandayus nenday* (Black-hooded conure) | Early 1980s, 2001 + | Escape | Not established | South America | 117 |
| *Cyanoliseus patagonus* (Patagonian conure) | 1999 + | Escape | Not established | South America | 117 |
| *Forpus passerinus* (Blue-winged parrotlet) | Early 1870s + | Escape | Not established | South America, Caribbean | 117 |
| *Amazona aestiva* (Blue-fronted parrot) | 1989 + | Escape | Not established | South America | 117 |

| Species | Date | Pathway | Status | Native range | Ref. |
|---|---|---|---|---|---|
| *Criniferoides leucogaster* (White-bellied go-away-bird) | | Escape | Not established | Northeast Africa | |
| *Crinifer piscator* (Western gray plaintain-eater) | | Escape | Not established | West Africa | |
| *Musophaga violacea*[a] (Violet turaco) | 1994, 1995 ++ | Frequent escape | Possibly local established in GP | West Africa | 117 |
| *Columba livia* (Rock dove/feral pigeon) | 1652 | Communication, hunting, pet bird trade | Irreversibly established and abundant | Eurasia, north Africa | 56,108 |
| *Columbina inca* (Inca dove) | 1992 + | Escape | | North America, central America | 117 |
| *Streptopelia decaocto*[a] (Eurasian collared-dove) | ++ | Frequent escape | Small, localized populations | Eurasia | 117 |
| *Geopelia cuneata*[a] (Diamond dove) | ++ | Frequent escape | | Australia | |
| *Gallinula nesiotis* (Tristan moorhen) | 1893 | Escape or ship-assisted | Extinct | Tristan da Cunha Island | 117 |
| *Fulica Americana* (American coot) | 1891 | Escape or ship-assisted | Extinct | North America, central America | 117 |
| *Eudocimus ruber* (Scarlet ibis) | 2000 + | Escape | Not established | Northern South America, Caribbean | 117 |
| *Dendrocitta vagabunda* (Rufous treepie) | 1997 | Escape | Not established | Southeast Asia | 117 |
| *Corvus frugilegus* (Rook) | Late 1890s | Introduced: aesthetic | Extinct | Eurasia | 108 |

*(Continued)*

*Table 15.3*  Alien Bird Species Recorded in South Africa (*Continued*)

| Species name (common name)[a] | Introduction date[b] | Pathway | Status | Native range[c] | References |
|---|---|---|---|---|---|
| *Corvus splendens* (Indian house crow) | Early 1970s | Self-introduced (ship-assisted) | Well-established in Durban and Cape Town; occasional sightings elsewhere | Southcentral Asia | 108 |
| *Turdus merula* (Eurasian blackbird) | Late 1890s | Introduced: aesthetic | Extinct | Eurasia, North Africa | 108 |
| *Turdus philomelos* (Song thrush) | Late 1890s | Introduced: aesthetic | Extinct | Western Eurasia, North Africa | 108 |
| *Luscinia megarhynchos* (Common nightingale) | Late 1890s | Introduced: aesthetic | Extinct | Europe, Africa | 108 |
| *Sturnus vulgaris* (Common/European starling) | Late 1890s | Introduced: aesthetic | Established: abundant and spreading | Western Eurasia, North Africa | 108 |
| *Acridotheres tristis* (Common/indian myna) | 1888, 1902, 1930s | Escape | Established: abundant and spreading | Southcentral Asia | 108,119 |
| *Lamprotornis iris*[a] (Emerald starling) | 1993 + | Escape | Possibly local breeding populations | West Africa | 117 |
| *Lamprotornis purpuropterus* (Rüppell's long-tailed starling) | 2001 + | Escape | Not established | Northeast Africa | 117 |
| *Lamprotornis superbus* (Superb starling) | 1995 + | Escape | Not established | Northeast Africa | 117 |
| *Pycnonotus jocosus* (Red-whiskered bulbul) | 1992 + | Escape | Not established | Southeast Asia | 117 |
| *Leiothrix argentauris* (Silver-eared mesia) | 2002 + | Escape | Not established | Southeast Asia | 117 |

| Species | Date | Introduction | Status | Native range | Ref |
|---|---|---|---|---|---|
| *Melanocorypha bimaculata* (Bimaculated lark) | 1930 | Escape | Not established | Asia, northeast Africa | 117 |
| *Ploceus nigerrimus* (Vieillot's black weaver) | 2001–02 + | Escape | | West Africa, central Africa | 117 |
| *Amandava amandava* (Red munia) | | Escape | | Asia | 117 |
| *Estrilda melpoda* (Orange-cheeked waxbill) | | Escape | | West Africa, central Africa | |
| *Euodice cantans* (African silverbill) | 2005 + | Escape | Possibly localized populations | Northcentral Africa | 117 |
| *Taeniopygia guttata* (Zebra finch) | 1984 ++ | Frequent escape | Probably small local populations | Australia | 117 |
| *Padda oryzivora* (Java sparrow) | + | Escape | | Southeast Asia | 117 |
| *Passer domesticus* (House sparrow) | 1880–90 | Multiple introductions | Established: abundant, but numbers decreasing | Eurasia | 56,108 |
| *Carduelis carduelis* (European goldfinch) | 1891, 1900s + | Escape | | Western Eurasia, North Africa | 117 |
| *Fringilla coelebs* (Common chaffinch) | Late 1890s | Introduced: aesthetic | Small but stable population in WC | Western Eurasia, North Africa | 108 |
| *Paroaria coronata* (Red-crested cardinal) | ca.1950s ++ | Frequent escape | | South America | 117 |
| *Paroaria dominicana* (Red-cowled cardinal) | 1960 + | Escape | Not established | South America | 117 |

*Note:* WC = Western Cape Province. GP = Gauteng Province.

[a] Invasion potential due to numbers and spread; warrants monitoring.
[b] Plus (+) symbol indicates species populations proven or suspected to be augmented by additional introduction events after first known introduction date. Double plus (++) symbol indicated species that experience regular additional introduction events.
[c] Including migratory range.

bird species have become extinct (or persecuted to extinction) after short-lived expatriate existences of a few seasons, even in cases where they managed to form conspecific populations and attempted breeding. Given the diversity of pet, aviary, and show birds kept in South Africa, the true number of escapee species is likely to be at least triple the figures presented here. The origins of the remaining 24 species are either uncertain, or a result of aesthetic, commercial, or hunting purposes.

## 15.4.2  *Extent of the problem—range expansion and nonstatic environment*

The most successful alien bird species in South Africa are currently in varying stages of dispersal as superimposed on their inherent invasion potential, rate of mobility, and climatic and physical tolerances. In the cases of the house sparrow and the rock dove, founding populations were assisted on multiple occasions by additional independent introductions and spread fairly rapidly after an initial lag period.[117] Both species are now ubiquitous throughout southern Africa occurring wherever there are human settlements in urban, suburban, or rural landscapes. The two alien starlings (the common myna and the common starling) were both introduced just over a century ago at the harbor cities of Cape Town and Durban, respectively.[118] Subsequently, both species have undergone major range extensions, and today their largely exclusive ranges (see Figures 15.3 and 15.4) cover all but the most arid regions of the country and marginally extend into neighboring countries.[118,119] For mallards, the situation is more complex and less directional, with two core populations established in major metropolitan regions of the Western Cape and Gauteng Provinces, and numerous small but unchecked populations scattered between the two provinces. Similarly, the invasive status of the rose-ringed parakeet is also unclear, but localized populations of this popular cage bird are well established around the three major cities of Pretoria, Johannesburg, and Durban.[118] In addition, widely scattered sight records (possibly of additional escapees rather than wanderers) coupled with increases in population density and local range in urban strongholds suggest that the species has significant invasion potential. Information on the Indian house crow is presented in Box 15.4.

Although the distinct or synergetic roles of climate change and land-cover modification are known to be important drivers in governing the initial establishment, rapid population growth, and widespread dispersal success of alien species,[109] species-specific studies are often lacking to quantify their importance. Nonetheless, some general trends are apparent. For example, urbanization (and the concurrent establishment of irrigated, shady gardens—see, e.g., reference[113]), commercial afforestation and the spread of hardy alien vegetation[118] have allowed many northerly species to expand their ranges southwards into the originally treeless grassland (mostly the Highveld), Karoo, and Fynbos biomes, sometimes by using natural corridors such as riparian strips. Likewise, the provision of nesting structures in the form of electricity pylons has allowed indigenous crows to occur at high densities in the Karoo dwarf shrublands where they may assist in seed dispersal and geographic spread of an alien plant, the undesirable *Opuntia cactus*.[120] Provision of livestock drinking troughs and farm dams has certainly assisted many aquatic and wetland species to spread to the arid interior.[118] Furthermore, irrigated agricultural land has created rich artificial conditions that, ironically, create an abundant food source for invasive crop pests such as the red-billed quelea.[121] Although difficult to demonstrate empirically, the influence of climate change, as a function of temperature and rainfall patterns, on avian distributions is certainly, at the least, a contributing factor. For example, it has been suggested that the southwesterly spread of common mynas is curbed by low temperatures in

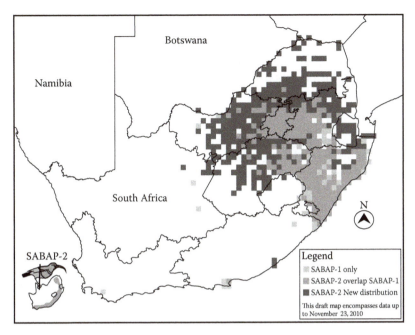

**Figure 15.3** The common myna's *Acridotheres tristis* distribution in South Africa and Lesotho at the quarter-degree scale. The first South African Bird Atlas Project (SABAP-1; 1987–1991)[118] data are indicated in light gray and SABAP-2 (2007–2010) data in dark gray. (Data from http://sabap2.adu.org.za.)

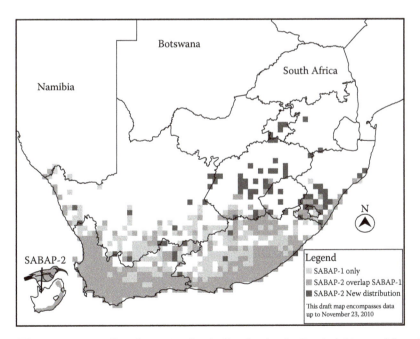

**Figure 15.4** The common starling *Sturnus vulgaris* distribution in South Africa and Lesotho at the quarter-degree scale. The first South African Bird Atlas Project (SABAP-1; 1987–1991)[118] data are indicated in light gray and SABAP-2 (2007–2010) data in dark gray. (Data from http://sabap2.adu.org.za.)

## BOX 15.4   INDIAN HOUSE CROW (*CORVUS SPLENDENS*)

Although the alien Indian house crow has been present in South Africa for only 50 years,[56] it is firmly established in two major metropolitan areas, Durban and Cape Town, and continues to spread with occasional sightings elsewhere along the coast-lines of both South Africa[118] and Mozambique.[136] Its first occurrence is attributed to birds making use of shipping traffic across the Indian Ocean.[56,137] This facilitated, but not deliberate, introduction is unique for established South African alien birds. Like other aliens that are exclusively commensal with humans, impacts related to Indian house crows are most severe in urban environments. Here, this small-bodied corvid is able to exploit a predatory niche from which the larger indigenous crow species are excluded, although it may compete with the introduced common myna (*Acridotheres tristis*) and common starling (*Sturnus vulgaris*) as well as indigenous species such as Burchell's coucal (*Centropus burchelli*) and red-winged starling (*Onychognathus morio*). As is typical for a globally successful colonist,[108,138] Indian house crows have a wide dietary tolerance. It has been recorded raiding heron and ibis colonies for eggs, and also preys on the eggs, nestlings, and adults of smaller birds, chicks of domestic chickens, small vertebrates, offal, and carrion. It is an effective kleptoparasite and may damage fruit and cereal crops to a limited extent.[117] Massive communal roosts distress local residents due to the nuisance factor, noise level, and fouling by birds.[139] Such roosts or other situations where crows occur in close association with people also pose a potential health risk to humans. Breeding birds are aggressive near their nests; they often mob and harass humans, domestic animals, and pets to obtain food and nesting material.[117,140] The considerable economic and environmental costs posed by this species in South Africa have prompted local eradication efforts, which achieved moderate success but failed to exterminate the species completely. For example, in 1989 the Durban population was estimated at ca. 1000 birds and by 1991, was reduced to ca. 150 birds following control operations,[141,142] and currently there are ca. 20 indi-viduals remaining (L. J. Smith, pers. comm., Invasive Alien Species Programme). A 6-year eradication plan currently underway in the city of Cape Town has a proposed budget of ZAR 3.7 million (approximately US$495,000).[140] A similar eradication plan is also planned for the Richards Bay metropolitan area. Such control efforts typically focus on communal roosting trees, or the poisoning of crows by baiting with tainted meat. Given the Indian house crow's current small local populations, its agricultural, environmental, and economic impacts on a national level are limited. However, judg-ing by the densities and apparent detrimental effects of this species in, for example, East Africa, further eradication efforts are warranted as soon as possible.

the Karoo,[108] although examining the wide climatic tolerance in its native range suggests[122] that this might be an oversimplification.

### 15.4.3   *Impacts or potential impacts—agricultural,*
### *environmental and economic*

From the rather scanty data available on the detrimental effects of birds on the South African agricultural industry, it seems that indigenous birds, including extralimital invaders, constitute a greater burden compared to alien species, with the red-billed quelea being a case in point (see Box 15.5). From an alien perspective, the wide dietary tolerance of the two starlings makes them potential pests of fruit crops and cereals of economic importance.[123] However, the detrimental effects of starlings on South African crops are believed to be less than in other regions of the world.[117] Internationally, starlings have also been shown to contribute toward secondary damage often found in crops in the form of fungal growths, bacterial infections, and facilitated insect damage.[123] Whether this is a major concern in South Africa remains to be established. Likewise, the effect of house sparrows on South African agriculture remains unknown, but elsewhere they have been reported to be minor pests of cereals and fruit crops.[117] However, given their general association with urban areas this seems minimal. In their native range, and to a lesser extent in their introduced range elsewhere in the world, rose-ringed parakeets are major pests of grains and minor pests of commercial fruit.[124]

As far as current expert opinion[117] is concerned, impacts (mostly interference and resource exploitation competition) related to alien birds on indigenous bird assemblages are, in general, believed to be of little consequence. This is because the distribution ranges of alien bird species are mostly restricted to urban areas dominated by an already alien bird community. Empirical studies attempting to quantify these impacts are, however, lacking both geographically and from a species-specific perspective, making it difficult for conservation and land managers to implement evidence-based conservation policies.

The continued spread of common mynas and common starlings in South Africa and neighboring countries[119] has led to public concern, particularly regarding interspecific competition and predation. Moreover, compared to the study by Brooke et al.[56] indicating that no undisturbed natural habitat in South Africa has been invaded by alien birds during the mid-1980s, the recent arrival of common mynas in many important conservation areas[119] is likely a matter of concern and warrants further monitoring. Common mynas and common starlings, as well as many indigenous species (e.g., the Cape crow [*Corvus capensis*] in the Karoo[120] and the African olive-pigeon [*Columba arquatrix*] spreading bugweed [*Solanum mauritianum*] in KwaZulu-Natal Province[127]), have the potential to assist in the spread of undesirable alien plants.[123] The possible spread of pathogens and parasites in situations where the congregation (mostly in large numbers) of postbreeding flocks of common starlings and common mynas often occur in proximity to humans is also of concern[128,129]; their accumulated droppings have the potential to act as a growth medium for fungi, prompting concerns about human health.[130]

The same applies to rock doves and house sparrows, in which further economic impacts stem primarily from the physical and chemical damage that their droppings cause to buildings and monuments, particularly in urban environments. Sparrow nests may also cause blockages and fire hazards to buildings. Additional concerns are collisions with aircraft[125,126] and aggression toward humans near nesting sites. Rock doves commonly occur in association with the indigenous speckled pigeon (*Columba guinea*) and occasionally hybridize with that species.[117] Nonetheless, this does not constitute a significant threat, and breeding opportunities and foraging behavior of the two species largely preclude competition. Hybridization is also the primary cause for concern in the case of mallard ducks, feral populations of which originate from domestic stock bred for ornamental collections as well as feathers, meat, and egg production.[131] Globally, the mallard is known to hybridize with about 50 other duck and goose species, more than any other duck in the world.[132] Mixed pairings of mallards and indigenous waterfowl such as the yellow-billed

## BOX 15.5   RED-BILLED QUELEA (*QUELEA QUELEA*)

Sub-Saharan Africa's red-billed quelea is regarded as the most abundant terrestrial bird species in the world, with an estimated breeding population of 1.5 billion.[143] At least 190 million are estimated to reside in South Africa[144] (depending on rainfall and nomadic movements). Its extreme sociability makes this 20-g passerine the world's most severe agricultural pest, with flocks of several million birds not uncommon. It has been suggested that an average-sized flock of 400,000 queleas can consume as much as 1.6 tons of grain per day in South Africa,[145] amounting to a daily loss of ZAR 960,000 (ca. US$128,000). This places an immense burden on South African agriculture, in addition to existing threats such as droughts, floods, weed infestation, and insect pests. Since 1957, South African government departments have, with the cooperation of landowners, implemented control measures to reduce quelea numbers in sensitive agricultural areas, particularly for crops such as wheat, sorghum, millet, manna, and to a lesser extent, sunflowers, peas, buckwheat, rice, and oats.[146] The two main control methods employed are aerial spraying of avicide poisons and ignition of combustible liquids in roosting colonies. In environmentally sensitive areas, less severe control efforts include the harvesting of nestlings, trapping with mist nets, chasing queleas to alternative roosts, planting lure roosts away from sensitive areas, manipulating roosting and foraging habitat before quelea infestations occur, application of repellent chemicals and agronomic practices such as planting of resistant cultivars, and varying planting schedules. Up to 190 control actions are undertaken per year,[147] resulting in the mortality of more than 100 million queleas annually. Despite high mortality rates, control operations do not appear to have any long-term effects on quelea populations,[117] partly due to their rapid reproduction and population augmentation by influxes of migratory birds. Repercussions of quelea control may include the death of large numbers of nontarget species, mainly waterbirds and terrestrial scavengers, effects on human well-being in the form of food security, contamination by poisons, and accidents with explosives, as well as significant expenditure on labor, training, and the purchase of materials and equipment.[134,135] From an invasion biology perspective, a current concern is the expansion of the species' range into previously unsuitable habitat outside its natural range. In particular, the provision of artificial water points, breeding habitat provided by bush encroachment and artificial wetlands, and abundant food stocks associated with farming practices have allowed queleas to penetrate areas such as the Karoo, Eastern Cape, KwaZulu-Natal midlands and recently, the Western Cape Province.[121] Its increasing presence in these areas may pose significant impacts, especially to agriculture in, for example, the Western Cape's wheat belt.

duck (*Anas undulate*) and African black duck (*Anas sparsa*) produce fertile offspring, thereby reducing the genetic purity of the indigenous waterfowl pool. Local control measures at the municipal level, for example, in Cape Town and Richards Bay, have, however, been put in place to reduce mallard population numbers. However, their population is regularly augmented by escapees and deliberate introductions. As in other alien hole nesters, the rose-ringed parakeet competes with indigenous species for nest cavities.[133] Nevertheless, given the rose-ringed parakeet's currently small population numbers and preference for urban environments coupled with its habit of frequently breeding in localized colonies, this is of limited concern.

### 15.4.4 Dealing with the problem

Quantitative data on and monetary costs of the economic impact of alien birds in South Africa are currently very limited, complicating decision-making processes and the addressing of public concerns. Similarly, and as was reported by Brooke[56] and Richardson,[109] data on the impacts of alien bird species on indigenous fauna and flora are also limited. Bird ringing should be seen as a valuable tool to improve our general understanding of avian biology and, consequently, improve on the management practices of pest species. For example, in view of the impact of red-billed queleas on grain crops, and the impact of control operations on nontarget species (see Box 15.5 for more information), an understanding of the movements and survival of the red-billed quelea based on bird ring recoveries is of particular importance for food security and for bird species conservation.[134,135]

Environmental costs are more difficult to express in monetary terms, but are likely to be insignificant due to the limited range restrictions of major invaders in urban environments. Interspecific competition and displacement are poorly studied, and prevailing opinions are often based on anecdotal observations, on which broad decisions are based. Nevertheless, preemptive action is essential to prevent population explosions in those species with as yet limited populations such as mallard, Indian house crow (see Box 15.4 for more information on this species), and rose-ringed parakeet.

## 15.5 Mammals

Dian Spear and Christian T. Chimimba

### 15.5.1 Introduction

Mammals have been intentionally introduced into South Africa for a number of different reasons including for wool, food (sheep [*Ovis aries*]), fur (nutria [*Myocastor coypus*] and American mink [*Mustela vison*]), pest control (European wild boar [*Sus scrofa*]), as pets (e.g., feral cats [*Felis catus*]), for hunting, and as ornamentals (e.g., fallow deer [*Dama dama*]; Table 15.4). They have also been unintentionally introduced as stowaways on boats (e.g., brown rat [*Rattus norvegicus*]) and have escaped from captivity (e.g., Himalayan tahr [*Hemitragus jemlahicus*]).[56] A number of introduced mammal species have also reached the wild in South Africa through the release of surplus animals from zoological gardens to private landowners.[9]

Apart from being deliberately introduced for food (e.g., European rabbit [*Oryctolagus cuniculus*]) and for biological control and/or as pets (e.g., feral cat), the significant driver of some small mammals seems to be through human-mediated introductions such as stowaways on ships (e.g., house mouse [*Mus musculus*] and roof rat [*Rattus rattus*]; Table 15.4).[153,156–159] More recent sampling efforts[160] for invasive rodents have yielded the brown rat (*Rattus norvegicus*) and the oriental house rat (*R. tanezumi*) (see references[161,162]) at or near O.R. Tambo International Airport near Johannesburg (northcentral South Africa). Currently, it is not known whether these rats were introduced to the area as aircraft stowaways from a variety of international destinations and further investigation is required at O.R. Tambo and other South African airports.

**Table 15.4** Mammal Species That Have Been Introduced into South Africa, Their Estimated Date of Introduction and Status (e.g., Invasive, Extinct) if Known and Reason for Introduction

| Species name (common name) | Introduction date | Pathway | Status | References |
|---|---|---|---|---|
| *Addax nasomaculatus* (Addax) | <1985 | Hunting | Unknown | 9 |
| *Ammotragus lervia* (Barbary sheep) | <1965 | Hunting | Unknown | 150,151,168 |
| *Antilope cervicapra* (Blackbuck) | <1985 | Hunting | Unknown | 9 |
| *Axis axis* (Axis deer) | <1985 | Hunting | Unknown | 169 |
| *Axis porcinus* (Hog deer) | <1985 | Hunting | Unknown | 169 |
| *Bison bison* (American bison) | <2005 | Hunting | Unknown | 168 |
| *Bos frontalis* (Gaur) | <2005 | Hunting | Unknown | |
| *Bos javanicus* (Banteng) | <1995 | Hunting | Unknown | |
| *Boselaphus tragocamelus* (Nilgai) | <1965 | Hunting | Unknown | 150,151,169 |
| *Bubalus bubalis* (Indian water buffalo) | <1965 | Hunting | Unknown | 9 |
| *Camelus dromedarius* (Dromedary camel) | <1945 | Transport | Unknown | 149,150,152,169 |
| *Canis familiaris* (Dog [feral]) | | Pet | Invasive | 149 |
| *Capra hircus* (Goat [feral]) | | Food | Invasive | |
| *Capra ibex* (Ibex) | <1985 | Hunting | Unknown | 152 |
| *Capra nubiana* (Nubian ibex) | <1985 | Hunting | Unknown | 152 |
| *Cervus elaphus* (Red deer) | 1895 | Hunting | Unknown | 148,149,152 |
| *Cervus nippon* (Sika deer) | 1897 | Hunting | Unknown | 148,149,170 |
| *Dama dama* (Fallow deer) | 1869 | Ornamental | Invasive | 151,153,169 |
| *Elaphurus davidianus* (Pere David's deer) | <1985 | Hunting | Unknown | 9 |
| *Equus asinus* (Donkey [feral]) | <1985 | Transport | Established | 9 |
| *Felis catus* (Cat [feral]) | | Pet | Invasive | 153 |
| *Hemitragus jemlahicus* (Himalayan tahr) | 1937 | Zoo escape | Eradication[a] | 149,154,171 |
| *Hydrochaeris hydrochaeris* (Capybara) | <1985 | Hunting | Unknown | 9 |
| *Hylochoerus meinertzhageni* (Giant forest hog) | <1985 | Hunting | Unknown | 152 |
| *Kobus kob* (Kob) | <2005 | Hunting | Unknown | |
| *Kobus leche* (Lechwe) | <1965 | Hunting | Invasive | 139,150,152 |
| *Kobus megaceros* (Nile lechwe) | <2005 | Hunting | Unknown | |

| Species | Date | Pathway | Status | Ref. |
|---|---|---|---|---|
| *Kobus vardonii* (Puku) | <2005 | Hunting | Unknown | |
| *Lama glama* (Llama) | <1965 | Wool, hunting | Unknown | 150–152 |
| *Madoqua kirkii* (Damara dik-dik) | <2005 | Hunting | Unknown | |
| *Mus musculus* (House mouse) | Seventeenth century | Stowaway | Invasive | 9 |
| *Mustela vison* (Mink) | 1950s | Fur | Extinct | 149 |
| *Myocastor coypus* (Nutria) | 1950s | Fur | Extinct | 149 |
| *Oryctolagus cuniculus*b (European rabbit) | Seventeenth century | Food | Invasive | |
| *Oryx beisa* (Beisa oryx) | <2005 | Hunting | Unknown | 9 |
| *Oryx dammah* (Scimitar-horned oryx) | <1985 | Hunting | Unknown | 9 |
| *Oryx leucoryx* (Arabian oryx) | <2000 | Hunting | Unknown | 168 |
| *Ovis aries* (Sheep) | <First century | Food | Unknown | 9 |
| *Ovis aries* (Mouflon) | <1970 | Hunting | Unknown | 9 |
| *Panthera tigris* (Tiger) | <1990 | Breeding | Unknown | |
| *Potamochoerus porcus* (Red river hog) | <2005 | Hunting | Unknown | |
| *Rattus norvegicus* (Brown rat) | Eighteenth century | Stowaway | Invasive | 153 |
| *Rattus rattus* (Roof rat) | First century | Stowaway | Invasive | 153 |
| *Rattus tanezumi* (Oriental house rat) | <2005 | Stowaway | Unknown | 161,162 |
| *Redunca redunca* (Bohor reedbuck) | <2005 | Hunting | Unknown | |
| *Rusa unicolor* (Sambar deer) | 1897 | Hunting | Unknown | 149,170 |
| *Sciurus carolinensis* (Gray squirrel) | 1890s | Ornamental | Invasive | 9,153 |
| *Sus scrofa* (European wild boar) | 1920s | Pest control | Invasive | |
| *Taurotragus derbianus* (Derby eland) | <2005 | Hunting | Unknown | 169 |
| *Tragelaphus eurycerus* (Bongo) | <2005 | Hunting | Unknown | |
| *Tragelaphus spekii* (Sitatunga) | <2005 | Hunting | Unknown | |

a Attempted eradication, individuals recently observed.
b On Robben Island.

The game industry in South Africa has been a significant driver of alien ungulate species introductions[2,9,56,148–153] due to the economic importance of some alien ungulate species for hunting.[154,155] For example, in 2008, auction prices were US$5300 for an addax (*Addax nasomaculatus*), US$3000 for a lechwe (*Kobus leche*), US$2900 for a scimitar-horned oryx (*Oryx dammah*), US$430 for a fallow deer (*Dama dama*), and US$160 for a red deer (*Cervus elaphus*). Alien ungulate species introductions by the game industry are not only species from outside South Africa but also species that have been moved outside their presumed historical distributional ranges (i.e., extralimital species) for hunting and/or ecotourism.

## 15.5.2   Extent of the problem

Of the 50 or so mammals that have been introduced into the wild in South Africa, only a small subset are known to be invasive, and there is little information available about whether many of the introduced species have established self-sustaining populations, although some species are thought to have gone locally extinct such as the nutria and mink.[56] Species that have become invasive include ornamentals, game animals, feral domestic animals, livestock, and species that have escaped from captivity (Table 15.4). Introduced mammals have been particularly invasive on islands. Fallow deer (*Dama dama*), feral cats, and European rabbits (*Oryctolagus cuniculus*) are invasive on Robben Island and feral cats and house mice on sub-Antarctic Marion Island. Fallow deer and red lechwe (*Kobus leche*) are among the most invasive mammals on mainland South Africa, and both species have spread into and require control in protected areas. Fallow deer have been introduced to at least 98 quarter-degree grid cells in the country.[9] Populations of feral pigs (*Sus scrofa*) occur in different regions of South Africa, with one population in the Western Cape Province estimated at over 1000 individuals. Populations of feral goats (*Capra hircus*) and donkeys (*Equus asinus*) are also thought to be established in the country.

Domestic and commensal animals have been introduced wherever people live. The genus *Rattus* represents the largest mammalian genus (ca. 66 species) of generally invasive and commensal murid rodents with a worldwide geographic distribution.[156,163–166] Until recently, the widely distributed roof rat (*R. rattus*) and the brown rat (*R. norvegicus*) were considered to be the only species of *Rattus* occurring in South Africa.[161,162] While the former species has traditionally been considered to be widely distributed throughout South Africa, the latter has conventionally been considered to be confined to coastal areas and harbors of the country.[156,163,164,167]

However, during a recent routine genetic screening of sibling rodents in South Africa using molecular data, Bastos et al.[161] reported on the first recorded occurrence in South Africa (and Africa) of an otherwise Asian endemic commensal and invasive sibling species, the oriental house rat (*R. tanezumi*), a species that is morphologically similar to *R. rattus* (i.e., both of the *R. rattus* species complex; Table 15.4).[160,162,165,166,168] This suggests that our current understanding of invasive rodents in South Africa (and beyond in Africa) is poor and in critical need of further investigation. Consequently, the recently recorded occurrence of *R. tanezumi* in South Africa subsequently led to intensive sampling and genetic identification initiatives of members of the genus *Rattus* as well as indigenous rodents, which apart from contributing to general small mammal studies in Africa, has implications in epidemiological, agricultural, biological conservation, and invasion biology research associated with problem rodents in the southern African subregion and beyond.

Some of the derived data so far show that all the three species of *Rattus* currently known to occur in South Africa exhibit a wide range of pelage color variation making morphological identification, particularly of the sibling species *R. rattus* and *R. tanezumi*, difficult. Recent molecular data[160] showed relatively low levels of intraspecific genetic variation in this group of rodents in South Africa that suggests a relatively recent introduction and that members of the genus *Rattus* in the country may have originated from the Far East. These data further suggest that *R. norvegicus* and *R. rattus* were introduced on multiple occasions in South Africa, while the occurrence of *R. tanezumi* in the country is based on a single, recent introduction, with the following suggested chronological sequence of single introductions: (1) *R. rattus*, (2) *R. norvegicus*, and (3) *R. tanezumi* that were followed by (4) two *R. rattus* introductions, and (5) one *R. norvegicus* introduction.

Of particular significance with the recent molecular data[160] is the strong suggestion that the traditionally considered geographic distributions of species of *Rattus* in South Africa as alluded to above are largely inaccurate and need to be investigated further by additional sampling, genotyping, and/or the use of ecological niche modeling together with ground-truthing initiatives. These recent geographic distribution data confirm that (1) *R. rattus* occurs over most of South Africa (except in the Karoo biome), (2) the newly recorded *R. tanezumi* has so far been sampled from localities in Limpopo, Gauteng, and KwaZulu-Natal Provinces of South Africa, and in Swaziland, while (3) *R. norvegicus*, which has traditionally been considered to be confined to South African coastal towns and harbors, has now been recorded at far inland localities in Gauteng Province.[160] Given that *R. tanezumi* was found to be sympatric with both *R. rattus* and *R. norvegicus* in South Africa and the lack of evidence of the co-occurrence of *R. rattus* and *R. norvegicus* in the country (unlike on Steward Island in New Zealand, where the two species are sympatric[169]),[160] there is a critical need to gain insights into the ecological dynamics between these three species of *Rattus* in South Africa, as well as how they interact with indigenous murid rodents.

Introductions of mammal species by the game industry have been extensive in South Africa[2,9] (see Figure 15.5). The country is second to the United States for the number of ungulate species introduced to any country, and all indigenous ungulate species have been translocated outside their presumed historical ranges leading to substantial range changes with extended ranges up to eight times larger than the presumed historical ranges for some species.[9] The introduction of alien species from outside South Africa has resulted in ungulate assemblages becoming more different (by about 4% for those areas

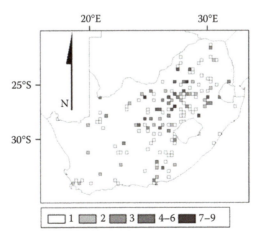

*Figure 15.5* The geographic spread of alien ungulate species that have been advertised for sale or hunting in *Landbou Weekblad* magazine in South Africa. (Data from Spear, D., and S. L. Chown, *Biol Conserv* 142:353, 2009.)

with introductions), whereas the movement of indigenous species extralimitally has made ungulate assemblages 6% more similar.[2]

## 15.5.3 *Impacts or potential impacts*

There is limited evidence of the impacts of alien mammals on biodiversity in South Africa.[9,170] Himalayan tahr (*Hemitragus jemlahicus*) have damaged vegetation and caused soil erosion on Table Mountain in Cape Town in the Western Cape Province of South Africa,[171] a protected area and key tourist attraction. Feral domestic and livestock animals have the potential to compete with or prey on indigenous species. For example, feral dogs kill indigenous mammals[149] and feral goats can have substantial impacts on vegetation. Feral pigs (*Sus scrofa*) have been shown to prey on indigenous critically endangered tortoises and small livestock[172] and to damage agricultural fields and untransformed habitat through rooting, wallowing, and feeding.[172,173] The pigs have also caused tens of thousands of US dollars damage to irrigation systems on private farms in the Western Cape.[173] There are no published data on the impacts of either fallow deer or lechwe on biodiversity in South Africa although concern has been raised that fallow deer are better adapted to vegetation high in $C_3$ grasses than indigenous species and they could be advantaged if vegetation becomes more $C_3$-dominated as a result of predicted climate change.[174]

Worldwide, members of the genus *Rattus* have been implicated in the epidemiology of a range of zoonotic diseases, some of which may be fatal to humans and other animals.[165,168,175–181] These rodents have also been reported to cause extensive damage to agricultural products (e.g., all crop stages and stored grain) and to human-made infrastructure (e.g., electrical installations) worldwide that result in economic losses,[165,168,177,179] the magnitude of which is unknown for South Africa due to lack of cost analysis and needs to be investigated. In South Africa, *Rattus* together with some indigenous murid rodent species of the genera *Mastomys* and *Gerbilliscus* (formerly *Tatera*) have been reported to cause leptospirosis, plague (caused by the bacillus *Yersinia pestis* transmitted by fleas), and toxoplasmosis in humans.[162] Although with a generally low prevalence, *Bartonella* and *Helicobacter*, both having a zoonotic potential, have recently been isolated in all three species of *Rattus* in South Africa, including the recovery for the first time of a species of *Bartonella* that otherwise occurs in *Rattus* worldwide.[160] Mostert[160] found the incidence of *Bartonella* and *Helicobacter* in South Africa to be higher in the filth-preferring, highly cautious, and therefore difficult to capture *R. norvegicus* than in *R. rattus* and *R. tanezumi*.

Other potential problems caused by these groups of rodents worldwide include murine typhus, rat-bite fever, and incidents of *Salmonella*-associated food poisoning,[182] which may also be the case for South Africa. A mid-1990s review estimated about 60 diseases worldwide (e.g., rodent-borne hemorrhagic fevers) and at least about 25 "new" viruses during the same period are associated with rodents and predicted that more await discovery,[183] particularly in the light of climate change scenarios. Although no data are available for South Africa, it is possible that the prevalence of these zoonotic diseases in humans may escalate due to immune systems compromised by the prevailing HIV/AIDS pandemic particularly in South African rural or township settings. In these communities, where *Rattus* occurs commensally, there could be an impact on worker productivity and health care costs, and hence considerable economic losses, the magnitude of which is unknown and needs to be investigated. Although no data are available for South Africa, members of the genus *Rattus* have otherwise also been reported to pose a threat to indigenous flora and fauna (e.g., other rodents, land birds, lizards, and weevils), particularly in island ecosystems.[184–190]

Although its invasion biology on mainland South Africa remains largely unknown, the more cosmopolitan and invasive house mouse (*Mus musculus*) is widespread in the country. Of critical concern is the occurrence of the northern European subspecies of the house mouse (*M. m. domesticus*) on the ecologically sensitive Southern Ocean sub-Antarctic Marion Island. This subspecies was introduced by sealers in the early 1800s[159,191] leading to the introduction of the cat (*Felis catus*) to the island as a means of biological control. With the subsequent eradication of the feral cat in the 1990s as well as changing climate scenarios, an increase in numbers of house mice may be leading to conservation concerns on the island with a significant impact on several indigenous species as well as on ecosystem functioning.[192–194] Direct and indirect impacts include (1) a decline in the biomass of invertebrate decomposers on Marion Island (compared to the nearby mouse-free Prince Edward Island[195,196]) due to a direct effect on terrestrial nutrient cycling,[194] (2) a suspected invertebrate body size reduction due to mouse-predation pressure,[192] (3) a decline in the populations of the lesser sheathbill (*Chionis minor*) which depends on invertebrates for its overwinter survival,[197] (4) mouse runways and burrows causing damage to plant communities especially the cushion plant *Azorella selago*,[198] and (5) being the main terrestrial mammal species on the island after the removal of the feral cat, the house mouse has been suspected of preying on chicks of the albatross and other seabirds as has been found on Gough Island.[199] Despite the negative impacts listed above, members of the genus *Rattus* and the albino strain of *M. musculus* have made positive contributions worldwide as model or experimental animals in immunology, pathology, epidemiology, genetics, physiology, and medicine, among other disciplines.[165,182,200] More recently, some members of the genus *Rattus* have made positive contributions toward studies that have attempted to trace historical human migration patterns.[201]

The domestic cat is considered to have been derived from the wild cat in Egypt around 4000 BC, and its human-assisted spread worldwide makes it the most widespread carnivore, and in South Africa, millions of domestic cats are kept as pets and thus it occurs throughout the country in areas where it is or has been associated with human settlements ranging from metropolitan and rural to protected areas and offshore and oceanic islands (e.g., Kruger National Park).[182] Impacts of the feral cat include: 1) threats to the genetic integrity of the African wild cat through hybridization with the feral domestic cat, and (2) feeding on rabbits and some birds on Dassen Island as was the case on Marion Island before its eradication in the 1990s.[182] Considered to be both a generalist predator (feeding on small mammals, birds, reptiles, and insects in rural and suburban areas) and a scavenger (feeding on garbage, rats, cockroaches, and human handouts in urban area),[182] the impacts of the feeding ecology of the feral cat in South Africa otherwise remains largely unknown.

Suspected of having been introduced into South Africa around 1900, the gray squirrel (*Sciurus carolinensis*) is indigenous to North America. Considered a game animal in the United States and an agricultural pest in the United Kingdom, the gray squirrel has not adapted well in South Africa where it is largely restricted to suburban areas with large seed-producing exotic trees such as oak, stone pines, and cluster pines.[182] Although it causes damage to pine nuts, vegetables, some deciduous fruits, grapes, and peaches, its impact is not considered to be economically detrimental,[182] and currently, there are no documented records of initiatives attempting to reduce the impact of the gray squirrel. However, the gray squirrel is believed to impact garden birds and pose a threat to indigenous squirrels.

Besides being popular as a pet, the European rabbit (*Oryctolagus cuniculus*) has generally not attained the status of a pest in Africa despite its breeding potential as was the case in Australia.[182] The only exception seems to be on Robben Island off the Atlantic coast of Cape Town in South Africa in recent years where unconfirmed newspaper reports otherwise suggest that the species is successfully being controlled. The European rabbit was initially introduced for food on Robben Island and subsequently but unsuccessfully, on almost all islands off the coast of southern Africa.[182] Apart from the Robben Island population, which is currently being controlled, the species also occurs on islands off the western (Namibian Possession Island, Jutten, Dassen, Vondeling, Schaapen), and eastern (Bird and Seal Islands near Port Elizabeth) coasts of southern Africa.[182] Although the impact of the European rabbit remains largely unknown and needs to be investigated further, its spread on the South African mainland seems to be limited by carnivore predators.[182]

## 15.5.4   Extralimital species

Extralimital species have the potential to have an impact on biodiversity (e.g., herbivory, hybridization, and competition) in the same way as any alien species. It has been suggested that nyala (*Tragelaphus angasii*) outcompetes bushbuck (*Tragelaphus scriptus*) in areas in KwaZulu-Natal where the nyala occurs extralimitally[202] and where it is also suspected of having substantial impacts on vegetation structure and could pose a risk of hybridizing with bushbuck. Giraffe (*Giraffa camelopardalis*) that have been introduced extralimitally for ecotourism have significantly altered tree abundance and composition.[203] Hybridization between blue wildebeest (*Connochaetes taurinus*) and black wildebeest (*Connochaetes gnou*) has resulted where *C. taurinus* has been introduced extralimitally,[204,205] and it has been suggested that the translocation of the springbok (*Antidorcas marsupialis*) extralimitally introduced a lungworm (*Bronchonema magna*), which then infected the bontebok (*Damaliscus pygargus pygargus*) contributing to mortality.[206] There are reports of extralimitally introduced warthog (*Phacochoerus africanus*) becoming problematic in the Eastern Cape Province, where they invaded a nature reserve and removal attempts were unsuccessful.[207] However, no impacts of warthog have been quantified in the scientific literature, and there appears to be no published data on the impacts of the most common and potentially harmful extralimital introductions such as that of impala (*Aepyceros melampus*). However, conservation authorities are concerned about impala invading *Acacia karroo* habitat along streams in the Karoo region and outcompeting with the reedbuck in the Eastern Cape Province where they have invaded reedbuck habitat (P. Lloyd, pers. comm.). In so far as small mammals are concerned, knowledge on the impact of extralimital species remains unknown and is in critical need of further investigation.

## 15.5.5   Dealing with the problem

In association with South Africa's legislation (see Section 15.6.3 "Research and capacity development" for more information on legislation), the feral pig (*Sus scrofa*) and Himalayan tahr (*Hemitragus jemlahicus*) have been listed as "controlled as part of invasive species control program." There have been substantial attempts to control the Himalayan tahr on Table Mountain in Cape Town in the past,[56] and most recently a complete eradication attempt was made. However, there have been sightings of Himalayan tahr on Table Mountain since the eradication initiative. It is important that the remaining Himalayan tahr individuals are eradicated to prevent re-invasion by this species (bearing in mind that the problem began with a pair of individuals). Feral pigs are hunted in farming areas by recreational hunters, and a much needed coordinated control program is being planned.

Despite the potential problems that these species pose to biodiversity in South Africa, fallow deer, lechwe, and nyala are classified as "controlled by area" in the draft list of invasive species.[209] Lechwe are economically valuable to the game industry, and this classification is a compromise between the game industry's preference for unrestricted movement and conservation authorities' preference for controlling the species. Research is required to determine what impacts this species is having on biodiversity in South Africa. The species is thought to have the potential to hybridize with other *Kobus* species, and other concerns include competition, disease transmission, and impacts on vegetation. If lechwe are shown to have impacts on biodiversity in South Africa, a case can be built for strict control of the species in the country. The same may also apply to the fallow deer and some indigenous species that are moved extralimitally.

The nyala is also important to the game industry but is thought to pose a substantial threat as a competitor to bushbuck and as a transformer of vegetation. For the numerous alien species in South Africa for which little is documented, it is important that their invasion status be determined, that is, which of these species are established, how big are populations, and are they spreading? It would be advisable to evaluate the extent of occurrence, rate of spread, and biodiversity impacts of all species to ascertain whether control measures should be put in place, remembering that prevention or rapid response at an early stage of invasion is more efficient than waiting for the problem to get out of control. The control of alien mammals in South Africa should not be restricted to wild animals but should include domestic (e.g., feral cats) and livestock (e.g., feral goats) animals that can substantially affect biodiversity. An example of relevant regulation is seen in the Kruger National Park, where employees who live in the park are not allowed to keep cats and dogs to reduce the incidence of problem feral animals in the park. This policy also exists for cats on Robben Island.

Most importantly, alien species should not be introduced to areas where they can hybridize and produce fertile offspring with indigenous species. This is covered by national legislation, but effective enforcement is lacking. Alien herbivores should not be introduced into areas with rare plants that are not adapted to herbivory, and likewise for alien predators that are particularly destructive on islands ecosystems. If landowners are going to keep alien species, they should have sufficient fencing to prevent the movement of alien species from their properties. In so far as small mammals are concerned, there is a critical need to include a legislative focus and implementation initiatives that can accommodate small mammals, which currently is generally lacking.

## 15.6   Legislation and policy

Berndt J. van Rensburg and Sarah J. Davies

### 15.6.1   Establishing a regulatory framework

Legislation governing the importation and movement of alien and extralimital species in South Africa falls under NEM:BA (Act No. 10, 2004).[208] Chapter 5 of the Act focuses on species and organisms posing potential threats to biodiversity. The development of regulations for implementing this chapter was initiated in 2004 and is still in progress.[3,210] Several versions of the regulations have been produced, and a revised draft was published for public comment in 2009.[209]

NEM:BA requires the listing of alien and invasive species[208] as prohibited species (under Section 67) or, under Section 70, as species that require compulsory control (Category 1a), control as part of an invasive species control program (Category 1b), regulation by area (Category 2), or regulation by activity (Category 3).[209] Prohibited species are invasive species that may not be imported, kept, bred, moved, released, or allowed to spread into South

Africa.[209] For all other alien and invasive species, a risk assessment is required to inform the issue of permits to undertake the above-mentioned restricted activities.

The effective control of alien and extralimital introductions through species-based legislation presents significant challenges to policy makers and managers. Introductions are largely driven by vectors and pathways rather than by the species themselves, a concern which has been raised for the United States.[107,211] The number of animals entering a given pathway needs to be controlled through integrated management that differentiates between intentional introductions, which are the focus of the pathway (e.g., species in the pet trade, excluding their parasites and pathogens), and accidental introductions, which are incidental to the true purpose of the pathway (e.g., hitchhikers in the nursery trade). The importance of establishing vector- or pathway-based legislation is becoming more evident as the genetic aspects of invasive species management and control are better understood (e.g., reference[212]), and because of the prevalence of accidental introductions of alien species, despite significant management interventions to limit such introductions (e.g., reference[213] for the Antarctic region, becomes clear).

NEM:BA emphasizes the management of individuals and species, rather than pathways and vectors of invasion. Although the act provides a sound foundation for the implementation of effective management strategies, lack of quantitative data on the impacts of many species makes thorough and objective risk assessment difficult. A review of U.S. invasive species legislation by the Ecological Society of America[211] revealed a number of priorities for invasive species legislation and its implementation, several of which are applicable in the South African context. These include (1) a focus on management of pathways to reduce introductions, (2) rigorous implementation of quantitative risk assessment, (3) the use of technology to improve surveillance and communication about invasive species, and (4) the creation of or mandating an institution that is responsible for prevention, early detection, and rapid response for emerging invasions.

### 15.6.2   Policy implementation

The management of invasive species happens in a complex multiagency environment where mandates and levels of operation of different institutions are critical factors in determining the success or failure of management initiatives. In order for legislated policy to effectively address conservation goals, it is essential that the best current science be incorporated into the policy environment.[3,214] Dedicated commitment is needed from all role players to take what is often an uninviting leap across the research-management gap. Some strategies to mitigate this problem involve defining and agreeing on the main objectives, raising sufficient funds, and, as suggested by Terborgh,[215] combining good science with strong institutions.[216]

Since 2004, many South African researchers spanning a variety of research institutions have worked in collaboration with the South African Department of Science and Technology—National Research Foundation (DST-NRF) Centre of Excellence for Invasion Biology (C·I·B), a nationwide research and capacity-building network funded principally by the national Departments of Science and Technology and Water Affairs and Stellenbosch University. The South African government's support for the C·I·B indicates substantial interest in the impacts of biological invasions on the economy and society.

To date, the integration and implementation of the best current science in the policy environment has been most successful in widely recognized areas of biodiversity importance, such as the Cape Floristic Region (see reference[217]), and in regions characterized by a smaller number of stakeholders, such as the Antarctic region.[3] The reasons for this include the constraints faced by policy makers in staying up to date with the most recent scientific developments and a lack of opportunities for communication between scientists and decision makers.[3]

### 15.6.3    Research and capacity development

Management of invasive species in South Africa is effected by national, provincial, and local government, nongovernment organizations, science councils, and the private sector (e.g., pest management companies, nursery industry, and pet trade), supported by research emerging from higher education institutions and science councils. While South Africa has an energetic corps of researchers working on aspects of biological invasions, there is a need to develop scientific capacity in many areas, in particular specialist expertise on vertebrate taxa (e.g., reptiles, small mammals) through local capacity development and strengthening of collaborations at the regional, continental, and intercontinental levels. Gaps in current knowledge include (1) assessing which of the species currently being kept and traded in South Africa as part of the pet, aquaculture, or game farming sectors pose a potential biodiversity threat; such species should be noted for inclusion on lists of prohibited species; (2) establishing the extent of geographic distributions and migratory pathways of invasive species, particularly those that have been unintentionally introduced, such as invasive small mammals; and (3) characterizing cryptic invasive species using a range of systematic techniques (e.g., molecular and cytogenetic data) as recently applied in the genus *Rattus*.[161,162] Integration of invasive species management plans into local development plans should take place through the integrated development planning (IDP) process. In this way, for example, municipal waste management practices may allow manipulation of the ecological dynamics of members of the genus *Rattus* that have an affinity for unhygienic conditions.

Little is known about how the status and trends of specific invasive species in South Africa may change with ongoing anthropogenic changes in land use, human activity, climate and atmospheric composition, and related changes in plant communities. Niche models go some way toward projecting the potential impacts of global change, but planning for change will be a priority.

## References

1. Chown, S. L., B. J. van Rensburg, K. J. Gaston, A. S. L. Rodrigues, and A. S. van Jaarsveld. 2003. Energy, species richness, and human population size: Conservation implications at a national scale. *Ecol Appl* 13:1233.
2. Spear, D., and S. L. Chown. 2008. Taxonomic homogenization in ungulates: Patterns and mechanisms at local and global scales. *J Biogeogr* 35:1962.
3. Chown, S. L., D. Spear, J. E. Lee, and J. D. Shaw. 2009. Animal introductions to southern systems: Lessons for ecology and for policy. *Afr Zool* 44:248.
4. Hulme, P. E. et al. 2008. Grasping at the routes of biological invasions: A framework for integrating pathways into policy. *J Appl Ecol* 45:403.
5. Blackburm, T. M., J. L. Lockwodd, and P. Cassey. 2009. *Avian Invasions. The Ecology and Evolution of Exotic Birds*. Oxford: Oxford University Press.
6. Leprieur, F., O. Beauchard, S. Blanchet, T. Oberdorff, and S. Brosse. 2008. Fish Invasions in the world's river systems: When natural processes are blurred by human activities. *PLoS Biol* 6:0404.
7. Robinson, T. B., C. L. Griffiths, C. D. McQuaid, and M. Rius. 2005. Marine alien species of South Africa—status and impacts. *Afr J Mar Sci* 27:29.

8. Griffiths, C. L., T. B. Robinson, and A. Mead. 2009. The status and distribution of marine alien species in South Africa. In *Biological Invasions in Marine Ecosystems*, ed. G. Rilov and J. A. Crooks, 393. Berlin: Springer.

9. Spear, D., and S. L. Chown. 2009. The extent and impacts of ungulate translocations: South Africa in a global context. *Biol Conserv* 142:353.

10. Elton, C. S. 1958. *The Ecology of Invasions*. Chicago: University of Chicago Press.

11. Stohlgren, T. J. et al. 1999. Exotic plant species invade hotspots of native plant diversity. *Ecol Monogr* 69:25.

12. Richardson, D. M., P. Pyšek, M. Rejmánek, M. G. Barbour, F. D. Panetta, and C. J. West. 2000. Naturalization and invasion of alien plants: Concepts and definitions. *Divers Distrib* 6:93.

13. Williamson, M., and A. Fitter. 1996. The varying success of invaders. *Ecology* 77:1661.

14. Skelton, P. H., and J. Skead. 1984. Early reports and paintings of freshwater fishes in the Cape province. *Afr Notes Rec* 26:29.

15. De Moor, I. J., and M. N. Bruton. 1988. Atlas of the alien and translocated indigenous aquatic animals in southern Africa. South African National Scientific Programmes, Report 144. Pretoria, South Africa: Foundation for Research Development, CSIR.

16. Cambray, J. A., and R. A. Jubb. 1977. Dispersal of fishes via the Orange-Fish tunnel, South Africa. *J Limnological Soc South Afr* 3:33.

17. Scott, L. E. P., P. H. Skelton, A. J. Booth, L. Verheust, R. Harris, and J. Dooley. 2006. *Atlas of Southern African Freshwater Fishes*. Smithiana, Monograph 2. Grahamstown, South Africa: SAIAB.

18. National Environmental Management. 2004. *Biodiversity Act, 2004*. Republic of South Africa. Government Gazette, 467, No. 26436. Cape Town: Government of the Republic of South Africa.

19. Cambray, J. A. 2003. The need for research and monitoring on the impacts of translocated sharptooth catfish, *Clarias gariepinus*, in South Africa. *Afr J Aquat Sci* 28:191.

20. Van der Waal, B. C. W., and R. Bills. 2000. *Oreochromis niloticus* (Teleostei: Cichlidae) now in the Limpopo River system. *S Afr J Sci* 96:47.

21. Bertasso, A., and A. Avenant-Oldewage. 2005. Aspects of the ecology of the Asian tapeworm, *Bothriocephalus acheilognathi* Yamaguti, 1934 in yellowfish in the Vaal Dam, South Africa. *Onderstepoort J Vet Res* 72:207.

22. Skelton, P. H. 1988. A taxonomic revision of the redfin minnows (Pisces, Cyprinidae) from southern Africa. *Ann Cape Prov Mus Nat Hist* 16:201.

23. Swartz, E. R., A. F. Flemming, and P. L. F. N. Mouton. 2004. Contrasting genetic patterns and population histories in three threatened redfin species (Cyprinidae) from the Olifants River system, western South Africa. *J Fish Biol* 64:1153.

24. Cambray, J. A. 2003. The global impact of alien trout species—a review; with references to their impact in South Africa. *Afr J Aquat Sci* 28:61.

25. Lowe, S. R., D. J. Woodford, D. N. Impson, and J. A. Day. 2008. The impact of invasive fish and invasive riparian plants on the invertebrate fauna of the Rondegat River, Cape Floristic Region, South Africa. *Afr J Aquat Sci* 33:51.

26. Mulder, P. F. S. 1973. Aspects on the ecology of *Barbus kimberleyensis* and *Barbus holubi* in the Vaal River. *Zool Afr* 8:1.

27. Moralee, R. D., F. H. Van Der Bank, and B. C. W. van Der Waal. 2000. Biochemical genetic markers to identify hybrids between the endemic *Oreochromis mossambicus* and the alien species, *O. niloticus* (Pisces: Cichlidae). *Water SA* 26:263.

28. Canonico, G. C., A. Arthington, J. K. McCrary, and M. L. Thieme. 2005. The effects of introduced tilapias on native biodiversity. *Aquat Conserv Mar Freshw Ecosyst* 15:463.

29. Lowe, S., M. Browne, S. Boudjelas, and M. DePoorter. 2000. *100 of the World's Worst Invasive Alien Species. A Selection from the Global Invasive Species Database*. Invasive Species Specialist Group (ISSG), a specialist group of the Species Survival Commission (SSC) of the World Conservation Union (IUCN), 12. Auckland, New Zealand.

30. Tweddle, D., R. Bills, E. Swartz, W. Coetzer, L. Da Costa, J. Engelbrecht, J. Cambray et al. 2009. The status and distribution of freshwater fishes. In *The Status and Distribution of Freshwater Biodiversity in Southern Africa*, ed. W. R. T. Darwall, K. G. Smith, D. Tweddle, P. H. Skelton, 21–37. Gland, Switzerland: IUCN and Grahamstown, South Africa: SAIAB.

31. Ramoejane, M., E. Swartz, and O. L. F. Weyl. 2010. The genetic integrity of *Labeo* species (Cyprinidae) in South Africa in relation to inter-basin water transfer schemes. In *Proceedings: The Annual Conference of the Southern African Sosciety of Aquatic Scientists "Aquatic biodiversity and climate change—an arid region perspective", 13–15 June, Augrabies Falls National Park*, Southern African Society of Aquatic Scientists. 37. South Africa.

32. Leibold, M., and C. J. van Zyl. 2008. The economic impact of sport and recreational angling in the republic of South Africa, 2007: Extensive report. In *Report of Project to Scientifically Determine the Overall Economic Impact and Strategic Value of Sport & Recreational Angling in the Republic of South Africa*. Development Strategies International PTY LTD. Consultants to Business and Government, 49. Cape Town: Unpublished consulting report for the South African Deep Sea Angling Association.

33. Ellender, B. R., O. L. F. Weyl, H. Winker, H. Stelzhammer, and G. R. L. Traas. 2010. Estimating angling effort and participation in a multi-user, inland fishery in South Africa. *Fish Manag Ecol* 17:19.

34. Ellender, B. R., O. L. F. Weyl, H. Winker, and A. J. Booth. 2010. Quantifying annual fish harvests from South Africas largest freshwater reservoir. *Water SA* 36:45.

35. Britz, P. J., B. Lee, and L. Botes. 2009. *AISA 2009 Aquaculture Benchmarking Survey: Primary Production and Markets*. AISA report. Cape Town: Aquaculture Institute of South Africa.

36. von Holdt, B., M. T. T. Davies, and L. J. H. Olyott. 2002. The role of the Federation of Southern African Flyfishers as the representative body of flyfishing in southern Africa. In *Regional Experiences for Global Solutions. The Proceedings of the 3rd World Recreational Fishing Conference 21–24 May 2002, Northern Territory, Australia. Fisheries Report 67*, ed. A. P. M. Coleman. 213–215. Darwin, Australia: Fisheries Group, Department of Business, Industry and Resource Development.

37. Davies, M. T. T. 2002. The development of flyfishing as a recreational sport in southern Africa. In *Regional Experiences for Global Solutions. The Proceedings of the 3rd World Recreational Fishing Conference 21–24 May 2002, Northern Territory, Australia. Fisheries Report 67*, ed. A. P. M. Coleman. 209–211. Darwin, Australia: Fisheries Group, Department of Business, Industry and Resource Development.

38. Abell, R., J. D. Allan, and B. Lehner. 2007. Unlocking the potential of protected areas for freshwater. *Biol Conserv* 134:48.

39. Weyl, O. L. F., T. Stadtlander, and A. J. Booth. 2009. Establishment of translocated populations of smallmouth yellowfish, *Labeobarbus aeneus* (Pisces: Cyprinidae) in lentic and lotic habitats within the Great Fish River system, South Africa. *Afr Zool* 44:93.

40. Fritts, T. H., and G. H. Rodda. 1998. The role of introduced species in the degradation of island ecosystems: A case history of Guam. *Annu Rev Ecol Syst* 29:113.

41. Freeland, W. J. 1984. Cane Toads: A Review of Their Biology and Impact in Australia. Technical report no 19. Darwin: Conservation Commission of the Northern Territory.

42. Fritts, T. H. 2002. Economic costs of electrical system instability and power outages caused by snakes on the Island of Guam. *Int Biodeterior Biodegradation* 49:93.

43. Kraus, F. 2009. *Alien Reptiles and Amphibians: A Scientific Compendium and Analysis*. Dordrecht: Springer.

44. Alexander, G. J., J. A. Harrison, D. H. Fairbanks, and R. A. Navarro. 2004. Biogeography of the frogs of South Africa, Lesotho and Swaziland. In *Atlas and Red Data Book of the Frogs of South Africa, Lesotho and Swaziland*, ed. L. R. Minter, M. Burger, J. A. Harrison, H. H. Braack, P. J. Bishop, and D. Kloepfer, 31. Washington, DC: Smithsonian Institution.

45. Poynton, J. C. 1964. The amphibian of southern Africa: A faunal study. *Ann Natal Mus* 17:1.

46. Branch, W. R. 2001. *A Photographic Guide to Snakes and Other Reptiles of Southern Africa*. Cape Town: Struik Publishers.

47. Uetz, P., J. Hallermann, and B. Baker. 2008. The JCVI/TIGR Reptile Database. http://www.reptile-database.org (accessed January 29, 2008).

48. Minter, L. R., M. Burger, J. A. Harrison, H. H. Braack, P. J. Bishop, and D. Kloepfer. 2004. *Atlas and Red Data Book of the Frogs of South Africa, Lesotho and Swaziland*. Washington, DC: Smithsonian Institution.

49. de Villiers, A. L. 2006. Geographical distribution: *Bufo gutturalis*. *Afr Herp News* 40:28.

50. Bishop, P. J. 2004. *Hyperolius marmoratus* Rapp, 1842. In *Atlas and Red Data Book of the frogs of South Africa, Lesotho and Swaziland,* ed. L. R. Minter, M. Burger, J. A. Harrison, H. H. Braack, P. J. Bishop, and D. Kloepfer, 141. Washington, DC: Smithsonian Institution.

51. Tolley, K. A., S. J. Davies, and S. L. Chown. 2008. Deconstructing a controversial local range expansion: Conservation biogeography of the painted reed frog (*Hyperolius marmoratus*) in South Africa. *Divers Distrib* 14:400.

52. Du Preez, L., and V. Carruthers. 2009. *A Complete Guide to the Frogs of Southern Africa.* Cape Town: Struik Nature.

53. Bruton, M. N., and J. van As. 1986. Faunal invasions of aquatic ecosystems in southern Africa, with suggestions for their management. In *The Ecology and Management of Biological Invasions in Southern Africa,* ed. I. A. W. Macdonald, F. J. Kruger, and A. A. Ferrar, 47. Cape Town: Oxford University Press.

54. Measey, G. J., and A. Channing. 2003. Phylogeography of the genus *Xenopus* in southern Africa. *Amphib-reptil* 24:321.

55. Jacobsen, N. H. G. 1995. *The Herpetofauna of Gauteng Province—Localities and Distribution Maps.* Johannesburg: Chief Directorate of Nature and Environmental Conservation.

56. Brooke, R. K., P. H. Lloyd, and A. L. de Villiers. 1986. Alien and translocated terrestrial vertebrates in South Africa. In *The Ecology and Management of Biological Invasions in Southern Africa,* ed. I. A. W. Macdonald, F. J. Kruger, and A. A. Ferrar, 63. Cape Town: Oxford University Press.

57. de Villiers, A. L. 2006. Geographical distribution: *Lygodactylus capensis capensis. Afr Herp News* 40:29.

58. Newbery, R. 1984. The American red-eared terrapin in South Africa. *Afr Wildl* 38:186.

59. Measey, G. J. 2004. *Xenopus laevis* (Daudin, 1802). In *Atlas and Red Data Book of the Frogs of South Africa, Lesotho and Swaziland,* ed. L. R. Minter, M. Burger, J. A. Harrison, H. H. Braack, P. J. Bishop, and D. Kloepfer, 264. Washington, DC: Smithsonian Institution.

60. Lobos, G., and G. J. Measey. 2002. Invasive populations of *Xenopus laevis* (Daudin) in Chile. *Herpetol J* 12:163.

61. Evans, B. J., J. C. Morales, M. D. Picker, D. J. Melnick, and D. B. Kelley. 1998. Absence of extensive introgression between *Xenopus gilli* and *Xenopus laevis laevis* (Anura: Pipidae) in southwestern Cape Province, South Africa. *Copeia* 1998:504–509.

62. van Wilgen, N. J., D. M. Richardson, and E. H. W. Baard. 2008. Alien reptiles and amphibians in South Africa: Towards a pragmatic management strategy. *S Afr J Sci* 104:13.

63. van Wilgen, N. J., J. R. Wilson, J. Elith, B. A. Wintle, and D. M. Richardson. 2009. Alien invaders and reptile traders: What drives the live animal trade in South Africa? *Anim Conserv* 13(Suppl. 1):1.

64. Rawlinson, P. A., R. A. Zann, S. Balen, and I. W. B. Thornton. 1992. Colonization of the Krakatau islands by vertebrates. *GeoJournal* 28:225.

65. Bruton, M. N., and S. V. Merron. 1985. *Alien and Translocated Aquatic Animals in Southern Africa: A General Introduction, Checklist and Bibliography.* Pretoria: Foundation for Research Development.

66. Passmore, N. I., and V. C. Carruthers. 1979. *South African Frogs.* Johannesburg: Witwatersrand University Press.

67. Passmore, N. I., and V. C. Carruthers. 1995. *South African Frogs, A Complete Guide.* Johannesburg: Southern Book Publishers Pty. Ltd., Witwatersrand University Press.

68. Channing, A. 2001. *Amphibians of Central and Southern Africa.* Pretoria: Protea Book House.

69. Westphal, M. I., M. Browne, K. MacKinnon, and I. Noble. 2008. The link between international trade and the global distribution of invasive alien species. *Biol Invasions* 10:391.

70. Rodriguez-Labajos, B., R. Binimelis, and I. Monterroso. 2009. Multi-level driving forces of biological invasions. *Ecol Econ* 69:63.

71. Bomford, M., F. Kraus, M. Braysher, L. Walter, and L. Brown. 2005. *Risk Assessment Model for the Import and Keeping of Exotic Reptiles and Amphibians.* Canberra: Australian Government Bureau of Rural Sciences.

72. Dennison, D. R. 2007. Ghost house snakes. *Reptile Exot Pet Mag Ultimate Exotics* 1:6.

73. Branch, W. R., and H. Erasmus. 1984. Captive breeding of pythons in South Africa, including details of an interesting hybrid (*Python sebae natalensis* X *Python molurus bivittatus*). *J Herpetol Assoc Afr* 30:1.

74. Phillips, B. L., and R. Shine. 2006. Spatial and temporal variation in the morphology (and thus, predicted impact) of an invasive species in Australia. *Ecography* 29:205.

75. Rodda, G. H., M. J. McCoid, T. H. Fritts, and E. W. Campbell. 1999. An overview of the biology of the brown treesnake (*Boiga irregularis*), a costly introduced pest on Pacific Islands. In *Problem Snake Management: The Habu and Brown Treesnake,* ed. G. H. Rodda, Y. Sawai, D. Chiszar, and H. Tanaka, 44. Ithaca, NY: Comstock Public.

76. Phillips, B. L., G. P. Brown, and R. Shine. 2003. Assessing the potential impact of cane toads on Australian snakes. *Conserv Biol* 17:1738.

77. Wells, K. D. 2007. *The Ecology and Behaviour of Amphibians.* Chicago: University of Chicago Press.

78. Kiesecker, J. M., and A. R. Blaustein. 1998. Effects of introduced bullfrogs and smallmouth bass on microhabitat use, growth, and survival of native red-legged frogs (*Rana aurora*). *Conserv Biol* 12:776.

79. Santos-Barrera, G., G. Hammerson, B. Hedges, R. Joglar, S. Inchaustegui, L. Kuangyang, C. Wenhao et al. 2006. *Lithobates catesbeianus*, 2008 IUCN Red List of Threatened Species. http://www.iucnredlist.org (accessed March 01, 2009).

80. Kiesecker, J. M., A. R. Blaustein, and C. L. Miller. 2001. Potential mechanisms underlying the displacement of native red-legged frogs by introduced bullfrogs. *Ecology* 82:1964.

81. Chivers, D. P., E. L. Wildy, J. M. Kiesecker, and A. R. Blaustein. 2001. Avoidance response of juvenile Pacific treefrogs to chemical cues of introduced predatory bullfrogs. *J Chem Ecol* 27:1667.

82. Kupferberg, S. J. 1997. Bullfrog (*Rana catesbeiana*) invasion of a California river: The role of larval competition. *Ecology* 78:1736.

83. Daszak, P., A. Strieby, A. A. Cunningham, J. E. Longcore, C. C. Brown, and D. Porter. 2004. Experimental evidence that the bullfrog (*Rana catesbeiana*) is a potential carrier of Chytridiomycosis, an emerging fungal disease of amphibians. *Herpetol J* 14:201.

84. Lever, C. 2003. *Naturalized Reptiles and Amphibians of the World.* New York: Oxford University Press.

85. Nagano, N., S. Oana, Y. Nagano, and Y. Arakawa. 2006. A severe *Salmonella enterica* serotype Paratyphi B infection in a child related to a pet turtle, *Trachemys scripta elegans. Jpn J Infect Dis* 59:132.

86. Cadi, A., and P. Joly. 2003. Competition for basking places between the endangered European pond turtle (*Emys orbicularis galloitalica*) and the introduced red-eared slider (*Trachemys scripta elegans*). *Can J Zool Rev Can de Zool* 81:1392.

87. Cadi, A., and P. Joly. 2004. Impact of the introduction of the red-eared slider (*Trachemys scripta elegans*) on survival rates of the European pond turtle (*Emys orbicularis*). *Biodivers Conserv* 13:2511.

88. Tinsley, R. C., and M. J. McCoid. 1996. Feral populations of *Xenopus* outside Africa. In *The Biology of Xenopus,* ed. R. C. Tinsley, and H. R. Kobel, 81. Oxford: Oxford University Press.

89. Lobos, G., and F. M. Jaksic. 2005. The ongoing invasion of African clawed frogs (*Xenopus laevis*) in Chile: Causes of concern. *Biodivers Conserv* 14:429.

90. Fouquet, A., and G. J. Measey. 2006. Plotting the course of an African clawed frog invasion in western France. *Anim Biol* 56:95.

91. Weldon, C., L. H. du Preez, A. D. Hyatt, R. Muller, and R. Speare. 2004. Origin of the amphibian chytrid fungus. *Emerg Infect Dis* 10:2100.

92. Pounds, J. A., M. R. Bustamante, L. A. Coloma, J. A. Consuegra, M. P. L. Fogden, P. N. Foster, E. La Marca et al. 2006. Widespread amphibian extinctions from epidemic disease driven by global warming. *Nature* 439:161.

93. Lips, K. R., J. Diffendorfer, J. R. Mendelson, and M. W. Sears. 2008. Riding the wave: Reconciling the roles of disease and climate change in amphibian declines. *PLoS Biol* 6:441.

94. Retallick, R. W. R., H. McCallum, and R. Speare. 2004. Endemic infection of the amphibian chytrid fungus in a frog community post-decline. *PLoS Biol* 2:1965.

95. Beard, K. H., and E. M. O'Neill. 2005. Infection of an invasive frog *Eleutherodactylus coqui* by the chytrid fungus *Batrachochytrium dendrobatidis* in Hawaii. *Biol Conserv* 126:591.

96. Schumacher, J., E. R. Jacobson, B. L. Homer, and J. M. Gaskin. 1994. Inclusion-body disease in boid snakes. *J Zoo Wildl Med* 25:511.

97. Oliver, J. H., M. P. Hayes, J. E. Keirans, and D. R. Lavender. 1993. Establishment of the foreign parthenogenetic tick *Amblyomma-rotundatum* (Acari, Ixodidae) in Florida. *J Parasitol* 79:786.

98. Burridge, M. J., L. A. Simmons, and S. A. Allan. 2000. Introduction of potential heartwater vectors and other exotic ticks into Florida on imported reptiles. *J Parasitol* 86:700.

99. Daszak, P., A. A. Cunningham, and A. D. Hyatt. 2000. Wildlife ecology—Emerging infectious diseases of wildlife—Threats to biodiversity and human health. *Science* 287:443.

100. Robert, J., L. Abramowitz, J. Gantress, and H. D. Morales. 2007. *Xenopus laevis*: A possible vector of ranavirus infection? *J Wildl Dis* 43:645.

101. Hoffman, W. A., and J. L. Janer. 1941. *Bufo marinus* as a vector of helminth eggs on the island of Puerto Rico. *PR J Public Health Trop Med* 16:505.

102. Responsible Exotic Animal Ownership (REXANO). 2009. USA: Captive constrictor (non-venomous) snakes statistics: Human fatalities. http://www.rexano.org/Statistics/Constrictor_Captive_Snake_Fatality.pdf (accessed 14 January, 2010).

103. Associated Press. 2009. Child dies after being strangled by pet python in Florida, *Fox News*. July 2, Florida. http://www.foxnews.com/story/0,2933,529709,00.html (accessed May 5, 2010).

104. Auliya, M. 2003. Hot trade in cool creatures: A review of the live reptile trade in the European Union in the 1990s with a focus on Germany. TRAFFIC Europe, Brussels. http://www.traffic.org/species-reports/traffic_species_reptiles2.pdf (accessed May 5, 2010).

105. Meshaka, W. E. J., B. P. Butterfield, and J. B. Hauge. 2004. *The Exotic Amphibians and Reptiles of Florida*. Malabar: Krieger Publishing Company.

106. Duggan, I. C., C. A. M. Rixon, and H. J. MacIsaac. 2006. Popularity and propagule pressure: Determinants of introduction and establishment of aquarium fish. *Biol Invasions* 8:377.

107. Kraus, F. 2003. Invasion pathways for terrestrial vertebrates. In *Invasive Species: Vectors and Management Strategies*, ed. G. Ruiz and J. Carlton, 292. Washington, DC: Island Press.

108. Dean, W. R. J. 2000. Alien birds in southern Africa: What factors determines their success? *S Afr J Sci* 96:9.

109. Richardson, D. M., W. J. Bond, W. R. J. Dean, S. I. Higgins, G. F. Midgley, S. J. Milton, L. W. Powrie, M. C. Rutherford, M. J. Samways, and R. E. Schulze. 2000. Invasive alien species and global change: A South African perspective. In *Invasive Species in a Changing World*, ed. H. A. Mooney and R. J. Hobbs, 303. Washington, DC: Island Press.

110. Van Rensburg, B. J., B. F. Erasmus, A. S. van Jaarsveld, K. J. Gaston, and S. L. Chown. 2004. Conservation during times of change: Interactions between birds, climate, and people in South Africa. *S Afr J Sci* 100:266.

111. Joubert, L. 2009. *Invaded —The Biological Invasion of South Africa*. Johannesburg: Wits University Press.

112. Kark, S., A. Iwaniuk, A. Schalimtzek, and E. Banker. 2007. Living in the city: Can anyone become an "urban exploiter"? *J Biogeogr* 34:638.

113. Van Rensburg, B. J., D. S. Peacock, and M. P. Robertson. 2009. Biotic homogenization and alien bird species along an urban gradient in South Africa. *Landsc Urban Plan* 92:233.

114. Okes, N. C., P. A. R. Hockey, and G. S. Cumming. 2008. Habitat use and life history as predictors of bird responses to habitat change. *Conserv Biol* 22:151.

115. Crawford, R. J. M., A. J. Tree, P. A. Whittington, J. Visagie, L. Upfold, K. J. Roxburg, A. P. Martin, and B. M. Dyer. 2008. Recent distributional changes of seabirds in South Africa: Is climate having an impact? *Afr J Mar Sci* 30:189.

116. Long, J. L. 1981. *Introduced Birds of the World*. London: David and Charles.

117. Hockey, P. A. R., W. R. J. Dean, and P. G. Ryan. 2005. *Roberts—Birds of Southern Africa*. 7th ed. 972. Cape Town: The Trustees of the John Voelcker Bird Book Fund.

118. Harrison, J. A., D. G. Allan, L. G. Underhill, M. Herremans, A. J. Tree, V. Parker, and C. J. Brown. 1997. *The Atlas of Southern African Birds*. Johannesburg: BirdLife South Africa.

119. Peacock, D. S., B. J. van Rensburg, and M. P. Robertson. 2007. The distribution and spread of the invasive alien common myna, *Acridotheres tristis* L. (Aves: Sturnidae), in southern Africa. *S Afr J Sci* 103:465.

120. Dean, W. R. J., and S. J. Milton. 2000. Directed dispersal of *Opuntia* species in the Karoo, South Africa: Are crows the responsible agents? *J Arid Environ* 45:305.

121. Oschadleus, H. D., and L. G. Underhill. 2006. Range expansion of the red-billed quelea, *Quelea quelea*, into the Western Cape, South Africa. *S Afr J Sci* 102:12.

122. Martin, W. K. 1996. The current and potential distribution of the common myna *Acridotheres tristis* in Australia. *Emu* 96:166.

123. Feare, C., A. Craig, B. Croucher, and C. Shields. 1998. *Starlings and Mynas.* London: Christopher Helm.

124. Khan, H. A. 1998. Rose-ringed parakeet, *Psittacula krameri*, a serious threat to agriculture. *Eclectus* 5:39.

125. Blackwell, B. F., T. L. DeVault, E. Fernández-Juricicb, and R. A. Dolbeerc. 2009. Wildlife collisions with aircraft: A missing component of land-use planning for airports. *Landsc Urban Plan* 93:1.

126. Byron, J., and C. T. Downs. 2002. Bird presence at Oribi Airport and recommendations to avoid bird strikes. *S Afr J Wildl Res* 32:49.

127. Oatley, T. B. 1984. Exploitation of a new niche by the Rameron pigeon *Columba arquatrix* in Natal. In ed. J. A. Ledger, *Proceedings of the 5th Pan-African Ornithological Congress*, p.323–330. Johannesburg: Southern African Ornithological Society.

128. Liversidge, R. 1975. Beware the exotic bird. *Bokmakierie* 27:86.

129. Pienaar, K. 2006. *Common or Indian Mynah, Acridotheres tristis.* Polokwane: Limpopo Department of Economic Development, Environment & Tourism.

130. Kirkpatrick, W., and A. Woolnough. 2007. *Common Starling. Pestnote 253.* Government of Western Australia. Perth, Australia: Department of Agriculture and Food.

131. Pienaar, K. 2006. *Mallard Duck, Anas platyrhynchos—A real "ugly duckling."* Limpopo Department of Economic Development, Environment & Tourism: Environmental Affairs. South Africa: Nylstroom, Limpopo province.

132. McCarthy, E. M. 2006. *Handbook of Avian Hybrids of the World.* Oxford: Oxford University Press.

133. Newton, I. 1994. The role of nest sites in limiting the numbers of hole-nesting birds: A review. *Biol Conserv* 70:265.

134. Bruggers, R. L., and C. C. H. Elliott. 1989. *Quelea quelea Africa's Bird Pest.* Oxford: Oxford University Press.

135. Mundy, P., and M. J. F. Jarvis. 1989. *Africa's Feathered Locust.* Harare: Baobab Books.

136. Parker, V. 1999. *The Atlas of the Birds of Sul do Save, Southern Mozambique*, 224. Cape Town and Johannesburg: Avian Demography Unit and Endangered Wildlife Trust.

137. Cheke, A. 2008. Seafaring behaviour in house crows *Corvus splendens*—a precursor to ship-assisted dispersal? *Phelsuma* 16:65.

138. Williamson, M. H., and A. Fitter. 1996. The characters of successful invaders. *Biol Conserv* 78:163.

139. Stafford, L. 2009. *Strategic Implementation Plan for Eradicating the Indian House Crow (Corvus splendens) Population in the City of Cape Town 2009–2015.* Cape Town, South Africa.

140. Stafford, L. 2009. *Report Cape Town House Crow Eradication Project for the Period July 2008–27 February 2009.* Cape Town, South Africa.

141. Berruti, A., and G. Nichols. 1991. Alien birds in southern Africa: The crow must go. *Bird South Afr* 43:52.

142. Berruti, A. 1997. House crow *Corvus splendens*. In *The Atlas of Southern African Birds Vol. 2: Passerines*, ed. J. A. Harrison, D. G. Allan, L. G. Underhill, M. Herremans, A. J. Tree, V. Parker, and C. J. Brown, 108. Johannesburg: BirdLife South Africa.

143. Perrins, C. M. 1990. *The Illustrated Encyclopaedia of Birds.* London: Headline Book Publishing.

144. Kweleabeleidskomitee [Quelea policy committee]. 1994. *Beleid vir die bestuur van die Rooibekkweleaprobleem* [Management policy for the Red-billed quelea problem]. Pretoria: Ministry of Agriculture.

145. Directorate: Land Use & Management. 2009. *Rooibekkwelea —'n landbouplaag* [Red-billed quelea—an agricultural plague]. Pretoria, South Africa: Department of Agriculture, Forestry and Fisheries.

146. National Department of Agriculture. 1999. *Redbilled Quelea—Control Manual.* Pretoria, South Africa: National Department of Agriculture.

147. Lötter, L., and M. Kieser. 2001. *Evaluation of the Non-Target Impact from the Existing Quelea Control Database: GIS and Fieldwork.* Pretoria, South Africa: ARC/LNR Plant Protection Research Institute.

148. Bigalke, R. C., and J. A. Bateman. 1962. On the status and distribution of ungulate mammals in the Cape province, South Africa. *Ann Cape Prov Mus* 2:85.

149. Siegfried, W. R. 1962. Introduced vertebrates in the Cape province. *Cape Dep Nat Conserv Annu Rep* 19:80.

150. van Ee, C. A. 1962. The distribution of ungulates in the Orange Free State. *Ann Cape Prov Mus* 2:53.

151. Chapman, N. G., and D. I. Chapman. 1980. The distribution of fallow deer: A world-wide review. *Mammal Rev* 10:61.

152. Lloyd, P. H., and J. C. G. Millar. 1983. A questionnaire survey (1969–1974) of some of the larger mammals of the Cape Province. *Bontebok* 3:1.

153. Smithers, R. H. N. 1983. *The Mammals of the Southern African Subregion*. Pretoria: University of Pretoria.

154. Castley, J. G., A. F. Boshoff, and G. I. H. Kerley. 2001. Compromising South Africa's natural biodiversity—inappropriate herbivore introductions. *S Afr J Sci* 97:344.

155. Lindsey, P. A., R. Alexander, L. G. Frank, A. Mathieson, and S. S. Romañach. 2006. Potential of trophy hunting to create incentives for wildlife conservation in Africa where alternative wildlife-based land uses may not be viable. *Anim Conserv* 9:283.

156. Skinner, J. D., and R. H. N. Smithers. 1990. *The Mammals of the Southern African Subregion*. 2nd ed. Pretoria: University of Pretoria.

157. Hänel, C., and S. L. Chown. 1999. Fifty years at Marion and Prince Edward Islands: A bibliography of scientific and popular literature. *S Afr J Sci* 95:87.

158. Bester, M. N., J. P. Bloomer, P. A. Bartlett, D. D. Muller, M. van Rooyen, and H. Büchner. 2000. Final eradication of feral cats from sub-Antarctic Marion Island, southern Indian Ocean. *S Afr J Wildl Res* 30:53.

159. Van Vuuren, B., and S. L. Chown. 2007. Genetic evidence confirms the origin of the house mouse on sub-Antarctic Marion Island. *Polar Biol* 30:327.

160. Mostert, M. E. 2009. *Molecular and Morphological Assessment of Invasive, Inland Rattus (Rodentia: Muridae) Congenerics in South Africa and Their Reservoir Host Potential with Respect to Helicobacter and Bartonella*. MSc. (Zoology) thesis. Pretoria: University of Pretoria.

161. Bastos, A. D. S., C. T. Chimimba, E. von Maltitz, F. Kirsten, and S. Belmain. 2005. Identification of rodent species that play a role in disease transmission to humans in South Africa. In *Proceedings of the South African Society for Veterinary Epidemiology and Preventive Medicine*, 78. Pretoria: South African Society for Veterinary Epidemiology and Preventive Medicine.

162. Taylor, P. J., L. Arntzen, M. Hayter, M. Iles, J. Frean, and S. Belmain. 2008. Understanding and managing sanitary risks due to rodent zoonoses in an African city: Beyond the Boston model. *Integr Zool* 3:38.

163. De Graaff, G. 1981. *The Rodents of Southern Africa: Notes on Their Identification, Distribution, Ecology and Taxonomy*. Durban: Butterworth.

164. Meester, J. A. J., I. L. Rautenbach, N. J. Dippenaar, and C. M. Baker. 1986. Classification of Southern African mammals. *Transvaal Mus Monogr* 5:1.

165. Aplin, K. P., T. Chesser, and J. ten Have. 2003. Evolutionary biology of the genus *Rattus*: Profile of an archetypal rodent pest. In *Rats, Mice and People: Rodent Biology and Management*, ed. G. R. Singleton, L. A. Hinds, C. J. Krebs, and D. M. Spratt, 487. Canberra: The Australian Centre for International Agricultural Research (ACIAR).

166. Musser, G. G., and M. D. Carleton. 2005. Family muridae. In *Mammal Species of the World: A Taxonomic and Geographic Reference*, ed. D. E. Wilson and D. M. Reeder, 894. Baltimore: Johns Hopkins University Press.

167. Apps, P. 2000. *Smithers's Mammals of Southern Africa. A Field Guide*. Cape Town: Struik Publishers.

168. Chinen, A. A., H. Suzuki, K. P. Aplin, K. Tsuchiya, and S. Suzuki. 2005. Preliminary genetic characterization of two lineages of black rats (*Rattus rattus sensu lato*) in Japan, with evidence for introgression at several localities. *Genes Genet Syst* 80:367.

169. Harper, G. A., K. J. M. Dickison, and P. J. Seddon. 2005. Habitat use by three rat species (*Rattus* spp.) on Stewart Island/Rakiura, New Zealand. *New Zealand J Ecol* 29:251.

170. Spear, D., and S. L. Chown. 2009. Non-indigenous ungulates as a threat to biodiversity. *J Zool Lond* 279:1.

171. Lloyd, P. H. 1975. *A Study of the Himalayan tahr* (Hemitragus jemlahicus) *on the Ecology of the Table Mountain Range*. Cape Town: Cape Department of Nature and Environmental Conservation.

172. Westdyk, J. H. 2000. *The Diet of Sus scrofa in the South Western Cape Province/Wellington Region.* BSc. (Honours) (Nature Conservation) thesis. Stellenbosch: University of Stellenbosch.

173. Hignett, D. L. 2006. *Feral Pigs (Sus scrofa) in the Western Cape Province: A Re-Evaluation.* MPhil (Environmental Management) thesis. Stellenbosch: University of Stellenbosch.

174. Codron, D., J. S. Brink, L. Rossouw, and M. Clauss. 2008. The evolution of ecological specialization in southern African ungulates: Competition or physical environmental turnover? *Oikos* 117:344.

175. Lazarus, A. B. 1989. Progress in rodent control and strategies for the future. In *Mammals as Pests,* ed. R. J. Putman, 53. London: Chapman and Hall Ltd.

176. Lodal, J., and M. Lund. 1989. Prevalence of *Pneumocystis carinii* and *Leptospira icterohaemorrhagiae* in Danish rodents. In *Mammals as Pests,* ed. R. J. Putman, 51. London: Chapman and Hall Ltd.

177. Putman, R. J. 1989. *Mammals as Pests.* London: Chapman & Hall Ltd.

178. Hugh-Jones, B. E., W. T. Hubbert, and H. V. Hagstad. 1995. *Zoonoses: Recognition, Control, and Prevention.* Ames: Iowa State University Press.

179. Mills, J. N. 1999. The role of rodents in emerging human disease: Examples from the hantaviruses and arenaviruses. In *Ecologically-Based Rodent Management,* ed. G. R. Singleton, H. A. Hinds, H. Leirs, and Z. Zhang, 134. Canberra: Australian Centre for International Agricultural Research.

180. Duplantier, J.-M., J. Catalan, A. Orth, B. Grolleau, and J. Britton-Davidian. 2003. Systematics of the black rat in Madagascar: Consequences for the transmission and distribution of plague. *Biol J Linn Soc* 78:335.

181. Hirano, M., X. Ding, T.-C. LI, N. Takeda, H. Kawabata, N. Koizumi, T. Kadosaka et al. 2003. Evidence for widespread infection of hepatitis E virus among wild rats in Japan. *Hepatol Res* 27:1.

182. Mills, M. G. L., and L. Hes. 1997. *The Complete Book of Southern African Mammals.* Cape Town: Struik Publishing Group (Pty) Ltd.

183. Singleton, G., L. Hinds, H. Leirs, and Z. Zhang. 1999. *Ecologically-Based Rodent Management.* Canberra: Australian Centre for International Agricultural Research.

184. Duplantier, J.-M., and D. Rakotondravony. 1999. The rodent problem in Madagascar: Agricultural pest and threat to human health. In *Ecologically-Based Rodent Management,* ed. G. Singleton, L. Hinds, H. Leirs and Z. Zhang, 441. Canberra: Australian Centre for International Agricultural Research.

185. Thorsen, M., R. Shorten, R. Lucking, and V. Lucking. 2000. Norway rats (*Rattus norvegicus*) on Fregate Island, Seychelles: The invasion; subsequent eradication attempts and implications for the island's fauna. *Biol Conserv* 96:133.

186. Amori, G., and M. Clout. 2003. Rodents on islands: A conservation challenge. In *Rats, Mice and People: Rodent Biology and Management,* ed. G. R. Singleton, L. A. Hinds, C. J. Krebs, and D. M. Spratt, 63. Canberra: Australian Centre for International Agricultural Research.

187. Aplin, K. P., and G. R. Singleton. 2003. Balancing rodent management and small mammal conservation in agricultural landscapes: Challenges for the present and the future. In *Rats, Mice and People: Rodent Biology and Management,* ed. G. R. Singleton, L. A. Hinds, C. J. Krebs, and D. M. Spratt, 80. Canberra: Australian Centre for International Agricultural Research.

188. Norris, A., and T. Low. 2005. Review of the management of feral animals and their impact on biodiversity in the Rangelands: A resource to aid NRM planning, Pest Animal Control. CRC Report, Pest Animal Control CRC. http://www.environment.gov.au/land/publications/rangelands-feral-animal.html (accessed January 17, 2011).

189. Harris, C. B., and D. W. MacDonald. 2007. Interference competition between introduced black rats and endemic Galápagos rice rats. *Ecology* 88:2330.

190. Harris, D. B. 2009. Review of negative effects of introduced rodents on small mammals on islands. *Biol Invasions* 11:1611.

191. Smith, V. R., N. L. Avenant, and S. L. Chown. 2002. The diet and impact of house mice on a sub-Antarctic island. *Polar Biol* 25:703.

192. Chown, S. L., and V. R. Smith. 1993. Climate change and the short-term impact of feral house mice at the sub-Antarctic Prince Edward Islands. *Oecologia* 96:508.

193. Van Aarde, R. J., S. M. Ferreira, T. D. Wassenaar, and D. G. Erasmus. 1996. With the cats away the mice may play. *S Afr J Wildl Res* 92:357.

194. Smith, V. R. 2002. Climate change in the sub-Antarctic: An illustration from Marion Island. *Clim Change* 52:345.

195. Crafford, J. E., and C. H. Scholtz. 1987. Quantitative differences between the insect faunas of sub-Antarctic Marion and Prince Edward Islands: A result of human intervention? *Biol Conserv* 40:255.

196. Chown, S. L., M. A. McGeoch, and D. J. Marshall. 2002. Diversity and conservation of invertebrates on the sub-Antarctic Prince Edward Islands. *Afr Entomol* 10:67.

197. Huyser, O., P. G. Ryan, and J. Cooper. 2000. Changes in population size, habitat use and breeding biology of lesser sheathbills (*Chionis minor*) at Marion Island: Impacts of cats, mice and climate change? *Biol Conserv* 92:299.

198. Avenant, N. L., and V. R. Smith. 2003. The microenvironment of house mice on Marion Island (sub-Antarctic). *Polar Biol* 26:129.

199. Cuthbert, R., and G. Hilton. 2004. Introduced House Mice *Mus musculus*: A significant predator of endangered and endemic birds on Gough Island, South Atlantic Ocean? *Biol Conserv* 117:483.

200. Schlick, N. E., M. I. Jensen-Seaman, K. Orlebeke, A. E. Kwitek, H. J. Jacob, and J. Lazar. 2006. Sequence analysis of the complete mitochondrial DNA in 10 commonly used inbred rat strains. *Am J Physiol Cell Physiol* 291:C1183.

201. Matisoo-Smith, E., and J. Robins. 2009. Mitochondrial DNA evidence for the spread of Pacific rats through Oceania. *Biol Invasions* 11:1521.

202. Coates, G. D., and C. T. Downs. 2005. Survey of the status and management of sympatric bushbuck and nyala in KwaZulu-Natal, South Africa. *S Afr J Wildl Res* 35:179.

203. Bond, W. J., and D. Loffell. 2001. Introduction of giraffe changes *Acacia* distribution in a South African savanna. *Afr J Ecol* 39:286.

204. Fabricius, C., D. Lowry, and P. van den Berg. 1988. Fecund black wildebeest x blue wildebeest hybrids. *S Afr J Wildl Res* 18:35.

205. Grobler, J. P., G. B. Hartl, N. Grobler, A. Kotze, K. Botha, and R. Tiedemann. 2005. The genetic status of an isolated black wildebeest (*Connochaetes gnou*) population from the Abe Bailey Nature Reserve, South Africa: Microsatellite data on a putative past hybridization with blue wildebeest (*C. taurinus*). *Mamm Biol* 70:35.

206. Meltzer, D. G. A. 1993. Historical survey of disease problems in wildlife populations: Southern Africa mammals. *J Zoo Wildl Med* 24:237.

207. Somers, M. J., and B. Fike. 1993. Aspects of the management of warthogs in the Andries Vosloo Kudu Reserve with implications for surrounding areas. *Pelea* 12:63.

208. Anonymous. Act No. 10 of 2004. *National Environmental Management: Biodiversity Act, 2004. 7 June 2004.* Republic of South Africa, Government Gazette, 467, No. 26436. Pretoria, South Africa: Staatskoerant.

209. Anonymous. Notice 347 of 2009. *Draft Alien and Invasive Species Regulations.* Department of Environmental Affairs and Tourism. Republic of South Africa, Government Gazette, 526, No. 32090. Pretoria, South Africa: Staatskoerant.

210. Macleod, F. 2006. *Aliens Invade South Africa.* Mail & Guardian. Johannesburg, South Africa: M&G Media.

211. Lodge, D. M., S. Williams, H. J. MacIsaac, K. R. Hayes, B. Leung, S. Reichard, R. N. Mack et al. 2006. Biological invasions: Recommendations for U.S. policy and management. *Ecol Appl* 16:2035.

212. Holsbeek, G., and R. Jooris. 2010. Potential impact of genome exclusion by alien species in the hybridogenetic water frogs (*Pelophylax esculentus* complex). *Biol Invasions* 12:1.

213. Rogan-Finnemore, M. 2008. *Non-Native Species in the Antarctic.* Special publication. Christchurch: Gateway Antarctica.

214. Ziman, J. 2007. *Science in Civil Society.* Exeter: Imprint Academic.

215. Terborgh, J. 1999. *Requiem for Nature.* Washington, DC: Island Press.

216. Dyson, F. 2006. *The Scientist as Rebel.* New York: New York Review of Books.

217. Cowling, R. M., R. L. Pressey, M. Rouget, and A. T. Lombard. 2003. A conservation plan for a global biodiversity hotspot—the Cape Floristic Region, South Africa. *Biol Conserv* 112:191.

*section eight*

# United States

# chapter sixteen

# Rodents and other vertebrate invaders in the United States

Michael W. Fall, Michael L. Avery, Tyler A. Campbell,
Peter J. Egan, Richard M. Engeman, David Pimentel,
William C. Pitt, Stephanie A. Shwiff, and Gary W. Witmer

## Contents

## 16.1   Introduction

All living organisms slowly, but constantly, increase or decrease their populations, alter their distribution, compete among each other for resources, and sometimes emerge as new species or become extinct. Most such changes in the distribution and abundance of species are invisible or undetectable on short time horizons, but changes in the ecological status of vertebrate species are often evident and sometimes demanding of human attention.

When species move (or are moved) from their natural ranges to new areas and become established, they are variously termed "nonindigenous," "alien," "introduced," or "invasive"; the technical or legal definitions for these terms are still in developmental stages. In the United States, an invasive species was legally defined by executive order in 1999 as "an alien species whose introduction does or is likely to cause economic or environmental harm or harm to human health. (p. 6186)" This limiting definition has apparently worked for legal purposes but has made the technical literature somewhat confusing. Some authors have used "invasive" and various synonyms to focus on injurious species not native to the United States; others include indigenous species introduced to new ecosystems with the potential to cause harm,[1,2] and still others have focused on all nonindigenous species, including those presumed innocuous or desirable.[3]

The presence of many invasive species has long been tolerated; they are often accepted by the public as naturally occurring—so ubiquitous, they may have long-standing local common names. Examples of such species in the United States include Norway rats (*Rattus norvegicus*), house sparrows (*Passer domesticus*), pigeons (*Columba livia*), European starlings (*Sturnus vulgaris*), European carp (*Cyprinus carpio*), and brown trout (*Salmo trutta*). The problems caused by invasive vertebrates, particularly those that have been established for a few years or have become widespread, are typically managed on a case-by-case basis. These problems are pervasive and formidable, including environmental degradation in terrestrial and aquatic ecosystems; economic damage to agriculture, commodities, property, and infrastructure; transmission of human and livestock diseases or acting as disease reservoirs; competing with desirable or native species; threats to public safety; and overall reduction of quality of life for rural, suburban, and urban populations.[4] Pimentel[5] (Chapter 17, this book) has estimated that introduced vertebrate species cause $46.7 billion in damages and control costs each year. (All monetary values in this chapter are expressed in US dollars.)

As in other countries, most of the problem invasive vertebrates arrived in the United States through activities of man, some through direct, purposeful introduction, many by accidental transport, and some by natural range expansion.[6] Invasive species, whether arriving by accident or purposefully introduced, are often not recognized as potentially harmful until they have become well established and problems begin to be recognized. Even then, some invasive species are championed by individuals or groups as having beneficial or redeeming features, uncertain futures, or rights to exist (e.g., starlings, nutria, or coqui frogs), generating sometimes rancorous exchanges, and what might have been a straightforward eradication effort becomes an ongoing program of wildlife damage management.[7-9]

In the past two decades, increased attention worldwide has been placed on the problems engendered by invasive species across a wide range of concerns, and these concerns are growing.[10,11] In the United States, hardly a day passes that media attention is not somehow focused on an invasive species issue, variously involving legislatures, mayors, governors, state and federal agencies, the courts, concerned citizens, and advocacy groups on all sides. These concerns range across the real and growing problems of economic damage, ecosystem degradation, and competition with rare or desirable native species and the real or perceived threats of such impacts. A growing technical literature—papers, books, online resources, and the associated nontechnical, educational resources—has helped fuel both scholarly interest and public concern. Much of the technical material to date is largely descriptive and speculative, although there are increasing numbers of analytical studies and examples of successful eradications, suppressions, or management programs for particular invasive vertebrate problems.

The organizational structure for addressing invasive species problems in the United States is still evolving, with new organizational resources, particularly at the state level, appearing frequently. The National Invasive Species Council (NISC), established by a 1999 executive order,[1] is cochaired by the secretaries of agriculture, commerce, and interior to assure coordination of federal programs involving 13 federal departments and agencies. NISC (www.invasivespecies.gov), guided by an external advisory committee, has developed and maintained a National Invasive Species Management Plan and provides extensive resources to states, local jurisdictions, and the public, as well as providing international coordination.[12] A number of states have established parallel invasive species councils and state plans to address invasive species problems or those of particular groups. Hawaii, for example, established the Coordinating Group on Alien Pest Species (CGAPS), a public and private partnership, in 1995,[13] and the Hawaii Invasive Species Council (HISC) by legislation in 2003 both with a variety of specific functions, including public information, strategy development, and policy coordination among state and federal agencies. The result has been a proactive program for invasive species identification and control, including a well-advertised invasive species hotline[13] and, when necessary, legislatively mandated access to private lands for control operations.

Depending on the definitions used and methods of compilation, at least 81 invasive mammal species, 99 bird species, 69 reptile species, 11 amphibian species,[2] and 533 nonindigenous fish species[3] (Table 16.1) are established in parts of the United States (including Alaska and Hawaii, but excluding territories). In this chapter, we provide a brief overview of invasive species problems in the United States, summarize species accounts of some important species of concern, and discuss issues related to present and future management of these problems.

*Table 16.1* Rodents and Other Invasive, Alien, or Nonindigenous Vertebrates Introduced in Parts of the United States

| | |
|---|---|
| Mammals[a] | 81 |
| Rodents | 18 |
| Marsupials | 2 |
| Primates | 6 |
| Insectivores | 2 |
| Lagomorphs | 6 |
| Carnivores | 14 |
| Ungulates | 33 |
| Birds[a] | 99 |
| Waterfowl | 5 |
| Herons | 1 |
| Gallinaceous birds | 25 |
| Pigeons and doves | 7 |
| Parrots | 20 |
| Owls | 1 |
| Perching birds | 40 |
| Reptiles[a] | 69 |
| Turtles | 6 |
| Crocodilians | 1 |
| Lizards | 56 |
| Snakes | 6 |

*(Continued)*

***Table 16.1*** Rodents and Other Invasive, Alien, or
Nonindigenous Vertebrates Introduced in Parts of
the United States (*Continued*)

| | |
|---|---|
| Amphibians[a] | 11 |
|   Frogs and toads | 10 |
|   Salamanders | 1 |
| Fish[b] | 533[b] |
|   Lampreys | 3 |
|   Sturgeons and paddlefish | 4 |
|   Gars | 4 |
|   Bowfin | 1 |
|   Bonytongues, mooneyes, and featherfin knifefishes | 4 |
|   Tarpons | 1 |
|   Bonefishes | 1 |
|   Freshwater eels | 4 |
|   Anchovies and herrings | 10 |
|   Milkfishes | 1 |
|   Minnows, suckers, and loaches | 156 |
|   Headstanders, trahiras, and characins | 18 |
|   Catfishes | 37 |
|   Pikes and mudminnows | 10 |
|   Smelts | 4 |
|   Trouts | 30 |
|   Trout-perches, pirate perches, and cavefishes | 3 |
|   Cods | 1 |
|   Mullets | 2 |
|   Rainbowfishes and silversides | 7 |
|   Ricefishes and needlefishes | 3 |
|   Rivulins, topminnows, live-bearers, splitfins, and pupfishes | 54 |
|   Sticklebacks | 4 |
|   Swamp eels | 1 |
|   Sculpins | 4 |
|   Snooks, basses, sunfishes, perches, roosterfishes, jacks, mojarras, grunts, drums, sea chubs, flagtails, cichlids, surfperches, wrasses, gobies, mackerels and tunas, butterfishes, gouramies, and snakeheads | 158 |
|   Flounders | 8 |

*Note:* These numbers, totaling 792, include species native to some parts of the country that have been introduced elsewhere outside their native ranges (mammals, birds, reptiles, and amphibians), and for fish, all established nonindigenous species.

[a] Estimated numbers of species for mammals, birds, reptiles, and amphibians summarized from Witmer et al.[2]

[b] Estimated numbers of fish species, hybrids, and unidentified forms summarized by order from Fuller et al.[3] Fuller et al.'s actual count from their database is 536 unique taxa.

## 16.2   *Assessing impacts of rodents and other vertebrate invaders*

The impacts and damage caused by vertebrate invaders in the United States have made this cluster of species a leading cause of environmental change and global biodiversity loss.[14–18] Invasions by nonindigenous species highlight the undeniable link between ecological and economic systems.[19,20] Ecological systems determine if the conditions are suitable for invasion and establishment of nonnative species; however, economic systems are affected by invasive species when the ecosystems are changed or diminished, when agricultural products are made unmarketable, or when public health and safety are compromised.[20–22] In general, the economic impacts of invasive species can be broken down into primary and secondary effects.

The primary negative economic effects most often caused by invasive species include disease transmission, predation, and/or destruction of environments.[23–25] Disease and predation cause mortality or morbidity in humans, companion animals, livestock, or wildlife,[26–28] while environmental destruction results from damaged ecosystems, crops, or property.[29–31] Valuation of the primary damage is usually accomplished by estimating the loss, market, and repair or restoration values associated with the affected resource. Loss values are often used in the case of death related to disease transmission, or predation to humans and companion animals, and in limited circumstances, wildlife. Market values are commonly used when monetizing the impact to livestock or crops.[17,20,22] In the case of property damage, repair costs are a typical method of valuation.[32,17] Finally, in the case of nonmarket ecosystems and wildlife, restoration values may be used to estimate the economic impact of damage to these resources.[33,34]

Primary effects can generate secondary effects due to interrelated economic factors that create linkages to established economic sectors.[25] For example, damage or destruction of an ecosystem is calculated by the number of acres damaged at the restoration price per acre. However, if the ecosystem damage also reduces tourism to the area, then the economic activity that would have been generated from tourist expenditures was also lost, representing a secondary impact.[25] Estimation of secondary effects usually requires the use of complex, computer-based input-output (I-O) simulation models. I-O modeling is an accepted methodology for estimating secondary impacts. This type of modeling attempts to quantify the impacts on output as a result of input changes in a regional economy based on the most current economic and demographic data available. An I-O model is developed by constructing a mathematical replica of a regional economy (city, county, state, etc.) that contains all the linkages between economic sectors (agricultural, manufacturing, and industrial) present in that economy. I-O models use the primary effect to generate the secondary effect, thereby calculating the resulting total effect on jobs and revenue in a specified regional economy.[35]

The challenge facing policy makers is to determine biologically effective and economically feasible methods of prevention, control, and damage mitigation of invasive species.[36] Accurate assessments of the economic impact of invasive species allow for the targeting of appropriate prevention and control methods.[37–39] The benefits and costs of all methods used to reduce the impacts of invasive species should be assessed to determine the most economically efficient techniques. Benefit–cost analysis is a common tool used by economists to evaluate programs and to determine the efficiency of management efforts; the monetary benefits and costs of program actions are identified and compared. A benefit–cost analysis is often used to value nonmarketed goods and services, such as environmental "goods."[40] The process of managing invasive species based on their environmental and economic impacts is an example. To estimate the values of such impacts, a number of measurement techniques have been developed.[41]

One accepted methodology to value nonmarket services is the damage-avoided method, which uses the value of resources protected as a measure of the benefits provided by a control program.[42] Benefits of any method to prevent, control, or mitigate the damage caused by invasive species are derived from the reduced burden associated with the impact of the species. Therefore, benefits are measured as cost savings resulting from diminished disease spread, predation, and/or environmental destruction. Costs are programmatic and derived from the labor and materials used to prevent, control, or mitigate invasive species damage. Total economic benefits of any control method are the summation of the primary and secondary effects "saved." The total economic benefits of the program can then be compared to total program costs to determine the economic efficiency. At least for the vertebrate invaders, economists have been slow to fully apply these tools to evaluate actual and potential impacts. Lack of such economic information is often cited at the political levels of government as a reason to tolerate invasion and establishment and to deal with problems that arise on an ad hoc basis.

## 16.3  Accounts of some important vertebrate invaders

Of the many invasive vertebrate species occurring in the United States, we provide species accounts of several that we view as particularly important because of economic losses, ecosystem impacts, or public interest.

### 16.3.1  Norway rat (Rattus norvegicus)

Introduced to North America about 1775 in trans-Atlantic shipping, the Norway rat[43,44] is now completely established in both rural and urban areas throughout the country, including Alaska and Hawaii. This species is one of the oldest and best-known invasive vertebrates in the United States and is responsible for a variety of types of damage to crops and stored commodities.[45,46] It spread rapidly and continuously across the country in shipped commodities, initially following wagon, riverboat, and rail routes. Areas with the least human traffic were the last to be reached.

The fur color of this rat is typically brown above and lighter brown gray below. The tail is sparsely haired and scaly, typically about the same length as the head and body. Its weight is about 500 g. One of the three common commensal rodent species on the North American continent, the Norway rat is closely tied to human settlements. Breeding may occur throughout the year. Populations expand rapidly when food, water, and habitat are available. Gestation is about 3 weeks, and animals reach sexual maturity about 3 weeks thereafter.[46]

In farm settings, damage to stored food and grains, damage to garden crops, and predation on eggs and baby chickens are common. Grain consumption and fecal contamination are common problems in commercial grain storage facilities.[45] Damage to roads, bridges, railroad track beds, and hydraulic structures may result from the burrowing activities and the associated soil loosening or flooding.[46] Structural damage in buildings results from gnawing and burrowing and may include damage to doors, window sills, and walls, as well as to pipes and wiring. Insulation may be damaged or removed in the course of nest building. In urban areas, Norway rat populations are commonly associated with poor sanitation or accumulation of trash and food refuse in inner-city areas, although outdoor feeding of pets and wildlife often support suburban populations as well. Norway rats serve as reservoirs of a number of diseases that may affect humans and domestic animals, most commonly salmonellosis, leptospirosis, and trichinosis.[44] In areas with high rat populations in close association with humans, rat bites may occur, particularly to babies and young children.

Davis[47] believed outdoor populations could be completely managed by environmental control and sanitation and demonstrated this repeatedly with experiments in Baltimore and New York City. However, Fall and Jackson[48] contended that the political and social impossibility of maintaining diligence by urban residents and sustained support by public and private sectors has allowed Norway rat problems to continue unabated. Numerous products are available commercially to property owners for Norway rat control, and extensive professional rodent control services are available through the structural pest control industry.[49]

## 16.3.2   *Roof rat* (Rattus rattus)

Roof rats, known also as black rats or ship rats, occur along port and shore areas in southeastern and western North America and throughout Hawaii and tropical Atlantic and Pacific Ocean islands. Although known most commonly as a commensal species closely tied to man, this species, particularly in warmer areas, readily establishes in landscape areas, including native forests in Hawaii and ocean islands. According to Brooks,[43] roof rats were well established in Virginia in the early 1600s. They were well established in North America's east coast areas by the 1800s. They occur sporadically in warmer inland areas but rarely persist. However, a recent infestation discovered in urban Phoenix, Arizona, raised concerns that the species could permanently establish in "islands" of suitable habitat and subsequently threaten crops and orchards.[50] In more temperate areas, they compete poorly with the larger and more aggressive Norway rats and occur mostly in port areas and generally indoors.[44]

The fur color is reddish brown, brown, or black, with the belly area being lighter or white. The tail is generally longer than the head and body. The weight of adults ranges from 150 to 250 g. As in Norway rats, breeding may occur throughout the year if resources are available, and the pattern of breeding is similar. Recently, a variant of *Rattus rattus*, the Asian house rat, has been separated taxonomically as *Rattus tanezumi*.[51] Animals of both species are generally similar in appearance; however, *Rattus tanezumi* appears more variable and has a somewhat shorter tail. A chief distinguishing feature is a differing number of chromosomes between the two species, but this is of course not evident without special study, and some authorities have not accepted the name change. *Rattus tanezumi* has recently been reported as a new invasive species in North America based on collections in California.[52]

Like the Norway rat, the roof rat invades homes and structures, causing damage and contamination of stored food and commodities. However, it readily adapts to field and forest habitats in tropical and semitropical areas causing damage to orchard, grain, and sugarcane crops. It preys on adult birds, nestlings, and eggs under some circumstances and is recognized worldwide as a likely cause of rare bird extinctions in many island areas, including Hawaii.[53–55]

This species is also a reservoir for a number of diseases of humans and animals but is most notorious for its role in bringing bubonic plague, the "Black Death," to fourteenth century Europe. The occurrence of bubonic plague in Hawaii during the period 1899–1958 was associated with this species,[56] as were the initial outbreaks in California in the early 1900s.[57]

Control methods and materials are the same or similar to those used for the Norway rat. However, this species has been a particular target of recent efforts, both in the United States and in many other countries, to eradicate them from islands where seabirds or other desirable species are threatened by rat predation.[58,59]

### 16.3.3   *Polynesian rat* (Rattus exulans)

The Polynesian rat (*Rattus exulans*) is a small tropical rat native to the Southeast Asia mainland, which spreads throughout islands in the Pacific in conjunction with human settlement of the region.[60] Although they do not occur on the United States mainland, they are well established on most tropical and semitropical islands (less than about 30° latitude) throughout the Pacific, including the Hawaiian Islands.[61] Polynesian rats are the smallest species (110–150 mm body length) in the genus *Rattus* and are slender (40–100 g), with relatively small feet and large ears. Like many rodent species, they are primarily nocturnal. Their fur is reddish brown on the dorsal surface and light gray on the belly area. Polynesian rats may breed throughout the year and have up to four litters annually with three to six young in each.[62] They are sexually mature at 2 months and may have a life expectancy of around 1 year.

Polynesian rats have adapted to a wide range of habitats from forests to grasslands to agricultural crops, such as sugarcane. They are good climbers but do not swim, so their dispersal to new islands is limited by human movement.[63,64,60] They are opportunistic omnivores, and their diet vary greatly by what is available and abundant by season and locale.[65–71] Predators of Polynesian rats include mongooses, cats, other larger rodents, and birds.[72] In addition, many Polynesian cultures consider rats to be a valuable food resource, and rodents may have been introduced into new areas intentionally for food.[64]

Polynesian rats are a significant agricultural pest throughout the Pacific region, and they damage a variety of crops, including rice, corn, macadamia nuts, sugarcane, coconut, cacao, pineapple, soybeans, and root crops.[73,69,70] Previous research documented the extensive effects of rat damage on sugarcane, but sugarcane production has largely been replaced by diversified agriculture in Hawaii.[55] Rat damage has now shifted to high-value seed crops (corn, soybean), and tropical fruits. Because Polynesian rats were spread through the Pacific Basin several thousand years ago, the impacts to the native flora and fauna are not readily apparent.[66,68,71,74] Polynesian rats are effective predators on seabirds, lizards, insects, and sensitive plant species that did not evolve with predation. Recent eradication efforts of Polynesian rats on islands have revealed these impacts as species recovery has occurred.[75]

A variety of methods have been employed to reduce the effects of Polynesian rats on agriculture and the environment. The primary successful methods have integrated rodenticides, alteration of cultural practices, and trapping.[70] Rodenticides have been effectively used to reduce agricultural damage, protect forest birds, and protect seabird colonies. Previous attempts to control rat damage biologically have been unsuccessful and deleterious to other species. The most frequently cited failure is the introduction of mongoose to Hawaii in 1883.[55]

### 16.3.4   *House mouse* (Mus musculus)

House mice are probably the most widespread mammalian species in the world next to humans. House mice originated in the grasslands of Central Asia and followed humans around the world. They are very good invaders and probably reached most parts of the world as stowaways on ships and cargo. House mice have remarkable abilities that have allowed them to be highly successful in many habitats around the world. Chief among these are their reproductive potential and their adaptability in different environments.

House mice are small, slender rodents with fur that is grayish brown above and gray to buff underneath. This small (about 20 g for adults) and highly prolific animal is a continuous breeder in many situations; a female can produce six to eight litters, each with four to seven young, per year. The young mature within about 3 weeks and soon become reproductively active. House mice are short lived (generally less than 1 year) and have high population turnover. In one study, 20 mice placed in an outdoor enclosure with abundant food, water, and cover became a population of 2000 in 8 months.[49]

House mice cause many types of damage.[76,77] A major concern is the consumption and contamination of stored foods; it has been estimated that substantial amounts of stored foods are lost each year in this manner. Mice also consume and contaminate large amounts of livestock feed at animal production facilities. Although mice generally live in proximity to humans,[49] sometimes feral populations occur. In these cases, the mice may damage many types of crops in the field, especially corn, cereal grains, and legumes. Australia has mouse "plagues" periodically resulting in enormous losses to stored crops and crops in the field.[78] In buildings, a mouse infestation can be a considerable nuisance because of the noise, odors, and droppings. More importantly, they damage insulation and wiring.[79] House fires have been caused by mice gnawing electrical wires; likewise, communication systems have been shut down for periods of time resulting in economic losses. Additionally, house mice are susceptible to a large number of disease agents and endoparasites. Consequently, they serve as reservoirs and vectors of disease transmission to humans, pets, and livestock.[80] Important among these diseases are leptospirosis, plague, salmonella, lymphocytic choriomeningitis, and toxoplasmosis. Finally, when introduced to islands, mice can cause significant damage to natural resources, including both flora and fauna. For example, on Gough Island, mice fed on nestling albatross chicks.[81]

A large number of methods and materials have been developed to help solve house mouse problems. In general, the use of multiple approaches and materials—integrated pest management—is more likely to reduce a mouse problem to a tolerable level. The tools available and their proper use have been extensively reviewed.[43,82,76,49]

## 16.3.5 *Nutria* (Myocastor coypus)

Nutria or coypu, semiaquatic rodents native to southern South America, are an invasive species having detrimental impacts mainly in the southern and eastern United States. Nutria were introduced into the United States in 1899 for fur farming and became established in several states.[83] Nutria dispersals resulted primarily from releases by fur farmers, escapes during hurricanes or rising floodwaters, or as translocations in an attempt to control nuisance aquatic vegetation. Some states, such as Louisiana, continue to recognize nutria as a beneficial natural resource for fur and food and manage populations for low densities—below presumed damage thresholds. In other situations, such as at the Blackwater National Wildlife Refuge in Chesapeake Bay, Maryland, where nutria have caused excessive marsh damage, government agencies have implemented an eradication strategy. Nutria and the damage they cause to crops, canals, and wetlands have been well described.[84,85]

Generally, nutria have dark brown fur and weigh about 5–9 kg. At first glance at a nutria swimming, they can be mistaken for a beaver or a muskrat, both rodents native to North America. Female nutria are polyestrous and are sexually mature at approximately 5 months old.[84] They are nonseasonal breeders capable of producing three to four litters a year with an average of four to five kits per litter. Nutria are voracious consumers of vegetation and are known to completely denude vegetation from areas where they feed before moving on. Their ease of mobility on land and in water makes them effective dispersers, posing significant challenges for resource managers.

The ravenous appetite of these herbivores can cause damage to agricultural crops and aquatic vegetation and can alter aquatic ecosystems. Crops damaged the most in the southeastern United States are rice and sugarcane, but other crops can be damaged as well: cereal grains, beets, peanuts, melons, and alfalfa.[84] In Louisiana, tens of thousands of acres of damaged marsh vegetation have been documented.[86] The areas damaged by nutria become permanent, open water ponds. Tidal and flooding impacts become more severe. The loss of marshland also removes habitat for native wildlife species such as waterfowl, wading birds, and muskrats. Finally, nutria burrowing habits can weaken irrigation structures and levees, and they are a host for some diseases.[84]

Nutria populations and damage have been controlled mainly by private hunters and trappers. When nutria fur prices declined in the 1980s, damage in many areas became a great concern. In Louisiana, a method was devised to manage nutria damage to supplement fur values with incentive payments to registered trappers and hunters of $4.00–$5.00 per nutria tail. Unlike classic bounty systems, the program is intensively managed to target specific areas for population reduction; in 2003–2004, 332,596 nutria tails were collected in designated harvest areas by 346 participants.[86] Rodenticides are rarely used for nutria control because of the concerns for hazards to nontarget animals and water quality. Research continues to develop new methods to control nutria populations, such as multiple-capture live traps[87] and improved attractants.[88]

## 16.3.6   *Gambian giant pouched rat* (Cricetomys gambianus)

Gambian giant pouched rats are native to a large area of central and southern Africa. They had become popular in the pet industry in some countries and became established on Grassy Key in the Florida Keys in 1999, following an escape or release by a pet breeder.[89] Despite a prolonged eradication effort, a free-ranging and breeding population remained on the island.[90] There is a concern that if this rodent reaches the mainland, there could be damage to the Florida fruit industry because Gambian rats are known to damage agricultural crops in Africa.[91] Imported Gambian rats may also pose risks as reservoirs of monkey pox and other diseases.[92] A climate-habitat modeling study suggested that their new range in North America could expand substantially were they to establish in the United States.[93]

Gambian rats are gray brown in color and can reach a considerable size—about 2.8 kg in weight and about 1 m long.[94] Females produce four young per litter. Because of their reproductive potential and their large size, they have been raised in captivity as a source of protein in Africa.[95] Since free-ranging Gambian rats are newcomers to North America, relatively little is known about their biology, habitat use, impacts, and interactions with native species or about the most effective means to capture or control these rodents. Hence, current efforts are concentrating on use of traditional live-trap capture methods and rodenticides in bait stations.[90] It will be important to develop additional tools to manage or eradicate this species and other rodent invaders from the United States.[96]

## 16.3.7   *Feral swine* (Sus scrofa)

Populations of feral swine have become established across the United States where their current range includes portions of at least 33 states.[97] Historical accounts of feral swine populations date to early European explorers, including Columbus, Cabot, Cortéz, and DeSoto.[98,99] These early populations, as well as the established and emerging populations, were founded from escaped domestic swine and through intentional releases of domestic swine, wild-caught feral hogs, and Eurasian wild boar;[99,100] today, feral swine populations are still comprised of these groups—now estimated at 4 million.[101]

Feral swine display high variability in body size, shape, and weight, as well as in pelage characteristics and coloration. Generally, their physical appearance is transitional between Eurasian wild boar, with a streamlined body and coarse dense hair, and domestic swine, with a round body and sparse hair.[101] While prevailing pelage color is black, other colors are common, including red, black and white spotted, brown, and roan.[99] Mean adult body weight for males and females from the southern United States range from 36 to 114 kg and from 34 to 92 kg, respectively.[101]

A major complicating factor in the management of feral swine populations is their exceptional reproductive potential. They are the most fecund, large, wild ungulate in North America.[102] Specifically, feral swine reach sexual maturity at a young age (5–10 months), have a large mean litter size (3.0–8.4), have the physiological capacity to breed year-round, and may produce two litters per year under favorable nutritional conditions.[101–103] These reproductive parameters suggest opportunity for management aimed at reducing birth rates to stabilize or reduce populations and related damage.[104]

Feral swine damage to property, agriculture, and natural resources often occurs as a result of their aggressive rooting, grubbing, plowing, and digging activities at and below the soil surface.[31] In sandy soils, feral swine may root to a depth of 1 m, causing extensive damage to crops, pastures, native plants, and farm equipment.[105] Rooting may also injure livestock and cause soil erosion.[101] Other sources of feral swine damage occur through wallowing, which may reduce water quality and disrupt sensitive wetland ecosystems,[29,30] destruction of livestock fencing,[105] and predation on young livestock and wildlife.[101] Feral swine are also present disease risks to both humans and livestock,[26] often carrying diseases, such as brucellosis, pseudorabies, and influenza.[27,28]

Methods that are available to control feral swine damage in the United States include exclusion fencing, trapping, and shooting.[31] Research efforts to develop immunocontraceptive vaccines and field-appropriate delivery systems,[106,107] coupled with improvements of existing control methodologies, are needed to formulate and implement comprehensive management programs for feral swine.

### 16.3.8 *Small Indian mongoose* (Herpestes javanicus)

The native range of the small Indian mongoose (*Herpestes javanicus*, synonymous with *H. auropunctatus*) extends from the Middle East through India, Pakistan, southern China, and parts of Indonesia.[108] During the late 1800s to early 1900s, mongooses were introduced with the hope of controlling rats and venomous snakes in sugarcane fields.[109,110] Thus, mongooses were introduced to many sugarcane-growing areas, including Puerto Rico, the Virgin Islands, and the Hawaiian Islands, in the hope of controlling rat damage.[111,112] Mongooses do eat rats, and early reports suggested that, while they were effective in reducing rat numbers, they were ineffective at exerting sufficient predatory impact to consistently reduce rodent damage.[113,114]

Mongooses are long (51–67 cm in length) and slender (300–900 g of weight) with short legs, small ears, a pointed nose, a bushy long tail, and short dark brown hair.[115] They are sexually dimorphic with males being larger than females. Mongooses have two to four offspring in a single litter each year but are capable of breeding year-round depending on food supply.[116] Further, where mongooses have been introduced, they have few predators or competitors to restrict their populations. Mongooses are found in a variety of habitats from tropical forests to open dry grasslands, marshes, and coastal sites, as long as adequate retreat sites are available. Mongooses are opportunistic omnivores and eat mammals, birds, reptiles, insects, and significant amount of plant material.[113,114,116] Home ranges are variable and may be small; however, the animals may move great distances to utilize unique or anthropogenic food sources.

Mongooses may have major economic impacts on local economies, agriculture, and natural resources, with damage estimates exceeding $50 million annually to Hawaii and Puerto Rico.[117] In addition, poultry, egg production, and game bird populations may be impacted unless significant efforts are made to exclude or control mongoose populations. In some areas, they may be reservoirs for several diseases threatening humans or livestock, including rabies and leptospirosis.[112,117,118] Mongooses are effective predators of ground-nesting birds and have been implicated in the extirpation of several species.[56,109,112,119,120] They have also had negative impacts on frogs, lizards, snakes, turtles, and small-mammal populations.[110,121–124]

Most mongoose control efforts have primarily related to attempts to protect ground-nesting birds or poultry operations using traps and toxicant baits.[125,126] Local control of mongoose populations has been most effective when kill traps, live traps, and toxicants are combined. Mongooses are very susceptible to toxicants including anticoagulant rodenticides. Large-scale eradication efforts and removal of incipient populations have proved to be difficult because of the availability of alternative foods and large foraging areas.[110,112,126]

### 16.3.9  *Rock pigeon* (Columba livia)

The rock pigeon, also known as the feral pigeon or rock dove, is a common sight in urban areas throughout the United States. Its native range extends from Britain to India, including northern Africa.[127] Although the species is most commonly associated with cities and towns, feral populations also inhabit rural and undeveloped areas throughout the range. Feral rock pigeons are highly variable in appearance due to frequent inter-breeding with domestic forms. Plumage of the wild form is generally blue-gray with a green and purple iridescence on the neck feathers. The wings have two black bars; the rump is white.[128]

The rock pigeon is likely the nonnative bird species with the longest tenure in the United States. It apparently arrived with early European colonists, along with poultry and livestock, in the early 1600s.[127,129] There have been many subsequent introductions, and the species has thrived in concert with human development and expansion. These pigeons now occur in all 50 states including Hawaii where it was introduced in 1796.[127]

The species is characterized by early sexual maturity; pigeons often breed before they are 1 year old.[128] In some locales, shortages of suitable nest sites might prevent all sexually mature birds from breeding. Depending on latitude, up to five clutches can be produced annually; clutch size is two. The average lifespan is about 2.5 years.[128]

Damage attributed to rock pigeons in the United States has been estimated to exceed $1 billion annually.[18] Negative impacts of pigeons include defacement and degradation of property and consumption or despoilment of grain and other food intended for livestock and humans. Pigeons are also associated with harboring or transmitting over 40 zoonotic diseases.[130] On the plus side, pigeons provide an important food source for urban-nesting peregrine falcons that were introduced to cities with high-rise buildings to help preserve threatened populations.[131] Furthermore, many people enjoy feeding pigeons in parks and urban centers. This recreational activity brings pleasure but also frustrates efforts to manage urban pigeon populations effectively. Methods for reducing pigeon populations and for addressing problems caused by pigeons are numerous and include exclusion, chemical repellents, toxicants, auditory and visual scare devices, trapping, and contraception.[132,133]

## 16.3.10   *House sparrow* (Passer domesticus)

According to Dawson,[134] the introduction of the English sparrow was "Without question the most deplorable event in the history of American ornithology . . . (p. 40)" The history of the house sparrow (often known as the English sparrow) in the United States dates from the early 1850s when at least 16 birds were released in New York City.[135] That event was followed by numerous subsequent releases, and the species eventually became established throughout the United States, including Alaska and Hawaii[127,136] The house sparrow is not migratory in the United States.

House sparrows have an extensive native range, and they are resident from the British Isles, Scandinavia, Russia, and Siberia south to northern Africa, Arabia, India, and Burma.[136] Introduced populations thrive in many other parts of the world.[127] Adult sparrows are about 16 cm long with bodyweights of about 28 g. Females and juveniles are a nondescript gray-brown in color. Males have a distinctive gray crown with a chestnut border, white cheek, and black throat and upper breast.[136] House sparrows have up to four clutches annually with four to six eggs per clutch.[137] Their diet is mostly grain and weed seeds, supplemented by insects and other invertebrates during the breeding season.[138] Bird seed (from feeders) in urban birds and commercial grains in rural birds were the principal food items recorded in one comparative study.[139] House sparrows readily consume livestock and poultry feed, and their droppings contaminate stored grain and create unsanitary conditions.

The house sparrow is primarily a commensal species in the United States, thriving in association with human activity and development. They frequent rural habitats around farms and dairies, as well as urban centers, where individual sparrows sometimes take residence inside large commercial buildings or stores or in airport terminals. They select nest sites in nooks and crevices on buildings and other structures, and they will readily usurp nest boxes intended for native hole-nesting species.[140] Very detailed guidance on how to protect bluebirds and other native species from house sparrows are available at http://www.sialis.org/hosp.htm.

House sparrows are associated with the transmission of at least 25 diseases potentially affecting humans or livestock.[130] Furthermore, they can harbor numerous types of ectoparasites. Because house sparrows often build their nests on buildings occupied by people, some of the parasites (e.g., the bedbug *Cimex lectularius*) can prove injurious to humans.[130] Management of problems caused by house sparrows is best accomplished using a combination of methods, including trapping, nest destruction, and exclusion. Many trap designs and other products are available.[141]

In the United States, the house sparrow population is in the midst of a sustained downward trend that reflects the trend throughout North America.[142] The reasons for this decline are unclear but might be related to the increasing conversion of American agriculture to large monoculture operations that renders formerly ideal rural farm habitats unsuitable.

## 16.3.11   *European starling* (Sturnus vulgaris)

The starling is a very successful invader worldwide and is on the list of 100 of the World's Worst Invasive Alien Species (http://www.issg.org/database/species/search.asp?st=100ss). In the United States, the starling became established following releases of 60 birds in 1890 and 40 more in 1891 in New York's Central Park.[127] During the next 50 years, the species spread across the country reaching the West Coast in the 1940s.[143] Currently, starlings are year-round residents from southern Alaska and Canada to northern Mexico with an estimated population of over 200 million.[144]

The starling is a stocky, compact bird, averaging 80–90 g body weight. It has a glossy black plumage, short, squared tail, pointed wings, and long bill. The plumage has a green/ purple iridescence. The sexes look alike. During the nonbreeding season, the head and body feathers have whitish tips which give the bird an overall spotted look. These spots wear off during the winter and are virtually gone by the next breeding season.[144]

Under favorable conditions, starlings can produce two clutches per year, with—four to six eggs per clutch. Starlings nest in natural and man-made cavities. Although aggressive and territorial in the breeding season, the starling is highly gregarious during the nonbreeding season. Starlings form large feeding flocks and share huge multispecies roosts in the winter. Both migratory and nonmigratory individuals can occur in the same population.[143]

The starling is responsible for an estimated $800 million of agricultural damage in the United States annually.[18] This includes depredations to fruit (cherries, blueberries, grapes) and grain crops, as well as feed consumption and fecal contamination at livestock feedlots and dairies, which may harbor huge numbers of birds. Starlings have been associated with transmission of at least 25 diseases, including toxoplasmosis, chlamydiosis, and salmonellosis.[130]

Beyond agricultural impacts, starling flocks cause numerous nuisance and damage problems through defecations on buildings, vehicles, and public spaces. Starlings are one of the most numerous bird species found in electric substations.[145] Because of their solid build and flocking tendency, starlings are a major risk to aircraft safety. According to the U.S. Federal Aviation Administration, there have been over 2000 reported collisions between starlings and civil aircraft since 1990.[146] Furthermore, starlings have impacted numerous native species through harassment and competition for cavity nest sites.[144] After many years of expansion, the trend of the European starling population has in recent years exhibited a broad decline throughout the United States.[142] The reason for this decline is unknown.

### 16.3.12   *Monk parakeet* (Myiopsitta monachus)

The monk parakeet (also known as the Quaker parrot) is a medium-sized parrot, approximately 30 cm long, weighing 100–120 g. These gregarious parrots have bright green backs and tails, and flight feathers with bluish cast. In contrast, their faces, throats, and breasts are pale gray. The sexes appear identical.

Their native range includes temperate and subtropical lowland regions of Argentina, Uruguay, Paraguay, and Brazil where they inhabit croplands, savannahs, and woodlots, including nonnative *Eucalyptus* stands planted as windbreaks. Monk parakeets have been introduced in many countries including the United States, Canada, Israel, Bahamas, Belgium, Italy, England, and Spain.

Thousands of monk parakeets were imported to the United States in the 1960s for the pet trade. The first free-flying birds were observed in New York in 1967[147] and Florida in 1969.[148] Since then, monk parakeet populations have been documented in over a dozen states.[149] The largest populations occur in south and southwest Florida, but thriving populations also exist in Texas, Connecticut, New York City, and Chicago among others. Unlike their selection of habitats in their native range, monk parakeets in the United States occupy habitats in urban and suburban areas. While the monk parakeet population in the United States exhibited exponential growth for many years, since 2005 the population has been steadily declining.

Monk parakeets have a variable diet comprised of seeds, fruits, berries, nuts, flowers, and leaf buds. In many locations, they obtain much of their food from backyard bird feeders. In South America, the species is considered a major crop pest,[149] but only isolated crop damage has been reported in the United States.[150]

Monk parakeets are unique among psittacines in that they do not nest in cavities but instead construct a large nest of sticks and branches. In the United States, they nest in trees and on radio towers, light poles, and electrical utility structures. The nest structure is maintained year-round, and as the structure is enlarged, other pairs of birds add nest chambers so that eventually a single, large structure can hold several nesting pairs.[151]

In the United States, their greatest economic impact is from nest construction on electric substations, transmission towers, and distribution poles. When nest materials get wet, the nests can cause short circuits, disrupt power, and damage sensitive equipment. No reliable, effective measures are available to prevent parakeet nesting on electric utility structures.[152] Although only temporarily effective, removal of nests from sensitive utility facilities is the most common method employed to prevent problems. In south Florida, during a recent 5-year period, costs associated with nest removal were estimated at \$1.3–\$4.7 million.[153] Long-term management options include the use of contraceptives to lower parakeet population levels.[153,154]

## 16.3.13 *Brown tree snake* (Boiga irregularis)

Brown tree snakes (*Boiga irregularis*), native to Australia, New Guinea and adjacent archipelagos, and the Solomon Islands, were probably introduced to the island of Guam as stowaways from Manus in the Admiralty Islands, north of New Guinea, shortly after World War II.[155,156] In 30 years, snakes spread throughout the island of Guam and attained extremely high population densities of 50–100 snakes per hectare in some areas.[155] Brown tree snakes are slender colubrid snakes, light brown in color, mildly venomous, and typically less than 2 m in length. The snakes are nocturnal and primarily arboreal, although they may be found on the ground.

As snakes spread across the island, their diverse food habits and the high population densities caused extirpation or drastic population reductions of Guam's resident bird species and native lizards and negative impacts to other wildlife.[156–160] Snakes have also had a significant impact on Guam's economy, agriculture, and human safety.[25] Because brown tree snakes are rear-fanged snakes and must chew to inject venom, their bites are unlikely to be fatal to adults. However, infants and small children (less than 10 years) are at risk for fatal bites, and bites to children and adults may exceed 170 bites per year.[25] Brown tree snakes are arboreal and use the tree canopies to move through landscapes; utility poles and wires provide ideal travel corridors. Snakes may ground electric systems when they move from grounded utility poles to electrified wires, thus producing power outages. In Guam, these power outages happen about every 2–9 days and cause millions of dollars in damage annually.[25,161] Additional effects of the high snake populations include loss of pets to predation, poultry industry losses, and concerns with the potential for diminished tourism.[25,162]

Brown tree snake management is focused on reducing the risk of snakes leaving the island and becoming established in other areas and on reducing their impacts to Guam. The risk of importation to Hawaii, with its many endangered species, and its thriving tourism industry is of particular concern. A combination of live trapping, hand capture from spotlight searches of fence lines, barriers, and the use of trained detector dogs are the primary interdiction methods to restrict snakes from aircraft and cargo leaving the island.[163–165] The trapping and spotlight searches are used to reduce the number of snakes in active transportation areas, then trained detector dogs are employed to search all cargoes leaving the island. This multitiered program has thus far been effective in preventing snakes from establishing in other Pacific island areas. To reduce the impacts to Guam, methods are being developed to deliver snake toxicants over large areas.[166] Other potential methods to control snakes include the use of repellents, fumigants, reproductive inhibition, and barriers, but these have yet to be deployed over large areas for population suppression or eradication.[163,167]

### 16.3.14 *Burmese python* (Python molurus bivittatus)

The Burmese python is a large invasive constrictor that has been entrenched in southernmost Florida for over a quarter-century.[168] The species' invasion pathway in south Florida has been attributed to illegal pet releases, although the highly destructive Hurricane Andrew in 1992 may also have released snakes from damaged captive breeding and holding facilities.[169,170] The origins of Burmese pythons entering the pet trade, and hence arriving in Florida, came from a portion of its native range, primarily Thailand near Bangkok (initially) and subsequently from Vietnam near Ho Chi Minh City after 1994.[171] Recent genetic testing of Burmese pythons showed little genetic differentiation among specimens captured in south Florida, but these specimens are genetically distinct from Vietnamese specimens; comparisons to Thai pythons were not conducted.[172] One possible consequence of minimal genetic variation among Florida pythons may be reduced ecological flexibility to adapt to significant changes in climatic conditions.

Burmese pythons are popular in the pet trade because of their relative docility and attractive coloration, with black-bordered dark brown blotches on a lighter background. The species is among the world's largest snakes, growing to over 7 m and weighing 90 kg,[173] with the largest Florida specimen reaching about 5 m. Burmese pythons are generalist feeders, consuming primarily mammals and birds but also reptiles, amphibians, and fish.[174] Their ecological impacts in south Florida continue to be identified, with documented predation on many native species, including endangered species.[174,175]

Burmese pythons are primarily a tropical lowland species strongly associated with water, with the vast majority of their native habitat below 200 m in altitude,[176,177] making south Florida an ideal locale for their establishment. Burmese python observations in south Florida have been expanding in recent years, including southward to the nearest island, Key Largo.[170,175] The potential range for python population expansion in the United States has been the subject of considerable controversy.[171,17,178] Projections of potential range expansions have used means such as climate matching with information from within the native range of the Burmese python and the closely related Indian python (*Python molurus molurus*),[178] and ecological niche modeling using the Burmese python's native range information.[179] Few empirical physiological and behavioral data are available on the Burmese python response to cooler mean temperatures and prolonged cold spells in the United States, but the available information seems to place doubt on whether this tropical species could establish sustained breeding populations beyond the warm climate and wetland habitats of south Florida.[180]

Smith et al.[181] have attempted analysis of the economic benefits for addressing the problem of snake depredations. Development of control tools and strategies for Burmese pythons is in its infancy and will likely follow similar conceptual approaches used for brown tree snakes (*Boiga irregularis*) on Guam.[163] Research has been initiated by several state and federal agencies and university scientists on technologies and strategies for controlling this invasive predator, including capture mechanisms, detection methods, reproductive vulnerabilities, baits and chemical cues, and toxicants.[170,182,183]

### 16.3.15   *Coqui frog* (Eleutherodactylus coqui)

The terrestrial tree frog *Eleutherodactylus coqui* was introduced to Hawaii from Puerto Rico via the horticultural trade in the late 1980s.[184] Since their introduction, coqui frogs have spread to six of the eight main Hawaiian Islands and are most widespread on the islands of Hawaii and Maui. Coqui frog populations occur in wet lowland forests and cover about 16,000 ha on the island of Hawaii and several hundred hectares on Maui. Due to extensive control efforts, only a few populations remain on Kauai, and incipient populations have been removed from Oahu, Lanai, and Molokai. Small populations and individual frogs have been reported from Florida, California, Guam, and the Virgin Islands.[185]

The most distinguishing feature of the coqui frog is the loud, two-note "ko-kee" mating call of the male—nearly 80 dB at 5 m. Coqui frogs are small tree frogs (30–50 mm in length) that vary in size across their native range, with females slightly larger than males.[185] Although numerous color and pattern variations have been described, frogs are typically gray to brown and may have a lighter colored stripe (or stripes) on the dorsal surface.[186] Coqui frogs are not dependent on standing water for reproduction; small froglets hatch directly from eggs.[187] These frogs reproduce up to four times a year and average 28 eggs per clutch.[188]

In Hawaii, coqui frog populations may exceed 50,000 frogs per hectare,[189,190] and because of these high densities, frogs may reduce native invertebrate populations, compete with native birds, alter food webs, and increase the nutrient cycling.[190,191,192] The loud frog calls have made people reluctant to purchase property or products infested with frogs, thus affecting agriculture, real estate, and the local economy.[191,193,194] Additionally, new quarantine and treatment procedures for minimizing the spread of frogs have increased the costs to the floriculture industry. Although the chytrid fungus *Batrachochytrium dendrobatidis* has been implicated in worldwide amphibian declines, chytrid has little effect on coquis, and because they are carriers, they could spread the organism, thus contraindicating its use as a biological control agent.[195,196]

A wide variety of techniques have been investigated to control frogs, but chemical control has been the most effective and safest option for large populations. A 16% solution of citric acid, a common food additive, classified as a "minimal-risk pesticide" exempt from federal registration requirements, has been approved by the Hawaii Department of Agriculture and is effective in controlling frog populations with minimal nontarget effects.[197] Other chemical options have been effective, but no other chemical control is currently registered for frog control. In addition to chemical control, hot-water treatments have been an effective means to kill frogs in plant shipments.[198] Mechanical controls, including traps and hand capture, and vegetation management have been effective on smaller-scale infestations.

### 16.3.16   *Sea lamprey* (Petromyzon marinus)

The sea lamprey (*Petromyzon marinus*) is a primitive boneless, jawless fish with a cartilaginous skeleton. They are native to the Atlantic Ocean, spawning in freshwater rivers,[3] but have long been a textbook example of a vertebrate invader causing great economic damage to commercial and sport fisheries by their establishment in the Great Lakes.[199] Adults are about 30–50 cm long and weigh about 225–370 g. They are grayish blue-black with metallic violet on the sides and silver-white coloration on the undersides. The body of the lamprey has smooth, scaleless skin with two dorsal fins (but no paired fins), no lateral line, no vertebrae, and no swim bladder.[200,201] Lampreys have a sucking disk mouth with sharp teeth surrounding a file-like tongue. They attach to fish and rasp into the soft tissues, feeding on body fluids, usually with lethal effects.

Sea lampreys were first observed in Lake Ontario in the 1830s, having entered through locks and canals. Niagara Falls served as a natural barrier, blocking lamprey entrance into the upper lakes until modifications to the Welland Canal in 1919 allowed the species to invade and spread through the rest of the Great Lakes system, where they were fully established by 1938. Establishment of sea lampreys was a principal factor in the collapse of the lake trout, whitefish, and chub populations during the 1940s and 1950s. Canadian and United States fisheries had harvested about £15 million of lake trout each year from the upper lakes, but by the 1960s, the harvest had fallen to £300,000.[200]

An ongoing integrated management program has resulted in a 90% reduction of sea lamprey populations.[201] Lampricides, aquatic pesticides selective for sea lamprey larvae, developed in the 1950s, are applied to some of the streams and tributaries where lamprey spawn every few years.[202] Other control methods include barriers constructed on streams to selectively block the migration of spawning lamprey while allowing other fish to pass with minimal disruption—including electric barriers that repel lampreys, velocity barriers that target lampreys' poor swimming ability, and adjustable-crest barriers that are inflated during the spawning run and deflated the rest of the year. Trapping of lampreys moving to spawning areas is also used, often in conjunction with barriers. In some areas, particularly the St. Mary's River, trapped male lampreys are sterilized and released into streams to compete with normal males.[201] Extensive technical resources and bibliographies, including many of the early unpublished reports, are maintained by the Great Lakes Fishery Commission (http://www.glfc.org).

## 16.3.17   *European and Asian carp* (Cyprinidae)

European carp (*Cyprinus carpio*) were introduced to the United States at some time during the middle to late 1800s (early date records for identification apparently conflict), originally as a food fish that was stocked throughout the country by the U.S. Fish Commission.[3] The species is now widely distributed in inland waters in at least 45 states, including Hawaii, and causes substantial damage to wetlands and aquatic ecosystems by habitat destruction and increased turbidity, competition for food with more desirable native species, and predation.[3] Few control efforts have been successful and, although the species is now often shunned as a food fish because of its association with polluted waters, it is sometimes managed by sport fishing and bow hunting.

More recently, concern has focused on the introduction of Asian cyprinids, including the grass carp (*Ctenopharyngogon idella*), the black carp (*Mylopharyngodon piceus*), the bighead carp (*Hypophthalmichthys nobilis*), and the silver carp (*Hypophthalmichthys molitrix*). These species were introduced to the United States in the 1960s and 1970s, originally for use in aquaculture, and escaped into lakes and rivers during flooding or through intentional movement of stocks. The latter two species, first found in native waters in the 1980s, are well established in the Mississippi River drainages.[203] These species grow rapidly and may weigh up to about 45 kg as adults. Scientists have speculated they are the most abundant large fish in the lower Mississippi River and describe impacts to include hazards to boaters and water skiers from the large, jumping fish, as well as direct competition with native fish for food and space, and predation on larva of native forms.[204]

These two carp species are of specific current concern because of potential dispersal from the Mississippi River Basin into the Great Lakes. Establishment in the Great Lakes would disrupt food chains supporting native fish and pose significant threats to Great Lakes ecosystems.[205] Since both commercial fishing and sport fishing, as well as tourism, boating, and water sports, are major facets of the shoreline and lake island economies, public concern and political action have increased as the carp issues became more widely known. The problem of Asian carp management has been described in detail, and the variety of research needs and potential control methods has been outlined.[206] Control and prevention efforts, including use of netting, the fish toxicant rotenone, and construction of an electric barrier in the Chicago Sanitary and Ship Canal, have appeared to protect the Great Lakes from invasion for an extended period. More recently, continuing concerns related to new detection methods brought the issue of closing navigation locks as a further means of keeping fish from entering the Great Lakes ecosystem from the Mississippi River Basin to the U.S. Supreme Court, which denied the request.[207] Containing the spread of Asian carp continues to be an important environmental issue.

## 16.4   Offshore threats

While the rodents and other vertebrate invaders already established in country are a serious concern, the potential for continuing invasions in the future must also be considered. These threats deserve careful analysis because on one hand, managers and decision makers should be alerted to potential primary (the invasive species themselves) and secondary impacts (including the associated diseases and parasites[208]) of vertebrate invasions, while on the other hand, the scientific and political disagreements this may entail are confusing to everyone and may sometimes be counterproductive, particularly if alarmist approaches make their way into mass media.[171,178,179,193]

The future threats of potential rodent invasions are illustrative. The source populations of the three species of commensal rodents (*Rattus norvegicus*, *Rattus rattus*, and *Mus musculus*) established throughout the United States were likely established in the European ports with substantial ship traffic bound for the New World colonies. Spread of these species throughout the country was fostered by their close association with human settlements and their propensity to infest commodity shipments and household goods moving by wagon, ship, riverboat, and railroad. They now occupy a number of sylvatic and agrarian habitats on the fringes of settlements, as well as the urban and suburban areas where they are ubiquitous. They have also established on a number of islands where they may occupy a wider range of habitats and a show wider range of food selection, causing myriad problems in these fragile ecosystems, particularly impacts on endangered species.

A number of other rodent species, most commonly recognized as agricultural pests in different parts of the world, sometimes occur as local commensals and may be found in port or shipping areas. These include *Apodemus agrarius* in East Asia, *Apodemus sylvaticus* in Europe,[209] *Bandicota bengalensis* and a number of smaller species in South Asia,[210,211] *Rattus exulans* in Hawaii and Pacific islands (where it is known as the Polynesian rat) and Southeast Asia (where it is known variously as the little Burmese or little Asian house rat),[211] *Rattus tanezumi* (a newly redesignated Asian variant of *Rattus rattus*)[51] that occurs in temperate and tropical areas of Asia and has already reached North America,[52] *Mastomys* [*Praomys*] *natalensis*,[211] and *Arvicanthis niloticus* in east and west Africa.[212] With the globalization of trade, containerized shipping, and the emphasis on increasing the range of trading partners, the probability of accidental transport of such species to the United States has clearly increased. The pet trade, which through accidental escapes or intentional releases of imported exotic animals, has been another principal route of the invasion and establishment of potentially injurious animals into the United States.[213] Increased efforts to identify and predict potential vertebrate invaders and to catalog potential primary and secondary problems, as well as to develop information on ecology and effective control methods in native ranges, should prove valuable, if not essential, in the future.[21,214]

Although the introduction, and at least local establishment, of some invasive species has happened in relatively short time spans (e.g., Asian carp, brown tree snakes, nutria, Gambian giant pouched rats), other species took decades to establish and move throughout the country (e.g., Norway rats, rock pigeons, house sparrows). The difficulties of rapid detections of invasions (or other events that occur in low frequencies) are well known. This may be a particular problem with invasive rodent species that are mostly nocturnal and similar in appearance, especially without specimens in hand.

Detection of invader propagules that may lead to incipient invasive populations holds the same challenges as monitoring populations for management or eradication. The detection of rare events (including the occurrence of animals in low numbers) can obviously be enhanced by increased sampling effort. Sampling effort is maximized within available resources if the sampling methods and observations are easily implemented and understood. Furthermore, quantification of animal population status can be greatly improved by applying detection methods that take advantage of behavioral characteristics, which increase the probability of observation, and by using measuring methods that are continuous rather than binary.[215–219] Public involvement in early detection efforts can be an important component, and a number of state and local programs have begun proactive efforts of public education using print and broadcast media, posters and displays, and in some cases establishing telephone "hotlines." Hawley[220] described a particularly well-organized and comprehensive public involvement program in Saipan, using modern marketing tactics to attempt to prevent importation and establishment of brown tree snakes.

## 16.5   Discussion

Worldwide, rodents and other invasive vertebrates have had devastating effects on the human enterprise and quality of life. Along with habitat loss and human activity, vertebrate invaders have been a principal cause of extinctions and continuing risks to endangered and threatened species in many areas. Infectious diseases and parasites carried with these invaders have amplified their direct effects on humans, domestic animals, wildlife, and the environment. The United States has been no exception to these effects.

Awareness of invasive species problems has greatly increased during the past several decades from the time when Norway rats, pigeons, starlings, and sea lampreys were the major vertebrate invaders that biologists studied and the public encountered. The prospects of needing to address the arrival and establishment of new vertebrate invaders appear almost certain. A central problem for biologists (and for politicians and the public) is to avoid thinking and planning only for the short term when analyzing the risks of vertebrate invasions and to consider longer time-horizons, best said as "ecological time,"[221] that involve time scales of decades or generations rather than years.

The questions of eradication versus management, particularly of new invaders, also require careful consideration in long-term planning since these actions may require substantially different strategies. Eradication is clearly feasible for founder populations and on small scales, based on the recent successes with island populations of rodents, carnivores, and ungulates. But the rapid response, persistent sustained action, and continuing surveillance required for successful eradication present both ecological and political challenges, and delays make any effort progressively more complex, expensive, and difficult to sustain. At some point, any eradication effort if not demonstrably successful in the short term—particularly if the invader, the invaded environment, or the impacts do not attract public concern—can easily slip into the case-by-case management mode typical of how vertebrate–human–wildlife conflicts are handled. In the long run (ecological time), continued faunal mixing,[222] with range changes, the loss of some species, and the addition of new nonindigenous species, including some that are injurious, seems inevitable. Many of these changes will happen slowly, and many will be undetected despite our best efforts. The social effect of the shifting baseline syndrome (in which each generation bases its ecological expectations on its own life experience rather than on historical patterns)[223,224] works against attempts to fend off these changes, making the case-by-case strategy of managing the specific problems caused by invasive species the most likely way to successfully mitigate impacts. Obtaining the ecological information and developing the range of vertebrate management tactics needed to accomplish this effectively are among the national challenges for the future.

## Acknowledgments

In preparing this chapter, we have drawn heavily on the papers, posters, and discussions from the 2007 symposium, *Managing Vertebrate Invasive Species*, sponsored by the U.S. Department of Agriculture/Animal and Plant Health Inspection Service/Wildlife Services, at the National Wildlife Research Center in Fort Collins, Colorado, to coincide with the inauguration of a new Invasive Species Research Building. We are grateful to our colleagues who participated in the symposium and to our agencies for their continuing commitments to addressing the problems of vertebrate invaders. We are also grateful to Peter Savarie, Robert Sugihara, Laura Driscoll, and Rogelio Doratt for helping to draw together information on invasive species problems in Hawaii and the Pacific Islands.

## References

1. Clinton, W. J. 1999. Executive order 133112 invasive species. *Fed Regist* 64:6183.
2. Witmer, G. W., P. W. Burke, W. C. Pitt, and M. L. Avery. 2007. Management of invasive vertebrates in the United States: An overview. In *Managing Vertebrate Invasive Species: Proceedings of an International Symposium*, ed. G. W. Witmer, W. C. Pitt, and K. A. Fagerstone, 127. Fort Collins, CO: USDA/APHIS/Wildlife Services, National Wildlife Research Center.
3. Fuller, P. L., L. G. Nico, and J. D. Williams. 1999. Nonindigenous fishes introduced into inland waters of the United States. Spec. Publ. 27, Bethesda, MD: American Fisheries Society.
4. Fall, M. W., and W. B. Jackson. 2000. Future technology for managing vertebrate pests and overabundant wildlife—an introduction. *Int Biodeterior Biodegradation* 45:93.
5. Pimentel, D. in press. Environmental and economic costs associated with alien invasive species in the United States. In *Biological Invasions: Economic and Environmental Costs of Alien Plant, Animal, and Microbe Species*, 2nd ed. ed. D. Pimentel. Boca Raton, FL: CRC Press.
6. Cox, G. W. 1999. *Alien Species in North America and Hawaii*. Washington, DC: Island Press.
7. Conover, M. 2002. *Resolving Human Wildlife Conflicts—The Science of Wildlife Damage Management*. Boca Raton, FL: Lewis Publishers.
8. Rodda, G. H., Y. Sawai, D. Chizar, and H. Tanaka. 1999. Snake management. In *Problem Snake Management—The Habu and the Brown Treesnake*, ed. G. H. Rodda, Y. Sawai, D. Chizar, and H. Tanaka, 1. Ithaca, NY: Comstock Publishing Associates.
9. Simberloff, D. 2002. Today Tiritiri, tomorrow the world! Are we aiming too low in invasives control? In *Turning the Tide: the Eradication of Invasive Species*, ed. C. R. Veitch, and M. N. Clout, 4. Gland: IUCN SSC Invasive Species Specialist Group, International Union for Conservation of Nature and Natural Resources.

10. Witmer, G. W., W. C. Pitt, and K. A. Fagerstone, eds. 2007. *Managing Vertebrate Invasive Species: Proceedings of an International Symposium*. Fort Collins, CO: USDA/APHIS/Wildlife Services, National Wildlife Research Center.

11. Veitch, C. R., and M. N. Clout, eds. 2002. *Turning the Tide: The Eradication of Invasive Species*. Gland: IUCN SSC Invasive Species Specialist Group, International Union for Conservation of Nature and Natural Resources.

12. Williams, L. 2007. Invasive species: A national perspective and the need for a coordinated response. In *Managing Vertebrate Invasive Species: Proceedings of an International Symposium*, ed. G. W. Witmer, W. C. Pitt, and K. A. Fagerstone, 9. Fort Collins, CO: USDA/APHIS/Wildlife Services, National Wildlife Research Center.

13. Martin, C. 2007. Promoting awareness, knowledge and good intentions. In *Managing Vertebrate Invasive Species: Proceedings of an International Symposium*, ed. G. W. Witmer, W. C. Pitt, and K. A. Fagerstone, 57. Fort Collins, CO: USDA/APHIS/Wildlife Services, National Wildlife Research Center.

14. Wilcove, D. S., D. Rothestein, J. Dubow, A. Phillips, and E. Losos. 1998. Quantifying threats to imperiled species in the United States. *BioScience* 48:607.

15. Mack, R. N., D. Simberloff, W. M. Lonsdale, H. Evans, M. Clout, and F. A. Bazzaz. 2000. Biotic invasions: Causes, epidemiology, global consequences and control. *Ecol Appl* 10:689.

16. Sala, O. E. et al. 2000. Biodiversity scenarios for the year 2100. *Science* 287:1774.

17. Pimentel, D. et al. 2002. Economic and environmental threats of alien plant, animal and microbe invasions. In *Biological Invasions: Economic and Environmental Costs of Alien Plant, Animal, and Microbe Species*, ed. D. Pimentel, 307. Boca Raton, FL: CRC Press. Reprinted from *Agriculture, Ecosystems, and Environment* 84:1–20, 2000.

18. Pimentel, D., R. Zuniga, and D. Morrison. 2005. Update on the environmental and economic costs associated with alien-invasive species in the United States. *Ecol Econ* 52:273.

19. Perrings, C. et al. 2002. Biological invasion risks and the public good: An economic perspective. *Conserv Ecol* 6:1.

20. Julia, R., D. W. Holland, and J. Guenther. 2007. Assessing the economic impact of invasive species: The case of yellow starthistle (*Centaurea solsitialis L.*) in the rangelands of Idaho, U. S. A. *J Environ Manage* 85:876.

21. Cook, D. C., M. B. Thomas, S. A. Cunningham, D. L. Anderson, and P. J. DeBarro. 2007. Predicting the economic impact of invasive species on an ecosystem service. *Ecol Appl* 17:1832.

22. Pineda-Krch, M., J. M. O'Brien, C. Thunes, and T. E. Carpenter. 2010. Potential impact of introduction of foot-and-mouth disease from wild pigs into commercial livestock premises in California. *Am J Vet Res* 71:82.

23. Williamson, M. 1996. *Biological Invasions*. London: Chapman and Hall.

24. Jay, M. T. et al. 2007. *Escherichia coli* O157:H7 in feral swine near spinach fields and cattle, central California coast. *Emerg Infect Dis* 13:1908.

25. Shwiff, S. A., K. Gebhardt, K. N. Kirkpatrick, and S. S. Shwiff. 2010. Potential economic damage from the introduction of the brown tree snake, *Boiga irregularis* (Reptilia: Colubridae), to the islands of Hawai'i. *Pac Sci* 64:1.

26. Witmer, G. W., R. B. Sanders, and A. C. Taft. 2003. Feral swine: Are they a disease threat to livestock in the United States? *Wildl Damage Manag Conf* 10:316.

27. Campbell, T. A., R. W. DeYoung, E. M. Wehland, L. I. Grassman, D. B. Long, and J. Delgado-Acevedo. 2008. Feral swine exposure to selected viral and bacterial pathogens in southern Texas. *J Swine Health Prod* 16:312.

28. Hall, J. S. et al. 2008. Influenza exposure in feral swine from the United States. *J Wildl Dis* 44:362.

29. Kaller, M. E., and W. E. Kelso. 2006. Swine activity alters invertebrate and microbial communities in a Coastal Plain watershed. *Am Midl Nat* 156:163.

30. Engeman, R. M., A. Stevens, J. Allen, J. Dunlap, M. Daniel, D. Teague, and B. Constantin. 2007. Feral swine management for conservation of an imperiled wetland habitat: Florida's vanishing seepage slopes. *Biol Conserv* 134:440.

31. Campbell, T. A., and D. B. Long. 2009. Feral swine damage and damage management in forested ecosystems. *For Ecol Manage* 257:2319.

32. Bergman, D. L., M. D. Chandler, and A. Locklear. 2002. The economic impact of invasive species to Wildlife Services' cooperators. In *Human Wildlife Conflicts: Economic Considerations*, ed. L. Clark, J. Hone, J. A. Shivik, R. A. Watkins, K. C. VerCauteren, and J. K. Yoder, 169. Fort Collins, CO: USDA/APHIS/Wildlife Services, National Wildlife Research Center.

33. Engeman, R. M., S. A. Shwiff, F. Cano, and B. Constantin. 2003. An economic assessment of the potential for predator management to benefit Puerto Rican parrots. *Ecol Econ* 46:283.

34. Engeman, R. M., H. T. Smith, S. A. Shwiff, B. Constantin, J. Woolard, M. Nelson, and D. Griffin. 2003. Prevalence and economic value of feral swine damage to native habitats in three Florida state parks. *Environ Conserv* 30:319.

35. Miller, R., and P. D. Blair. 1985. *Input-Output Analysis: Foundations and Extensions*. Englewood Cliffs, NJ: Prentice-Hall.

36. Burnett, K. M., S. D'Evelyn, B. A. Kaiser, P. Nantamanasikarn, and J. A. Roumasset. 2008. Beyond the lamppost: Optimal prevention and control of the brown tree snake in Hawaii. *Ecol Econ* 67:66.

37. McNeely, J. A., H. A. Mooney, L. E. Neille, P. Schei, J. K. Waage, eds. 2001. *Global Strategy on Invasive Alien Species*. Gland, Switzerland: International Union for Conservation of Nature and Natural Resources (on behalf of the Global Invasive Species Programme).

38. National Invasive Species Council. 2001. *Meeting the Invasive Species Challenge: National Invasive Species Management Plan*. Washington DC: National Invasive Species Council.

39. National Invasive Species Council. 2008. *2008–2012 National Invasive Species Management Plan*. Washington DC: National Invasive Species Council.

40. Shwiff, S. A., R. T. Sterner, and K. N. Kirkpatrick. 2008. Economic evaluation of a Texas oral rabies vaccination program for control of a domestic dog-coyote rabies epizootic: 1995–2006. *J Am Vet Med Assoc* 233:1736.

41. Zerbe, R. O., and D. D. Dively. 1994. *Benefit-Cost Analysis: In Theory and Practice*. New York: HarperCollins College Publishers.

42. King, D. M., and M. Mazzotta. 2000. Damage cost avoided, replacement cost, and substitute cost methods. Ecosystem Valuation. http://www.ecosystemvaluation.org/cost_avoided.htm (accessed December 1, 2006).

43. Brooks, J. E. 1973. A review of commensal rodents and their control. *CRC Crit Rev Environ Control* 3:405.

44. Meehan, A. P. 1984. *Rats and Mice*. East Grinstead: Rentokil Ltd.

45. Jackson, W. B. 1977. Evaluation of rodent depredations to crops and stored products. *Eur Plant Prot Organ Bull* 7:439.

46. Timm, R. M. 1994. Norway rats. In *Prevention and Control of Wildlife Damage*, ed. S. E. Hygstrom, R. M. Timm, and G. E. Larson, B105. Lincoln, NE: University of Nebraska Cooperative Extension.

47. Davis, D. E. 1953. The characteristics of rat populations, *Q Rev Biol* 28:373.

48. Fall, M. W., and W. B. Jackson. 1998. A new era of vertebrate pest control? An introduction. *Int Biodeterior Biodegradation* 42:85.

49. Corrigan, R. M. 2001. *Rodent Control: A Practical Guide for Pest Management Professionals*. Richfield, MN: GIE, Inc.

50. Nolte, D. L., D. Bergman, and J. Townsend. 2003. Roof rat invasion of an urban desert island. In *Rats, Mice, and People: Rodent Biology and Management*, ed. G. R. Singleton, L. A. Hinds, C. J. Krebs, and D. M. Spratt, 481. Canberra, Australia: Australian Centre for International Agriculture Research.

51. Musser, G. G., and M. D. Carleton. 2005. Superfamily Muroidea. In *Mammal Species of the World: A Taxonomic and Geographic Reference*, 3rd ed. D. E. Wilson and D. M. Reeder, 894. Baltimore, MD: Johns Hopkins University Press.

52. James, D. K. 2006. New rat species in North America. *Vector Ecol Newsletter* 37:5.

53. Munro, G. C. 1945. Tragedy in bird life. *Elapaio* 5:48.

54. Atkinson, I. A. E. 1977. A reassessment of factors, particularly *Rattus rattus* L. that influenced the decline of endemic forest birds in the Hawaiian Islands. *Pac Sci* 31:109.

55. Pitt, W. C., and G. W. Witmer. 2007. Invasive predators: A synthesis of the past, present, and future. In *Predation in Organisms—a Distinct Phenomenon*, ed. A. M. T. Elewa, 265. Heidelberg, Germany: Springer Verlag.

56. Tomich, P. Q. 1986. *Mammals in Hawaii*. 2nd ed. Honolulu, HI: Bishop Museum Press.

57. Witmer, G. W. 2004. Rodent ecology and plague in North America. In *19th International Congress of Zoology*. Beijing, China: China Zoological Society.

58. Howald, G., C. J. Donlan, J. P. Galvan, J. C. Russell, J. Parkes, A. Samaniego, Y. Wang et al. 2007. Invasive rodent eradication on islands. *Conserv Biol* 21:1258.

59. Witmer, G. W., F. Boyd, and Z. Hillis-Starr. 2007. The successful eradication of introduced roof rats (*Rattus rattus*) from Buck Island using diphacinone, followed by an irruption of house mice (*Mus musculus*). *Wildl Res* 34:108.

60. Matisoo-Smith, E., and J. H. Robins. 2004. Origins and dispersals of Pacific peoples: Evidence from mtDNA phylogenies of the Pacific rat. *Proc Natl Acad Sci* 101:9167.

61. Roberts, M. 1991. Origin, dispersal routes, and geographic distribution of *Rattus exulans*, with special reference to New Zealand. *Pac Sci* 45:123.

62. Jackson, W. B. 1965. Litter size in relation to latitude in two murid rodents. *Am Midl Nat* 73:245.

63. McCartney, W. C. 1970. Arboreal behavior of the Polynesian rat (*Rattus exulans*). *BioScience* 20:1061.

64. Spenneman, D. H. R. 1997. Distribution of rat species (*Rattus* spp.) on the atolls of the Marshall Islands: Past and present dispersal. *Atoll Res Bull* 446:1.

65. Kami, H. T. 1966. Foods of rodents in the Hamakua District, Hawaii. *Pac Sci* 20:367.

66. Kepler, C. B. 1967. Polynesian rat predation on nesting Laysan albatrosses and other Pacific seabirds. *Auk* 84:426.

67. Fall, M. W., A. B. Medina, and W. B. Jackson. 1971. Feeding patterns of *Rattus rattus* and *Rattus exulans* on Eniwetok Atoll, Marshall Islands. *J Mammal* 52:69.

68. Cook, I. G. 1973. The tuatara, *Sphenodon punctatus* gray, on islands with and without populations of the Polynesian rat, *Rattus exulans* (Peale). *Proc New Zealand Ecol Soc* 20:115.

69. Tobin, M. E., and R. T. Sugihara. 1992. Abundance and habitat relationships of rats in Hawaiian sugarcane fields. *J Wildl Manag* 56:815.

70. Sugihara, R. T. 1997. Abundance and diets of rats in two native Hawaiian forests. *Pac Sci* 51:189.

71. Rufaut, C. G., and G. W. Gibbs. 2003. Response of a tree weta population (*Hemideina crassidens*) after eradication of the Polynesian rat from a New Zealand island. *Restor Ecol* 11:13.

72. Marshall Jr., J. T. 1962. Predation and natural selection. In *Pacific Island Rat Ecology: Report on a Study Made on Ponape and Adjacent Islands 1955–1958*, ed. T. I. Storer, 177. Honolulu, HI: Bernice P. Bishop Museum.

73. Strecker, R. L. 1962. Economic relations. In *Pacific Island Rat Ecology: Report on a Study Made on Ponape and Adjacent Islands 1955–1958*, ed. T. I. Storer, 200. Honolulu, HI: Bernice P. Bishop Museum.

74. Meyer, J. Y., and J. F. Butaud. 2009. The impacts of rats on the endangered native flora of French Polynesia (Pacific Islands): Drivers of plant extinction or coup de grâce species? *Biol Invasions* 11:1569.

75. Gibbs, G. W. 2009. The end of an 80-million year experiment: A review of evidence describing the impact of introduced rodents on New Zealand's 'mammal-free' invertebrate fauna. *Biol Invasions* 11:1587.

76. Timm, R. M. 1994. House mice. In *Prevention and Control of Wildlife Damage*, ed. S. E. Hygstrom, R. M. Timm, and G. E. Larson, B31. Lincoln, NE: University of Nebraska Cooperative Extension.

77. Witmer, G., and S. Jojola. 2006. What's up with house mice? A review. *Vertebrate Pest Conf* 22:124.

78. Brown, P. R., M. Davies, G. Singleton, and J. Croft. 2004. Can farm-management practices reduce the impact of house mouse populations on crops in an irrigated farming system? *Wildl Res* 31:597.

79. Hyngstrom, S. E. 1995. House mouse damage to insulation. *Int Biodeterior Biodegradation* 36:143.

80. Gratz, N. G. 1994. Rodents as carriers of disease. In *Rodent Pests and Their Control*, ed. A. P. Buckle and R. H. Smith, 85. Wallingford, UK: CAB International.

81. Cuthbert, R., and G. Hilton. 2004. Introduced house mice: A significant predator of threatened and endemic birds on Gough Island, South Atlantic Ocean? *Biol Conserv* 117:483.

82. Prakash, I. 1988. *Rodent Pest Management*. Boca Raton, FL: CRC Press.

83. Carter, J., and B. P. Leonard. 2002. A review of the literature on the worldwide distribution, spread of, and efforts to eradicate the coypu (*Myocastor coypus*). *Wildl Soc Bull* 30:162.

84. LeBlanc, D. J. 1994. Nutria. In *Prevention and Control of Wildlife Damage*, ed. S. E. Hygstrom, R. M. Timm, and G. E. Larson, B71. Lincoln, NE: University of Nebraska Cooperative Extension.

85. Bounds, D. L., M. H. Sherfy, and T. A. Mollett. 2003. Nutria. In *Wild Mammals of North America: Biology, Management, and Conservation*, ed. G. A. Feldhamer, B. C. Thompson, and J. A. Chapman, 1119. Baltimore, MD: John Hopkins University Press.

86. Marx, J., E. Mouton, and G. Linscombe. 2004. Nutria harvest distribution 2003–2004 and a survey of nutria herbivory damage in coastal Louisiana in 2004. In *Fur and Refuge Division, Louisiana Department of Wildlife and Fisheries/Coastwide Nutria Control Program*. Baton Rouge, LA: CWPPRA Project (LA-03b).

87. Witmer, G. W., P. W. Burke, S. Jojola, and D. L. Nolte. 2008. A live trap model and field trial of a nutria multiple capture trap. *Mammalia* 72:352.

88. Jojola, S., G. W. Witmer, and P. W. Burke. 2009. Evaluation of attractants to improve trapping success of nutria on Louisiana coastal marsh. *J Wildl Manag* 73:1414.

89. Perry, N. et al. 2006. New invasive species in southern Florida: Gambian rat. *J Mammal* 87:262.

90. Engeman, R. M. et al. 2007b. The path to eradication of the Gambian giant pouched rat in Florida. In *Managing Vertebrate Invasive Species: Proceedings of an International Symposium*, ed. G. W. Witmer, W. C. Pitt, and K. A. Fagerstone, 305. Fort Collins, CO: USDA/APHIS/Wildlife Services, National Wildlife Research Center.

91. Fiedler, L. A. 1994. Rodent pest management in Eastern Africa. FAO Plant Production and Protection Paper, Food and Agriculture Organization of the United Nations, Rome.

92. Enserink, M. 2003. U.S. monkey pox outbreak traced to Wisconsin pet dealer. *Science* 300:1639.

93. Peterson, A. et al. 2006. Native-range ecology and invasive potential of *Cricetomys* in North America. *J Mammal* 87:427.

94. Kingdon, J. 1974. *East African Mammals, Vol. 2, Part B (Hares and Rodents)*. Chicago, IL: University of Chicago Press.

95. Ajayi, S. 1975. Observations on the biology, domestication, and reproductive performance of the African giant rat *Cricetomys gambianus* waterhouse in Nigeria. *Mammalia* 39:343.

96. Witmer, G., Snow, N., and P. Burke. 2010. Potential attractants for detecting and removing invading Gambian giant pouched rats (*Cricetomys gambianus*). *Pest Manag Sci* 66:412.

97. Southeastern Cooperative Wildlife Disease Study. 2010. National Feral Swine Mapping System, University of Georgia, Athens. http://128.192.20.53/nfsms/ (accessed January 20, 2010).

98. Towne, C. W., and E. N. Wentworth. 1950. *Pigs from Cave to Cornbelt*. Norman, OK: University of Oklahoma.

99. Mayer, J. J., and I. L. Brisbin Jr., 1991. *Wild Pigs of the United States: Their History, Morphology, and Current Status*. Athens: University of Georgia.

100. Wood, G. W., and T. E. Lynn Jr., 1977. Wild hogs in southern forests. *South J Appl For* 1:12.

101. Sweeney, J. R., J. M. Sweeney, and S. W. Sweeney. 2003. Feral hog (*Sus scrofa*). In *Wild Mammals of North America: Biology, Management, and Conservation*, ed. G. A. Feldhamer, B. C. Thompson, and J. A. Chapman, 1164. Baltimore, MD: John Hopkins University Press.

102. Taylor, R. B., E. C. Hellgren, T. M. Gabor, and L. M. Ilse. 1998. Reproduction of feral pigs in southern Texas. *J Mammal* 79:1325.

103. Gabor, T. M., E. C. Hellgren, R. A. Van Den Bussche, and N. J. Silvy. 1999. Demography, socio-spatial behaviour and genetics of feral pigs (*Sus scrofa*) in a semi-arid environment. *J Zool (London)* 247:311.

104. Campbell, T. A., M. R. Garcia, L. A. Miller, M. A. Ramirez, D. B. Long, J. Marchand, and F. Hill. 2010. Immunocontraception of male feral swine with a recombinant GnRH vaccine. *J Swine Health Prod* 18:118–124.

105. Mapston, M. E. 2004. Feral hogs in Texas. Texas AgriLife Extension, Publication B-6149. Texas A&M System, College Station.

106. Campbell, T. A., S. J. Lapidge, and D. B. Long. 2006. Using baits to deliver pharmaceuticals to feral swine in southern Texas. *Wildl Soc Bull* 34:1184.

107. Long, D. B., T. A. Campbell, and G. Massei. 2010. Evaluation of feral swine-specific feeder systems. *Rangelands* 32:8.

108. Corbet, G. B., and J. E. Hill. 1992. *Mammals of the Indomalayan Region: A Systematic Review*. Oxford: Oxford University Press.

109. Gorman, M. L. 1975. The diet of feral *Herpestes auropunctatus* (Carnivora: Viverridae) in the Fijian Islands. *J Zool (London)* 175:273.

110. Sugimura, K., F. Yamada, and A. Miyamoto. 2005. Population trend, habitat change and conservation of the unique wildlife species on Amami Island, Japan. *Glob Environ Res* 6:79.

111. Nellis, D. W., and C. O. R. Everard. 1983. The biology of the mongoose in the Caribbean Islands. *Studies Fauna Curacao other Caribbean Islands* 64:1.

112. Long, J. L. 2003. *Introduced Mammals of the World*. Canberra: CSIRO Publishing.

113. Baldwin, P. H., C. W. Schwartz, and E. R. Schwartz. 1952. Life history and economic status of the mongoose in Hawaii. *J Mammal* 33:335.

114. Pimentel, D. 1955. Biology of the Indian mongoose in Puerto Rico. *J Mammal* 36:62.

115. Nellis, D. W. 1989. *Herpestes auropunctatus. Mamm Species* 342:1.

116. Nowak, R. M. 1991. *Walker's Mammals of the World*. 5th ed. vol. 2. Baltimore, MD: Johns Hopkins University Press.

117. Pimentel, D., L. Lach, R. Zuniga, and D. Morrison. 2000. Environmental and economic costs associated with non-indigenous species in the United States. *BioScience* 50:53.

118. Everard, C. O., and J. D. Everard. 1988. Mongoose rabies. *Rev Infect Dis* 10:S610.

119. Baker, R. H., and C. A. Russell. 1979. Mongoose predation on a nesting nene. *Elapaio* 40:51.

120. Stone, C. P., M. Dusek, and M. Aeder. 1994. Use of an anticoagulant to control mongooses in Nene breeding habitat. *Elepaio* 54:73.

121. Seaman, G. A., and J. E. Randall. 1962. The mongoose as a predator in the Virgin Islands. *J Mammal* 43:544.

122. Nellis, D. W., and V. Small. 1983. Mongoose predation on sea turtle eggs and nests. *Biotropica* 15:159.

123. Coblentz, B. E., and B. A. Coblentz. 1985. Control of the Indian mongoose *Herpestes auropunctatus* on St. John, U.S. Virgin Islands. *Biol Conserv* 33:281.

124. Vilella, F. J. 1998. Biology of mongoose (*Herpestes javanicus*) in a rain forest in Puerto Rico. *Biotropica* 30:120.

125. Smith, D. G., J. T. Polhemus, and E. A. VanderWerf. 2000. Efficacy of fish-flavored diphacinone bait blocks for controlling small Indian mongooses (*Herpestes auropunctatus*) populations in Hawaii. *Elepaio* 60:47.

126. Roy, S., C. Jones, and S. Harris. 2002. An ecological basis for control of the mongoose in Mauritius: Is eradication possible? In *Turning the Tide: The Eradication of Invasive Species*, ed. C. R. Veitch and M. N. Clout, 266. Gland, Switzerland: IUCN SSC Invasive Species Specialist Group, International Union for Conservation of Nature and Natural Resources.

127. Long, J. L. 1981. *Introduced Birds of the World*. New York: Universe Books.

128. Johnston, R. F. 1992. Rock dove (*Columba livia*). In *The Birds of North America Online*, ed. A. Poole, Cornell Laboratory of Ornithology. http://bna.birds.cornell.edu/bna/species/013 (accessed January 25, 2010).

129. Johnston, R. F., and M. Janiga. 1995. *Feral Pigeons*. New York: Oxford University Press.

130. Weber, W. J. 1979. *Health Hazards from Pigeons, Starlings and English Sparrows*. Fresno, CA: Thomson Publications.

131. White, C. M., N. J. Clum, T. J. Cade, and W. G. Hunt. 2002. Peregrine falcon (*Falco peregrinus*). In *The Birds of North America Online*, ed. A. Poole, Cornell Laboratory of Ornithology. http://bna.birds.cornell.edu/bna/species/660 (accessed January 25, 2010).

132. Williams, D. E., and R. M. Corrigan. 1994. Pigeons (rock doves). In *Prevention and Control of Wildlife Damage*, ed. S. E. Hygstrom, R. M. Timm, and G. E. Larson, E87. Lincoln, NE: University of Nebraska Cooperative Extension.

133. Avery, M. L., K. L. Keacher, and E. A. Tillman. 2008. Nicarbazin bait reduces reproduction by pigeons (*Columba livia*). *Wildl Res* 35:80.

134. Dawson, W. L. 1903. *The Birds of Ohio*. Columbus, OH: Wheaton Publishing Company.

135. Moulton, M. P., W. P. Cropper Jr., M. L. Avery, and L. E. Moulton. 2010. The earliest house sparrow introductions to North America. *Biol Invasions* 12:2955–2958, http://www.springerlink.com/content/1771077353t22353 (accessed January 14, 2010).

136. Lowther, P. E., and C. L. Cink. 2006. House sparrow (*Passer domesticus*). In *The Birds of North America Online*, ed. A. Poole, Cornell Laboratory of Ornithology. http://bna.birds.cornell.edu/bna/species/012 (accessed January 25, 2010).
137. McGillivray, W. B. 1983. Intraseasonal reproductive costs for the house sparrow (*Passer domesticus*). *Auk* 100:25.
138. Kalmbach, E. R. 1940. Economic status of the English sparrow in the United States. *US Dep Agric Tech Bull* 711.
139. Gavett, A. P., and J. S. Wakeley. 1986. Diets of house sparrows in urban and rural habitats. *Wilson Bull* 98:137.
140. Jackson, J. A., and J. Tate Jr., 1974. An analysis of nest box use by purple martins, house sparrows and starlings in eastern North America. *Wilson Bull* 86:435.
141. Fitzwater, W. D. 1994. House sparrows. In *Prevention and Control of Wildlife Damage*, ed. S. E. Hygstrom, R. M. Timm, and G. E. Larson, E101. Lincoln, NE: University of Nebraska Cooperative Extension.
142. Sauer, J. R., J. E. Hines, and J. Fallon. 2008. The North American breeding bird survey, results and analysis 1966–2007. Version 5.15.2008, U.S. Geological Survey Patuxent Wildlife Research Center, Laurel, Maryland.
143. Kessel, B. 1953. Distribution and migration of the European starling in North America. *Condor* 55:49.
144. Cabe, P. R. 1993. European starling (*Sturnus vulgaris*). In *The Birds of North America Online*, ed. A. Poole, Cornell Laboratory of Ornithology. http://bna.birds.cornell.edu/bna/species/48 (accessed January 25, 2010).
145. James, J. B., E. C. Hellgren, and R. E. Masters. 1999. Effects of deterrents on avian abundance and nesting density in electrical substations in Oklahoma. *J Wildl Manag* 63:1009.
146. Dolbeer, R. A., S. E. Wright, J. Weller, and M. J. Begier. 2009. Wildlife strikes to civil aircraft in the United States 1990–2008. Serial Report Number 15, Federal Aviation Administration National Wildlife Strike Database, Washington, DC.
147. Neidermyer, W. J., and J. J. Hickey. 1977. The monk parakeet in the United States, 1970–75. *Am Birds* 31:273.
148. Owre, O. T. 1973. A consideration of the exotic avifauna of southeastern Florida. *Wilson Bull* 85:491.
149. Spreyer, M. F., and E. H. Bucher. 1998. Monk parakeet (*Myiopsitta monachus*). In *The Birds of North America Online*, ed. A. Poole, Cornell Laboratory of Ornithology. http://bna.birds.cornell.edu/bna/species/322 (accessed January 26, 2010).
150. Tillman, E. A., A. Van Doorn, and M. L. Avery. 2001. Bird damage to tropical fruit in south Florida. *Wildl Damage Manag Conf* 9:47.
151. Eberhard, J. R. 1998. Breeding biology of the monk parakeet. *Wilson Bull* 110:463.
152. Avery, M. L., J. R. Lindsay, J. R. Newman, S. Pruett-Jones, and E. A. Tillman. 2006. Reducing monk parakeet impacts to electric utility facilities in south Florida. In *Advances in Vertebrate Pest Management*, vol. IV, ed. C. J. Feare and D. P. Cowan, 125. Furth, Germany: Filander Verlag.
153. Avery, M. L., C. A. Yoder, and E. A. Tillman. 2008. Diazacon inhibits reproduction in invasive monk parakeet populations. *J Wildl Manag* 72:1449.
154. Pruett-Jones, S., J. R. Newman, C. M. Newman, M. L. Avery, and J. R. Lindsay. 2007. Population viability analysis of monk parakeets in the United States and examination of alternative management strategies. *Hum Wildl Confl* 1:35.
155. Rodda, G. H., T. H. Fritts, and P. J. Conry. 1992. Origin and population growth of the brown tree snake, Boiga irregularis, on Guam. *Pac Sci* 46:46.
156. Savidge, J. A. 1987. Extinction of an island forest avifauna by an introduced snake. *Ecology* 68:660.
157. Rodda, G. H., T. H. Fritts, and D. Chiszar. 1997. The disappearance of Guam's wildlife; new insights for herpetology, evolutionary ecology, and conservation. *BioScience* 47:565.
158. Savidge, J. A. 1988. Food habits of *Boiga irregularis*, an introduced predator on Guam. *J Herpetol* 22:275.
159. Wiles, G. J. 1987. The status of fruit bats on Guam. *Pac Sci* 41:148.

160. Wiles, G. J., J. Bart, R. E. Beck Jr., C. F. Aguon. 2003. Impacts of the brown tree snake: Patterns and species persistence in Guam's avifauna. *Conserv Biol* 17:1350.

161. Fritts, T. H. 2002. Economic costs of electrical system instability and power outages caused by snakes on the island of Guam. *Int Biodeterior Biodegradation* 49:93.

162. Fritts, T. H., and M. J. McCoid. 1991. Predation by the brown treesnake on poultry and other domesticated animals in Guam. *Snake* 23:75.

163. Engeman, R. M., and D. S. Vice. 2001. Objectives and integrated approaches for the control of brown tree snakes. *Int Pest Manag Rev* 6:59.

164. Vice, D. S., and M. E. Pitzler. 2002. Brown tree snake control: Economy of scales. In *Human Wildlife Conflicts: Economic Considerations*, ed. L. Clark, J. Hone, J. A. Shivik, R. A. Watkins, K. C. VerCauteren, and J. K. Yoder, 127. Fort Collins, CO: USDA/APHIS/Wildlife Services, National Wildlife Research Center.

165. Vice, D. S., and D. L. Vice. 2004. Characteristics of brown tree snakes *Boiga irregularis* removed from Guam's transportation network. *Pac Conservation* 10:216.

166. Savarie, P. J., J. A. Shivik, G. C. White, and Clark, L. 2001. Use of acetaminophen for large scale control of brown tree snakes. *J Wildl Manag* 65:356.

167. Savarie, P. J., S. Wood, G. Rodda, R. L. Bruggers, and R. M. Engeman. 2005. Effectiveness of methyl bromide as a cargo fumigant for brown tree snakes (*Boiga irregularis*). *Int Biodeterior Biodegradation* 56:40.

168. Meshaka Jr., W. E., W. F. Loftus, and T. Steiner. 2000. The herpetofauna of Everglades National Park. *Fla Sci* 63:84.

169. Bilger, B. 2009. The Natural World, "Swamp Things". *The New Yorker*, April 20, 80.

170. Snow, R. W., K. L. Krysko, K. M. Enge, L. Oberhofer, A. Walker-Bradley, and L. Wilkins. 2007. Introduced populations of boa constrictor (Boidae) and *Python molurus bivittatus* (Pythonidae) in southern Florida. In *The Biology of Boas and Pythons*, ed. R. W. Henderson and R. Powell, 416. Eagle Mountain, UT: Eagle Mountain Publishing.

171. Barker, D. G., and T. M. Barker. 2008. Comments on a flawed herpetological paper and an improper and damaging news release from a government agency. *Bull Chic Herpetol Soc* 43:45.

172. Collins, T. M., B. Freeman, and S. Snow. 2008. Genetic characterization of populations of the nonindigenous Burmese python in Everglades National Park. Final Report for the South Florida Water Management District. Department of Biological Sciences, Florida International University, Miami.

173. Minton, S. A., and M. R. Minton. 1973. *Giant Reptiles*. New York: Charles Scribner's Sons.

174. Snow, R. W., M. L. Brien, M. S. Cherkiss, L. Wilkins, and F. J. Andazzotti. 2007. Dietary habits of the Burmese python, *Python molurus bivittatus*, in Everglades National Park, Florida. *Herpetol Bull* 101:5.

175. Greene, D. U., J. M. Potts, J. G. Duquesnel, and R. W. Snow. 2007. Geographic distribution: *Python molurus bivittatus* (Burmese python). *Herpetol Rev* 38:355.

176. Barker, D. G., and T. M. Barker. 2008. The distribution of the Burmese python, *Python molurus bivittatus*. *Bull Chic Herpetol Soc* 43:33.

177. Pope, C. H. 1961. *The Giant Snakes*. New York: Alfred A. Knopf.

178. Rodda, G. H., C. S. Jarnevich, and R. N. Reed. 2009. What parts of the U.S. mainland are climatically suitable for invasive alien pythons spreading from Everglades National Park? *Biol Invasions* 11:241.

179. Pyron, R. A., F. T. Burbrink, and T. J. Guiher. 2008. Claims of potential expansion throughout the U.S. by invasive python species are contradicted by ecological niche models. *PLoS ONE* 3:e293, http://www.plosone.org/article/info%3Adoi%2F10.1371%2Fjournal.pone.0002931 (accessed January 22, 2010).

180. Barker, D. G. 2008. Will they come out in the cold? Observations of large constrictors in cool and cold conditions. *Bull Chic Herpetol Soc* 43:93.

181. Smith, H. T., A. Sementelli, W. E. Meshaka Jr., and R. M. Engeman. 2007. Reptilian pathogens of the Florida everglades: The associated costs of Burmese pythons. *Endanger Species Update* 24:63.

182. Engeman, R., B. U. Constantin, S. Hardin, H. T. Smith, and W. E. Meshaka Jr., 2009. "Species pollution" in Florida: A cross-section of invasive vertebrate issues and management responses. In *Invasive Species: Detection, Impact and Control*, ed. C. P. Wilcox and R. B. Turpin, 179. Hauppauge, NY: Nova Science Publishers.

183. Mauldin, R. E., and P. J. Savarie. 2010. Acetaminophen as an oral toxicant for Nile monitor lizards (*Varanus niloticus*) and Burmese pythons (*Python molurus bivitattatus*). *Wildl Res* 37:215–222.

184. Kraus, F., E. W. Campbell, A. Allison, and T. Pratt. 1999. *Eleutherodactylus* frog introductions in Hawaii. *Herpetol Rev* 30:21.

185. Beard, K. H., E. A. Price, and W. C. Pitt. 2009. Biology and impacts of Pacific islands invasive species, 5. *Eleutherodactylus coqui*, the coqui frog (Anura: Leptodactylidae). *Pac Sci* 63:297.

186. Woolbright, L. L. 2005. A plot-based system of collecting population information on terrestrial breeding frogs. *Herpetol Rev* 36:139.

187. Townsend, D. S., M. M. Stewart, F. H. Pough, and P. H. Brussard. 1981. Internal fertilization in an oviparous frog. *Science* 212:469.

188. Townsend, D. S., and M. M. Stewart. 1994. Reproductive ecology of the Puerto Rican frog *Eleutherodactylus coqui*. *J Herpetol* 28:34.

189. Woolbright, L., A. H. Hara, C. M. Jacobsen, W. J. Mautz, and F. J. Benevides. 2006. Population densities of the coqui, *Eleutherodactylus coqui* (Anura: Leptodactlylidae) in newly invaded Hawaii and in native Puerto Rico. *J Herpetol* 40:122.

190. Beard, K. H., E. R. Al-Chokhachy, N. C. Tuttle, and E. M. O'Neil. 2008. Population density and growth rates of *Eleutherodactylus coqui* in Hawaii. *J Herpetol* 42:626.

191. Beard, K. H., and W. C. Pitt. 2005. Potential consequences of the coqui frog invasion in Hawaii. *Divers Distrib* 11:427.

192. Sin, H., K. H. Beard, and W. C. Pitt. 2008. An invasive frog, *Eleutherodactylus coqui*, increases new leaf production and leaf litter decomposition rates through nutrient cycling in Hawaii. *Biol Invasions* 10:335.

193. Kraus, F., and E. W. Campbell III. 2002. Human-mediated escalation of a formerly eradicable problem: The invasion of Caribbean frogs in the Hawaiian Islands. *Biol Invasions* 4:327.

194. Kaiser, B., and K. Burnett. 2006. Economic impacts of coqui frogs in Hawaii. *Interdiscip Environ Rev* 8:1.

195. Beard, K. H., and E. M. O'Neil. 2005. Infection of an invasive frog *Eleutherodactylus coqui* by the chytrid fungus *Batrachochytrium dendrobatidis* in Hawaii. *Biol Conserv* 126:591.

196. Carey, C., and L. Livo. 2008. To use or not to use the chytrid pathogen, *Batrachochytrium dendrobatidis*, to attempt to eradicate coqui frogs from Hawaii. In *1st International Conference on the Coqui Frog*, 7–9 February 2008, Hilo. Coqui Frog Working Group, Island of Hawaii and University of Hawaii, College of Tropical Agriculture and Human Resources. http://www .ctahr.hawaii.edu/coqui/WEBCCareyFICCF.pdf.pdf (accessed January 28, 2010).

197. Sin, H., and A. Radford. 2007. Coqui frog research and management efforts in Hawaii. In *Managing Vertebrate Invasive Species: Proceedings of an International Symposium*, ed. G. W. Witmer, W. C. Pitt, and K. A. Fagerstone, 157. Fort Collins, CO: USDA/APHIS/Wildlife Services, National Wildlife Research Center.

198. Hara, A. H., C. M. Jacobsen, S. R. Marr, and R. Y. Niino-DuPonte. 2010. Hot water as a potential disinfestation treatment for an invasive anuran the coqui frog, *Eleutherodactylus coqui* Thomas (Anura: Leptodactilidae), on plotted plants. *Int J Pest Manag.* 56:255.

199. Hickman, C. P. 1955. *Integrated Principles of Zoology*. St. Louis, MO: C. V. Mosby.

200. Great Lakes Fishery Commission. 2000. *Sea Lamprey—A Great Lakes Invader*. Fact Sheet 3. Ann Arbor, MI: Great Lakes Fishery Commission. http://www.glfc.org/pubs/FACT_3.pdf (accessed January 25, 2010).

201. Minnesota Sea Grant. 2009. *Sea Lamprey—The Battle Continues*. Duluth, MN: University of Minnesota. http://www.seagrant.umn.edu/ais/sealamprey_battle (last modified March 6, 2009; accessed January 25, 2010).

202. Great Lakes Fishery Commission. 2004. *Lampricides and Facts about Stream Treatments*. Fact Sheet 4a. Ann Arbor, MI: Great Lakes Fishery Commission. http://www.glfc.org/pubs/FACT_4a. pdf (accessed January 25, 2010).

203. Chick, J. H., and M. A. Pegg. 2001. Invasive carp in the Mississippi River Basin. *Science* 292:2250.

204. Columbia Environmental Research Center. 2004. *Facts about Bighead and Silver Carp*. Colombia, MO: U.S. Geologic Survey. http://www.cerc.usgs.gov/pubs/center/pdfDocs/Asian_carp-2-2004.pdf (accessed January 25, 2010).

205. Great Lakes National Program Office. 2009. *Asian Carp and the Great Lakes*. Chicago, IL: U.S. Environmental Protection Agency. http://www.epa.gov/glnpo/invasive/asiancarp/index.html (accessed January 25, 2010).

206. Conover, G., R. Simmonds, and M. Whalen, eds. 2007. Management and control plan for bighead, black, grass, and silver carps in the United States. Asian Carp Working Group, Aquatic Nuisance Species Task Force, Washington, DC.

207. Kendall, B. 2010. Supreme Court rejects bid to close waterway in carp case. *The Wall Street Journal*, January 19, 2010.

208. Pavlin, B. I., L. M. Schloegel, and P. Daszak. 2009. Risk of importing zoonotic diseases through wildlife trade, United States. *Emerg Infect Dis* 15:1721.

209. Lund, M. 1988. Rodent problems in Europe. In *Rodent Pest Management*, ed. I. Prakash, 29. Boca Raton, FL: CRC Press.

210. Prakash, I., and R. P. Mathur. 1988. Rodent problems in Asia. In *Rodent Pest Management*, ed. I. Prakash, 67. Boca Raton, FL: CRC Press.

211. Lund, M. 1994. Commensal rodents. In *Rodent Pests and Their Control*, A. P. Buckle and R. H. Smith, 23. Wallingford, CT: CAB International.

212. Fiedler, L. A. 1988. Rodent problems in Africa. In *Rodent Pest Management*, ed. I. Prakash, 35. Boca Raton, FL: CRC Press.

213. Jenkins, P. T. 2007. The failed regulatory system for animal imports into the United States and how to fix it. In *Managing Vertebrate Invasive Species: Proceedings of an International Symposium*, ed. G. W. Witmer, W. C. Pitt, and K. A. Fagerstone, 85. Fort Collins, CO: USDA/APHIS/Wildlife Services, National Wildlife Research Center.

214. Christy, M. T., A. A. Y. Adams, G. H. Rodda, J. A. Savidge, and C. L. Tyrell. 2010. Modelling detection probabilities to evaluate management and control tools for an invasive species. *J Appl Ecol* 47:106.

215. Engeman, R. M. 2005. A methodological and analytical paradigm for indexing animal populations applicable to many species and observation methods. *Wildl Res* 32:203.

216. Engeman, R. M., and D. A. Whisson. 2006. Using a general indexing paradigm to monitor rodent populations. *Int Biodeterior Biodegradation* 58:2.

217. Engeman, R. M., D. A. Whisson, J. Quinn, F. Cano, and T. White Jr. 2006. Monitoring invasive mammalian predator populations sharing habitat with the critically endangered Puerto Rican parrot *Amazona vittata*. *Oryx* 40:95.

218. Engeman, R. M. et al. 2006. Rapid assessment for a new invasive species threat: The case of the Gambian giant pouched rat in Florida. *Wildl Res* 33:439.

219. Whisson, D. A., R. M. Engeman, and K. Collins. 2005. Developing relative abundance techniques (RATS) for monitoring rodent populations. *Wildl Res* 32:239.

220. Hawley, N. B. 2007. Custom trucks, radio snake jingles, and temporary tattoos: An overview of a successful public awareness campaign related to brown treesnakes in the Commonwealth of the Northern Mariana Islands. In *Managing Vertebrate Invasive Species: Proceedings of an International Symposium*, ed. G. W. Witmer, W. C. Pitt, and K. A. Fagerstone, 53. Fort Collins, CO: USDA/APHIS/Wildlife Services, National Wildlife Research Center.

221. Rosenzweig, M. L. 1995. *Species Diversity in Space and Time*. Cambridge: Cambridge University Press.

222. Knopf, F. L. 1992. Faunal mixing, faunal integrity, and the biopolitical template for diversity conservation, *Trans N Am Wildl Nat Resour Conf* 57:330.

223. Pauly, D. 1995. Anecdotes and the shifting baseline syndrome of fisheries. *Trends Ecol Evol* 10:430.

224. Rosenzweig, M. L. 2003. *Win-Win Ecology—How the Earth's Species Can Survive in the Midst of Human Enterprise*. New York: Oxford University Press.

# chapter seventeen

# Environmental and economic costs associated with alien invasive species in the United States

*David Pimentel*

## Contents

## 17.1  Introduction

Approximately 50,000 alien invasive (nonnative) species have been introduced into the United States throughout its history. Introduced species, such as corn, wheat, rice, and other food crops, and cattle, poultry, and other livestock, now provide more than 98% of the U.S. food system at a value of approximately $800 billion per year.[1] (All monetary values in this chapter are expressed in US dollars.) Other exotic species have been used for landscape restoration, biological pest control, sport, pets, and food processing. Some nonindigenous species, however, have caused major economic losses in agriculture, forestry, and several other segments of the U.S. economy, in addition to harming the environment. Recent studies report $100–$200 billion in damages from the exotic species introduced to the United States.[2,3]

Estimating the full extent of environmental damage caused by exotic species and the number of indigenous species extinctions they have caused is difficult because little is known about the estimated 750,000 species in the United States—half of these species have not been described.[4] Nonetheless, about 400 of the 958 species listed as threatened or endangered under the Endangered Species Act are at risk primarily because of competition with and predation by nonindigenous species.[5] In other regions of the world, as many as 80% of the endangered species are threatened and at risk due to the pressures of nonnative species.[6] Many other species not listed are also negatively affected by alien species and/or ecosystem changes caused by alien species. Estimating the economic impacts associated with nonindigenous species in the United States is also difficult; nevertheless, enough data are available to quantify some of the impacts on agriculture, forestry, and public health. In this chapter, we assess, as much as possible, the magnitude of the environmental impacts and economic costs associated with the diverse invasive species that have become established within the United States. Although moving native species to other national locations where they did not exist previously can also have significant impacts, this assessment is limited to alien species that did not originate within the United States or its territories.

## 17.2    *Environmental damages and associated control costs*

Most plant and vertebrate animal introductions have been intentional, whereas most invertebrate animal and microbe introductions have been accidental. In the past 40 years, the rate of and the risk associated with biotic invaders have increased enormously because of human population growth, rapid movement of people, and alteration of the environment. In addition, more goods and materials are being traded among nations than ever before, thereby creating opportunities for unintentional introductions.[1,7]

Some of the approximately 50,000 species of plants and animals that have invaded the United States cause a wide array of damages to managed and natural ecosystems (Table 17.1). Some of these damages and control costs are assessed in Sections 17.2.1 through 17.2.7.

### 17.2.1    *Plants*

Most alien plants now established in the United States were introduced for food, fiber, and/or ornamental purposes. An estimated 5,000 plant species have escaped and now exist in U.S. natural ecosystems,[8] compared with a total of about 17,000 species of native U.S. plants.[9] In Florida, of the approximately 25,000 alien plant species imported mainly as ornamentals for cultivation, more than 900 have escaped and become established in surrounding natural ecosystems.[10–12] More than 3,000 plant species have been introduced into California.[13]

Most of the 5000 alien plants established in U.S. natural ecosystems have displaced one or more native plant species.[8] Alien weeds are spreading and invading approximately 700,000 ha/y of U.S. wildlife habitat.[14] One of these pest weeds is the European purple loosestrife (*Lythrum salicaria*), which was introduced in the early nineteenth century as an ornamental plant.[15] It has been spreading at a rate of 115,000 ha/y and is changing the basic structure of most of the wetlands it has invaded.[16] Competitive stands of purple loosestrife have reduced the biomass of 44 native plants and endangered wildlife, like the bog turtle and several duck species dependent on these native plants.[17] Loosestrife now occurs in 48 states and costs $45 million per year in control costs and forage losses.[18] There has been significant success in the biological control of purple loosestrife employing several insect species.[19]

Many introduced plant species established in the wild are having an effect on U.S. national parks.[20] In Great Smoky Mountains National Park, 400 of approximately

***Table 17.1*** Estimated Annual Costs Associated with Some Alien Species Introduction in the United States[a] (× Millions of Dollars)

| Category | Nonindigenous species | Losses and damages | Control costs | Losses + costs |
|---|---|---|---|---|
| Plants | 25,000 | | | |
| Purple loosestrife | | — | — | 45 |
| Aquatic weeds | | 10 | 100 | 110 |
| Melaleuca tree | | NA | 3–6 | 3–6 |
| Crop weeds | | 17,500 | 3000 | 20,500 |
| Weeds in pastures | | 1000 | 5000 | 6000 |
| Weeds in lawns, gardens, golf courses | | NA | 1500 | 1500 |
| Mammals | 20 | | | |
| Wild horses and burros | | 5 | NA | 5 |
| Feral pigs | | 1000 | 0.5 | 1000.5 |
| Mongooses | | 50 | NA | 50 |
| Rats | | 19,000 | NA | 19,000 |
| Cats | | 18,000 | NA | 18,000 |
| Dogs | | 425 | NA | 425 |
| Birds | 97 | | | |
| Pigeons | | 2200 | NA | 2200 |
| Starlings | | 800 | NA | 800 |
| Reptiles and amphibians | 53 | | | |
| Brown tree snake | | 2.6 | 13.6 | 16.2 |
| Fish | 150 | 5400 | NA | 5400 |
| Arthropods | 4500 | | | |
| Imported fire ant | | 1200 | 800 | 2000 |
| Formosan termite | | 1000 | NA | 1000 |
| Green crab | | 44 | NA | 44 |
| Gypsy moth | | NA | 11 | 11 |
| Crop pests | | 10,400 | 500 | 10,900 |
| Pests in lawns, gardens, golf courses | | NA | 1500 | 1500 |
| Forest pests | | 2100 | NA | 2100 |
| Mollusks | 88 | | | |
| Zebra mussel | | — | — | 2000 |
| Asian clam | | 1000 | NA | 1000 |
| Shipworm | | 205 | NA | 205 |
| Microbes | 20,000 | | | |
| Crop plant pathogens | | 18,000 | 400 | 18,400 |

(*Continued*)

*Table 17.1* Estimated Annual Costs Associated with Some Alien Species Introduction in the
United States[a] (× Millions of Dollars) (*Continued*)

| Category | Nonindigenous species | Losses and damages | Control costs | Losses + costs |
|---|---|---|---|---|
| Plant pathogens in lawns, gardens, golf courses | | NA | 2000 | 2000 |
| Forest plant pathogens | | 2100 | NA | 2100 |
| Dutch elm disease | | NA | 100 | 100 |
| Livestock diseases | | 9,000 | NA | 9,000 |
| Human diseases | | NA | 91,631 | 91,631 |
| **TOTAL** | | | | **$219,049** |

[a] See text discussion for details and sources.

1500 vascular plant species are exotic, and 35 of these are currently displacing and threatening other species in the park.[20,21] Hawaii has a total of 2690 plant species, 946 of which are alien species.[22] About 800 native species are endangered, and more than 218 endemic species are believed to be extinct because of alien species.[23]

Sometimes one nonindigenous plant species competitively overruns an entire ecosystem. For example, in California, yellow starthistle (*Centaurea solstitalis*) now dominates more than 4 million ha of northern California grassland, resulting in the total loss of this once productive grassland.[24] European cheatgrass (*Bromus tectorum*) is dramatically changing the vegetation and fauna of many natural ecosystems. This annual grass has invaded and spread throughout the shrub-steppe habitat of the Great Basin in Idaho and Utah, predisposing the invaded habitat to fires.[25-27] Before the invasion of cheatgrass, fire burned once every 60–110 years and shrubs had a chance to become well established. Now, fires occur about every 3–5 years; shrubs and other vegetations are diminished, and competitive monocultures of cheatgrass now exist on 5 million ha in Idaho and Utah.[28] The animals dependent on the shrubs and other original vegetations have been reduced or eliminated.

An estimated 138 alien tree and shrub species have invaded native U.S. forest and shrub ecosystems.[29] Introduced trees include salt cedar (*Tamarix pendantra*), eucalyptus (*Eucalyptus* spp.), Brazilian pepper (*Schinus terebinthifolius*), and Australian melaleuca (*Melaleuca quinquenervia*).[30-32] Some of these trees have displaced native trees, shrubs, and other vegetation types, and populations of some associated native animal species have been reduced in turn.[30] For example, the melaleuca tree is competitively spreading at a rate of 11,000 ha/y throughout the vast forest and grassland ecosystems of the Florida Everglades,[24] where it damages the natural vegetation and wildlife.[30]

Exotic aquatic weeds in the Hudson River basin of New York number 53 species.[33] In Florida, exotic aquatic plants, such as hydrilla (*Hydrilla verticillata*), water hyacinth (*Eichhornia crassipes*), and water lettuce (*Pistia stratiotes*), are altering fish and other aquatic animal species, choking waterways, altering nutrient cycles, and reducing recreational use of rivers and lakes. Active control measures of these aquatic weeds have become necessary. For instance, Florida spends about $20 million each year on hydrilla control.[34] Despite this large expenditure, hydrilla infestations in just two Florida lakes have prevented their recreational use, causing $10 million annually in losses.[35] In the United States, a total of $100 million is invested annually in alien species aquatic weed control.[30]

## 17.2.2 Mammals

About 28 species of mammals have been introduced into the United States; these include dogs, cats, horses, burros, cattle, sheep, pigs, goats, and deer.[36] Several of these species have escaped or were released into the wild; many have become pests by preying on native animals, grazing on vegetation, or intensifying soil erosion. For example, goats (*Capra hircus*) introduced on San Clemente Island, California, are responsible for the extinction of eight endemic plant species and the endangerment of eight other native plant species.[25]

A number of small mammals have also been introduced into the United States. These species include a number of rodents, such as the European black or tree rat (*Rattus rattus*), Asiatic Norway or brown rat (*Rattus norvegicus*), house mouse (*Mus musculus*), and European rabbit (*Oryctolagus cuniculus*), as well as the domestic cat (*Felis catus*) and dog (*Canis familiaris*).[36]

Some introduced rodents have become serious pests on farms, in industries, and in homes.[36] On farms, rats and mice are particularly abundant and destructive. On poultry farms, there is approximately one rat per five chickens[37] (D. Pimentel, unpublished data, 1951). Using this ratio, the total rat population on U.S. poultry farms may easily number more than 1.4 billion.[38] Assuming that the number of rats per chicken has declined because of improved rat control since these observations were made, we estimate that the number of rats on poultry and other farms is approximately 1 billion. With an estimated one rat per person in the United States,[39] there are an estimated 280 million rats in U.S. urban and suburban areas.[1]

If we assume, conservatively, that each adult rat consumes and/or destroys stored grains[40,41] and other materials valued at $15 per year, then the total cost of destruction by introduced rats in the United States is more than $19 billion per year. In addition, rats cause fires by gnawing electric wires, pollute foodstuffs, and act as vectors of several diseases, including salmonellosis and leptospirosis, and, to a lesser degree, plague and murine typhus.[42] They also prey on some native invertebrate and vertebrate species like birds and bird eggs.[43]

The Indian mongoose (*Herpestes auropunctatus*) was first introduced into Jamaica in 1872 for biological control of rats in sugarcane.[44] It was subsequently introduced to Puerto Rico, other West Indian Islands, and Hawaii. The mongoose controlled the Asiatic rat but not the European rat and preyed heavily on native ground nesting birds.[44,45] It also preyed on beneficial native amphibians and reptiles, causing a minimum of 7–12 amphibian and reptile extinctions in Puerto Rico and other islands of the West Indies.[46] In addition, the mongoose emerged as the major vector and reservoir of rabies and leptospirosis in Puerto Rico and other islands.[47] Based on public health damages, killing of poultry, extinctions of amphibians and reptiles, and destruction of native birds, we estimate that the mongoose is causing approximately $50 million in damages each year in Puerto Rico and the Hawaiian Islands (D. Pimentel, unpublished data; David Foote, pers. comm., Hawaii Volcanoes National Park, 1998).

There are an estimated 88 million pet cats in the United States,[48,49] plus 50–60 million feral cats.[48,49] Cats prey on native birds[50] as well as on small native mammals, amphibians, and reptiles.[51] Estimates are that feral cats in Wisconsin and Virginia kill more than 3 million birds in each state per year.[52] Assuming eight birds killed per feral cat per year and about 60 million feral cats,[49] 480 million birds are killed per year in the nation. Each adult bird is valued at $30; this cost per bird is based on the literature that reports that a bird watcher spends $0.40 per bird observed, a hunter spends $216 per bird shot, and specialists spend $800 per bird reared for release; in addition, note that the

U.S. Environmental Protection Agency fines polluters $10 per fish killed, including small, immature fish.[53] Therefore, the total damage to U.S. bird population is approximately $18 billion per year. This cost does not include the number of birds killed by pet or urban cats, a figure reported to be similar to the number killed by feral cats,[49] nor does the cost include the many small mammals, amphibians, and reptiles that are killed by feral and pet cats.[51]

Most dogs introduced into the United States were introduced for domestic purposes, but some have escaped into the wild. There are 75 million pet dogs and 30 million feral dogs.[48] Some of these wild dogs run in packs and kill deer, rabbits, and domestic cattle, sheep, and goats. Carter[54] reported that feral dog packs in Texas cause more than $5 million in livestock losses each year. Dog packs have also become a serious problem in Florida.[55] In addition to the damages caused by dogs in Texas, and assuming $5 million for all damages for the other 49 states combined, total losses in livestock kills by dogs per year would be approximately $10 million per year.

There are 15–20 deadly dog attacks per year in the United States.[56] An estimated 4.7 million people are bitten by feral and pet dogs annually, with 800,000 cases requiring medical treatment.[57] Centers for Disease Control estimates medical treatment for dog bites costs $165 million per year, and the indirect costs, such as lost work, increase the total costs of dog bites to $250 million per year.[58,59] Most dog attack victims are small children.[60]

## 17.2.3   Birds

About 97 of the 1000 bird species in the United States are exotic.[61,62] Of the 97 introduced bird species, only 5% are considered beneficial, while most (56%) are pests.[61] However, several species, including chickens and pigeons, were introduced into the United States for agricultural purposes.

In Hawaii, 35 of 69 alien birds introduced between 1850 and 1984 are still extant on the islands.[63-65] The common myna (*Acridotheres tristis*), introduced into Hawaii, helped control pest cutworms and armyworms in sugarcane.[25] However, it became the major disperser of seeds of the introduced pest weed *Lantana camara*. To cope with the weed problem, Hawaii resorted to the introduction of insects as biocontrol agents.[25]

The English or house sparrow (*Passer domesticus*) was introduced into the United States intentionally in 1853 to control the cankerworm.[66,67] By 1900, the birds were considered pests because they damage plants around homes and public buildings and consume wheat, corn, and the buds of fruit trees.[66] Furthermore, English sparrows harass robins, Baltimore orioles, yellow-billed cuckoos, and black-billed cuckoos, and displace native bluebirds, wrens, purple martins, and cliff swallows.[66-68] They are also associated with the spread of about 29 diseases of humans and livestock.[69]

The exotic common pigeon (*Columba livia*) exists in most cities of the world, including those in the United States.[70] Pigeons are considered a nuisance because they foul buildings, statues, cars, and sometimes people, and feed on grain.[68,71] The control costs of pigeons are at least $9 per pigeon per year.[72] Assuming there are two pigeons per hectare in urban areas[73] or approximately one pigeon per person in urban areas, and using potential control costs as a surrogate for losses, pigeons cause an estimated $2.2 billion per year in damages. These damage costs do not include the environmental damages associated with their serving as reservoirs and vectors for over 50 diseases, including parrot fever, ornithosis, histoplasmosis and encephalitis.[68,69]

## 17.2.4 Amphibians and reptiles

Amphibians and reptiles introduced into the United States number about 53 species. All these alien species occur in states where it seldom freezes; Florida is now host to 30 species and Hawaii to 12.[74,75] The negative ecological impacts of a few of these exotic species have been enormous.

The brown tree snake (*Boiga irregularis*) was accidentally introduced to snake-free Guam immediately after the Second World War when military equipment was moved onto Guam.[76] Soon the snake population reached densities of 100 per hectare and dramatically reduced native bird, mammal, and lizard populations. Of the 13 species of native forest birds originally found on Guam, only 3 species still exist;[77] of the 12 native species of lizards, only 3 retain the possibility of surviving.[77] The snake eats chickens, eggs, and caged birds, which causes major problems to small farmers and pet owners. This snake crawls up trees and utility poles causing power outages on the island. One island-wide power outage caused by the snake cost the power utility more than $6 million. Local outages that affect business are estimated to cost about $10,000 per commercial customer.[78,79] With about 133 outages per year,[80] our estimate of the cost of snake-related power outages is conservatively $2.6 million per year.

In addition, the brown tree snake is slightly venomous and has caused public health problems, especially when it has bitten children. At one hospital emergency room, about 26 people per year are treated for snake bites.[30] Some bitten infants require hospitalization and intensive care; this is estimated to cost $25,000 per year (T. Fritts, U.S. Geological Survey, 1998, pers. comm.). The total costs of endangered species recovery efforts, environmental planning related to snake containment on Guam, and other programs directly stemming from the snake's invasion of Guam constitute costs in excess of an additional $1 million per year; in addition, up to $2 million per year is invested in research and control of this serious pest (1998 federal budget, T. Fritts, U.S. Geological Survey, 1998, pers. comm.). Hawaii's concern about the snake has prompted the federal government to invest $1.6 million per year in brown tree snake control.[81] The total cost associated with the snake is more than $16.2 billion per year (Table 17.1).

## 17.2.5 Fish

A total of 150 alien fish species has been introduced into the United States.[82–84] Most of these introduced fish species have been established in states with mild climates, like Florida (50 species)[85] and California (56 species).[86] In Hawaii, 33 alien freshwater fish species have become established.[87] Forty-four native species of fish are threatened or endangered by alien invasive fish species.[88] An additional 27 native fish species are also negatively affected by introductions.[88]

Introduced fish species frequently alter the ecology of aquatic ecosystems. For instance, the grass carp (*Ctenopharyngodon idella*) reduces natural aquatic vegetation, while the common carp (*Cyprinus carpio*) reduces water quality by increasing turbidity. These changes have caused the extinctions of some native fish species.[84,89]

Although some native fish species are reduced in numbers, forced to extinction, or hybridized by alien fish species, alien fish do provide some benefits in the improvement of sportfishing. Sportfishing contributes $69 billion to the economy of the United States.[90,91] However, based on more than 40 alien invasive species that have negatively affected native fishes and other aquatic biota, and considering the fact that sports fishing is valued at

$69 billion per year,[91] the conservative economic losses due to exotic fish is $5.4 billion per year (D. Pimentel, unpublished data).

## 17.2.6    Arthropods and other invertebrates

Approximately 4500 arthropod species (2582 species in Hawaii and more than 2000 in the continental United States) have been introduced. Also, 11 earthworm species[92] and nearly 100 aquatic invertebrate species have been introduced.[30] More than 95% of these introductions were accidental, with many species gaining entrance through plants or through soil and water ballast from ships.

The introduced balsam woolly adelgid (*Adelges piceae*) inflicts severe damage in balsam-fir natural forest ecosystems.[93] According to Todd Wilkinson,[94] aphids, fungi, and rust are destroying old-growth spruce-fir, dogwoods, and whitebark pine trees in many regions. Over about a 20-year period, the pests have spread throughout the southern Appalachians and destroyed most of the trees mentioned. Alsop and Laughlin[95] report the loss of two native bird species and the invasion by three other species as a result of adelgid-mediated forest death.

Other introduced insect species have become pests of livestock and wildlife. For example, the red imported fire ant (*Solenopsis invicta*) kills poultry chicks, lizards, snakes, and ground nesting birds.[96] A 34% decrease in swallow nesting success and a decline in the northern bobwhite quail populations was reported due to these ants.[97] The estimated damage to livestock, wildlife, and public health caused by fire ants in Texas is estimated to be $300 million per year. Two people were killed by fire ants in Mississippi in 2002. An additional $200 million is invested in control per year.[98,99] Assuming equal damages in other infested southern states, the fire ant damages total approximately $2 billion per year.[100] The Formosan termite (*Coptotermes formosanus*) is reported to cause structural damages of approximately $1 billion per year in southern United States, especially in the New Orleans region.[101]

The European green crab (*Carcinus maenas*) has been associated with the demise of the soft-shell clam industry in New England and Nova Scotia.[102] It also destroys commercial shellfish beds and preys on large numbers of native oysters and crabs.[102] The annual estimated economic impact of the green crab is $44 million per year.[102]

## 17.2.7    Mollusks

Eighty-eight species of mollusks have been introduced and established in U.S. aquatic ecosystems.[30] Three of the most serious pests are the zebra mussel (*Dreissena polymorpha*), Asian clam (*Corbicula fluminea*), and quagga mussels (*Dreissena bugensis*).[103]

The European zebra mussel was first found in Lake St. Clair after gaining entrance via ballast water released in the Great Lakes from ships that had traveled over from Europe.[104] The zebra mussel has spread into most of the aquatic ecosystems in the eastern United States and is expected to invade most freshwater habitats throughout the nation.[104]

Another related mussel species, the quagga mussel, is spreading rapidly and, in some cases, displacing zebra mussels (C. A. Stepien, Cleveland State University, 2003, pers. comm.). Large mussel populations reduce food and oxygen for native fauna. In addition, zebra mussels have been observed completely covering native mussels, clams, and snails, thereby further threatening their survival.[104,105] Mussel densities have been recorded as high as 700,000/m$^2$.[106] Zebra and quagga mussels also invade and clog water intake pipes, water filtration, and electricity generating plants; it is estimated that they cause $2 billion per year in damages and associated control costs per year.[107]

Though the Asian clam (*Corbicula fluminea*) grows and disperses less rapidly than the zebra mussel, it too causes significant fouling problems and threatens native species. Costs associated with its damage are about $1 billion per year.[30,108]

The introduced shipworm (*Teredo navalis*) in the San Francisco Bay has also caused serious damage since the early 1990s and also on the East Coast.[109] Currently, damages are estimated to be about $205 million per year on the West Coast and about $200 million on the East Coast.[109]

## 17.3    Crop, pasture, and forest losses and associated control costs

Many weeds, pest insects, and plant pathogens are biological invaders causing several billion dollars in losses to crops, pastures, and forests annually in the United States. In addition, several billion dollars are spent on pest control.

### 17.3.1    Weeds

In crop systems, including forage crops, an estimated 500 introduced plant species have become weed pests; many of these were actually introduced as crops and then became pests.[110] Most of these weeds were accidentally introduced with crop seeds from ship-ballast soil or from various imported plant materials; among them were yellow rocket (*Barbarea vulgaris*) and Canada thistle (*Cirsium arvense*).

In U.S. agriculture, weeds cause a reduction of 12% in crop yields. In economic terms, this represents about $24 billion in lost crop production annually, based on the crop potential value of all U.S. crops, more than $200 billion per year.[1] Based on the estimate that about 73% of the weeds are alien,[111] it follows that about $17.5 billion of these crop losses are due to introduced weeds. Note, alien invasive weeds are more serious pests than native weeds; thus, $17.5 billion is likely to be a conservative estimate. In addition, approximately $4 billion in herbicides are applied to U.S. crops,[112] of which about $3 billion is used for control of alien invasive weeds. Therefore, the total cost of introduced weeds to the U.S. economy is about $20.5 billion per year.

In pastures, 45% of weeds are alien species.[111] U.S. pastures provide about $10 billion in forage crops annually,[113] and the estimated losses due to weeds is approximately $2 billion.[114] Since about 45% of the weeds are alien invasives,[111] the forage losses due to these nonindigenous weeds are nearly $1 billion per year.

Some introduced weeds are toxic to cattle and wild ungulates, like leafy spurge (*Euphorbia esula*).[115] In addition, several alien thistles have replaced desirable native plant species in pastures, rangelands, and forests, thus reducing cattle grazing.[116] According to former Interior Secretary Bruce Babbitt,[14] ranchers spend about $5 billion each year to control invasive alien weeds in pastures and rangelands, yet these weeds continue to spread.

Management of weed species in lawns, gardens, and golf courses is a significant proportion of their total management costs of about $36 billion per year.[1] In addition, Templeton et al.[117] estimated that about $1.3 billion of the $36 billion is spent just on residential weed, insect, and disease pest control each year. Because a large proportion of these weeds are exotics, we estimate that $500 million is spent on residential exotic weed control, and an additional $1 billion is invested in alien invasive weed control on golf courses. Weed trees also have an economic impact, from $3 to $6 million per year is being spent in efforts to control only the melaleuca tree in Florida (Curtis J. Richardson, Duke University, 1998, pers. comm.).

## 17.3.2   Vertebrate pests

Horses (*Equus caballus*) and burros (*Equus asinus*) released in the western United States have attained wild populations of approximately 50,000 animals.[118] These animals graze heavily on native vegetation, allowing alien invasive annuals to displace native perennials.[119] Furthermore, burros inhabiting the northwestern United States diminish the primary food sources of native bighorn sheep and seed-eating birds, thereby reducing the abundance of these native animals.[25] In general, the large populations of introduced wild horses and burros cost the nation an estimated $5 million per year in forage losses.[120]

Feral pigs (*Sus scrofa*), native to Eurasia and North Africa, have been introduced into some U.S. parks for hunting, including the California coastal prairie and Hawaii, and have substantially changed the vegetation in these parks.[121] In Hawaii, more than 80% of the soil is bare in regions inhabited by pigs.[25] This disturbance allows annual plants to invade the overturned soil and intensifies soil erosion. Pig control per park in Hawaii (~1500 pigs per park)[122] costs about $150,000 per year. Assuming that the three parks in Hawaii have similar pig control problems, the total cost is $450,000 per year (R. Zuniga, Cornell University, 1999, pers. comm.).

Feral pigs have also become a serious problem in Florida, where their population has risen to more than 500,000[55]; similarly, in Texas, their number ranges from 1 to 1.5 million (J. P. Bach, Texas A & M University, 1999, pers. comm.). In Florida, Texas, and elsewhere, pigs damage grain, peanut, soybean, cotton, hay, and various vegetable crops, and the environment.[123] Pigs also transmit and are reservoirs for serious diseases of humans and livestock, like brucellosis, pseudobrucellosis, and trichinosis.[124]

Nationwide, there are an estimated 4 million feral pigs. Based on environmental and crop damages of about $200 per pig (one pig can cause up to $1000 of damages in one night; J. P. Bach, Texas A & M University 1999, pers. comm.), and assuming 4 million feral pigs inhabit the United States, the yearly cost is about $1 billion per year. This estimate is conservative because pigs cause significant environmental damages and diseases that cannot be easily translated into dollar values.

European starlings (*Sturnus vulgaris*) are serious pests and are estimated to occur at densities of more than 1 per hectare in agricultural regions.[125] Starlings are capable of destroying approximately $2000 per hectare worth of cherries.[126] In grain fields, starlings consume about $6 per hectare of grain.[126] Conservatively, assuming $5 per hectare for all damages to agriculture crop production in the United States, the total loss due to starlings would be approximately $800 million per year. In addition, these aggressive birds have displaced numerous native birds, like bluebirds, woodpeckers, wood ducks, purple martins, and others.[127] Starlings have also been implicated in the transmission of 25 diseases, including parrot fever and other diseases of humans.[66,69]

## 17.3.3   Insect and mite pests

Approximately 500 alien insect and mite species are pests in crops. Hawaii has 5246 identified native insect species and an additional 2582 introduced insect species.[10,22,128] Introduced insects account for 98% of the pest insects in the state.[129] In addition to Florida's 11,500 native insect species, 949 introduced species have invaded the state (42 species were introduced for biological control).[130] In California, the 600 introduced species are responsible for 67% of all crop losses.[13]

Each year, pest insects destroy about 13% of potential crop production representing a value of about $26 billion in U.S. crops.[1] Considering that about 40% of the pests were introduced,[111] we estimate that these pests cause about $10.4 billion in crop losses each year. In addition, about $1.2 billion in pesticides are applied for all insect control each year.[112] The portion applied against introduced pest insects is approximately $500 million per year. Therefore, the total cost for introduced invasive insect pests is approximately $10.9 billion per year. In addition, based on the earlier discussion of management costs of lawns, gardens, and golf courses, we estimate the control costs of pest insects and mites in lawns, gardens, and golf courses to be at least $1.5 billion per year.

About 360 alien insect species have become established in American forests.[131,132] Approximately 30% of these are now serious pests. Insects cause the loss of approximately 9% of forest products, amounting to a cost of $7 billion per year.[1,133] Because 30% of the pests are alien pests, annual losses attributed to alien invasive species is about $2.1 billion per year.

The emerald ash borer was introduced into the United States in about 2000 and is having a devastating impact on all types of the 7–10 billion ash trees in the United States. White-ash wood is favored for producing wood bats in American baseball. The beetle probably came from China in wood packing material. The total value of trees, including wood for bats and other uses, probably total $1 billion in the United States. There is no known control of the beetle, which girdles the tree and eventually causes the death of the ash tree. The spread of the beetle occurs by beetle flight, but the rapid spread is by humans distributing firewood long distances.

There is no known effective control. Woodpeckers feed on the larvae and pupae in the ash tree, but their effect is relatively minor. Insecticides do not work because the beetle larvae bore into the wood. Eventually, when the ash tree density is reduced to about one ash tree per square mile the beetle population will be reduced and the remaining ash trees will probably survive at a very low density along with a very low density of the emerald ash borer population.

The gypsy moth (*Lymantria dispar*), intentionally introduced into Massachusetts in the 1800s, has developed into a major pest of U.S. forests and ornamental trees, especially oaks.[134] The U.S. Forest Service currently spends about $11 million annually on gypsy moth control.[134,135]

## 17.3.4 Plant pathogens

There are an estimated 50,000 parasitic and nonparasite diseases of plants in the United States and most of these are fungi species.[136] In addition, there are more than 1300 species of viruses that are plant pests in the United States.[136] Many of these microbes were introduced inadvertently with seeds and other parts of host plants and have become major crop pests in the United States.[111] Including the introduced plant pathogens plus other microbes, we estimate that more than 20,000 species of microbes have invaded the United States.

U.S. crop losses to all plant pathogens total approximately $16 billion per year.[1,112] Approximately 65%,[111] or an estimated $10.4 billion per year of losses are attributable to alien plant pathogens. In addition, $0.72 billion is spent annually for fungicides,[112] with approximately $0.6 billion for the control of alien plant pathogens. This brings the costs of damage and control of alien invasive plant pathogens to about $11 billion per year. In addition, based on the earlier discussion of pests in lawns, gardens, and golf courses, we estimate the control costs of plant pathogens in lawns, gardens, and golf courses to be at least $2 billion per year.

In forests, more than 20 alien species of plant pathogens attack woody plants.[131,133] Two of the most serious plant pathogens are the chestnut blight fungus (*Cryphonectria parasitica*) and Dutch elm disease (*Ophiostoma ulmi*). Before the introduction of chestnut blight, approximately 25% of the eastern U.S. deciduous forest consisted of American chestnut trees.[24] Elm tree removal costs about $100 million per year.[134]

Approximately 9%, or $7 billion, of forest products are lost each year due to plant pathogens.[91,133] Assuming that the proportion of introduced plant pathogens in forests is similar to that of introduced insects (about 30%), approximately $2.1 billion in forest products are lost each year to alien invasive plant pathogens in the United States.

### 17.3.5   Livestock pests

Similar to crops, exotic microbes (e.g., calf-diarrhea-rotavirus) and parasites (e.g., face flies, *Musca autumnalis*) were introduced when livestock were brought to the United States.[137,138] In addition to the hundreds of microbes and parasites that have already been introduced and are pests, there are more than 60 microbes and parasites that could invade and become serious pests to U.S. livestock.[139] A conservative estimate of the losses to U.S. livestock from exotic microbes and parasites is approximately $9 billion per year (Kelsey Hart, College of Veterinary Medicine, Cornell University, 2001, pers. comm.).

### 17.3.6   Human diseases

The alien diseases now having the greatest impact are acquired immune deficiency syndrome (AIDS), syphilis, and influenza.[120,140] In 1993, there were 103,533 cases of AIDS with 37,267 deaths.[141] The total federal funding of health care costs for the treatment of AIDS had reached about $11 billion in 2004.[142]

New influenza strains, originating in the Far East, quickly spread to the United States. These are reported to cause 5%–6% of all deaths in 121 U.S. cities.[143,144] Costs of influenza in the United States can exceed $80 billion per year.[145] In addition, each year there are approximately 53,000 cases of syphilis; to treat only newborn children infected with syphilis costs the nation $18.4 million per year.[146]

In total, AIDS, influenza, and syphilis take the lives of more than 40,000 people each year in the United States, and treatment costs for these diseases and syphilis total over $90 billion per year and does not include the other exotic diseases. Also, West Nile virus is a new invading disease of humans, birds, horses, and other animals. Approximately 4200 humans were infected by the West Nile virus in 2003, with 284 deaths[147]; the estimated public health costs are $631 million per year (Table 17.1). An increasing threat of exotic diseases exists because of rapid transportation, encroachment of civilization into new ecosystems, and growing environmental degradation.

## 17.4   Conclusion

With more than 50,000 alien invasive species in the United States, the fraction that is harmful does not have to be large to inflict significant damage to natural and managed ecosystems and cause public health problems. There is a suite of ecological factors that may cause alien invasive species to become abundant and persistent. These include the lack of controlling natural enemies (e.g., purple loosestrife and imported fire ant), the development of new associations between alien parasite and host (e.g., the AIDS virus in humans and the gypsy moth in U.S. oaks), effective predators in a new ecosystem (e.g., brown tree snake and feral cats), artificial or disturbed habitats that provide favorable invasive ecosystems for the aliens (e.g., weeds in crop and lawn habitats), and invasion by some highly adaptable and successful alien species (e.g., water hyacinth, and zebra and quagga mussels).

Our study reveals that economic damages associated with alien invasive species effects and their control amount to approximately $219 billion per year. The Office of Technology Assessment (OTA)[30] reported costs of $1.1 billion per year ($97 billion averaged over 85 years) for 79 species. The reason for our higher estimate is that we included more than 10 times the number of species in our assessment and found higher costs reported in the literature than OTA[30] for some of the same species. For example, for the zebra mussel, OTA reported damages and control costs of slightly more than $300,000 per year; we used an estimate of $1 billion per year.[107]

Although we reported specific total economic damages and associated control costs, precise economic costs associated with some of the most ecologically damaging exotic species are not available. The brown tree snake, for example, has been responsible for the extinction of dozens of bird and lizard species on Guam. Yet for this snake, only minimal cost data are known. In other cases, such as the zebra mussel and feral pigs, only combined damage and control cost data are available. These are considered low when compared with the extensive environmental damages these species cause and do not include indirect costs. If we had been able to assign monetary values to species extinctions and losses in biodiversity, ecosystem services, and aesthetics, the costs of destructive alien invasive species would undoubtedly be several times higher than $219 billion per year. Yet even this understated economic loss indicates that alien invasive species are exacting a significant toll.

We recognize that nearly all our crop and livestock species are aliens and have proven essential to the viability of our agriculture and economy. Although certain alien crops (e.g., corn and wheat) are vital to agriculture and the U.S. food system, this does not diminish the enormous negative impacts of other nonindigenous species (e.g., mussels and exotic weeds).

The true challenge lies not in determining the precise costs of the impacts of exotic species but in preventing further damage to natural and managed ecosystems. Formulation of sound prevention policies needs to take into account the means through which alien species gain access to and become established in the United States. Since the invasions vary widely, we should expect that a variety of strategies would be needed for prevention programs. For example, public education, sanitation, and effective prevention programs at airports, seaports, and other ports of entry will help reduce the chances for biological invaders becoming established in the United States.

Fortunately, the problem is gaining the attention of policy makers. On February 2, 1999, President Clinton issued an executive order allocating $28 million and creating an interagency Invasive Species Council to produce a plan within 18 months to mobilize the federal government to defend against alien species invasions. In addition, a Federal Interagency Weed Committee has been formed to help combat nonindigenous plant species invasions.[148] The objective of this interagency committee is education, formation of partnerships among concerned groups, and stimulation of research on the biological invader problem. Former Secretary Bruce Babbitt[149] also established an Invasive Weed Awareness Coalition to combat the invasion and spread of nonnative plants.

While these policies and practices may help prevent accidental and intentional introduction of potentially harmful exotic species, we have a long way to go before the resources devoted to the problem are in proportion to the risks. We hope that this environmental and economic assessment will advance the argument that investments made now to prevent future introductions will be returned many times over in the preservation of natural ecosystems, diminished losses to agriculture and forestry, and lessened threats to public health.

## Acknowledgments

We wish to express our sincere gratitude to the Cornell Association of Professors Emeriti for the partial support of our research through the Albert Podell Grant Program. We thank the following people for reading an earlier draft of this article and for their many helpful suggestions: D. Bear, Council on Environmental Quality, Executive Office of the President, Washington, DC; J. W. Beardsley, University of Hawaii; A. J. Benson, U.S. Geological Survey, Gainesville, FL; B. Blossey, Cornell University; C. R. Bomar, University of Wisconsin; F. T. Campbell, Western Ancient Forest Campaign, Springfield, VA; R. Chasen, Editor, *BioScience*; P. Cloues, Geologic Resources Division, Natural Resource Program Center, Lakewood, CO; W. R. Courtenay, Florida Atlantic University; R. H. Cowie, Bishop Museum, Honolulu, HI; D. Decker, Cornell University; R. V. Dowell, California Department of Food and Agriculture; T. Dudley, University of California, Berkeley; H. Fraleigh, Colorado State University; H. Frank, University of Florida; T. Fritts, U.S. Geological Survey, Washington, DC; E. Groshoz, University of New Hampshire; J. Jenkins, Forest Service, USDA, Radnor, PA; J. N. Layne, Archbold Biological Station, Lake Placid, FL; J. Lockwood, University of Tennessee; J. D. Madsen, U.S. Army Corps of Engineers, Vicksburg, MS; R. A. Malecki, N. Y. Cooperative Fish and Wildlife Research Unit, Ithaca, NY; E. L. Mills, Cornell University; S. F. Nates, University of Southwestern Louisiana; H. S. Neufeld, Appalachian State University; P. J. O'Connor, Colorado State University; B. E. Olson, Montana State University; E. F. Pauley, Coastal Carolina University; M. Pimentel, Cornell University; S. Pimm, University of Tennessee; W. J. Poly, Southern Illinois University; W. Roberts, Rainbow Beach, Australia; M. Sagoff, Institute for Philosophy and Public Policy, University of Maryland, College Park; B. Salter, Maryland Department of Natural Resources; D. L. Scarnecchia, University of Idaho; D. Simberloff, University of Tennessee; G. S. Rodrigues, Empresa Brasileira de Pesquisa Agropecuaria, Brazil; J. N. Stuart, University of New Mexico; S. B. Vinson, Texas A & M University; L. A. Wainger, University of Maryland; J. K. Wetterer, Columbia University; and C. E. Williams, Clarion University of Pennsylvania.

## References

1. USBC. 2009. *Statistical Abstract of the United States 2009*. Washington, DC: U.S. Census Bureau, U.S. Government Printing Office.
2. National Aeronautics and Space Administration. 2003. Invasive species management: 2003–2007. http://aiwg.gsfc.nasa.gov/esappdocs/oldpplans/2003/IV2003ProgramPlan.doc (accessed January 7, 2010).
3. National Invasive Species Council. 2004. FY 2004 interagency invasive species performance budget. http://www.docstoc.com/docs/4141379/National-Invasive-Species-Council-fy-interagency-invasive-species-performance (accessed January 21, 2010).
4. Raven, P. H., and G. B. Johnson. 1992. *Biology*. 3rd ed. St. Louis: Mosby Year Book.
5. Wilcove, D. S., D. Rothstein, J. Bubow et al. 1998. Quantifying threats to imperiled species in the United States. *BioScience* 48:607.
6. Armstrong, S. 1995. Rare plants protect Cape's water supplies. *New Sci* 1964:8.
7. Bryan, R. T. 1996. Alien species and emerging infectious diseases: Past lessons and future applications. In *Proceedings of the Norway/UN Conference on Alien Species*, ed. G. T. Sandlund, P. J. Schel, and A. Viken, 74. Trondheim: Norwegian Institute for Nature Research.
8. Morse, L. E., J. T. Kartesz, and L. S. Kutner. 1995. Native vascular plants. In *Our Living Resources: A Report to the Nation on the Distribution, Abundance, and Health of U.S. Plants, Animals and Ecosystems*, ed. E. T. LaRoe, G. S. Farris, C. E. Puckett, P. D. Doran, and M. J. Mac, 205. Washington, DC: U.S. Department of the Interior, National Biological Service.

9. Morin, N. 1995. Vascular plants of the United States. In *Our Living Resources: A Report to the Nation on the Distribution, Abundance, and Health of U.S. Plants, Animals and Ecosystems*, ed. E. T. LaRoe, G. S. Farris, C. E. Puckett, P. D. Doran, and M. J. Mac, 200. Washington, DC: U.S. Department of the Interior, National Biological Service.

10. Frank, J. H., and E. D. McCoy. 1995. Introduction to insect behavioral ecology: The good, the bad and the beautiful: Non-indigenous species in Florida. *Fla Entomol* 78:1.

11. Frank, J. H., E. D. McCoy, H. G. Hall et al. 1997. Immigration and introduction of insects. In *Strangers in Paradise*, ed. D. Simberloff, D. C. Schmitz, and T. C. Brown, 75. Washington, DC: Island Press.

12. Simberloff, D., D. C. Schmitz, and T. C. Brown. 1997. *Strangers in Paradise*. Washington, DC: Island Press.

13. Dowell, R. V., and C. J. Krass. 1992. Exotic pests pose growing problem for California. *Calif Agric* 46:6.

14. Babbitt, B. 1998. Statement by secretary of the interior on invasive alien species. In *Proceedings, Natlional Weed Symposium*, April 8–10, 1998. http://web4.audubon.org/bird/at_home/pdf/GFLchap1.pdf (accessed January 27, 2011).

15. Malecki, R. A., B. Blossey, S. D. Hight et al. 1993. Biological control of purple loosestrife. *BioScience* 43:680.

16. Thompson, D. G., R. L. Stuckey, and E. B. Thompson. 1987. Spread, impact, and control of purple loosestrife (*Lythrum salicaria*) in North American wetlands. U.S. Fish and Wildlife Service, Fish and Wildlife Research 2, Washington, D.C. http://www.npwrc.usgs.gov/resource/plants/loosstrf/index.htm (accessed February 25, 2010).

17. Gaudet, C. L., and P. A. Keddy. 1988. Predicting competitive ability from plant traits: A comparative approach. *Nature* 334:242.

18. National Wildlife Refuge Association. 2007. Protecting America's Wildlife: Invasive Species. http://www.refugenet.org/New-issues/invasives.html#TOC02 (accessed February 10, 2011).

19. USDA. 2005. *Agricultural Statistics*. Washington, DC: U.S. Department of Agriculture.

20. Hiebert, R. D., and J. Stubbendieck. 1993. *Handbook for Ranking Exotic Plants for Management and Control*. Denver: U.S. Department of Interior, National Park Service.

21. National Park Service. 2009. Invasive species. … What are they and why are they a problem? http://www.nature.nps.gov/biology/invasivespecies/ (accessed January 2010).

22. Eldredge, L. G., and S. E. Miller. 1997. Numbers of Hawaiian species: Supplement 2, including a review of freshwater invertebrates. *Bishop Mus Occas Pap* 48:3.

23. Vitousek, P. M. 1988. Diversity and biological invasions of Oceanic Islands. In *Biodiversity*, ed. E. O. Wilson and F. M. Peter, 181. Washington, DC: National Academy of Sciences.

24. Campbell, F. T. 1994. Killer pigs, vines, and fungi: Alien species threaten native ecosystems. *Endanger Species Tech Bull* 19:3.

25. Kurdila, J. 1995. The introduction of exotic species into the United States: There goes the neighborhood. *Environ Aff* 16:95.

26. Vitousek, P. M., C. M. D'Antonio, L. L. Loope et al. 1996. Biological invasions as global environmental change. *Am Sci* 84:468.

27. Vitousek, P. M., C. M. D'Antonio, L. L. Loope et al. 1997. Introduced species: A significant component of human-caused global change. *N Z J Ecol* 21:1.

28. Whisenant, S. G. 1990. Changing fire frequencies on Idaho's Snake River plain: Ecological and management implications. In *Proceedings–Symposium on Cheatgrass Invasion, Shrub Die-Off, and Other Aspects of Shrub Biology and Management*, ed. E. D. McArthur, E. M. Romney, E. M. Smith, and P. T. Tueller. Ogden, UT: U.S. Forest Service, Intermountain Research Station, INT-276, 4–10.

29. Campbell, F. T. 1998. "Worst" invasive plant species in the conterminous United States. Report, Western Ancient Forest Campaign, Springfield, VA.

30. OTA. 1993. *Harmful Non-Indigenous Species in the United States*. Washington, DC: Office of Technology Assessment, United States Congress.

31. Miller, J. H. 1995. Exotic plants in southern forests: Their nature and control. *Proc South Weed Sci Soc* 48:120.

32. Randall, J. M. 1996. Weed control for the preservation of biological diversity. *Weed Technol* 10:370.

33. Mills, E. L., M. D. Scheuerell, J. T. Carlton, and D. L. Strayer. 1997. Biological invasions of the Hudson River basin. *New York State Mus Circ* 57:1.

34. Koschnick, T. 2007. You thought milfoil was tough: Options for and challenges associated with Hydrilla control. In *Midwest Aquatic Plant Management Society 27th Annual Conference*, 2007 March 4, Milwaukee: Midwest Aquatic Plant Management Society. http://dnr.wi.gov/invasives/classification/pdfs/LR_Hydrilla_verticillata.pdf (accessed February 9, 2011).

35. Center, T. D., J. H. Frank, and F. A. Dray. 1997. Biological control. In *Strangers in Paradise*, ed. D. Simberloff, D. C. Schmitz, and T. C. Brown, 245. Washington, DC: Island Press.

36. Schmidly, D. J., and W. B. Davis. 2004. *The Mammals of Texas*. Revised ed. Austin: University of Texas Press.

37. Smith, R. 1984. Producers need not pay startling "rodent tax" losses. *Feedstuffs* 56:13.

38. USDA. 2007. *Agricultural Statistics*. Washington, DC: U.S. Department of Agriculture.

39. Wachtel, S. P., and J. A. McNeely. 1985. Oh rats. *Int Wildl* 15:20.

40. Chopra, G. 1992. Poultry farms. In *Rodents in Indian Agriculture*, ed. I. Prakash and P. K. Ghosh, 309. Jodhpur: Scientific Publishers.

41. Ahmed, E., I. Hussain, and J. E. Brooks. 1995. Losses of stored foods due to rats at grain markets in Pakistan. *Int Biodeterior Biodegradation* 36:125.

42. Richards, C. G. J. 1989. The pest status of rodents in the United Kingdom. In *Mammals as Pests*, ed. R. J. Putman, 21. London: Chapman and Hall.

43. Amarasekare, P. 1993. Potential impact of mammalian nest predators on endemic forest birds of western Mauna Kea, Hawaii. *Conserv Biol* 7:316.

44. Pimentel, D. 1955. The control of the mongoose in Puerto Rico. *Am J Trop Med Hyg* 41:147.

45. Vilella, F. J., and P. J. Zwank. 1993. Ecology of the small Indian mongoose in a coastal dry forest of Puerto Rico where sympatric with the Puerto Rican nightjar. *Caribb J Sci* 29:24.

46. Henderson, R. W. 1992. Consequences of predator introductions and habitat destruction on amphibians and reptiles in the post-Columbus West Indies. *Caribb J Sci* 28:1.

47. Everard, C. O. R., and J. D. Everard. 1992. Mongoose rabies in the Caribbean. *Ann N Y Acad Sci* 653:356.

48. Humane Society of the United States. 2009. U.S. pet ownership statistics, October 22, 2009. http://www.humanesociety.org/issues/pet_overpopulation/facts/pet_ownership_statistics .html (accessed December 17, 2009).

49. McKay, G. M. 1996. Feral cats in Australia, origins and impacts. In *Unwanted Aliens? Australia's Introduced Animals*, ed. G. M. McKay, 9. The Rocks: Nature Conservation Council of NSW.

50. Fitzgerald, B. M. 1990. Diet of domestic cats and their impact on prey populations. In *The Domestic Cat: The Biology of Its Behavior*, ed. D. C. Turner and P. Bateson, 123. Cambridge: Cambridge University Press.

51. Dunn, E. H., and D. L. Tessaglia. 1994. Predation of birds at feeders in winter. *J Field Ornithol* 65:8.

52. Luoma, J. R. 1997. Catfight. *Audubon* 99:85.

53. Pimentel, D., and A. Greiner. 1997. Environmental and socio-economic costs of pesticide use. In *Techniques for Reducing Pesticide Use: Economic and Environmental Benefits*, ed. D. Pimentel, 51. Chichester: John Wiley & Sons.

54. Carter, C. N. 1990. Pet population control: Another decade without solutions? *J Am Vet Med Assoc* 197:192.

55. Layne, J. N. 1997. Nonindigenous mammals. In *Strangers in Paradise*, ed. D. Simberloff, D. C. Schmitz, and T. C. Brown, 157. Washington, DC: Island Press.

56. Wedro, B. C., and M. C. Stoppler. 2010. Dog bite treatment. http://www.medicinenet.com/dog_ bite_treatment/article.htm (accessed January 21, 2010).

57. Sacks, J. J., M. Kresnow, and B. Houston. 1996. Dog bites: How serious a problem? *Inj Prev* 2:52.

58. Colburn, D. 1999. Dogs take a big bite out of health care costs. *Washington Post* February 2.

59. Quinlan, K. P., and J. J. Sacks. 1999. Hospitalizations for dog bite injuries. Centers for Disease Control. http://www.cdc.gov/ncipc/duip/hospital.htm (accessed February 23, 1999).

60. Centers for Disease Control (CDC). 1997. Dog-bite-related fatalities—United States, 1995–1996. *Mortal Morb Wkly Rep Commun Dis Cent* 46:463.

61. Temple, S. A. 1992. Exotic birds, a growing problem with no easy solution. *Auk* 109:395.

62. Birding.com. 2010. Bird checklists of the United States. http://www.birding.com/checklists.asp (accessed September 1, 2010).

63. Moulton, M. P., and S. L. Pimm. 1983. The introduced Hawaiian avifauna: Biogeographic evidence for competition. *Am Nat* 121:669.

64. Pimm, S. L. 1991. *The Balance of Nature?* Chicago: The University of Chicago Press.

65. Keitt, T. H., and P. A. Marquet. 1996. The introduced Hawaiian avifauna reconsidered: Evidence for self-organized criticality. *J Theor Biol* 182:161.

66. Laycock, G. 1966. *The Alien Animals*. New York: Natural History Press.

67. Roots, C. 1976. *Animal Invaders*. New York: Universe Books.

68. Long, J. L. 1981. *Introduced Birds of the World: The Worldwide History, Distribution, and Influence of Birds Introduced to New Environments*. New York: Universe Books.

69. Weber, W. J. 1979. *Health Hazards from Pigeons, Starlings and English Sparrows: Diseases and Parasites Associated with Pigeons, Starlings, and English Sparrows which Affect Domestic Animals*. Fresno: Thomson Publications.

70. Robbins, C. S. 1995. Non-native birds. In *Our Living Resources: A Report to the Nation on the Distribution, Abundance, and Health of U.S. Plants, Animals and Ecosystems*, ed. E. T. LaRoe, G. S. Farris, C. E. Puckett, P. D. Doran, and M. J. Mac, 437. Washington, DC: U.S. Department of the Interior, National Biological Service.

71. Smith, R. H. 1995. Rodents and birds as invaders of stored-grain ecosystems. In *Stored-Grain Ecosystems*, ed. D. S. Jayas, N. D. G. White, and W. E. Muir, 289. New York: Marcel Dekker, Inc.

72. Haag-Wackernagel, D. 1995. Regulation of the street pigeon in Basel. *Wildl Soc Bull* 23:256.

73. Johnston, R. F., and M. Janiga. 1995. *Feral Pigeons*. New York: Oxford University Press.

74. McCoid, M. J., and C. Kleberg. 1995. Non-native reptiles and amphibians. In *Our Living Resources: A Report to the Nation on the Distribution, Abundance, and Health of U.S. Plants, Animals and Ecosystems*, ed. E. T. LaRoe, G. S. Farris, C. E. Puckett, P. D. Doran, and M. J. Mac, 433. Washington, DC: U.S. Department of the Interior, National Biological Service.

75. Lafferty, K. D., and C. J. Page. 1997. Predation of the endangered tidewater goby, *Eucyclogobius newberryi*, by the introduced African clawed frog, *Xenopus laevis*, with notes on the frog's parasites. *Copeia* 3:589.

76. Fritts, T. H., and G. H. Rodda. 1995. Invasions of the brown tree snake. In *Our Living Resources: A Report to the Nation on the Distribution, Abundance, and Health of U.S. Plants, Animals and Ecosystems*, ed. E. T. LaRoe, G. S. Farris, C. E. Puckett, P. D. Doran, and M. J. Mac, 454. Washington, DC: U.S. Department of the Interior, National Biological Service.

77. Rodda, G. H., T. H. Fritts, and D. Chiszar. 1997. The disappearance of Guam's wildlife. *BioScience* 47:565.

78. Coulehan, K. 1987. Powerless again. About your partners in business: Snakes and GPA. *Guam Bus News* January 1987:13.

79. Bergman, D. L., M. D. Chandler, and A. Locklear. 2002. The economic impact of invasive species to wildlife services' coordinators. In *Human Conflicts with Wildlife: Economic Considerations, USDA NWRC Symposia*, ed. L. Clark, Fort Collins: National Wildlife Research Center. http://digitalcommons.unl.edu/nwrchumanconflicts169. http://digitalcommons.unl.edu/nwrchumanconflicts/21/ (accessed February 24, 2010).

80. Fritts, T. H. 2002. Economic costs of electrical system instability and power outages caused by snakes on the island of Guam. *Int Biodeterior Biodegradation* 49:93.

81. Holt, A. 1997–1998. Hawaii's reptilian nightmare. *World Conserv* 4/97–1/98:31.

82. Courtenay, W. R. Jr., D. P. Jennings, and J. D. Williams. 1991. Appendix 2: Exotic Fishes. In *Common and Scientific Names of Fishes From the United States and Canada*, ed. C. R. Robins, R. M. Bailey, C. E. Bond, J. R. Brooker, E. A. Lachner, R. N. Lea, W. B. Scott, 97. Bethesda, MD: American Fisheries Society.

83. Courtenay, W. R. 1993. Biological pollution through fish introductions. In *Biological Pollution: The Control and Impact of Invasive Exotic Species*, ed. B. N. McKnight, 35. Indianapolis: Academy of Science.

84. Pimentel, D. 2005. Aquatic nuisance species in the New York state canal and Hudson River system and the great lakes basin: An economic and environmental assessment. *Environ Manage* 35:692.

85. Courtenay, W. R. 1997. Nonindigenous fishes. In *Strangers in Paradise*, ed. D. Simberloff, D. C. Schmitz, and T. C. Brown, 109. Washington, DC: Island Press.

86. Dill, W. A., and A. J. Cordone. 1997. History and status of introduced fishes in California, 1871–1996. Fish *Bull* 178:1.

87. Maciolek, J. A. 1984. Exotic fishes in Hawaii and other islands of Oceania. In *Distribution, Biology, and Management of Exotic Fishes*, ed. W. R. Courtenay and J. R. Stauffer, 131. Baltimore: Johns Hopkins University Press.

88. Wilcove, D. S., and M. J. Bean. 1994. *The Big Kill: Declining Biodiversity in America's Lakes and Rivers*. Washington, DC: Environmental Defense Fund.

89. Taylor, J. N., W. R. Courtenay, and J. A. McCann. 1984. Known impacts of exotic fishes in the continental United States. In *Distribution, Biology, and Management of Exotic Fishes*, ed. W. R. Courtenay and J. R. Stauffer, 322. Baltimore: Johns Hopkins University Press.

90. Bjergo, C., C. Boydstun, M. Crosby et al. 1995. Non-native aquatic species in the United States and coastal water. In *Our Living Resources: A Report to the Nation on the Distribution, Abundance, and Health of U.S. Plants, Animals and Ecosystems*, ed. E. T. LaRoe, G. S. Farris, C. E. Puckett, P. D. Doran and M. J. Mac, 428. Washington, DC: U.S. Department of the Interior, National Biological Service.

91. USBC. 2001. *Statistical Abstract of the United States 2001*. Washington, DC: U.S. Bureau of the Census, U.S. Government Printing Office.

92. Hendrix, P. F. 1995. *Earthworm Ecology and Biogeography*. Boca Raton: Lewis Publishers.

93. Jenkins, J. C. 1998. "Measuring and Modeling Northeaster Forest Response to Environmental Stresses." PhD diss. University of New Hampshire, Durham, NH.

94. Wilkinson, T. 2002. Losing the forests and the trees. *Natl Parks* 76:18.

95. Alsop, F. J., and T. F. Laughlin. 1991. Changes in the spruce-fir avifauna of Mt. Guyot, Tennessee, 1967–1985. *J Tenn Acad Sci* 66:207.

96. Vinson, S. B. 1994. Impact of the invasion of *Solenopsis invicta* (Buren) on native food webs. In *Exotic Ants: Biology, Impact, and Control of Introduced Species*, ed. D. F. Williams, 241. Boulder, CO: Westview Press.

97. Allen, C. R., R. S. Lutz, and S. Demarais. 1995. Red imported fire ant impacts on northern bobwhite populations. *Ecol Appl* 5:632.

98. Vinson, S. B. 1992. The economic impact of the imported fire ant infestation on the state of Texas. Report, Texas A & M University, College Station.

99. TAES. 1998. Texas imported fire ant research and management plan. Report, Texas Agricultural Extension Service, Texas A & M University, College Station.

100. Congressional Research Service. 1999. Harmful non-native species: Issues for Congress VI. http://ncseonline.org/NLE/CRSreports/Biodiversity/biodv-26e.cfm (accessed January 2010).

101. Corn, M. L., E. H. Buck, J. Rawson et al. 1999. *Harmful Non-Native Species: Issues for Congress*. Washington, DC: Congressional Research Service, Library of Congress.

102. Lafferty, K. D., and A. M. Kuris. 1996. Biological control of marine pests. *Ecology* 77:1989.

103. Khalanski, M., F. Bergot, and E. Vigneux. 1997. Industrial and ecological consequences of the introduction of new species in continental aquatic ecosystems: The zebra mussel and other invasive species. *Bulletin Francais de la Peche et de la Pisciculture* 0 (344–345):385.

104. Benson, A. J., and Boydstun, C. P. 1995. Invasion of the zebra mussel into the United States. In *Our Living Resources: A Report to the Nation on the Distribution, Abundance, and Health of U.S. Plants, Animals and Ecosystems*, ed. E. T. LaRoe, G. S. Farris, C. E. Puckett, P. D. Doran, and M. J. Mac, 445. Washington, DC: U.S. Department of the Interior, National Biological Service.

105. Keniry, T., and J. E. Marsden. 1995. Zebra mussels in Southwestern Lake Michigan. In *Our Living Resources: A Report to the Nation on the Distribution, Abundance, and Health of U.S. Plants, Animals and Ecosystems*, ed. E. T. LaRoe, G. S. Farris, C. E. Puckett, P. D. Doran, and M. J. Mac, 445. Washington, DC: U.S. Department of the Interior, National Biological Service.

106. Griffiths, D. W., D. W. Schloesser, J. H. Leach et al. 1991. Distribution and dispersal of the zebra mussel (*Dreissena polymorpha*) in the Great Lakes Region. *Can J Fish Aquat Sci* 48:1381.

107. U.S. Army. 2002. Economic impacts of zebra mussel infestation. http://www.wes.army.mil/el/ zebra/zmis/zmishelp/economic_impacts_of_zebra_mussel_infestation.htm (accessed December 2002).

108. Isom, B. G. 1986. *ASTM (American Society for Testing and Materials) Special Technical Publication, 894. Rationale for Sampling and Interpretation of Ecological Data in the Assessment of Freshwater Ecosystems.* Philadelphia: American Society for Testing and Materials.

109. Cohen, A. N., and J. T. Carlton. 1995. *Nonindigenous Aqautic Species in a United States Estuary: A Case Study of the Biological Invasions of the San Francisco Bay and Delta.* Washington, DC: United States Fish and Wildlife Service.

110. Pimentel, D., M. S. Hunter, J. A. LaGro, R. A. Efronymson et al. 1989. Benefits and risks of genetic engineering in agriculture. *BioScience* 39:606.

111. Pimentel, D. 1993. Habitat factors in new pest invasions. In *Evolution of Insect Pests – Patterns of Variation,* ed. K. C. Kim and B. A. McPheron, 165. New York: John Wiley & Sons.

112. Pimentel, D. 1997. *Techniques for Reducing Pesticides: Environmental and Economic Benefits.* Chichester: John Wiley & Sons.

113. USDA. 1998. *Agricultural Statistics.* Washington, DC: U.S. Department of Agriculture.

114. Pimentel, D. 1991. *Handbook on Pest Management in Agriculture, vol. 1, 2, and 3.* Boca Raton, FL: CRC Press.

115. Trammel, M. A., and J. L. Butler. 1995. Effects of exotic plants on native ungulate use of habitat. *J Wildl Manage* 59:808.

116. Kadrmas, T., and W. S. Johnson. 2002. Managing yellow and dalmatian toadflax, fact sheet FS-09-96, University of Nevada, Cooperative Extension, Reno. http://www.unce.unr.edu/ publications/FS02/FS0296.pdf (accessed August 18, 2003).

117. Templeton, S. R., D. Zilberman, and S. J. Yoo. 1998. An economic perspective on outdoor residential pesticide use. *Environ Sci Technol* 2:416A.

118. Pogacnik, T. 1995. Wild horses and burros on public lands. In *Our Living Resources: A Report to the Nation on the Distribution, Abundance, and Health of U.S. Plants, Animals and Ecosystems,* ed. E. T. LaRoe, G. S. Farris, C. E. Puckett, P. D. Doran and M. J. Mac, 456. Washington, DC: U.S. Department of the Interior, National Biological Service.

119. Rosentreter, R. 1994. Displacement of rare plants by exotic grasses. In *Proceedings—Ecology and Management of Annual Rangelands,* ed. S. B. Monsen and S. G. Kitchen, 170. Washington, DC: USDA, Forest Service, Rocky Mountain Research Station.

120. Pimentel, D., L. Lach, R. Zuniga et al. 2000. Environmental and economic costs associated with introduced non-native species in the United States. *BioScience* 50:53.

121. Kotanen, P. M. 1995. Responses of vegetation to a changing regime of disturbance: Effects of feral pigs in a California coastal prairie. *Ecography* 18:190.

122. Stone, C. P., L. W. Cuddihy, and T. Tunison. 1992. Response of Hawaiian ecosystems to removal of pigs and goats. In *Alien Plant Invasions on Native Ecosystems in Hawaii: Management and Research,* ed. C. P. Stone, C. W. Smith, and J. T. Tunison, 666. Honolulu: University of Hawaii Cooperative National Park Studies Unit.

123. Rollins, D. 1997. *Statewide Attitude Survey on Feral Hogs in Texas.* College Station: Texas Agricultural Extension Service, Texas A&M University.

124. Davis, D. S. 1998. *Feral Hogs and Disease: Implications for Humans and Livestock.* College Station: Texas A & M University, Department of Veterinary Pathology.

125. Moore, N. W. 1980. How many wild birds should farmland support? In *Bird Problems in Agriculture,* ed. E. N. Wright, I. R. Inglis, and C. J. Feare, 2. Croydon: The British Crop Protection Council.

126. Feare, C. J. 1980. The economics of starling damage. In *Bird Problems in Agriculture,* ed. E. N. Wright, I. R. Inglis, and C. J. Feare, 39. Croydon: The British Crop Protection Council.

127. Johnson, S. A., and W. Givens. 2009. Florida's introduced birds: European starling (Sturnus vulgaris). University of Florida IFAS Extension. Publication #WEC 255. http://edis.ifas.ufl.edu/ uw300 (accessed December 22, 2009).

128. Howarth, F. G. 1990. Hawaiian terrestrial arthropods: An overview. *Bishop Mus Occas Pap* 30:4.

129. Beardsley, J. W. 1991. Introduction of arthropod pests into the Hawaiian Islands. *Micronesia Suppl* 3:1.
130. Frank, J. H., and E. D. McCoy. 1995. Precinctive insect species in Florida. *Fla Entomol* 78:21.
131. Liebold, A. M., W. L. MacDonald, D. Bergdahl et al. 1995. Invasion by exotic forest pests: A threat to forest ecosystems. *For Sci* 41:1.
132. Haack, R. A., and J. W. Byler. 1993. Insects and pathogens: Regulators of forest ecosystems. *J For* 91:32.
133. Hall, J. P., and B. Moody. 1994. Forest depletions caused by insects and diseases in Canada 1982–1987. Forest Insect and Disease Survey Information Report ST-X-8, Forest Insect and Disease Survey, Canadian Forest Service, Natural Resources Canada, Ottawa.
134. Campbell, F. T., and S. E. Schlarbaum. 1994. *Fading Forests: North American Trees and the Threat of Exotic Pests*. New York: Natural Resources Defense Council.
135. Moore, B. A. 2005. *Alien Invasive Species: Impacts on Forests and Forestry—A Review*. Forest Health and Biosecurity Working Paper 8, Forest Resources Development Service Working Paper FBS/8E. Forestry Department, Forest Resources Division, Food and Agriculture Organization of the United Nations, Rome. http://www.fao.org/docrep/008/j6854e/J6854E06.htm (accessed December 22, 2009).
136. USDA. 1960. *Index of Plant Diseases in the United States*. Washington, DC: Crop Research Division, ARS, U.S. Department of Agriculture.
137. Drummond, R. O., G. Lambert, H. E. Smalley et al. 1981. Estimated losses of livestock to pests. In *Handbook of Pest Management in Agriculture*, ed. D. Pimentel, 111. Boca Raton, FL: CRC Press, Inc.
138. Morgan, N. O. 1981. Potential impact of alien arthropod pests and vectors of animal diseases on the U.S. livestock industry. In *Handbook of Pest Management in Agriculture*, ed. D. Pimentel, 129. Boca Raton, FL: CRC Press, Inc.
139. USAHA. 1984. *Foreign Animal Diseases: Their Prevention, Diagnosis and Control*. Richmond, VA: Committee on Foreign Animal Diseases of the United States Animal Health Association.
140. Newton-John, H. 1985. Exotic human diseases. In *Pests and Parasites as Migrants*, ed. A. J. Gibbs and H. R. C. Meischke, 23. Sydney: Cambridge University Press.
141. Centers for Disease Control (CDC). 1996. Summary of notifiable diseases, United States, 1995. *Mortal Morb Wkly Rep Commun Dis Cent* 44:1.
142. Bartlett, J. G. 2006. A look back at the cost of HIV/AIDS care—then and now. Webscape from WebMD. http://www.medscape.com/viewarticle/547640 (accessed January 6, 2010).
143. Kent, J. H., L. E. Chapman, L. M. Schmeltz et al. 1992. Influenza surveillance—United States, 1991–92. *MMWR Surveill Summ* 41(SS-5):35.
144. Chapman, L. E., M. A. Tipple, L. M. Schmeltz et al. 1992. Influenza—United States, 1989–90 and 1990–91 seasons. *MMWR Surveill Summ* 41(SS-3):35.
145. Molinari, N.-A. M., I. R. Ortega-Sanchez, M. L. Messonier et al. 2007. Annual impact of seasonal influenza in the US: Measuring disease burden and costs. *Vaccine* 25:5086.
146. Bateman, D. A., C. S. Phibbs, T. Joyce et al. 1997. The hospital cost of congenital syphilis. *J Pediatr* 130:752.
147. Centers for Disease Control (CDC). 2003. West Nile Virus: Summer is coming and so are the mosquitoes. Centers for Disease Control. http://www.cdc.gov/ncidod/dvbid/westnile/Westnilespotlight2003.htm (accessed September 1, 2010).
148. FICMNEW. 2010. Federal Interagency Committee for the Management of Noxious and Exotic Weeds. http://www.fs.fed.us/ficmnew/index.shtml (accessed May 6, 2010).
149. Babbitt, B. 1999. *Weed Coalition Announces National Strategy to Combat the Spread of Non-Native Invasive Plants*. Washington, DC: U.S. Department of the Interior, Wednesday, March 10.

# Index

Lightning Source UK Ltd.
Milton Keynes UK
UKHW030620151019
351624UK00017B/245/P